BIOMEDICINE EXAMINED

CULTURE, ILLNESS, AND HEALING

Editors:

MARGARET LOCK

Departments of Anthropology and Humanities and Social Studies in Medicine, McGill University, Montreal, Canada

ALLAN YOUNG

Department of Anthropology, Case Western Reserve University, Cleveland, Ohio, U.S.A.

BIOMEDICINE EXAMINED

Edited by

MARGARET LOCK

*Departments of Humanities and Social Studies
in Medicine and Anthropology, McGill University, Quebec, Canada*

and

DEBORAH GORDON

*Program in Medical Anthropology,
University of California, San Francisco, U.S.A.*

KLUWER ACADEMIC PUBLISHERS

DORDRECHT / BOSTON / LONDON

Library of Congress Cataloging in Publication Data

Biomedicine examined.

  (Culture, illness, and healing)
  Includes indexes.
  1. Social medicine--Miscellanea.   I. Lock,
Margaret M.   II. Gordon, Deborah R.   III. Series.
[DNLM: 1. Culture--essays.   2. Medicine--essays.
3. Social Conditions--essays.   WB 9 B615]
RA418.B53   1988   362.1'042   88-21633

ISBN 1-55608-071-9
ISBN 1-55608-072-7 (pbk.)

The human mind has at no period accepted a moral chaos.

> George Eliot
> Middlemarch, 1871

The conclusions that we seek to draw from the likeness of events are unreliable, because events are always unlike. There is no quality so universal in the appearance of things as their diversity and variety.

> Michel De Montaigne
> Essays: On Experience, 1580
> (trans. Cohen, 1958)

If the cultural influences upon science can be detected in the humdrum minutiae of a supposedly objective, almost automatic quantification, then the status of biological determinism as a social prejudice reflected by scientists in their own particular medium seems secure.

> Stephen Jay Gould
> The Mismeasure of Man, 1981

Perfect health, like perfect beauty, is a rare thing;
and so it seems, is perfect disease.
>   Peter Mere Latham 1789-1875
>   Collected Works, Book I, ch. 443

A model is by definition that in which nothing has
to be changed, that which works perfectly whereas
reality, as we see clearly, does not work and constantly falls to pieces; so we must force it, more or
less roughly, to assume the form of the model.
>   Italo Calvino
>   Mr. Palomar, 1983

. . . values and knowledge are always and necessarily associated in action just as in discourse . . . *the
very definition of "true" knowledge reposes in the
final analysis upon an ethical postulate.*
>   Jacques Monod
>   Le Hazard et la Nécessité, 1970

TABLE OF CONTENTS

PART I

THE SOCIAL SCIENCES AND BIOMEDICINE

MARGARET LOCK

INTRODUCTION

The culture of contemporary medicine is the object of investigation in this book; the meanings and values implicit in biomedical knowledge and practice and the social processes through which they are produced are examined through the use of specific case studies. The essays provide examples of how various facets of 20th century medicine, including education, research, the creation of medical knowledge, the development and application of technology, and day to day medical practice, are pervaded by a value system characteristic of an industrial-capitalistic view of the world in which the idea that science represents an objective and value free body of knowledge is dominant.

The authors of the essays are sociologists and anthropologists (in almost equal numbers); also included are papers by a social historian and by three physicians all of whom have steeped themselves in the social sciences and humanities. This co-operative endeavor, which has necessitated the breaking down of disciplinary barriers to some extent, is perhaps indicative of a larger movement in the social sciences, one in which there is a searching for a middle ground between grand theory and attempts at universal explanations on the one hand, and the context-specific empiricism and relativistic accounts characteristic of many historical and anthropological analyses on the other.

For many years social scientists left unquestioned the dominant ideology of their time; scientific "facts" were reified, assumed to be pristine and beyond the realm of social analysis. Anthropologists were particularly blind in this respect, and while they blithely examined the exotic healing ceremonies and rituals of other cultures and situated them in local cosmologies, they stubbornly ignored modern medicine, assuming it to have evolved beyond the superstition, religion, and value laden beliefs so clear to them in traditional medicine. Anthropologists, of course, have been careful to show how local medical knowledge is not arbitrary, but grounded in the "seamless web" of local culture; scurrilous words such as "superstition" are studiously avoided and use is made of the more elegant term "belief", which has nevertheless, until recently, been assumed to be an explanatory system of an entirely different order than one grounded in science. Despite this major limitation, the heritage of ethnomedicine and the more encompassing comparative study of healing systems, is a rich one in which the methods of symbolic anthropology have been dominant.

3

M. Lock and D. R. Gordon (eds.), Biomedicine Examined, 3–10.
© *1988 by Kluwer Academic Publishers.*

Keesing states that symbolic anthropology is an "exploration, an exca-vation, of the cumulated, embodied symbols of other peoples, a search for meanings, for hidden connections, for deeper saliences than those presented by the surface evidence of ethnography" (1987:161). In medi-cal anthropology, the task becomes one of describing local beliefs about the causes and prevention of distress, diagnosis, classification and labell-ing of illnesses, and the performance of healing rituals, and then attempt-ing an interpretation of these beliefs usually by demonstrating the way in which they are related to and act upon other aspects of culture. This enterprise has been described as dependent upon a "virtuosity in seeing hidden meanings enciphered as tropes" (Keesing 1987:161), and in cases where it is well executed has provided us with stimulating accounts of the medical beliefs of other cultures (see for example, Good 1977; Ngubane 1977; Sindzingre and Zempléni 1981).

It is perhaps in connection with beliefs about causation that medical anthropology has made one of its most important contributions to the study of medical systems. A comparative analysis of causal explanations for distress is a constant and forceful reminder that by confining one's attention to the human body and to encounters between healers and patients, major distortions are introduced into the analytic procedure. In most cultures the social origins of illness and distress are of overriding concern (Lewis 1975; Manning and Fabrega 1973) and medically related activities are much more encompassing than a visit to a healer (Janzen 1978).

Worsley, in a review of the comparative study of medical systems, has argued that the "treating of bodily ills takes place in *any* culture within a "metamedical" framework of thought", an overarching philosophy which guides the basic features of medical knowledge, its organization and prac-tice (1982:315, emphasis added). The characteristic settings of doctor's office, clinic, and hospital, and the battery of medical professionals and auxilliaries associated with the modern "health-care complex" are specific products of industrial society. Similarly, the very idea of a bounded med-ical system, reasonably autonomous and clearly distinct from other social institutions, is a cultural construct, as is the "belief" (superstition?) that diseases are "real" entities and that their elimination crystallizes the essence of what medicine is all about.

One strength of the interpretive approach in anthropology is that, although the researcher records an account of the medical beliefs of other peoples, the exercise is done with a sensitivity to the fact that the interpretation is itself a product of particular historical and cultural deter-minants. Ideally, no assumption is made of a privileged analytical view-point which is value free. This attitude, when combined with rich and detailed descriptive data, has furnished us with invaluable context-specific accounts of medical beliefs, more of which would be a welcome addition to the storehouse of information on comparative medical systems. However, in portraying the complex and subtle shades of meanings which are thought to form the basis for everyday life in other cultures, two very important arenas have been almost totally neglected: firstly, the control, distribution, and dynamics of the actual application of medical knowledge, and secondly the relationship of knowledge and its application not only to social and symbolic healing, but also to the reduction of individual pain and suffering.

Cultures, including those of industrial-capitalistic societies, form systems of meanings which provide explanations of how the world works, of what is thought of as "real" and what is designated as "natural" and inevitable. These meanings link people to one another and form the basis for social action. There has been a tendency among symbolic anthropologists to emphasise the shared signification and meaning created by a culture in a positive light, to consider it only as a unifying whole, and as providing the *raison d'etre* for human existence in social groups. On the other hand, the way in which culture constitutes not only meanings but also an ideology has been largely ignored. In all societies cultural ideologies serve to institute and legitimate certain political and economic realities. Inequalities in the distribution of knowledge, power, and privilege are accepted as cosmically ordained relationships: "Cultures are webs of mystification as well as signification" (Keesing 1987:161). A balanced description is needed in the case of medical beliefs, not only of ideas about causation and healing, but of who creates and defines these meanings and what significance this has for the allocation of responsibility for the occurrence of illness and for the restoration and maintenance of the social order. An account of the control of medical knowledge and the way in which it is selectively applied can demonstrate "the manner in which social interest becomes seamlessly incorporated in the set of tacit assumptions about reality" (Comaroff 1982:50).

Medical sociologists, unlike anthropologists, have traditionally paid attention to the unequal distribution of power and to the way in which medicine can act as an institution of social control (Zola 1972). They have also shown how medicine in capitalist societies reflects the values of society at large, including those of class, race, gender, and age (Ehrenreich 1978; Frankenberg 1980; Stark 1982). However, sociologists, in common with their anthropological colleagues, have until recently rarely made the content of modern medical knowledge itself a subject for analysis, and have tended to rest assured that, because it is supposedly grounded in science, it is not, therefore, a subject for sociological enquiry.

In the stimulating and important precursor to this present volume,

The Problem of Medical Knowledge (1982), Wright and Treacher discuss why some of the traditional assumptions about biomedicine have recently been called into question. They point out that the strategy of the modern medical profession has been to claim that it functions with knowledge which has a special technical status and hence is not contestible in the same way as are say, religion or ethics. Power and knowledge are clearly linked in this claim which has increasingly been criticized from a number of angles: that the self-interest of the medical profession influences at times the generation of medical knowledge and its practice (Jackson 1970; Lewin and Olesen 1985); a lack of autonomy of medicine in other cultural domains (Bastien 1985; Kleinman 1980; Lock 1980); and, most notably under the influence of Foucault, how the language of medicine does not merely describe a pre-existing biological reality, but instead creates its own objects of analysis (Foucault 1975, 1979; Armstrong 1983). Wright and Treacher are quick to point out that what is being demonstrated in research of this kind is not that medicine is "unscientific" because it is permeated by social forces: but, in contrast, that both medicine and science are essentially social enterprises (1982:7). This shift in thinking is not merely with respect to medicine, but is part of a larger reorientation in connection with science in general (Latour and Woolgar 1979; Mulkay 1979) and has entailed the incorporation of a perspective in which more attention is given to the interrelationships of "the researcher, the scientific community of which he is a member, the knowledge which the community shares, and the broader religious, social and political currents within which the community exists" (Wright and Treacher 1982:8). The way has gradually been paved for the examination of medicine as a social and cultural construction and for a simultaneous rapprochement and refinement of the methods used by sociologists, anthropologists and social historians in its analysis. The volume edited by Wright and Treacher was one of the first to make its focus an analysis of medical knowledge, and emphasis was placed in several essays on the way in which various topics related to medicine are perceived in different historical periods and how they change through time and space. A second book, *Physicians of Western Medicine* (1985), edited by Hahn and Gaines, focuses on modern medical settings and shows very clearly the way in which medical knowledge and practice is not the product of a monolithic autonomous institution but rather is made up of numerous sub-specialities, interest groups, and individuals who bring a variety of perspectives to their work. Use is made of the ethnographic approach, and the opinions and experiences of physicians are incorporated into the analysis. The present volume is heavily indebted to these and other earlier endeavors and has tried to develop more fully some of the arguments implicit in them. Most notably, an effort is made systematically to uncover some of the cultural assumptions, the ideology, in which modern medicine is grounded, and to show how these assumptions are embedded in the creation and transmission of medical knowledge and practice. It is generally accepted by the authors that medical knowledge is not an isolatable autonomous body of information, but that it is rooted in and is sustained by practice. In the majority of the essays, rich ethnographic or textual accounts locate the data in specific contexts, which are also often analysed in terms of their implications for relationships of power and control.

Our purpose is to demonstrate the social and cultural character of *all* medical knowledge, but by so doing we are not denying the existence of real, painful stress and suffering. There is, of course, a biological reality, but the moment that efforts are made to explain, order, and manipulate that reality, then a process of contextualization takes place in which the dynamic relationship of biology with cultural values and the social order has to be considered.

The relationship of medical knowledge and practice to the sick human body is a topic which so far has not been taken up very seriously by researchers using either interpretive or political analyses. It has generally been accepted by anthropologists, for example, that a shared belief system can be mobilized in order to aid in the healing process, largely though its effect on the psyche (Levi Strauss 1963) or because it stimulates a restructuring of the social order (Turner 1968). The reader is rarely informed if and when biological healing occurs as the result of healing rituals; in common with the shamans who are so often featured in their analyses, anthropologists have tended to relegate the body physical to the background in their scramble to expose metaphorical and metonymical links between the various discourses on the cosmos, society, gender, and the body (*Culture and Depression* [1985] edited by Kleinman and Good is a notable exception to this tendency; see also Lock [1986]). Similarly, the biological status of the human body has not been made problematic in most sociological analyses, although the work of Bourdieu and Foucault has stimulated a move in this direction (see, for example, Turner 1984).

An examination of the social and cultural construction of biomedicine forces a questioning of a privileged status for its knowledge and at the same time reveals how the practice of medicine is laden with moral evaluation. In theory the biomedical model of disease causation, by focusing attention on the neutral terrain of the physical body, serves to "depoliticize" the medical encounter (Habermas 1971). In practice the question of ultimate causation, of why one particular person becomes sick at one particular time, is not readily cast aside, and patients, families, and healers generate and exchange explanations in which implications about responsibility are embedded. In their various ways, statements which focus on social, familial, psychological, or biological explanations for disease causation all highlight contradictions and tensions between the postulated relationships of nature, the individual, and society. An analysis of these statements exposes the way in which beliefs in connection with health and ill-health are reinforced by the basic ideology of a culture, and also reveals how images of selfhood and the creation of social identities are intimately linked to ideas about a moral and healthy body generated both within medicine and society at large.

As the practice of modern medicine becomes increasingly a technical enterprise, it is more encumbent upon us than ever to recognize that the human body is not a machine, that health and illness are not merely biological states, but rather that they are conditions which are intimately related to and constituted by the social nature of human life. The study of health, illness, and medicine provides us with one of the most revealing mirrors for understanding the relationship between individuals, society, and culture; it is an exciting task which has only just begun.

REFERENCES

Armstrong, David
    1983   Political Anatomy of the Body: Medical Knowledge in Britain in the Twentieth Century. Cambridge: Cambridge University Press.
Bastien, Joseph
    1985   Qollahuaya - Andean Body Concepts: A Topographical-Hydraulic Model of Physiology. American Anthropologist 87:595-611.
Comaroff, Jean
    1982   Medicine: Symbol and Ideology. In P. Wright and A. Treacher (eds.), The Problem of Medical Knowledge: Examining the Social Construction of Medicine. Pp. 49-68. Edinburgh: University of Edinburgh Press.
Ehrenreich, John (ed.)
    1978   The Cultural Crisis of Modern Medicine. New York: Monthly Review Press.
Foucault, Michel
    1975   The Birth of the Clinic: An Archeology of Medical Perception. New York: Vintage.
    1979   Discipline and Punish: The Birth of the Prison. New York: Vintage.
Frankenberg, Ronald
    1980   Medical Anthropology and Development: A Theoretical Perspective. Social Science and Medicine 14B:197-207.
Good, Byron
    1977   The Heart of What's the Matter: The Semantics of Illness in Iran. Cul-

ture, Medicine and Psychiatry 1:25-58.

Habermas, Jürgen
  1971   Toward a Rational Society. London: Heinemann.

Hahn, Robert and Atwood Gaines (eds.)
  1985   Physicians of Western Medicine: Anthropological Approaches to Theory
          and Practice. Dordrecht, Holland: D. Reidel Publishing Co.

Jackson, J.A. (ed.)
  1970   Professions and Professionalization. Cambridge: Cambridge University
          Press.

Janzen, John
  1978   The Quest for Therapy in Lower Zaire. Berkeley: University of Califor-
          nia Press.

Keesing, Roger
  1987   Anthropology as Interpretive Quest. Current Anthropology 28:161-169.

Kleinman, Arthur
  1980   Patients and Healers in the Context of Culture: An Exploration in the
          Borderland Between Anthropology, Medicine, and Psychiatry. Berke-
          ley: University of California Press.

Kleinman, Arthur and Byron Good
  1985   Culture and Depression: Studies in the Anthropology and Psychiatry of
          Cross-Cultural Affect and Disorder. Berkeley: University of California
          Press.

Latour, Bruno and Steve Woolgar
  1979   Laboratory Life: The Social Construction of Scientific Facts. Beverly
          Hills, California: Sage.

Levi-Strauss, Claude
  1963   The Sorcerer and His Magic. In Structural Anthropology. Pp. 167-185.
          New York: Basic Books.

Lewin, Ellen and Virginia Olesen
  1985   Women, Health and Healing: Toward a New Perspective. London:
          Tavistock.

Lewis, Gilbert
  1975   Knowledge of Illness in a Sepik Society. London: Athlone.

Lock, Margaret
  1980   East Asian Medicine in Urban Japan: Varieties of Medical Experience.
          Berkeley: University of California Press.
  1986   Ambiguities of Aging: Japanese Experience and Perceptions of Meno-
          pause. Culture, Medicine and Psychiatry 10:223-46.

Manning, Peter and Horatio Fabrega
  1973   The Experience of Self and Body: Health and Illness in the Chiapas
          Highlands. In G. Psathas (ed.), Phenomenological Sociology. Pp.
          59-73. New York: Wiley.

Mulkay, Michael
    1979   Science and the Sociology of Knowledge. London: George Allen and
            Unwin.
Ngubane, Harriet
    1977   Body and Mind in Zulu Medicine: An Ethnography of Health and Dis-
            ease in Nyuswa - Zulu Thought and Practice. London: Academic Press.
Sindzingre, Nicole and Andras Zempléni
    1981   Modèles et Pragmatique, Activation et Répétition: Réflexions sur la
            Causalité de la Maladie chez les Senufo de Côte d'Ivoire. Social Sci-
            ence and Medicine 15B:279-293.
Stark, E.
    1982   Doctors in Spite of Themselves. International Journal of Health Servi-
            ces 12:419-457.
Turner, Bryan
    1984   The Body and Society: Explorations in Social Theory. Oxford: Basil
            Blackwell.
Turner, Victor
    1968   The Drums of Affliction: A Study of Religious Processes Among the
            Ndembu of Zambia. Oxford: Clarendon.
Worsley, Peter
    1982   Non-Western Medical Systems. Annual Review of Anthropology
            11:315-348.
Wright, Peter W.G. and Andrew Treacher (eds.)
    1982   The Problem of Medical Knowledge: Examining the Social Construction
            of Medicine. Edinburgh: University of Edinburgh Press.
Zola, Irving
    1972   Medicine as an Institution of Social Control. Sociological Review
            20:487-504.

MARGARET LOCK AND DEBORAH R. GORDON

RELATIONSHIPS BETWEEN SOCIETY, CULTURE, AND BIOMEDICINE: AN INTRODUCTION TO THE ESSAYS

The essays in this volume demonstrate the interdependence of biomedicine, society and culture. The authors, for the most part using an ethnographic approach, show how language, values, metaphor, ritual practices, institutions, and social organization, among other variables, contribute to the creation of ideas about contemporary medical theory and to clinical practice. In particular these essays demonstrate how cultural and social variables function to support certain assumptions about the natural order, the "reality" of disease, and their "rational" management, very frequently through the use of technological intervention. Although the argument in most of the articles is developed around specific issues or case studies, several common themes, which we will briefly consider, recur frequently throughout the essays.

The volume begins with Gordon's exploration of several ideas derived from the Western philosophical and cultural heritage which may account for the tenacity of certain medical assumptions and practices in biomedicine today. She highlights parallels between the assumed autonomy of nature in both the natural science and biomedical paradigms and the assumed autonomy of the individual in a Western understanding of personhood, society, morality and religion. As a corollary she emphasizes how in biomedical practice there is an assertion of the autonomy of the individual from a social and cultural context.

THE CONSTRUCTION OF "PSYCHE" AND "SOMA"

As numerous writers on Western medicine have noted, both in biomedical knowledge and practice, a separation is usually assumed between the working of mind and body. A number of articles in this volume illustrate the cultural construction of this separation, how it is created and perpetuated, and the kinds of discourse and practices which actually contribute to the credibility of this dichotomy.

"Mind" and "body" are metaphors, Kirmayer argues, polarizing two distinct constellations of feelings and values. Their dualist relationship and the meanings associated with them are reproduced through the specialty of "psychosomatics" which protects the biomedical model by siphoning off cases with which it is unsuccessful. Blame for intransigent

M. Lock and D. R. Gordon (eds.), Biomedicine Examined, 11–16.
© *1988 by Kluwer Academic Publishers.*

illness is placed on the patient's irrational emotions instead of inadequate medical theory, diagnosis, therapy, or difficult social relationships. Further, psychosomatic therapy itself reaffirms the triumph of reason over both the "irrational" and the body.

Similarly, Helman finds that physicians increasingly label patients' emotions or personality characteristics as "pathological" and cite them as causal in "non-organic" illnesses. In response, patients then "reify" these pathogenic emotions, traits, or body parts, separate them from their "ideal social self," and then blame these isolated factors for the occurence of their illnesses. Psychological explanations such as these compensate for the limits of the biomedical model in diagnosis, treatment, and explanations of illness.

Several other papers provide insight into how the separation of mind and body, body and emotions, and medicine and emotions are brought about. Observing the rite of the anatomy lesson, Lella and Pawluch trace changing relationships among attitudes towards emotions, cadaver dissection, and historical periods. They argue that the matter-of-fact, non-emotional attitude toward the anatomy lesson that has characterized its teaching during this century is changing. Today, as was once the case in the past, the emotional and philosophical meanings of cadaver dissection for medical students are increasingly being given attention.

Another emotional situation many physicians must face is telling patients "bad news." As an example, Taylor observed surgeons while they were informing patients about a diagnosis of breast cancer. Although different physicians adopted different strategies, they all routinized the strategy they adopted, rather than adapting their behavior to each patient. The contrast between the patients' obvious emotionality and many physicians' discomfort with emotions probably accounts in part for these physicians' experiencing disclosure of diagnosis as a very stressful part of their work.

Two papers question the dominance of the natural science model of explanation in medicine, in which the observer (medical practioner) is thought to sustain a detached attitude. Kirmayer suggests that it is important to arrive at an understanding of the metaphoric basis upon which explanations in modern medicine are built. Such an understanding will, he suggests, alert us to the involvement of both physician and patient emotions in medical practice. Similarly, Gordon discusses how knowledge becomes "embodied" through experience, such that "intuition" comes to supersede calculative reasoning. The common dichotomy of mind and body limits our understanding not only of illness, but also of what we accept as knowledge.

Kaufert highlights another dimension that contributes to the separation of body from emotions and experience in arguing that the medical construction of menopause is strongly influenced by the demands of the scientific research paradigm. This paradigm favors the study of phenomena as discrete events rather than as a process. Kaufert notes the emphasis which is placed on the physical and biological changes associated with the mid-life transition to near exclusion of the actual experiences of women.

HOW MEDICINE PERPETUATES ITSELF

The transmission of medical knowledge and values is another theme which is taken up in several of the papers. Atkinson studied the teaching strategies used to reproduce clinical medicine. He shows how senior physicians teach physicians-in-training to construct normal or abnormal pictures of patient situations and to integrate and apply theory in practice. Indirectly, he also shows how physicians make use of patients while teaching students how to arrive at a diagnosis and that this is often accomplished by means of the Socratic method.

Sankar, in watching the reactions of senior medical students during and after visits to the homes of patients, discovered the power of the *context* of the clinic/hospital. This environment not only is the "turf" of the physicians, but also contributes to what information physicians will or will not have to confront. As one student put it:

In the hospital . . . you have social problems where you want them, you don't have to deal with them. There is no social . . . Whereas at home, there are all kinds of interactions that you really have no control over (Sankar, this volume).

MEDICINE ADAPTING

Medicine is continually undergoing change, adaptation, and expansion, another theme taken up in several of the essays. Armstrong traces changes in ideas about space and time in British medical practice. The location of practice in the health center made possible a new analysis of illness in terms of time (process) and "community." The rise of medical concern for chronic illness and the "biography" (the patient as a person) reflects these changes. A temporal model of illness is also revealed in the incorporation of "prevention" as the concern of general medical practitioners, a topic discussed by Williams and Boulton. How, in fact, this new domain is interpreted in theory and practice is left much to the discretion of individual physicians. Their interpretations reflect personal biography as well as age, subspecialty, or institutional affiliation. Wright also traces the incorporation of a new domain into the medical sphere — that of babyhood. He analyzes the complex culmination of interacting factors that contributed to infancy being reconceptualized in a metaphorically novel way. This "generative metaphor" allowed babyhood to be considered a medical problem and thus subject to scientific and social control. Lock discusses the transformation of school refusal into a syndrome in Japan, and analyses the social, cultural, and political reasons as to why this particular problem creates a potent concern which it is assumed the medical profession should help to resolve.

Gordon describes a contemporary movement to make medical practice more scientific, by further developing scientific guidelines for practice and by making the judgement and decision making process of physicians more explicit, formal, and "rational." She critiques this movement by considering an alternative paradigm of knowledge and clinical expertise.

BIAS OF A TECHNOLOGICAL KIND

Several papers in this volume reconstruct the processes that contribute to a bias for the use of technology in medical care. Muller and Koenig demonstrate how physicians can usually find something more to do for "terminally ill patients." By continuing to try whatever possible technological interventions remain for a particular case, they not only postpone labeling a patient as "dying," but also appear to express their commitment to patient care.

On a more impersonal level, however, one finds a frightening collusion implied here. New technology needs to be tested and improved; while untested it is considered "high risk." As terminally ill patients worsen, as the likelihood of their dying *increases*, the *relative* risk incurred through the use of a new technological intervention *decreases* and becomes the only possible salvation. The "postponement of dying" that Muller and Koenig describe allows for the trial of new experimental technology on patients who, it is thought, have little to lose.

"Routinization" is the key process that moves "experimental" technology into normal procedure which, effectively, creates a "moral imperative" (Koenig). Barley shows us how radiological technicians often approach computer technology ineffectively and how they use a series of strategies, including magical thinking, superstitions, and anthropomorphism, to cope with computer breakdowns while still using a "mechanical" model. The technicians' experiences highlight how the use of technology is not neutral but must be made sense of by those involved. Koenig describes nurses managing the problems of new technology and discusses how "routinization" contributes to a sense of the inevitability of the use of technology. Both essays illustrate how this apparent routinization is constructed.

While radiological technicians and nurses are busy managing day to day problems with technology, physicians often construct research findings based on the very same technology (Koenig). In fact, the research perspective in medicine, that which is closest to pure science, shows up repeatedly in several of these papers as a powerful force in medical practice (see Taylor, Kaufert, Koenig, Gordon).

BIOMEDICINE IN VARIOUS SOCIAL AND CULTURAL CONTEXTS

In contrast to the universalism assumed in biomedical and naturalist ideology, biomedicine -- both in theory and practice -- is very much a product of particular social and cultural conditions. This situation is highlighted in several of the papers which show how biomedicine takes form in different social and cultural contexts. Low, for example, discusses how physicians respond to *nervios* -- a "folk" symptom and syndrome in Costa Rica -- how the discourse between patient and physician produces diagnoses and treatments that are adapted to local understandings and social dynamics. Lock presents a picture of how biomedicine is used to support core values in Japan. Atkinson, Wright, Armstrong, Williams and Boulton write on past and present periods in the United Kingdom; while the rest of the essays focus on contemporary North America.

DEPERSONALIZATION AND THE PRESERVATION OF MEDICINE

Several of the essays offer illustrative examples of the production of depersonalization and distancing from everyday events, for which biomedicine has often been criticized: through research designs (Kaufert); rites, such as the anatomy lesson (Lella and Pawluch); the context of health care (Sankar, Armstrong); the use of patients as teaching objects (Atkinson); an exclusive focus on "therapeutic possibility" (Muller and Koenig); and a division of labor that allows physicians to distance themselves from problems related to technology, patients, and other health professionals (Koenig, Barley). We also see some of the ways in which biomedicine protects and preserves itself when its limits are challenged: by "psychologizing" patients' "psychosomatic" problems (Helman, Kirmayer); by routinizing stressful medical events (Taylor); and by continuing to treat dying patients (Muller and Koenig). In many cases these responses to the limits of biomedical theory and practice serve the interests of physicians more than patients.

Finally, several papers show medicine responding to social and professional changes, demands, and needs: by attending more to physicians' emotions and philosophy (Lella and Pawluch); by responding to newly recognized social problems (Lock, Wright); by incorporating diagnosis and treatment of "folk" illnesses such as *nervios* (Low); by expanding into new dimensions of the life cycle (Armstrong, Williams and Boulton, Kaufert, Wright); and by trying to improve medical practice by making it more "scientific" (Gordon).

Together these papers, themselves constructions, further our understanding of both the changes and the tenacity that characterize biomedicine.

ACKNOWLEDGEMENTS

The editors gratefully acknowledge the contribution made by Naomi Adelson in the final preparation of this manuscript.

PART II

MIND, BODY, VALUES, AND SOCIETY

DEBORAH R. GORDON

TENACIOUS ASSUMPTIONS IN WESTERN MEDICINE

> "To return to things themselves is to
> return to that world which precedes
> knowledge, of which knowledge always
> *speaks . . ."* (Merleau-Ponty 1962:ix).

> ". . . I believe that a deeper examina-
> tion must show that the struggle
> between rival approaches in the sci-
> ence of man . . . is no mere question
> of the relative efficacy of different
> methodologies, but is rather one facet
> of a clash of moral and spiritual out-
> looks. And I believe that we can only
> make even the first halting steps
> towards resolving it if we can give
> explicit recognition to this fact"
> (Charles Taylor 1985a:114).

INTRODUCTION

While biomedicine has successfully created and hoarded a body of technical knowledge to call its own, its knowledge and practices draw upon a
background of tacit understandings that extend far beyond medical
boundaries. The biological reductionism by which modern medicine is
frequently characterized is more theoretical than actual; in its effects,
biomedicine speaks beyond its explicit reductionist reference through the
implicit ways it teaches us to interpret ourselves, our world, and the relationships between humans, nature, self, and society. It draws upon and
projects *cosmology* (ways of ordering the world), *ontology* (assumptions
about reality and being), *epistemology* (assumptions about knowledge and
truth), understandings of *personhood, society, morality,* and *religion*
(what is sacred and profane). Although biomedicine both constitutes and
is constituted by society, this interdependency is nevertheless denied by
biomedical theory and ideology which claim neutrality and universality.

Mirroring this ideology, social and historical studies traditionally have
considered western medicine as a "constant," the universal against which
other medical systems were mapped. But much as scientific facts have
recently lost some of their status as absolute, universal truths (both

19

M. Lock and D. R. Gordon (eds.), Biomedicine Examined, 19–56.
© *1988 by Kluwer Academic Publishers.*

within the world of science and without)[1] so too is western medicine increasingly being understood as one of many medicines -- "biomedicine" -- culturally and historically specific and far from universal. A growing body of literature demonstrates how biomedical knowledge and practices are eminently and irreducibly social and cultural.[2] Much as the patient has been the object of medical attention (what Foucault [1975] calls the "medical gaze"), made comprehensible and visible through language and technology, we may increasingly speak of a social scientific/historical gaze turned on medicine, describing hidden cultural scaffolding and social processes that shape practice and knowledge. These studies reveal more and more how biomedicine evolves through social choices rather than natural inevitability and how, in fact, a sense of "natural inevitability" or "givenness" is constructed out of social choice.

Biomedicine as a product of western culture and society draws on some of the dominant western philosophical traditions. I wish in this essay to synthesize some basic assumptions which show up as common themes in studies of biomedicine by addressing more directly the "background"[3] through which biomedicine operates: the "western" in "western medicine."

There are good reasons for studying this background.[4] For one, medicine projects a symbolic reality that is becoming more dominant. Once the exclusive model of physicians, the biomedical approach is being incorporated into the lay/folk model (Fabrega 1975). From a healing model it is rapidly expanding into a moral and engineering one, increasingly "remaking" humans not in nature's image but in its own (Engelhardt 1973).

Further, these background assumptions may account for the tenacity of some important medical practices. For despite many significant recent changes -- such as greater attention to patients' experiences, the whole life cycle, uncertainty and probability theory, a multi-causal view of disease, and greater partnership with patients -- many important approaches remain the same. The limits of mind/body dualism, for example, have been enumerated for years, yet, as Dossey notes:

It is a mistake to underestimate the force of Cartesian dualism in medicine today. In spite of a growing disaffection of a section of the populace with traditional approaches to health, the dualist philosophy is alive and well, the guiding light of almost all theoretical and clinical efforts of Western medicine (1984:15).

Assumptions based on seventeenth century physics continue to dominate, as "new physics" has made only a slight inroad into medicine. Ostensible reforms -- such as psychosomatic medicine, holistic health, the expanded "biopsychosocial" model (Engel 1977), "bioethics," and the patient autonomy movement -- often retain the basic values and assumptions of biomedicine and unwittingly perpetuate the "status quo."[5] Many of these same values and assumptions are also reproduced in the social science/ historical research that is being used increasingly in medical practice.

Clearly the tenacity of biomedical assumptions and practices derives from numerous sources -- the efficacy of much of medical practice, the extensive web of institutions, the political, economic and personal invest- ments in the current model and the successful response of many physi- cians to criticism (for example Arney 1982). Here I will focus on the cul- tural basis of this tenacity,[6] asking as W.T. Jones does, ". . . how might the world-view of practitioners and patients affect a medical system by limiting its capacity to change?" (1976:383). I will consider the back- ground of biomedicine primarily in terms of two major western traditions -- "naturalism" (usually referred to in the literature as "science") and "individualism" -- a complex of values and assumptions asserting the pri- macy of the individual and of individual freedom (Dumont 1977).

Charles Taylor puzzles over the enduring dominance of what he calls "naturalism," the view that humans are not only a part of nature but that this nature should be understood according to the canons of the Enlight- enment (1985a,b:2),[7] and asks why the human sciences, despite unargua- bly dubious results, persist in modeling themselves after the natural sci- ences? Epistemological explanations alone which claim that natural science produces better knowledge cannot explain the prestige and domi- nance of naturalism. Taylor argues that just as all human communities are defined and colored by their "strong evaluations" (values), the human sciences are also colored by modern values and aspirations:

. . . behind and supporting the impetus to naturalism . . . stands an attachment to a certain picture of the agent [actor]. This picture is deeply attractive to moderns, both flattering and inspiring. It shows us as capable of achieving a kind of disen- gagement from our world by objectifying it. We objectify our situation to the extent that we can overcome a sense of it as what determines for us our paradigm purposes and ends, and can come to see it and function in it as a neutral environ- ment, within which we can effect the purposes which we determine out of our- selves (Taylor 1985a:4).

Taylor believes that the force of the natural science paradigm derives "from the underlying image of the self . . . and the images of freedom, dignity and power which attach to it" (1985a:6). More specifically he claims that:

. . . the more we are led to interpret ourselves in the light of the disengaged pic- ture, to define our identity by this, the more the connected epistemology of natu- ralism will seem right and proper to us. Or otherwise put, a commitment to this

identity generates powerful resistances against any challenges to the naturalist outlook. In short, its epistemological weaknesses are more than made up for by its
moral appeal (Taylor 1985a:6).

Undoubtedly the dominance of the natural science paradigm in a human
practice like medicine benefits from extra support similar to that
described above. Although this paradigm has produced impressive
results in biomedicine, its limitations have been discussed for some time
(e.g. Powles 1973; Kleinman, Eisenberg and Good 1978). Yet its support
remains tacit and broad. Its demise would be radical and even painful,
as Dossey warns his medical colleagues:

To allow more than objects to enter our experience [as doctors] -- *really* enter --
would entail a painful reassessment of who we are. It would mandate a redefinition of our relationship with the world, a renunciation of the ordinary subject-ob-
ject way we habitually define ourselves (Dossey 1984:5).

Before elaborating on this topic, we should consider a few important
qualifications. First, I do not want to encourage thinking about biomedicine and western traditions in monolithic terms, much as we once spoke
about Primitive Medicine. If research has demonstrated anything it is the
plurality of biomedicines.[8] In fact, North American medicine receives
disproportionate attention here, a function of the relative abundance of
medical and social scientific literature and public discussion about many
biomedical issues and the relative exaggeration of themes that are less
obvious and acute in other countries. While the term "western" masks a
multitude of differences which we are only beginning to understand
(Gaines 1987), I assume the validity of referring to the modern western
world in terms of a set of traditions of which many societies partake to
different degrees (cf. Mauss 1938, 1985; Dumont 1970, 1977; Bellah, et
al. 1985). One such common tradition is naturalism and while I make
this my focus because of its continued dominance and intransigence, I in
no way equate naturalism with biomedicine (nor do most physicians).
Nor do I argue for a causal connection between naturalist assumptions
and biomedical practice but that these assumptions are sustained and
implied in many medical practices. My intent is to stimulate awareness
of and reflection on these assumptions and on how and perhaps why they
are sustained in practice, both by physicians and by western-trained
social scientists/historians. For the future, more comparative research
among western contexts which not only grounds or challenges some commonly ascribed assumptions about western culture, but which also shows
how medical practice reflects local culture, is needed.

Finally, what exactly do I mean by "background"? Like other aspects
of a society's culture, biomedical discourse and practices operate against

and through a background of understandings that are shared to varying degrees with other subcultures in western society. I do not, however, regard these as understandings or assumptions that people carry around in their heads (consciously or not) and then "apply" in different situations, that is, a cognitive "world view" or a "belief system." This view is, in fact, itself a tenacious western assumption of naturalism that I will question, and as such, it is difficult to find adequate language to express an alternative approach. Nevertheless my intention is to stress ways of being and feeling, more than a cognitive view.[9] I interpret practices as themselves embodying or supposing interpretations about what is real, what it is to be an object, a subject, a person, a body, a society, and so on. I have in mind flexible, undefined "taken-for-granteds" which allow practices to make sense (Dreyfus 1980), what Heidegger calls "foreknowledge" (1962). One participates *through* background assumptions, one does not consciously *see* them (cf. Heidegger 1962; Dreyfus 1979; Benner 1984b; Wrubel 1985). In fact it is this very embodiment in practices, which thus entails ways of being and not just explicit beliefs, that make the assumptions so tenacious.

I consider naturalism as a cultural construct in the sense that it takes language, practices, and institutions that have extensive societal support to maintain it. Following Taylor's lead, I focus particularly on the mutual support between naturalism, biomedicine, and individualism. In the first part of this essay I present the claims of naturalism to autonomy from the supernatural, the cultural, the social, the psychological, the moral, as well as from particular time and space. In the second part, I discuss values and assumptions about personhood, society, morality, and religion which are subsumed under the general term individualism. In the final section the assumed autonomy of naturalism and biomedicine from society is challenged by showing connections between a dominant value in Western society, that of autonomy (freedom) on the one hand, and biomedicine and naturalism on the other.

I. THE "AUTONOMY OF NATURE"

NATURALIST COSMOLOGY AND ONTOLOGY

Naturalism is founded on a number of important distinctions that assert the autonomy of "nature" from the "supernatural," from human consciousness, from "culture," "society," "morality," "psychology," and particular time and space. Naturalism also asserts a separation between cosmology/ontology on the one hand and epistemology on the other, a distinction I will maintain but which is historically and culturally specific

(Needham 1962).

Biomedical practitioners approach sickness as a natural phenomenon, legitimize and develop their knowledge using a naturalist method (scientific rationality) and see themselves as practicing on nature's human representative -- the human body. Thus many of the following assumptions salient in naturalism appear in biomedical and social scientific discourse and practices. It should be emphasized from the outset that I am presenting the traditional view of science because it remains dominant in medicine (Kleinman 1986).

1. "Nature" is Distinct from the "Supernatural": Matter is Opposed to Spirit

Enlightenment thinkers devoted themselves to disengaging nature from its previous metaphysical and spiritual connections.[10] Where once nature was seen as sacred, the reflection of a divine plan or the embodiment of Ideas (Taylor 1985a,b), the Enlightenment task was to "disenchant" nature and see "God's world" as a mechanism, composed of physical matter obeying natural mechanistic laws rather than spiritual ones. The key to the universe is its logical rather than spiritual order (Descartes 1950). In contrast to the somewhat capricious supernatural world, nature is considered to be orderly, lawful, and therefore predictable (King 1975).

Medicine repeats this claim of autonomy from religion, regarding illness as distinct from "misfortune" or divine punishment and disease not as sin but as mechanism. In fact, medicine's development took off when "disease broke away from the metaphysics of evil to which it had been related for centuries" (Foucault 1975:196). Not prayer, but "looking and seeing" what is wrong in the body machine and repairing it is the key. Medicine is usually described as "lacking a metaphysics" and failing to provide religious spiritual meaning.

In the naturalist view, the universe consists of discrete material essences, albeit often invisible, which are fixed and stable in their identity and follow the law of non-contradiction: Being "A" does not equal Being "non-A" (Lee 1959:131). Reality is directly proportional to materiality, which is considered lacking in spirit; the more physical, the more real. "Body" is distinct from "mind."

Medicine exemplifies materialism. "Real" illness corresponds to the degree to which physical traces show up in the body. Health and illness are defined in terms of materialist indicators, such as blood pressure, rather than "spirit," such as "feeling" healthy (Dossey 1984).

2. Nature is Autonomous from Human Consciousness

Another essential separation assumed in naturalism and embodied in the terms "objective" and "subjective" is between nature and the individual's experience of it. Nature is "given," "out there," distinct from the observer. Its meaning is its own, distinct from its meaning for society or for a particular person. Thus it must be understood on its own terms -- objectively -- free of society's need to project human traits such as will or purpose onto it. Nature is a thing-in-itself -- neutral, indifferent to human purpose and to human relationships. These are the meanings embodied in the term "object." Subjective meaning -- what something is *for someone*, such as hot or cold, good or bad -- is exterior to the being of objects. Connection or relationship is also separate from the identity of objects. Naturalist cosmology consists in part of "subjects" encountering "objects." It distinguishes between the primary quality of a thing-in-itself and the secondary, the meaning *for someone* (Lakoff and Johnson 1980:199).

Similarly, medicine distinguishes between "signs," objective indications in the patient's body, and "symptoms," the patient's complaints. Medical social scientists distinguish between "disease," biological abnormality, and "illness," the patient's experience. Health or illness is defined more through objective data offered by the body than the experience of the patient. They are defined as separate from the individual (one "has" good health).

Emotions are also not the right idiom for understanding nature, according to naturalism. They in fact impede our understanding, for nature is neither emotional nor psychological. In medicine it is assumed that emotions can certainly cause illness, but "psychosomatic" or "functional" illness, not "real" illness (see Kirmayer, Helman, this volume).

Nature is mechanism and thus to approach it correctly one must understand *how it works*, not *what it means* for humans, because it doesn't *mean* anything in terms of human purposes and goals. To "move" nature, one must not "think so," one must "do so," that is, manipulate it physically.

The "bias towards action" frequently noted in medicine (see Muller and Koenig, this volume) comes to mind. One can manipulate nature without "hurting" it; the corpse on which medical students learn anatomy, for example, doesn't feel anything, and according to the traditional ideal, neither should the dissecting physician (see Lella and Pawluch, this volume).

Human meaning is subjective in the naturalist understanding. Meaning, no longer seen to reside in the universe, in objects themselves, is now located within the mental confines of the individual and taken to be his/her personal private subjective property (Taylor 1985a,b). Much as objectivity and consciousness are sealed off from each other, subjectivity does not exist "out there" but rather "in there." In medicine the "subjectivity" of the patient (which is receiving more attention recently, Armstrong 1983) is often considered asocial, idiosyncratic, and given; it is the unique personal property of an individual, not of an historical context or situation.

3. Atomism: The Part is Independent of and Primordial to the Whole

Being is in essences, not in relationships or in process. The meaning or identity of a thing is given in itself alone rather than in the "living" context of the "whole" of which it is a part (Taussig 1977). The common scientific notion of "variable," for example, relies on this atomistic view as W.T. Jones points out:

> But what is a variable? It is an aspect of the situation that can be lifted out of its context and altered without affecting that context. In other words, underlying the notion of a variable is a vision of the universe as a collection . . . of relatively discrete and encapsulated items. The notion of a variable would only occur to, and seem plausible to, scientists with discreteness-bias (1976:388).

That parts of nature are considered autonomous "things-in-themselves" has three major consequences: (1) Relationships are derivative; the whole is determined by the sum of its parts, rather than the whole determining the parts (Taussig 1977:152). (2) Given that their identity is self-determined, the parts may be removed from their context without altering their identity. They may be "decontextualized." (3) Relations between parts are "external" not "internal" (Merleau-Ponty 1962; Taussig 1977:143), since parts interact across their distinct boundaries. Thus we speak of "causation" between parts, rather than "creation" or "production" of an identity.

Atomism of many sorts prevails in medicine: diseases are considered to have an identity separate from their specific hosts and are located and treated in the "atom" of society – the individual, his/her body divided into parts and parts which are approached as autonomous units. In fact, at times they are discussed as if, "all the organs just sit there doing their own thing and that from the concert of individual actions, the function of the whole organism occurs" (Cassell 1984:26-7). Health and illness, life and death are seen as discrete and opposed entities, not complementary parts of each other (as in the eastern view of ying/yang). Models of illness are usually cause and effect. Atomism also shows up in the tendency of social scientists to discuss a "medical system" as if it had an independent reality of its own (Comaroff 1983; Worsley 1982).

In fact, naturalism is characterized by a *unitary ontology* (Taylor 1985a), an assumption that while the world is composed of varying strata of complexity that can be ordered hierarchically, underneath it all the nature of being is basically the same; qualitative differences are secondary (Bhasker 1979). Different levels are not incommensurable (as in the model of the "Great Chain of Being," Gould 1983) but reducible. Little distinction is made between humans and things.

Medicine has thrived on biological reductionism. Even "systems theorists" and "new-age" anti-reductionists treat different dimensions of human life, such as "psychology" and "biology" as if they were commensurable (Gould, ibid). More recently, "subjective" data of the patient (such as changes in "body image") are treated much like other "objective" data. Patients' "values" are treated as though merely another "variable" in the complete picture. "Motivation," for example, is often discussed as if it were a physical force, the exclusive property of an individual.

4. Nature is Separate from Culture

Nature stands not only independent from culture but prior to it. Symbols/language/representations *depict* an independent empirical reality, rather than *constitute* it. Meaning is in the correspondence between representation and external reality. Disease taxonomy mirrors nature's "real" diseases.

Similarly "natural man" is distinct from "cultural man," the "animal of man" is distinguishable from the "human," and more specifically the "body" is distinguishable from the "mind." This animal/natural part is considered universal, the human part -- the mind, the spirit -- is where diversity shows up. Yet this diversity is really only "skin deep." In fact a common notion is that humans are animals with rationality and culture added (Geertz 1973; Straus 1963). This "stratigraphic approach" assumes that underneath all the exterior cultural coating -- specifically underneath anatomy at the levels of physiology and biochemistry -- all humans are basically the same (Geertz 1973). Culture is a superficial, "sometimes-thing."

Furthermore, culture is usually conceived of as applying more to groups than to individuals (Obeyesekere 1981), understood largely as external constraints that *limit* rather than enable or empower individuals (Fox and Swazey 1984). In fact, culture is often seen as essentially opposed to individuality. Increasingly culture is being approached in a biological image analogous to DNA -- as a code. Anthropologists "decode" culture, just as physicians "decode" symptoms.

The cultural dimensions in medicine are usually seen to be restricted to the superficial, to apply to patients' behavior and understandings, to exist primarily in "others" beliefs. Culture is basically external to biology and disease, just as biology is regarded as essentially universal. Medical taxonomy, it is commonly believed, describes not prescribes (Rawlinson 1982). In some extreme cases, however, disease is considered to be essentially all culture and little biology, referred to in the anthropological literature as "culture-bound syndromes" (Hahn 1986). Students of bio-medicine and science have been slow to explore the symbolic dimension of biomedicine's apparently technical, instrumental, and neutral activities (Habermas 1970; Powles 1973; Comaroff 1984; Gordon 1987a,b).

5. Nature is Separate from Morality

In naturalism it is assumed that nature is indifferent to the good and the bad -- to human values and morality in general. Similarly in medicine it is frequently assumed that illnesses "happen" to people and that sickness has no special attraction to virtue or vice. It is merely mechanism, not good, bad or righteous behavior that counts. Instead of judging, medicine diagnoses, explains "how," and treats. Disease processes are generally indifferent to the value of their inhabitant; in fact, they are basically democratic (Zempléni 1985). Yet as we learn more of the mechanisms of disease, responsibility for being sick is shifting to the individual, who through will, reason, and healthy "lifestyle" should be able to prevent it.

6. Nature is Autonomous From Society

The natural order is also separate from the social order according to naturalism. What is regarded as the truth (knowledge) is considered to be autonomous, separate from power (Foucault 1977, 1980b). In other words, what is "natural" is beyond the sphere of social influence. Many gender differences, for example, are "natural," neither human projections nor cultural products.

In medicine it is basically believed that disease follows its own rules, neither those of kings nor slaves. Disease is neither the fault of the individual nor of society. To be sure, different social classes become sick more than others, and this can be explained by differences in personal hygiene or "lifestyle." Disease is essentially an individual problem and is systematically abstracted from a social context. Even the placebo effect (which usually takes place in the context of a relationship) is described as "self healing" (Ferguson 1984).

7. Nature/Truth is Universal, Autonomous from Time or Space

Nature's truth is beyond any particular time or space. It is a reality that is omnipresent, universal, eternal, and absolute. In contrast to many other societies, a dominant assumption in western culture, particularly science, is that,

Time does not . . . enter into the inner nature of things; it merely carries things along with it, as a river carries with it the boats and ships that float on its surface -- they are not a part of it, and it is not a part of them (Jones 1976:389).

Events are connected more through the laws of nature rather than through the notion of individual lives (*ibid*.). Time is linear. There is one truth.

The history of medicine is frequently understood to be a cumulative progression towards the unfolding of truth about nature's diseases. The belief in primarily one truth, which is best captured in the neutral language of numbers, is strong. Physicians increasingly are expected to "tell the truth" to patients about their diagnosis and prognosis, assuming that there is *a* truth that can be told.

Objects are representations or manifestations of invisible laws and forces. These laws are described in abstract terms, such as "gravity," "atoms," "diseases." Jones calls this the "abstract bias," and while not exclusive in western society, it is an important dimension of naturalism:

The West has always been more interested in the class than in its individual members, in the "what" than in the "that," in "essence" than in "existence." That is, the aim of Western science has always been to discover and formulate "laws" . . . . It takes abstract-bias to conceive of a world of ideal forms, and to view these ideal forms as more real, and therefore more important than the objects encountered in perception (1976:388-89).

Medicine is propelled by abstractions which are taken as real. The "cause" of illness is often confused with the illness itself (Cassell 1976). The very concept of "disease" entities is based on the abstraction of diseases from individual patients; biology or theories of "stress" are based on an "hypothetical average man" (Young 1980) abstracted from society. Medicine's prejudice for diagnosis, that is, placing an event in a class, reflects this bias.

The approach to the body in biomedicine is an exemplar of naturalism in medicine.[11] The body is regarded as nature's representative in human beings: it is an "it," a physical object (and as such passive), with a stable identity (anatomy and physiology were "discovered" and are universal, Armstrong 1983); separate from the self ("I have a body, not I am a body," Wilber 1984); and bounded from others by skin ("my body"). Neither the body nor symptoms belong to a social field. As representative of nature, the body is distinct from and lower than the mind (Kirmayer, this volume; Turner 1984) and opposed to reason. It is a resource to cultivate, manipulate, train (Foucault 1977), that is ahistorical, acultural, asocial, amoral, non-emotional. As a "thing," the body is neither a person nor something sacred ("a thou"), but run by mechanisms and best approached objectively through the purest and most objective of languages -- numbers.

NATURALIST EPISTEMOLOGY: THE AUTONOMY OF RATIONALITY

Naturalist assumptions about knowledge and truth repeat the distinctions enumerated above. Naturalist truth is not supernatural or spiritual knowledge but the truth of matter, of mechanism. "Truth" is in the accurate explanation of material reality, not in the good, or the beautiful, or the spiritual. Naturalist knowledge should maintain the separation between culture and reality. Culture -- symbols, language -- are vehicles for knowing; they connect the outside reality with the internal knower. Ideally they should leave no trace on the knowledge, that is, they should depict rather than constitute. Similarly, society and knowledge are separate. Knowledge is autonomous yet accessible to those in power and it is separate *from*, not defined *by* power (as Foucault [1977] argues). Rationality is also separate from morality. In fact, it is supposed to be stripped of value and to present only "facts." Truth tells us about how things work "naturally," not ideologically. Finally, ideal truth is beyond time and space -- singular, universal, eternal, and neutral.

Knowing in Naturalist Epistemology

In keeping with the notion of a real "out there," the knower must first separate himself from the "known" and copy this object as accurately as possible onto his/her mental sketchpad. The important assumption here is that we apprehend the world indirectly through representation in our minds (Descartes 1950; Dreyfus in press; Wrubel 1985). Since objective meaning is believed to be independent from language, scientific knowledge seeks to make word correspond to object, designating and representing an autonomous reality. A more recent version of this view, the "information processing" model of cognition, has us picking out meaningless bits of isolated information in the world and then "processing" them, as the saying goes, whereupon we "assign" meaning to them. Perception and understanding are separate. Even if we are unaware, it is argued that we "process" through a set of rules we have already "internalized." Reason is regarded as calculation (see Gordon, this volume).

Mere representation, however, is not enough in the naturalist viewpoint. The knower does not just want to copy the object. To really know is to know theory, the rule, the object's mechanisms that explain the particular in terms of the universal. The meaning of the particular is not in the particular itself but in its representation of a class, as a symbol. The meaning of symptoms, for example, is not in the symptoms *per se*, but in the underlying disease process they may indicate (Good and Good 1980). Truth is behind things, not in them.

While knowledge is mental, it is acquired through the senses which are considered to be ignorant and neutral. Ideas change over time but sensual knowledge is separate from ideas and thus ahistorical – basically a passageway. Indeed, the body and emotions are seen to interfere with reason. In general, emotions are considered to detract from physicians' clear understanding of patient problems.

The agent of knowledge in naturalism is the individual. This "epistemological individualism" of the empirical test of the senses contrasts with other loci of knowing – specifically in tradition or in the authority of others. In fact, tradition and authority are seen as blinders or distorters of vision and it is through liberation from them that empirical reason can prevail (Shils 1981). Reason then is democratic, since it is open to all.

Consciousness is an essential condition of knowing in naturalist epistemology (Merleau-Ponty 1962), such that knowing that is not conscious is suspect. Consciousness is associated with the mental, and true knowledge derives from the mind, not the body (see Kirmayer, Gordon, this volume). The assumption is that we can arrive at an unbiased position or at least control of our suppositions through mathematical manipulations (Young 1978). Precision, clarity, and numbers do that job best.

Ideal Knowing Through Manipulation of Both Knower and Known

The ideal knowledge of naturalism is not natural, everyday knowledge. Rather it is based on manipulations of both the known and the knower. While the goal is to "see things for what they are," this is not a direct and easy process. The Enlightenment was revolutionary in part because 18th century scientists contrived and studied a "vexed nature" (Bacon cited in Berman 1984:17), rather than nature in its ordinary state. Knowledge was and is produced in artificial settings that practitioners create for nature to show itself. Manipulation thus is embedded within the scientific program as the very cornerstone of truth and is epitomized by the laboratory experiment and randomized clinical trials in clinical medicine.

It follows that in a universe made up of atoms, the predominant mode of knowing is *analytic*, that is, the breaking down of the whole into its constitutive parts. Entities are dissolved into discrete chains or units from which it is assumed that the whole, such as the human body, can then be understood by reconstructing the parts (Engel 1977:131).

The western philosophical and scientific traditions have long assumed that detachment provides the purest window to truth. It is by withdrawing from the noise and the crowds that one can truly understand what is happening. Thus to arrive at real knowledge we must alter ourselves -- move back, distance ourselves from values, local bias, and particular interest by taking a universal standpoint, one which is disengaged from everyday life. *Knowing* then is distinct from *being* according to naturalism. In philosophical terms, *ontology* is distinct from *epistemology*. This encourages the frequent separation of knowledge and practice. This separation of knowledge and practice is manifest in the division between the basic science and clinical years in medical education. The distancing from everyday understandings that naturalism encourages contributes to the distance between physicians and patients. It undoubtedly cuts physicians off from their own everyday knowledge and understanding (Baron 1981; Sankar, this volume; Mishler, et al. 1981; Duff and Hollingshead 1968).

"The Visual Metaphor"

Vision is the dominant sense in naturalist and medical epistemology, and the important role vision plays in western thinking is often referred to as the "visual metaphor." The characteristics of vision support naturalist cosmology and ontology (Jonas 1966; Keller and Grontkowski 1983; Richters 1985). Compared to other senses (smell, touch, hearing), vision presents the most static picture, revealing more structure than process (see Sacks 1982). On the other hand, it can record change most accurately in a simultaneous frame, thus supporting a belief in the eternal. Vision fosters the sense of separation between subject and object, particularly when compared to touch, and supports a physicalist interpretation of the elements of reality as things-in-themselves. It enhances a sense that the knower is autonomous, free to choose from everything he/she sees. Finally, vision allows for the quickest route to the mind and abstraction -- to "the mind's eye" (Keller and Grontkowski 1983).

In sum, vision supports an understanding of both nature and humans as autonomous, and through the numerous techniques for seeing that have been developed (Reiser 1978), naturalist cosmology and individualism (which we will discuss shortly) are mutually reinforced.

Biomedicine relies on naturalist epistemology as its official epistemology, using "science" as the measure of truth. In fact, medicine's burst of development came with the displacement of the criterion of truth from tradition and rationality to "look and see" (Foucault 1975; Arney and Bergen 1984).

Naturalism, then, is not very natural. Things are studied and manipulated in a contrived context in order to arrive at underlying laws, while the knower also manipulates him/herself to know them. And while naturalist rationality claims separation from morality, we must note that one of the initial goals of science was to further the cause of freedom — freedom from nature's whims, from social constraint, from the authority of tradition (Shils 1981). Scientific rationality places the individual as sovereign knower, but this is an individual who, through disengagement, is separate from society, culture, emotion, and particular time and place.

II. THE "AUTONOMY OF MAN"

> "Each patient is a free agent entitled to full explanation and full decision-making authority with regard to his medical care. John Stuart Mill expressed it as: 'Over himself, his own body and mind, the individual is sovereign'" (Ad Hoc Committee on Medical Ethics, American College of Physicians 1984:264).

The rest of this chapter challenges the claim made by naturalism and much of biomedicine to autonomy from the social, cultural, moral, and spiritual orders and simply to reflect the "way things are." In this section ideas about personhood, society, morality, and religion will be considered. I argue that biomedicine is a powerful mechanism for reproducing these distinctions and for perpetuating the ideal of autonomy.

I begin here by considering some of the cultural infrastructure necessary for biomedical and naturalistic practice to operate and to make sense -- in fact to be desirable. This means looking at one major constellation of values in western culture, which I refer to as "individualism." In particular I consider the commitment to specific interpretations of freedom and liberty which are in turn embodied in modern western identity.

Naturalism and biomedicine demand and project a particular image of humans in order to work. Humans must be capable of pulling themselves out of their particular local, emotional, and historical context in order to objectively reflect on nature and themselves in absolute and neutral terms. Such objectification requires an understanding of humans as outside of society and able to view the world free from human meaning (that is, able to deanthropomorphize, Taylor 1985a,b; O'Neill 1985).

These assumptions are corroborated by and sustain the political and humanist philosophy of individualism so dominant in western societies (Dumont 1970, 1977). In essence, Individual is to Society what Atom is to Nature.

THE INDIVIDUAL AND SOCIETY

The Sovereign Individual

If we were to write a creation story for contemporary western society, we would have to begin with, "In the beginning there was the individual . . ." The individual has distinct priority in the individual/society equation, in fact, the "individual" is a central symbol in western cosmology.[12]

In the most extreme version of this cosmology (as found in North America), the individual is seen not only as culturally distinct, but *prior* to society and culture, even in conflict with them. Indeed, one of the goals of life is to free oneself from social and cultural determination. The prominent western ideal of individual freedom is not the autonomous free person living outside of society (this, in fact, is found in other societies) but the "autonomous distinct individual living-in-society" (Shweder and Bourne 1982:128). Society in this view is derivative -- the product of an association of individuals. In analytic terms, the definition of the person is "ego-centric" as opposed to "socio-centric," the society is based on atomism (individualism) rather than holism (Dumont 1970; Douglas 1970), relationships are contractual rather than organic. Supporting such a vision is a moral code of "rights" rather than "duties" (Shweder and Bourne 1982), based on "equality" rather than "hierarchy" (Dumont 1977).

The Individual as a Distinct Unit

The person in the European Protestant tradition is distinguished from his social position or role (Geertz 1977:9; Gaines 1982), and is thought to have a self that is distinct, nameable, and bounded by its skin such that what is "self" and "not self" is relatively clear (Lee 1959). Shweder and

Bourne (1982:105) found, for example, that Americans in contrast to Indians tended to define persons separately from social context by abstracting a set of absolute dispositions they identified with the person ("ego-centric"). This contrasts with defining a person in terms of "cases" or "contexts" ("socio-centric") or social roles as is done in many other societies. Similarly people are characterized in terms of "personality traits," such as "Type A personality," which are understood as belonging to an individual, not also to situations or to society (Benner 1984b; Wrubel 1985; Helman 1987). This "referential" way of defining the person contrasts with the Mediterranean "indexical" (Crapanzano 1980, cited in Gaines 1982) or the Japanese "socially relative" ways (Lebra 1976) which are based more on context. Like the atom, the self is considered a thing (an essence, not a process), often composed of elements, such as needs, values, beliefs. Language describes the self; it does not constitute it.

The self is identified not with the physical (the body) but with the conscious self, the seat of control (Lee 1959:133). The self (person) is conceived as a "dynamic center of awareness, emotion, judgment, and action organized into a distinctive whole . . ." (Geertz 1977:9). In fact, it is difficult to think of power without a driver, of significance without intention, of strategies without strategists (Foucault 1977; Taylor 1985a). Physical or emotional states, such as symptoms, emanate from the individual (Scheman 1983), not from situations or social groups.

The Individual as Prior to Society and Culture

Caudill (1976) observed that American mothers tended to perceive their babies from the earliest moments as entities separate from themselves, and thus see their role as translators and interpreters of their infants. Japanese mothers, he found, experience their babies more as parts of themselves, and therefore act *for* their infants rather than try to interpret them. The recent demand for the "rights of fetuses" also exemplifies a belief that we are human even before we emerge from the womb. A conviction exists among some that underneath the cultural/social coating, a real, unique, "deep" natural self exists, one that is "given" to the individual. Society poses a threat to this "real self." To be socially determined is to be weak, trapped, limited. Shils writes:

There is the metaphysical dread of being encumbered by something alien to oneself. There is a belief, corresponding to a feeling, that within each human being there is an individuality lying in potentiality which seeks an occasion for realization but is held in the toils of the rules, beliefs, and roles which society imposes. In a more popular . . . form, the concern "to establish one's identity," "to discover oneself," or to "find out who one really is" has come to be regarded as a first obligation of the individual . . . To be "true to oneself" means, they imply, discovering what is contained in the uncontaminated self, the self which has been freed from the encumbrance of accumulated knowledge, norms, and ideals handed down by previous generations (1981:10-11).[13]

The Self as Object to Itself: The Objectification of Subjectivity

As Foucault points out, the individual in western culture is understood as having a double aspect: both object and subject to itself (Foucault 1973; Dreyfus and Rabinow 1983). The self can be autonomous and self-determining because it can distance itself from itself without being totally detached. It is not determined by what it has, by its values, its relationships with others or with tradition. Its identity fixed once and for all, the self is relatively invulnerable to experience. As Sandel characterizes the stance taken in the work of John Rawls:

. . . No commitment could grip me so deeply that I could not understand myself without it. No transformation of life purposes and plans could be so unsettling as to disrupt the contours of my identity . . . Given my independence from the values I have, I can always stand apart from them . . . (Rawls 1980:544-5, paraphrased by Sandel 1982:62).

Like the atoms discussed earlier, relationships are external to the self. Thus one speaks of "co-operation," "domination," "subordination," all terms that imply separate identities. The experience of the self is seen to be outside of identity. Illness, for example, does not "alter" the self. Doctors then can tell patients, "not to worry, 5 weeks after the operation you'll be your old self again" (Ferguson 1984). It is this very autonomy from circumstance and the ability to stand back and reason, the ability to disengage, that is considered the hallmark of the freedom of the individual.

"Possession" of things becomes a primary mode of being for the self -- related but distanced from itself at the same time: "It is mine rather than me. If I lose it, I am the same person as before the loss" (Sandel 1982). Health too becomes a possession, even a commodity that one purchases.

Recent years have witnessed an increase in the objectification not only of the body but of the self (the "self talking about the self," Foucault 1980a; Armstrong 1983; Arney and Bergen 1984). American reforms in patient care, for example, by encouraging patients to "express themselves" and by providing objective data on them and their bodies (through practices such as "informed consent," options to review their medical charts and X-rays), at the same time further the objectification of the self. Health is seen as a project for the self to work on.

The *ideal* modern identity is as free of traces of social and cultural determination as possible. It strives to be its own author, consciously choosing its path, able to disengage itself and step back and judge rationally what it will be, where it will go. Self-control by the modern identity is potentially unlimited.[14]

Society as Derivative and Potential Enemy of Freedom

Much as the "whole" in nature derives from its parts, society (according to this western tradition) is seen as the "association" of predefined individuals (Dumont 1970; Shweder and Bourne 1982; Sandel 1982), made up of their independently existing characteristics (Scheman 1983:232). The "more than individual" is spoken of in terms of "the common good," the "public interest" (Fox and Swazey 1984); society is "decomposable" (Young 1980). Social institutions are often understood as providing *means* for realizing the *ends* of individuals, of realizing the ideas in people's heads, rather than as expressive phenomena in themselves (Taylor 1979:26; Lukes 1973). This interpretation of social institutions reflects an understanding of meaning and spirit as located inside the individual, not outside in public, shared space (Taylor 1985a; Geertz 1973).

Social relationships are also seen as a potential threat to freedom. Emotional involvement with other people can deplete the self. "Burn-out" becomes a concern for health care providers, worried about "losing themselves" in "over-involvement" with patients (Benner 1984a). Involved relationships with others are less often understood as being a potential *source* of freedom, power, and resources (Benner 1984a[15]; Fox and Swazey 1984).

This dominant modern approach to the individual and society has important implications for health, illness, and medical care. It undoubtedly contributes to a sense of isolation, anxiety, and loneliness, again most extremely in North America (Yankelovich 1981). Some heart attacks may in fact be emotionally "broken hearts" (Lynch 1977). Patient autonomy is the dominant goal of patient rehabilitation (Kaufman and Becker 1986); emphasis is on the patient who must learn new skills in order to enter social life, less on the incorporation of the patient into a social unit (Corin 1985). Further, the emphasis on independence puts the act of caring in crisis -- best illustrated by the nursing profession -- as caring in a self-transcending way raises concern for subservience (Benner 1984a; Fox and Swazey 1984).

MORALITY: NATURALISM AND THE IDEAL OF FREEDOM

Reflecting the individual as a core value (Fox and Swazey 1984; Dumont 1977), western morality is dedicated to defending the sovereignty of the individual and to ensuring freedom from interference, equal treatment, and opportunity for development. This support is expressed through a language of "rights" and "justice." The freedom it seeks is absolute: "The ability to act on one's own, without outside interference or subordination to outside authority" (Taylor 1985a:5). And it is universal: "For us, every man is in principle an embodiment of humanity at large, and as such he is equal to every other man and free" (Dumont 1977:4), as opposed to societies based on "hierarchy" and "holism". In contrast to the individual, society receives little support in western morality. Morality that represents the interests of society we know as "duties" or "obligations" -- terms that convey a sense of burden. A morality that supports connection, caring and responsibility is poorly articulated (Gilligan 1983; Fox and Swazey 1984; Bellah, et al. 1985; Shweder and Bourne 1982). Paralleling the strong dichotomy between individual and society, private and public morality are dichotomized.

In such a context, ideal social relationships are those which rise out of consent and contract between autonomous individuals (Shweder and Bourne 1982:129), a model that is growing quickly in medicine (Quill 1983). Honesty and communication that support equality rather than hierarchy take on particular value (Bellah, et al. 1985). Information becomes essential, the "rightful" means for the individual to act autonomously. Indeed, medical morality, called "bioethics," is dominated by a concern for rights, interpreted predominantly in terms of protection of patient autonomy *against* encroachment ("paternalism") and discrimination (justice, equality). Rarely are rights *for* "caring" and "community" (Gilligan 1983; Fox and Swazey 1984) asserted, or are qualities such as "decency" and kindness formally considered morally essential (Zaner 1982).

With nature understood as being devoid of human purpose, moral responsibility has followed two paths: subjectivism, by which values are left to human construction; and naturalism or utilitarianism,[16] by which the good and the bad are defined in terms of what is "natural," what is "normal," and what "works."

The Individual as Moral Agent: The "Disengaged" Self

With values no longer anchored in the "more than human," one approach has been to leave them to the rational choice of individuals. MacIntyre describes (albeit in somewhat extreme terms) this approach to morality:

It is in this capacity of the self to evade any necessary identification with any particular contingent state of affairs that some modern philosophers . . . have seen the essence of moral agency. To be a moral agent is, on this view, precisely to be able to stand back from any and every situation in which one is involved, from any and every characteristic that one may possess, and to pass judgment on it from a purely universal and abstract point of view . . . it is in the self and not in social roles or practices that moral agency has to be located (1981:30).

This moral autonomy of the individual reflects the naturalist cosmology described earlier.

Only in a universe empty of *telos* . . . is it possible to conceive a subject apart from and prior to its purposes and ends. Only a world ungoverned by a purposive order leaves principles of justice open to human construction and conceptions of the good to individual choice (Sandel 1982:175).

This ideal of moral responsibility, then, is considered essentially asocial, universal, abstract, and disengaged, the epitomy of moral freedom. But it presents several problems for individual and social life. Social values are denied and come to be regarded as private attributes of the individual and of "mind," rather than of public life. As private, they are not considered open to public discourse. Morality then runs the risk of being regarded as *nothing but* personal preference (referred to varyingly as "subjectivism," "relativism" [Bernstein 1983] or "nihilism" [Dreyfus and Rabinow 1983]), what MacIntyre calls "emotivism":

. . . the doctrine that all evaluative judgments and more specifically all moral judgments are *nothing but* expressions of preference, expressions of attitude or feeling, insofar as they are moral or evaluative in character (1981:11).

Morality from Naturalism and Biomedicine

While naturalism separated nature from morality, naturalism itself serves as a basis for morality. To say something is natural provides a bedrock of support (Rosaldo 1983). To say something is "rational" is even stronger. Rationality is not only a dominant western value, it is where values are hidden (Young 1987). Yet it again excludes social responsibility from the picture. Further, substitution of mechanistic for teleological explanations of nature has generated an abundance of "means" with little guidance as to ends. Medicine's proliferating technology has become an end in itself. Bellah and his colleagues speak for many physicians: "For most of us, it is easier to think about how to get what we want than to know what exactly we should want" (1985:21). The moral dilemmas regularly encountered in contemporary medicine may best be understood in the context of the denial of shared values and the reframing of values in terms of rationality and the individual instead of culture and society (Young 1987).

Biomedicine has both created and stepped in to fill the apparent moral vacuum, replacing moral idioms with medical ones -- illness replaces sin, rehabilitation replaces corporal punishment, domination, or exclusion, normality/abnormality replace vice and virtue, and "truth" becomes power (Foucault 1975, 1977, 1980). In this way, medicine has inherited the former clientele of the church, the prison, the social periphery: "madness" is transformed into "mental illness," "teenage illegitimacy" becomes "teenage pregnancy," "anomalies" become "handicaps" (Foucault 1965; Arney and Bergen 1983, 1984).

In conclusion, the image of the modern self, capable of nearly absolute freedom from social determinacy, able to disengage in order to reach a higher level of truth, is a western ideal of what it means to be human. Thus the disengagement that naturalism prescribes is important not only as a means to an end -- a means to freedom through control and knowledge -- but also as an end in itself. To be disengaged means to be free to be one's own unique author; this attitude epitomizes the modern western identity particularly in North America.

SPIRITUAL FREEDOM: NATURALISM AS A MEANS BEYOND THE
"MERELY HUMAN"

This ideal of absolute freedom rises above the "merely human." Transcending time, place, contingency, bias, and emotion, it takes us beyond the vicissitudes of humanity (Taylor 1985a). Biomedicine in particular, through its technological rites, presents an ideal of life beyond the ravages of time -- beyond death (Comaroff 1984; Gordon 1987a). Are we not back into the supernatural camp which science aspired to leave? Is not our attachment to the scientific viewpoint also connected to our ideal vision of ourselves and our spiritual need to believe in an absolute truth, understood from an absolute, universal position (Taylor 1985a), not unlike that in many religions? Much as the western religious traditions sought to rise above the "merely human" and be free from human emotions, human cares, and the demands of social life, we may see the ideal modern free "subject" as reflecting a similar aspiration.

This ideal also envisions humans detaching themselves from the world -- through objectification -- and able to understand it from a universal, non-particular standpoint. It also provides a sense of spiritual freedom:

like pre-Enlightenment religion, it envisions a place for humans to stand outside the context of human emotions in order to determine what is truly important:

In one case, that of the tradition, this is seen as a larger order which is the locus of more than human significance; in the modern case, it is an order of nature which is meant to be understood free of any significance at all, merely naturalistically (Taylor 1985a:113).

But something of the same aspiration is evident in both, the aspiration for something more than the merely human. This desire is probably much too fundamental to human life to ever be eliminated. The paradox of modernity is that the search goes on even when denied (ibid.). In this way medicine, in presenting an image of life beyond time and space through rituals that dramatize the dominance of humans over nature, occupies similar religious space as the medical practices in more traditional societies.

III. THE MUTUAL SUPPORT AND REINFORCEMENT BETWEEN INDIVIDUALISM AND NATURALISM/BIOMEDICINE

Society Uses Naturalism and Biomedicine to Assert Values

Naturalism and particularly biomedicine sustain the viability of the ideal modern identity: independent of society and culture, owner of his or her own symptoms, increasingly able now to detach from self and body to observe, cultivate, and even contractually join the physician in rationally treating his or her own case. In general, medicine offers a strong sense that humans can overcome nature, no longer a victim, but in the omnipotent driver's seat. The radical autonomy projected in western society is a social construct, aided greatly by naturalism and biomedicine.

In fact, as indicated above, biomedicine is now a central stage in which the assertion of autonomy takes place. Rights are claimed for everyone and everything, from fetuses to wombs, to neonatals, to people in comas. Contracts between "informed partners" are becoming the means for increasing the "equality" of patients *vis-à-vis* physicians. While such contracts are an obvious corrective to years of medical domination, how much are we still seeing a rebellion against individuals' need for society, for dependence, or for nurturing? How much is biomedicine the central social space in which the causes of autonomy, individualism, and control over nature are celebrated, perhaps in a religious fashion (Comaroff 1984; Gordon 1987a)?

The Resemblance of Naturalism/Biomedicine and Individualism

Naturalism and individualism have much in common. Both take the atom as the model and have little use for society and culture. Both the naturalists (who regard humans as material beings) on the one hand, and the individualists (who treat the "essentially human being" as pure will and reason) on the other leave the space between subjects, objects, and atoms uncharted and impotent (Kass 1985). The one sees all unity underlying apparent diversity, the other sees diversity underlying apparent similarity. Neither recognizes the dignity of society or of the body (ibid.).

The symmetry between nature and the human world in naturalist/ biomedical cosmology and individualism is not unusual. In many societies conceptions of the natural order appear to be patterned after the social order (see Needham 1962; Lock 1980; Douglas 1970). This symmetry raises several possibilities: How much is naturalism a projective system for society, a wish fulfillment in that our ideal is projected onto nature (as a model for society)? Is the social world modelled after the natural (Kass 1985; Shweder and Bourne 1982:130)? Does this symmetry reflect a human need to belong to the universe (Gregory 1984)?

The Non-Autonomy of Naturalism and Biomedicine

To continue to blame biomedicine exclusively for its ills is to reproduce its own ideology -- that it is independent of society and has an exclusive relationship with nature. Biomedicine's practices take place against a particular background of what it means to be human. When calling it "dehumanizing" we do well to consider what it means in western society to be "human." Similarly as medicine strives to be more "holistic," we must ask whether real "holism" can exist in medicine in a society in which atomism prevails to such a profound extent? Or is holism the ideal to strive for? How much has biomedicine been doing its duty for society in supporting these views of the autonomous man? The social dimension of illness and medical systems extends beyond simply socializing the individual through disease and illness (Young 1982). Sickness expounds a truth about the order of the world as much as the body of the sick person.

Further, we must consider whether the adamant and relatively successful denial of the social dimension in medicine and naturalism is not paradoxically the exact evidence of the power of social assumptions and practices. To sustain as tenable an ideal of the autonomous, cultureless man, when from our first moments of life we exist in a social context, requires tremendous cultural and social support. Biomedicine and naturalism provide much of it.

ACKNOWLEDGEMENTS

I am very grateful to Margaret Lock for her continuous support and invaluable contribution throughout the entire development of this article. I also thank Victoria Kahn, Elisabeth Giansiracusa, Sharon Kaufman, Jessica Muller and Patricia Benner for their suggestions and editorial contributions.

NOTES

1. See for example the literature on the "new science" or "new physics" (Capra 1982; Dossey 1984; Wilber 1984); philosophical critique of traditional science, and social studies of science, such as Fleck 1935; Kuhn 1970; Mendelsohn and Elkana 1981; Knorr-Cetina 1981; Latour and Woolgar 1979; Barnes 1977; Durbin 1980; and feminist critiques of science (Harding and Hintikka 1983; Richters 1985) among many others.

2. Among earlier examples see Parsons 1951; Fox 1959; Becker, et al. 1961; Freidson 1970; Foucault 1965, 1975, 1980a; Powles 1973; Fabrega 1975; Hahn and Kleinman 1983, in press; Kleinman 1973, 1980; Hahn and Gaines 1985; Eisenberg and Kleinman 1981; Mishler, et al. 1981; Wright and Treacher 1982; Armstrong 1983; Young 1976, 1978, 1980; Figlio 1977; Comaroff 1978; Illich 1975. See also works cited in Note 4 below.

3. See following discussion for explanation of the use of "background" in this essay.

4. Some of works that already have addressed this topic include: (a) *on views of nature and ontology* by (Hahn 1985; Hahn and Kleinman 1983, in press); Douglas (1966, 1970); Wright 1979; Foucault 1973; Comaroff 1982, 1984; Powles 1973; Gaines 1987; Young 1978, 1980; Gordon 1980; (b) *on medical epistemology* by Foucault 1975, 1977; Young 1978, 1980; Kleinman 1973; Eisenberg 1977; Good and Good 1980, 1981; Wright 1979; and Wright and Treacher 1982; (c) *on personhood and the self* by Foucault 1975, 1980a; Armstrong 1983; Arney and Bergen 1984; Manning and Fabrega 1973; Hahn 1985; and Gaines 1982, 1985; (d) and *on morality* by Parsons, Fox, Lidz 1975; Fox 1974; Fox and Swazey 1978, 1984; Sontag 1978; Engelhardt Jr. 1974; Carlton 1978; Bosk 1979, as well as other works in the field of "medical ethics" (see Engelhardt and Erde 1980 for summary of philosophical ones).

5. See Kirmayer, Helman (this volume) on psychosomatic medicine; Guttmacher (1979), Berliner and Salmon (1980), Comaroff (1982), Dossey

(1984) on holistic health; Lock and Lella (1986) for a critique of Engel's model (1977); Fox and Swazey (1984) on bioethics; Armstrong (1983), Arney and Bergen (1984) on "patient autonomy."

6. While in no way asserting it as primary.

7. Specifically by avoiding "subjective" or "anthropomorphic" explanations, and by giving explanations in absolute terms (Taylor 1985a:2). This use of the term "naturalism," which I will follow in this paper, diverges from Barnes (1977) and Young (1978). They distinguish naturalism from "positivism"; while both approaches assume an objective reality, naturalism in their version allows for the social production of a given positive reality. In this essay, many of the traits of positivism outlined by Young are included in my use of the term naturalism.

8. International differences (see for example Lock, Low, this volume; Weisberg and Long 1984; Gaines 1987); generational differences (Good 1985; Williams and Boulton, this volume); specialty, subspecialty and institutional differences (Helman 1985; Gifford 1986; K. Taylor, Gordon, this volume); differences in social/cultural backgrounds (Gaines 1982; 1985; Good, et al. 1985) for examples.

9. Geertz (1975) refers to "common sense systems," Young to "tacit beliefs" (1978:100) that enter discourse and thinking as unexamined or unarticulated assumptions in contrast to explicit beliefs.

10. It is beyond the scope of this paper to trace the roots of what I am calling naturalism. Many extend back to the Greeks, others to the Enlightenment, and others to contemporary western society. Similarly, limited space precludes discussion of the particular historical conditions and "producers" of these assumptions, and the interests involved in their acceptance (for example see Wright, this volume, 1979; Wright and Treacher 1982; Figlio 1977; Armstrong 1983; Brandt 1985; Comaroff 1982; Taussig 1980; Osherson and AmaraSingham 1981).

11. In light of Kirmayer's article (this volume) and the number of recently published accounts on this subject, I am giving very little space to this topic (see Foucault 1975, 1977; Armstrong 1983; Turner 1984; Freund 1982; O'Neill 1985; Scheper-Hughes and Lock 1987; Comaroff 1985, among others).

12. As Gaines (1982) points out, at least two models of personhood dominate western culture – the Mediterranean and the Protestant European. As the latter is most associated with science and medicine I shall discuss only it in this paper. Further, the most extreme version of this approach is found in North America.

13. In addition to the impact of naturalism on the view of the individual, a second major force in the contemporary western view of the self came from the Romantic period of the 18th and 19th centuries. During this time the belief in the "radical freedom" of individuals was articulated. This approach to individualism is referred to as "expressive individualism" (Bellah, et al.

1985:333-334; see also Taylor 1979).

14. This version applies most to the North American context. Note 12 is also operative in this discussion.

15. Benner (1984a) illustrates, for example, the freedom a committed relationship between nurse and patient can allow and the "positive" power (for example, for transformation) of involved relationships. These and other issues are discussed in Gordon, "Alternatives to Tenacious Assumptions in Western Medicine," in progress.

16. For reasons of space, elaboration of these important topics is beyond the capacity of this essay.

REFERENCES

Ad Hoc Committee on Medical Ethics, American College of Physicians
    1984   American College of Physicians Ethics Manual. Part II: Research, Other Ethical Issues. Recommended Reading. Annals of Internal Medicine 101:263-274.
American Board of Internal Medicine
    1979   Clinical Competence in Internal Medicine. Annals of Internal Medicine 90:402-411.
Armstrong, David
    1983   Political Anatomy of the Body. Medical Knowledge in Britain in the Twentieth Century. Cambridge: Cambridge University Press.
Arney, William Ray
    1982   Power and the Profession of Obstetrics. Chicago: University of Chicago Press.
Arney, William Ray and Bernard J. Bergen
    1983   The Anomaly, the Chronic Patient and the Play of Medical Power. Sociology of Health and Illness 5:1-24.
    1984   Medicine and the Management of Living. Chicago: University of Chicago Press.
Barnes, Barry
    1977   Interests and the Growth of Knowledge. London: Routledge and Kegan Paul.
Baron, Richard J.
    1981   Bridging Clinical Distance: An Empathic Rediscovery of the Known. The Journal of Medicine and Philosophy 6:5-23.
Becker, Howard, et al.
    1961   Boys in White. Chicago: University of Chicago Press.
Bellah, Robert, et al.
    1985   Habits of the Heart. Berkeley: University of California Press.

Benner, Patricia
  1984a From Novice to Expert. Excellence and Power in Clinical Nursing
        Practice. Menlo Park, Addison-Wesley.
  1984b Stress and Satisfaction on the Job: Work Meanings and Coping of Mid-
        Career Men. New York: Praeger Scientific Press.
Berliner, Howard S. and Salmon, J. Warren
  1980 The Holistic Alternative to Scientific Medicine: History and Analysis.
       International Journal of Health Services 10:133-147.
Berman, Morris
  1984 The Re-Enchantment of The World. New York: Bantum.
Bernstein, Richard
  1983 Beyond Objectivism and Relativism: Science, Hermeneutics, and Praxis.
       Philadelphia: University of Pennsylvania Press.
Bhasker, Roy
  1979 The Possibility of Naturalism. Brighton: Harvester Press.
Bosk, Charles
  1979 Forgive and Remember. Chicago: University of Chicago Press.
Bourdieu, Pierre
  1977 Outline of a Theory of Practice. London: Cambridge University Press.
Brandt, Alan
  1985 No Magic Bullet. Cambridge: Harvard University Press.
Capra, Fritjof
  1982 The Turning Point. New York: Bantam Books.
Carlton, Wendy
  1978 "In Our Professional Opinion . . ." Notre Dame: University of Notre
       Dame Press.
Cassell, Eric J.
  1976 The Healer's Art. New York: Penguin Books.
  1984 How is the Death of Barney Clark to be Understood? *In* M. Shaw
       (ed.), After Barney Clark: Reflections on the Utah Artificial Heart Pro-
       gram. Pp. 25-41. Austin, Texas: University of Texas Press.
Caudill, William
  1976 The Cultural and Interpersonal Context of Everyday Health and Illness
       in Japan and America. *In* C. Leslie (ed.), Asian Medical Systems. Pp.
       159-183. Berkeley: University of California Press.
Comaroff, Jean
  1978 Medicine and Culture: Some Anthropological Perspectives. Social Sci-
       ence and Medicine 12B:247-254.
  1982 Medicine: Symbol and Ideology. *In* P. Wright and A. Treacher (eds.),
       The Problem of Medical Knowledge. Pp. 49-68. Edinburgh: Edinburgh
       University Press.
  1983 The Defectiveness of Symbols or the Symbols of Defectiveness? On
       the Cultural Analysis of Medical Systems. Culture, Medicine, and

Psychiatry 7:3-20.

1984   Medicine, Time and the Perception of Death. Listening: Journal of Religion and Culture 19:155-169.

1985   Body of Power, Spirit of Resistance: The Culture and History of a South African People. Chicago: University of Chicago Press.

Corin, Ellen

1985   Les Espaces Sociaux et Symboliques de la Réhabilitation de Patients Schizophrenes. *In* M. Pandolfi and A. Zempléni (eds.), Ethnopsychiatry Today (*Ethnopsichiatria Oggi*). *Psichiatria e Psicoterapia Analitica* IV:2:91-121.

Crapanzano, Vincent

1980   The Subject as Object: The Phenomenology of Encounter. Paper presented at the 24th Annual meeting of the American Academy of Psychoanalysis: San Francisco, May.

Descartes, René

1950   Discourse on Method. Indianapolis: Liberal Arts Press. Translation, Laurence J. Lafleur (original edition 1637).

Dossey, Larry

1984   Beyond Illness. Boulder, Colorado: New Science Library.

Douglas, Mary

1966   Purity and Danger. London: Routledge and Kegan Paul.

1970   Natural Symbols. Middlesex, England: Penguin.

Dreyfus, Hubert L.

1979   What Computers Can't Do. Revised Edition. New York: Harper and Row.

1980   Holism and Hermeneutics. Review of Metaphysics 34:3-23.

In press   Being-in-the-World. A Commentary on Heidegger's *Being and Time, Division I*. Cambridge, MIT Press.

Dreyfus, Hubert and Paul Rabinow

1983   Michel Foucault: Beyond Structuralism and Hermeneutics. Second Edition. Chicago: University of Chicago Press.

Duff, Raymond S. and Hollingshead, August B.

1968   Sickness and Society. New York: Harper and Row Publishers.

Dumont, Louis

1970   Homo Hierarchicus. London: Paladin.

1977   From Mandeville to Marx. Chicago: University of Chicago Press.

Durbin, Paul T. (ed.)

1980   A Guide to the Culture of Science, Technology, and Medicine. New York: The Free Press.

Eisenberg, Leon

1977   Disease and Illness: Distinctions Between Professional and Popular Ideas of Sickness. Culture, Medicine and Psychiatry 1:9-23.

Eisenberg, Leon and Arthur Kleinman (eds.)
    1981   The Relevance of Social Science for Medicine. Dordrecht, Holland: D. Reidel Publishing Co.

Engel, George
    1977   The Need For a New Medical Model: A Challenge For Biomedicine. Science 196:129-136.

Engelhardt, Jr, J.H. Tristram
    1973   The Philosophy of Medicine: A New Endeavor. Texas Reports on Biology and Medicine 31:445.
    1974   The Disease of Masturbation. Values and the Concept of Disease. Bulletin of the History of Medicine 48:234-248.

Engelhardt, Jr, J.H. Tristram and Edmund L. Erde
    1980   Philosophy of Medicine. In P. Durbin (ed.), A Guide to the Culture of Science, Technology and Medicine. Pp. 364-461. New York: The Free Press.

Fabrega, Horatio
    1975   The Need for an Ethnomedical Science. Science 189:969-975.

Ferguson, Marilyn
    1984   The Acquarian Conspiracy. New York: Basic Books.

Figlio, Karl
    1977   The Historiography of Scientific Medicine. An Invitation to the Human Sciences. Comparative Studies in Society and History 19:262-286.

Fiske, Donald and Richard Shweder (eds.)
    1986   Metatheory in Social Science. Chicago: University of Chicago Press.

Fleck, Ludwik
    1979   The Genesis and Development of a Scientific Fact. Chicago: University of Chicago Press (original edition 1935).

Foucault, Michel
    1965   Madness and Civilization. New York: Vintage.
    1973   The Order of Things. The Archaeology of the Human Sciences. New York: Vintage.
    1975   The Birth of the Clinic: The Archaeology of Medical Perception. New York: Vintage Books.
    1977   Discipline and Punish. New York: Vintage Books.
    1980a The History of Sexuality, Vol. 1. An Introduction. New York: Vintage Books.
    1980b Power/Knowledge. Selected Interviews and Other Writings, 1972-1977. Colin Gordon (ed.). New York: Pantheon.

Fox, Renée
    1959   Experiment Perilous. Glencoe, Illinois: The Free Press.
    1974   Ethical and Existential Developments in Contemporaneous American Medicine: Their Implications for Culture and Society. Milbank Memorial Fund Quarterly 52 (Fall): 445-483.

Fox, Renée C. and Judith P. Swazey
    1978   The Courage to Fail: A Social View of Organ Transplants and Dialysis. Chicago: University of Chicago Press.
    1984   Medical Morality is not Bioethics - Medical Ethics in China and the United States. Perspectives in Biology and Medicine 27:336-360.
Freidson, Eliot
    1970   The Profession of Medicine. New York: Dodd and Mead.
Freund, Peter E.S.
    1982   The Civilized Body. Philadelphia: Temple University Press.
Gaines, Atwood D.
    1982   Cultural Definitions, Behavior and the Person in American Psychiatry. In A.J. Marsella and G.M. White (eds.), Cultural Conceptions of Mental Health and Therapy. Pp. 167-192. Dordrecht, Holland: D. Reidel Publishing Co.
    1985   The Once- and the Twice-Born: Self and Practice Among Psychiatrists and Christian Psychiatrists. In R. Hahn and A. Gaines (eds.), Physicians of Western Medicine. Pp. 223-243. Dordrecht, Holland: D. Reidel Publishing Co.
    1987   Culture and Medical Knowledge in France and America. In A. Gaines and A. Young (eds.), The Social Origins of Biomedical Knowledge. Special Issue of Medical Anthropological Quarterly. Forthcoming.
Geertz, Clifford
    1973   The Interpretation of Cultures. New York: Basic Books.
    1975   Common Sense as a Cultural System. The Antioch Review 33 (Spring): 5-26.
    1977   On the Nature of Anthropological Understanding. In Annual Editions in Anthropology. Guilford, Conn.: Dushkin.
Gilligan, Carol
    1983   Do the Social Sciences Have an Adequate Theory of Moral Development? In N. Haan, et al. (eds.), Social Science as Moral Inquiry. Pp. 33-51. New York: Columbia University Press.
Good, Byron and Mary-Jo DelVecchio Good
    1980   The Meaning of Symptoms: A Cultural Hermeneutic Model for Clinical Practice. In L. Eisenberg and A. Kleinman (eds.), The Relevance of Social Science for Medicine. Pp. 65-196. Dordrecht, Holland: D. Reidel Publishing Co.
    1981   The Semantics of Medical Discourse. In E. Mendelsohn and Y. Elkana (eds.), Sciences and Cultures. Pp. 177-212. Dordrecht, Holland: D. Reidel Publishing Co.
Good, Byron, et al.
    1985   Reflexivity, Countertransference and Clinical Ethnography: A Case From a Psychiatric Cultural Consultation Clinic. In R.A. Hahn and A.D. Gaines (eds.), Physicians of Western Medicine. Pp. 193-221.

Dordrecht, Holland: D. Reidel Publishing Co.

Good, Mary-Jo DelVecchio
1985   Discourses on Physician Competence. *In* R. Hahn and A. Gaines (eds.), Physicians of Western Medicine. Pp. 247-267. Dordrecht, Holland: D. Reidel Publishing Co.

Gordon, Deborah R.
1980   Assumptions in American Medicine. Unpublished manuscript.
1987a Magico-Religious Dimensions of Western Medicine: The Case of the Artificial Heart. Paper presented at the First National Conference of the Cultural Anthropology of Complex Societies, Rome, Italy, May 27-30, 1987.
1987b Magical Aspects of Biomedicine (*Aspetti Magici in Biomedicina*). *Psichiatria e Psicoterapia Analitica* VI:I:93-98.
n.d.   Alternatives to Tenacious Assumptions in Western Medicine. In preparation.

Gould, Stephen Jay
1983   Review of F. Capra's, The Turning Point. New York Review of Books, March 3, 1983.

Gregory, Michael S.
1984   Science and Humanities: Toward a New Worldview. *In* D. Brock and A. Harward (eds.), The Culture of Biomedicine. Pp. 11-33. Newark: University of Delaware Press.

Guttmacher, Sally
1979   Whole in Body, Mind and Spirit. Holistic Health and the Limits of Medicine. Hastings Center Reports 9:15-21.

Haan, Norma, et al. (eds.)
1983   Social Science as Moral Inquiry. New York: Columbia University Press.

Habermas, Jürgen
1970   Toward a Rational Society. Boston: Beacon Press.

Hahn, Robert A.
1985   A World of Internal Medicine. *In* R. Hahn and A. Gaines (eds.), Physicians of Western Medicine. Pp. 51-111. Dordrecht, Holland: D. Reidel Publishing Co.
1986   Culture-Bound Syndromes Unbound. Social Science and Medicine 22:6:679-685.

Hahn, Robert A. and Arthur M. Kleinman
1983   Biomedical Practice and Anthropological Theory: Frameworks and Directions. Annual Review of Anthropology. Pp. 305-333. Palo Alto: Annual Review Press.
In press Biomedicine as a Cultural System. *In* Massimo Piattelli-Palmarini (ed.), Encyclopedia of the Social History of the Biomedical Sciences. Milan: Franco Maria Ricci Publishing Co.

Hahn, Robert A. and Atwood D. Gaines (eds.)
  1985 Physicians of Western Medicine. Dordrecht, Holland: D. Reidel Publishing Co.
Harding, Sandra and Merill B. Hintikka (eds.)
  1983 Discovering Reality. Dordrecht, Holland: D. Reidel Publishing Co.
Heidegger, Martin
  1962 Being and Time. New York: Harper and Row.
Helman, Cecil G.
  1985 Disease and Pseudo-Disease: A Case History of Pseudo-Angina. In R. Hahn and A. Gaines (eds.), Physicians of Western Medicine. Pp. 293-331. Dordrecht, Holland: D. Reidel Publishing Co.
  1987 Heart Disease and the Cultural Construction of Time: The Type A Behavior Pattern as a Western Culture-Bound Syndrome. Social Science and Medicine 25:969-979.
Illich, Ivan
  1975 Medical Nemisis. London: Maryon Boyars.
Jones, W.T.
  1976 World View and Asian Medical Systems. In C. Leslie (ed.), Asian Medical Systems. Pp. 383-404. Berkeley: University of California Press.
Jonas, Hans
  1966 The Nobility of Sight. In The Phenomenon of Life. Toward a Philosophical Biology. New York: Dell.
Kass, Leon R.
  1985 Thinking About the Body. The Hastings Center Report (Feb.):20-30.
Kaufman, Sharon and Gay Becker
  1986 Stroke: Health Care on the Periphery. Social Science & Medicine 22:983-989.
Keller, Evelyn Fox and Christine R. Grontkowski
  1983 The Mind's Eye. In S. Harding and M.B. Hintikka (eds.), Discovering Reality. Pp. 207-224. Dordrecht, Holland: D. Reidel Publishing Co.
King, Lester
  1975 Explanations of Disease. In H.T. Engelhardt, Jr. and S.F. Spicker (eds.), Evaluation and Explanation in the Biomedical Sciences. Pp. 141-150. Dordrecht, Holland: D. Reidel Publishing Co.
Kleinman, Arthur
  1973 Medicine's Symbolic Reality: On the Central Problem in the Philosophy of Medicine. Inquiry 16:206-213.
  1980 Patients and Healers in the Context of Culture. Berkeley: University of California Press.
  1986 Some Uses and Misuses of the Social Sciences in Medicine. In D. Fiske and R. Shweder (eds.), Metatheory in Social Science. Pp. 222-245. Chicago: University of Chicago Press.

Knorr-Cetina, Karin D.
    1981   The Manufacture of Knowledge. Oxford. Pergamon.
Kuhn, Thomas
    1970   The Structure of Scientific Revolutions. Chicago: University of Chi-
           cago Press.
Lakoff, George and Johnson, Mark
    1980   Metaphors We Live By. Chicago: University of Chicago Press.
Latour, Bruno and Steve Woolgar
    1979   Laboratory Life. Beverley Hills: Sage.
Lebra, Takie
    1976   Japanese Patterns of Behavior. Honolulu: University Press of Hawaii.
Lee, Dorothy
    1959   Freedom and Culture. Englewood Cliffs, N.J.: Prentice-Hall.
Lock, Margaret
    1980   East Asian Medicine in Urban Japan: Varieties of Medical Experience.
           Berkeley: University of California Press.
    1985   Models and Practice in Medicine: Menopause as Syndrome or Life
           Transition? In R. Hahn and A.D. Gaines (eds.), Physicians of Western
           Medicine. Pp. 115-140. Dordrecht, Holland: D. Reidel Publishing Co.
Lock, Margaret and Joseph Lella
    1986   Reforming Medical Education: Towards a Broadening of Attitudes. In
           S. McHugh and T.M. Vallis (eds.), Illness Behavior: A Multidisciplinary
           Model. Pp. 47-70. New York: Plenum Press.
Lukes, Steven
    1973   Individualism. Oxford: Basil Blackwell.
Lynch, James J.
    1977   The Broken Heart: The Medical Consequences of Loneliness. New
           York: Basic Books.
MacIntyre, Alasdair
    1981   After Virtue. Notre Dame: University of Notre Dame Press.
Manning, Peter and Horatio Fabrega
    1973   The Experience of Self and Body: Health and Illness in the Chiapas
           Highlands. In G. Psathas (ed.), Phenomenological Sociology. Pp.
           59-73. New York: Wiley.
Maretzki, Thomas
    1985   The Physician in Healer-Centered Research. In R. Hahn and A. Gaines
           (eds.), Physicians of Western Medicine. Pp. 23-47. Dordrecht, Hol-
           land: D. Reidel Publishing Co.
Mauss, Marcel
    1985   A Category of the Human Mind: The Notion of Person, The Notion of
           Self. In M. Carrithers et al. (eds.), The Category of the Person. Pp.
           1-25. Cambridge: Cambridge University Press (original edition 1938).

Mendelsohn, Everett and Yehuda Elkana (eds.)
1981 Sciences and Cultures. Anthropological and Historical Studies of the Sciences. Dordrecht, Holland: D. Reidel Publishing Co.
Merleau-Ponty, Maurice
1962 Phenomenology of Perception. London: Routledge & Kegan Paul.
Millman, Marcia
1978 The Unkindest Cut. New York: William Morrow.
Mishler, Elliot, et al.
1981 Social Contexts of Health, Illness, and Patient Care. Cambridge: Cambridge University Press.
Needham, Joseph
1962 Science and Civilization in China. Vol. 2. Cambridge: Cambridge University Press.
Obeyesekere, Gananath
1981 Medusa's Hair. Chicago: University of Chicago Press.
O'Neill, John
1985 Five Bodies. Ithaca: Cornell University Press.
Osherson, Samuel and Lorna AmaraSingham
1981 The Machine Metaphor in Medicine. In E. Mishler, et al., Social Contexts of Health, Illness, and Patient Care. Cambridge: Cambridge University Press.
Parsons, Talcott
1951 The Social System. Glencoe, Ill. The Free Press.
Parsons, Talcott, Renée Fox, and Litz V.
1972 The "Gift of Life" and its Reciprocation. Social Research 39:367-415.
Powles, John
1973 On the Limitations of Modern Medicine. Science, Medicine and Man 1:1-30.
Quill, Timothy E.
1983 Partnerships in Patient Care: A Contractual Approach. Annals of Internal Medicine 98:228-234.
Rawls, John
1980 Kantian Constructivism in Moral Theory. Journal of Philosophy 77:515-72.
Rawlinson, Mary
1982 Medicine's Discourse and the Practice of Medicine. In V. Kestenbaum (ed.), The Humanity of the Ill. Pp. 69-85. Knoxville: University of Tennessee Press.
Reiser, Stanley Joel
1978 Medicine and the Reign of Technology. Cambridge: Cambridge University Press.
Richters, Annemiek
1985 Perception and Apperception in Cultural Anthropology: A Feminist Cri-

tique. Unpublished manuscript.

Roberts, Cecilia M.
  1977   Doctor and Patient in the Teaching Hospital. Lexington, Mass: D.
         Heath Co.

Rosaldo, Michelle
  1983   Moral/Analytic Dilemmas Posed by the Intersection of Feminism and
         Social Sciences. *In* N. Haan, et al. (eds.), Social Science as Moral
         Inquiry. Pp. 76-95. New York: Columbia University Press.

Sacks, Oliver
  1982   Awakenings. London: Pan.

Sandel, Michael J.
  1982   Liberalism and the Limits of Justice. Cambridge: Cambridge University
         Press.

Scheman, Naomi
  1983   Individualism and the Objects of Psychology. *In* S. Harding and M.B.
         Hintikka (eds.), Discovering Reality. Pp. 225-244. Dordrecht, Hol-
         land: D. Reidel Publishing Co.

Scheper-Hughes, Nancy and Margaret M. Lock
  1987   The Mindful Body: A Prolegomenon to Future Work in Medical Anthro-
         pology. Medical Anthropology Quarterly 1:6-41.

Shils, Edward
  1981   Tradition. Chicago: University of Chicago Press.

Shweder, Richard A. and Edmund J. Bourne
  1982   Does the Concept of the Person Vary Cross-Culturally? *In* A.J. Mar-
         sella and G.M. White (eds.), Cultural Conceptions of Mental Health
         and Therapy. Pp. 97-137. Dordrecht, Holland: D. Reidel Publishing
         Co.

Sontag, Susan
  1978   Illness as Metaphor. New York: Vintage Books.

Starr, Paul
  1982   The Social Transformation of American Medicine. New York: Basic
         Books.

Straus, Erwin W.
  1963   Born to See, Bound to Behold: Reflections on the Function of Upright
         Posture in the Esthetic Attitudes. *In* S.F. Spicker (ed.), The Philosophy
         of the Body. Pp. 334-362. New York: Quadrangle.

Taussig, Michael
  1977   The Genesis of Capitalism Amongst a South American Peasantry: Dev-
         il's Labour and the Baptism of Money. Comparative Studies in Society
         and History 19:130-155.
  1980   Reification and the Consciousness of the Patient. Social Science and
         Medicine 14B:3-13.

Taylor, Charles
  1979 Hegel and Modern Society. Cambridge: Cambridge University Press.
  1985a Human Agency and Language. Philosophical Papers 1. Cambridge: Cambridge University Press.
  1985b Philosophy and the Human Sciences. Philosophical Papers 2. Cambridge: Cambridge University Press.
Turner, Bryan
  1984 The Body and Society. London: Basil Blackwell.
Weisberg, Daniel and S.O. Long (eds.)
  1984 Biomedicine in Asia: Transformations and Variations. Special Issue. Culture, Medicine, and Psychiatry 8 (2).
Wilber, Ken (ed.)
  1984 Quantum Questions. Boulder, Colorado: New Science Library.
Will, George F.
  1987 There is No "Right Way" to Rent a Human Womb. International Herald Tribune. January 23, 1987, p. 5.
Wittgenstein, Ludwig
  1958 Philosophical Investigations. New York: Basil Blackwell.
Worsley, Peter
  1982 Non-Western Medical Systems. Annual Review of Anthropology 11:315-248.
Wright, Peter W.G.
  1979 Some Recent Developments in the Sociology of Knowledge and Their Relevance to the Sociology of Medicine. Ethics in Science and Medicine 6:93-104.
Wright, Peter W.G. and Treacher, Andrew (eds.)
  1982 The Problem of Medical Knowledge. Edinburgh: Edinburgh University Press.
Wrubel, Judith C.
  1985 Personal Meanings and Coping Processes. Ph.D. Dissertation. Human Development and Aging, University of California, San Francisco.
Yankelovich, Daniel
  1981 New Rules: Searching for Self-Fulfillment in a World Turned Upside Down. New York: Bantam.
Young, Allan
  1976 Some Implications of Medical Beliefs and Practices for Social Anthropology. American Anthropologist 78:5-24.
  1978 Mode of Production of Medical Knowledge. Medical Anthropology 2:97-122.
  1980 The Discourse on Stress and the Reproduction of Conventional Knowledge. Social Science and Medicine 14B:133-146.
  1982 The Anthropologies of Illness and Sickness. Annual Review of Anthropology 11:257-285.

1987   Personal communication.
Zaner, Richard
    1982   Chance and Morality: The Dialysis Phenomenon. *In* V. Kestenbaum
            (ed.), The Humanity of the Ill. Pp. 39-68. Knoxville: University of
            Tennessee Press.
Zempléni, Andras
    1985   La Maladie et Ses Causes? Introduction. *In* A. Zempléni (ed.),
            Causes, Origines et Agents de la Maladie Chez Les Peuples Sans Ecri-
            ture. L'Ethnologie 96/97:13-44.

LAURENCE J. KIRMAYER

MIND AND BODY AS METAPHORS: HIDDEN VALUES IN BIOMEDICINE

INTRODUCTION

Although Western medicine has often been characterized, and criticized, as dualistic and reductionistic, contemporary biomedical physicians are largely unconcerned with the metaphysical "world-knot" of the mind-body problem. Science seems to be slowly untangling this knot, offering a multitude of empirical correspondances between physiology and behaviour that constrain philosophical speculation. Modern biology explains mindful action as an emergent property of the hierarchical organization of the nervous system. A more sophisticated version of this materialism recognizes that mind and consciousness are not simply functions of the isolated nervous system but can be better understood as emergent properties of social systems, that is, of interactions between many individual organisms (Bateson 1979; Harré 1984). However, biology leaves unexplored an aspect of the mind-body problem that is essentially ethical. This residual mind-body problem occurs because mind and body symbolize contrasting poles in human experience: the voluntary or intentional and the involuntary or accidental. It is because the contrast between willful action and impersonal accident is central to both the private sense of self and the public concept of the person that mind-body dualism persists in Western thinking about morally significant events like sickness and disability.

In this chapter I use the notion of metaphor to examine how values are hidden behind biomedicine's rhetoric of scientific rationality. Even the empirically tested categories of science rest on a choice of metaphor (Hesse 1978), but the high emotion of clinical practice ensures that the affective coloring and value orientations imparted by metaphor exert powerful effects on the discourse of medicine. When values are explicit they may be openly debated but rhetoric uses metaphor to smuggle values into discourse that proclaims itself rational, even-handed and value-free (Perelman 1982).

In the first section I review some ways in which biomedicine expresses the dualistic values of Western metaphysics through its emphasis on the rational mastery of the body. The dominant metaphor of biomedicine is THE BODY IS A BIOCHEMICAL MACHINE (Osherson and AmaraSingham 1981). The patient is the owner of the body-machine which is brought to the physician for repairs. A rational patient adheres to the rules of the

M. Lock and D. R. Gordon (eds.), Biomedicine Examined, 57–93.
© 1988 by Kluwer Academic Publishers.

sick-role: seeking out medical expertise, giving the body over to be examined and complying with the treatment regimen. When patients deviate at any step in this process they may be judged irrational or responsible for their illness. Even when the norms of illness behavior are strictly followed, if medicine cannot explain or alleviate the illness, the patient may be blamed for its failure. These maneuvers act to maintain the rationality and coherence of the biomedical world view even while they disqualify the patient's suffering or moral agency.

Psychosomatic approaches to medicine have been offered as a corrective to the depersonalized view of the patient in biomedicine. But, as examples in the second section of this chapter will illustrate, psychosomatic theory reproduces the same values as biomedicine. In the healing vision of psychosomatic medicine, mind and body are to be brought into harmony. Most often, however, this goal is described not as an equal marriage but as the reestablishment of the mind's dominance and control over the body and with it, of reason over emotion. Thus, psychosomatics expresses its holistic perspective in dualistic terms that ultimately invoke the same values of rational control and distance from passion and bodily-felt meaning that are part of the mechanistic world view of biomedicine.

If the efforts of holistic and psychosomatic medicine to transcend mind-body dualism have largely failed it is because these concepts are so deeply entrenched in Western experience. Although some form of dualism is not hard to find in disparate cultures, the specific nature of mind-body dualism in the West is a consequence of rational agency, individualism and the psychologization of experience. This does not originate in the ideology or practice of medicine but, in both its biological and psychological approaches to irrationality and sickness, Western medicine serves as a powerful reinforcer of duality. The concluding section then considers the potential for healing some of the painful splits in Western medical practice. If the dualism of Western culture is not primarily a problem of technical knowledge or metaphysical belief system but is rooted in the moral order and the social construction of the person, then focusing on revising the conceptual categories of medicine is a misplaced endeavour. It is the fundamental experiences of agency and accident, and their moral consequences, that must be addressed if medical practice is to respond sensitively to the emotional needs of patients and the social implications of distress.

THE RATIONALITY OF BIOMEDICINE

> "[I]n the medicine of the future the
> interdependence of mind and body
> will be more fully recognized, and the
> influence of one over the other may be
> exerted in a manner which is now not
> thought possible." Osler (1928)

Biomedicine was founded on a Cartesian division of man into a soulless mortal machine capable of mechanistic explanation and manipulation, and a bodyless soul, immortal, immaterial, and properly subject to religious authority, but largely unnecessary to account for physical disease and healing (Osherson and AmaraSingham 1981). This dualism arose from a struggle between religious tradition and science and carved out two domains that, initially, allowed each its separate province (Hankins 1985). With the waning of religious authority and the corresponding rise of scientific materialism, morality became a problem less of the relation of the soul to God, or man to fellow man, than of the proper conditions for the efficient working of body and mind. In place of dualism, science has come to favor a monistic materialism: the person is a physiochemical machine, all of whose functions can be described in biological terms and rationalized for efficiency. But beneath this monism, dualistic values persist in the metaphoric descriptions of disease, the symbolic classification of diagnoses and the organization of the health care system.

Sullivan (1986) details how the earlier Cartesian dualism which posited two distinct substances for mind and body was superseded by an "epistemological dualism" that emphasized two different ways of knowing -- subjective awareness and direct observation. This dualism, implicit in the work of the 19th century French physician Xavier Bichat, focused on visual inspection and the autopsy as the final arbiters of clinical truth. The patient's subjective account of distress was deemed unreliable and essentially irrelevant to the physical diagnosis. Thus, the conscious awareness of the patient was subordinated to the physician's privileged knowledge of the body acquired by direct examination. The body revealed its disease to the doctor without the need for the patient's self-interpretation. The real dualism in modern medicine, Sullivan argues, is not between two substances but between the physician as active knower and the patient as passive known. This duality is captured in the distinction between disease and illness (Eisenberg 1977). Disease stands for the biological disorder, or, more accurately, the physician's biomedical interpretation of disorder, while illness represents the patient's personal experience of distress. In biomedicine, these two aspects of distress are accorded different status and it is "real disease" that is viewed as the true object of medicine.

The physician's biological approach to disease is one expression of a broader scientific rationality that sets aside the emotional and moral dimensions of distress. Rationality has come to stand for the mental pole in the dualistic opposition of mind and body. In William Osler's vision of *Aequanimitas*, mind and body are united in a harmonious whole, but it is control of the body and the emotions that is medicine's ideal (Dearing 1980; Osler 1928). In Osler's advice to young physicians it is not difficult to recognize the old problem of subordinating the bodily to the spiritual -- here refigured as the rational faculties of physician's mind. Although the notion that medicine has a moral dimension -- explicit in medical writings and practices from ancient through renaissance times -- underwent a gradual transformation during the Enlightenment, the link between right living and good health has never been lifted completely out of the moral realm. Indeed, in the detached view of the body as a well-functioning machine, rationality itself becomes a moral good.

Sickness as a Threat to Rationality

Despite our mastery of many diseases, sickness often escapes scientific explanation and control. The sick person is a reminder to others of the limits of reason. Sickness is violent and capricious; it erodes the sense of self-control and threatens the exchanges that maintain family life and social structure. The ability to contain and control the expression of sickness is thus necessary to maintain confidence in rationality and individual will. Biomedicine applies rational techniques to sickness: focusing on the body as a machine by distancing from the emotional significance of illness; ordering the body according to a physiological scheme; and, finally, demanding that patients adopt the detached perspective of the scientist toward their own bodies.

Medical practice has evolved many obvious and some subtle ways for distancing from the body as person. The architecture of the hospital provides a series of barriers that separate the sick body from the social person. The hospital gown effaces individuality, leaving the body half exposed and available for quick examination. This minor loss of dignity marks a major change in social status: from free agent to docile patient, from actor to acted-upon. Patients are interviewed behind curtains that provide only the pretense of privacy. During bedside rounds, the patient's case is discussed and physical signs demonstrated on the body as though no person is present. Even the drapes and baffles that

surround the sterile surgical field serve more than a biological function –
they help to reduce the sleeping person to a technical problem of organs
and blood.

But it is language that draws the most subtle boundaries. Physicians
launder bodily language to reduce the affectivity of medical discourse.
The savage incursion of a heart attack becomes "a code" or an "MI." The
many clawed crab of cancer invading from within becomes "CA" or
"mitotic figures." Medical jargon abounds with acronyms whose convenience disguises the reality of sickness: CRF, CHF, RF – kidney, heart,
and lung failure. Arbitrary signs are substituted for words that would
reveal the body's disorder and bring the physician into a more intense
emotion-laden relationship with the suffering patient.

The emotional reaction to sickness displayed by the patient, and vicariously experienced by the physician, sets uncomfortable limits to technical expertise and scientific detachment. The biomedical physician's competence is largely defined in terms of the ability to explain and
manipulate the physiology of the body. The work of internal medicine is
directed toward "physiological integrity," an abstraction already one
remove from the fleshy substance of the body (Hahn 1985a). This integrity is restored when disease is cured or, when cure is not possible, when
the body's disorder is made to conform to physiological theory.

A thirty-four year old professional woman with low back pain continues working
despite her family doctor's prescription of strict bed rest. Her doctor describes
her as "a workaholic, a driven perfectionist." She endures one year of pain with little change in her life-style. Eventually, her pain worsens and she undergoes tests
which demonstrate a collapsed intervertebral disc. A neurosurgeon operates on
her spinal column to decompress the pinched nerve root. The patient feels better
immediately after the operation but within a few days has recurrent pain. Without
performing an examination, the surgeon tells the family doctor that he has corrected the back problem so there should be nothing wrong and suggests that the
problem is "behavioral." He refuses to see the patient and advises she seek psychiatric help. (It is subsequently found on myelogram X-ray that the patient has a new
nerve root compression from a re-extruded disc.)[1]

The neurosurgeon's unwillingness to re-examine the patient reflects his
certainty that the biological machine has been set right. He discounts the
patient's report because she has not behaved reasonably according to his
physical model of her disease: living with severe pain for a year and then
complaining of pain after he has fixed the real problem. The disease
revealed by physical examination and technical instruments is at once
more real and more important than the patient's subjective distress. The
rational order of medicine eclipses the bodily-felt reality of the patient.
If biology provides no rationale for suffering then medicine can wash its
hands of the patient who must be responsible for her own recalcitrant

problem.

Rational Cooperation and the Sick Role

Just as physiological theory provides standards for the rational ordering of disease, lay and professional norms provide prescriptions for behavior when one is ill. To the extent that patients follow the norms of the sick role they are judged rational. Parson's (1951) classical conception of the sick role is essentially a moral concept consisting, as it does, of both rights and obligations. People who are sick have the right to be excused from usual social duties and to not be blamed or held responsible for their illness. In return they are expected to try to get well by seeking out recognized medical help and following the doctor's orders. Parson's scheme expresses an important prototype in both lay and professional thinking about illness. However, the classic sick role model poorly describes both chronic and psychological, stress or lifestyle-related illness (Blackwell 1967; Segall 1976; Twaddle 1969). Thus, most illness occupies a morally ambiguous realm where social context and underlying attitudes can exert powerful effects on the physician's legitimation of distress.

Our commitment to rational mastery means that we would like to believe that if we do everything right, that is sensibly, we will not get sick. Sickness makes the patient's stewardship of the body suspect. The physician then appropriates the body and performs caretaking functions for the patient who has failed to protect the body and lacks the expert knowledge necessary to understand what is wrong. Physicians take charge of medical interviews to ensure that the technically important data are collected (although in the process they may miss relevant information about symptoms: Beckman and Frankel 1984). The crucial data are clues to the mechanical functioning of the body; the patient's life-world is largely ignored (Mishler 1985). Physicians' confidence that they know what is really important can result in a paternalistic attitude that extends to giving explanations and advice:

Almost all patients, regardless of intellectual capacity, are naive and simplistic when dealing with their own health problems. One should assume nothing, and start from basic facts, and build upward. A brilliant person is often a dull patient. A less endowed patient is often like a child (Tumulty 1970:23).

The body stands in relationship to the mind as a child to a parent. When the physician perceives the patient as a sick body it is natural that he take over the parenting functioning, bypassing the patient's own self-knowledge and self-care as demonstrably incompetent. Rationality in biomedicine then, is largely equivalent to the patient's willingness to abandon his body to professional care.

The physician's expertise is a valuable commodity not to be casually given away. Medical knowledge is the private property of the profession (Bologh 1981) but the patient may participate in rational discourse, to a limited extent, by treating his own body as an object. When he does so he is viewed as a "good patient" who cooperates with treatment. The sick-role is founded on a split between person and body so that the former need not be held responsible for the social consequences of the troubles of the latter (Parsons 1951). Sickness is something that just happens to bodies while the person is a sort of helpless bystander. But to maintain this blameless posture the patient must follow the physician's directives. Acceptance of medical authority is in itself a sign of competence and rationality. When patients do not accept physicians' directives or explanations they are labelled "noncompliant," "uncooperative," "difficult," and "irrational."

Physicians have exaggerated standards for rationality, based on their distancing from bodily-feeling and emotion. The physician's rational mastery of the emotional threat of illness is seen most dramatically in those instances where a doctor is sick. Many physicians react to their own illness by adopting the distant role of student of pathology (Hahn 1985b) and may respond to sick colleagues with a similar strategy.

A medical resident develops herpes zoster (shingles), a viral infection of the nervous system. She is examined by a dermatologist who asks if she has seen her own chest X-ray. When she says "no," he replies, in an enthusiastic tone usually reserved for teaching on hospital ward rounds: "Well I just thought with your dry skin, and zoster and if you had hilar lymph nodes [on chest X-ray] -- well that's a classic triad -- it's got to be Hodgkin's for sure." He thus raises the possibility of a serious malignancy, Hodgkin's lymphoma, with this patient as though she were a colleague discussing an interesting case.

The physician's attention is focused not on the patient but on the "it" of disease. His insensitivity can be seen as a defensive maneuver that protects him from the threat of a colleague's sickness. But the presence of a colleague -- even if she is a patient -- is an occasion to speak aloud the cognitive distancing from suffering that the physician is engaged in all day long.

In medical school, physicians may have experienced medical student's disease -- a sort of transient hypochondriasis associated with examination anxiety and a flood of information about pathology (Mechanic 1972). Physicians master this anxiety of uncertainty by accumulating technical knowledge. So it is that physicians expect that giving a biomedical explanation to the patient will reassure, calm and satisfy them. Physicians forget that it was not simply technical knowledge that reassured them (for when they had only a small amount of it they themselves suffered) but its articulation with a larger body of knowledge that forms a coherent biomedical world view and of their own investiture, in the role of physician, as the legitimate interpreter of that world view.

In the clinic, the physician's ability to explain and to cure becomes the power to control (Taussig 1980). The patient's refractoriness to diagnosis or treatment is a challenge to the physician's authority and to all that it stands for socially as well as personally. If the norms for illness behavior are contravened or pathophysiological explanations evaded, the patient may be blamed for diagnostic or treatment failures. Attributing the limitations of biomedical explanation and treatment to hidden motivations of the patient is convenient for the preservation of medical authority because it maintains an opposition with the physician, rationality and health on one side and the patient, irrationality and sickness on the other. The patient is seen to be choosing to be sick, rejecting the physician's help and, with it, the legitimate sick-role.

PSYCHOSOMATIC KNOWLEDGE

> *An eingeredte krenk is erger fun an emeser* (An illness you've talked yourself into is worse than a real one) -- Yiddish proverb.

Holistic or psychosomatic medicine has been offered as a corrective to the conceptual limitations of the biomedical view. One recent version of holism takes the form of a "biopsychosocial approach" that would give equal consideration in diagnosis and treatment to biological, psychological and social levels in a natural hierarchy of organization (Engel 1977). But it is not simply the lack of a comprehensive theoretical model that prevents physicians from integrating psychosocial information into clinical practice. Psychosocial "data" have affective and moral implications that pose significant problems for physicians.

The modern use of the term psychosomatic encompasses two old conceptions: the importance of psychological factors in causing disease (psychogenesis) and a holistic view of man (Lipowski 1984).[2] Whatever acceptance these concepts have gained in lay and professional circles, psychosomatic medicine has had little noticeable impact on the organization of biomedicine. For the most part, psychosomatic disorders have been incorporated into biomedicine as a class of not quite legitimate illnesses best handled by mental health practitioners. Psychiatrists working in medical settings have developed psychosomatic theory in an effort to make the emotional care of patients intelligible and important to physicians focused on bodily disease. Yet, hospital physicians initiate psychiatric consultations with considerable reluctance both because of the stigma attached to psychiatry and because they expect little benefit. In most hospitals, only extremes of emotional distress or erratic behavior precipitate psychiatric consultation and at this point it is often with the hope that the psychiatrist will take over the total management of the disturbing patient (Greenhill 1977).

Psychiatrists have promoted the view that all disease is psychosomatic, in the sense that it involves both mind and body, a view supported in a report of the World Health Organization (1964) which recommended eliminating "psychosomatic disorder" as a misleading diagnostic label. The current nosology of the American Psychiatric Association (1980) has replaced psychosomatic and psychophysiological disorders with a descriptive label, "psychological factors affecting physical condition," clearly applicable to any illness. Despite these efforts at reform, the term "psychosomatic disorder" continues to be widely used and psychosomatic illness retains the stigma associated with "purely" mental illness (Blackwell, 1967).

The diagnosis of a psychosomatic condition transforms the real into the imaginary, the innocent into the culpable. This is well brought out in the range of informal terms used to label psychosomatic patients and their conditions. Some of these are presented in Table I. They range from casual dismissal of the problem as not real ("It's all in your head," "imaginary," "pseudoseizures," "pseudoangina," and so on) to frankly moralistic labels. Even the ostensibly neutral categories of diagnosis have a pejorative quality. This they share with the psychological labels that are also often applied to psychosomatic patients.

Psychosomatic diagnosis creates the reality it intends to describe. Even where empirical evidence is lacking, the psychosomatic explanation forges a link between mind, as subject, and body as object. Helman (this volume) suggests that when physicians make psychosomatic diagnoses, socially objectionable parts of the self ("bad" emotions) are split off and reified as impersonal causal agents responsible for the patient's illness. In this way, doctor and patient join together in addressing the common problem "out there."

An Imaginary Malady

Hysteria or conversion disorder serves well to illustrate psychosomatic theory and practice because it is *defined* as a syndrome of somatic symptoms of psychological origin. Conversion disorder is diagnosed when: (1)

TABLE I.

Terms applied to patients with psychosomatic illness

Existential (Real/Not Real)
     All in your head
     Complaining
     Imaginary
     Exaggerating
     Factitious, fake
     Game (e.g. "pain game")
     Histrionic
     Malingering
     Pseudo (e.g. "pseudoangina," "pseudoepilepsy")

Evaluative (Good/Bad)
     Bad (e.g. "bad back")
     Cripple (e.g. "cardiac cripple")
     Crock
     Weak (e.g. "weak stomach," "weak nerves")

"Diagnostic"
     Abnormal illness behaviour
     Compensation neurosis
     Dysregulation syndrome
     Functional (versus organic)
     Green poultice syndrome (symptoms that are relieved when paper money
         is applied directly to the affected region)
     Hypochondriacal
     Neurasthenic
     Hysterical
     Somatizing
     Stress, strain
     Tension

Psychological
     Behavioral (e.g. "behavioral seizures")
     Crazy
     Emotional
     Psych case
     Psychophysiological
     Secondary gain
     Supratentorial (i.e. involving higher mental processes located above the
         tentorium cerebri in the cerebral hemispheres)

there are no organic lesions or dysfunctions that can account for physical symptoms that otherwise resemble a neurological disorder; (2) there is evidence for personal or social trauma that could incite psychological conflict and emotional distress; (3) the form of the somatic symptoms can be symbolically related to the nature of the psychological or social conflict (American Psychiatric Association 1980; Engel 1970). Conversion symptoms do not always follow such symbolic correspondences. Many conversion symptoms occur in individuals who have a pre-existing somatic illness or who have had exposure to a sick person. The symptom then follows the person's conscious model of illness more than unconscious symbolism.

Mrs. D, a twenty-three year old woman employed as a secretary and recently married, was admitted to hospital following the sudden onset of weakness affecting the left side of her body. She was unable to walk, had only slight movement of her left arm, and a noticeable droop of the left side of her mouth. A computerized tomography (CT) scan of her brain found evidence for a stroke affecting the right cerebral hemisphere. Over the two weeks of her hospitalization Mrs. D was examined numerous times by the neurology housestaff, however, no note was made of her emotional reaction or adjustment to her illness. When her physical condition was judged stable she was discharged to a convalescent hospital for physiotherapy and rehabilitation.

After two weeks in the convalescent hospital Mrs. D was scheduled to return home to continue physiotherapy as an outpatient. Instead, she began to notice numbness and weakness affecting the entire left half of her body. Fearing another stroke, she was readmitted to the neurology service of the acute care hospital. However, the neurology resident who performed Mrs. D's readmission physical examination found few signs of neurologic disease. Her loss of sensation was inconsistant and crossed the midline of the body unlike any known neuroanatomical pathway. A repeat CT scan provided no evidence for new brain injury.

During ward rounds the morning after readmission, Mrs. D's case was discussed at her bedside. She listened closely but the doctors spoke with jargon which they made no effort to explain to her. The attending neurologist conducted a thorough physical examination. He had Mrs. D walk on her heels and toes up and down the hospital corridor to test the motor function of her lower limbs. Although she stated she could not do this, when encouraged she faltered at first but performed well. While she struggled to follow the physician's directions the housestaff audience talked and joked among themselves. She began to suspect the doctors did not think she was really sick and were laughing at her. Angry and exposed but feeling powerless to protest, she began to cry. The neurologist then stated: "There's no real evidence of a new CVA [cerebrovascular accident, stroke]. This is probably a conversion symptom. Let's get psychiatry."

To Mrs. D the reference to psychiatry plainly meant the neurologists did not believe she was sick -- they thought she was crazy. The resident in neurology told her that she could go home once a psychiatrist saw her because there was nothing wrong with her neurologically. He arranged for a psychiatric consultation "as soon as possible" because this was now the only delay in discharging Mrs. D from hospital. At this point, the resident felt his work was done.

Mrs. D's two overlapping episodes of illness were similar in her mind but sharply distinguished by her doctors. Her initial loss of muscle power was terrifying and she recognized she was having a stroke. In the emergency room, the neurologist found the pattern of weakness consistent with a specific brain lesion and the CT scan supported this hypothesis. Although Mrs. D was young for a stroke, her history of migraine, heavy smoking and birth control pills put her at high risk. For the physicians, there was a satisfying fit between history, physical findings and a specific neurological diagnosis. In hospital, Mrs. D's course was described as "uneventful" -- her recovery accorded well with physiological expectations and her emotional distress was viewed as appropriate for the seriousness of her disease. Soon after plans were made for her to return home from the convalescent hospital, she experienced symptoms that resembled her initial stroke. On readmission to the acute care hospital, however, she had symptoms that followed the conventional lay person's representation of the body rather than established neuroanatomical pathways. Thus, it appeared that the problem involved Mrs. D's mind rather than her brain. Of course, the neurologists, if questioned, would state that her mental symptoms were ultimately caused by the brain. But hers was a disorder in brain function not structure, akin to learning bad habits, and of little neurological interest except as a difficult diagnosis to make.

Mrs. D's body was seen by the doctors as alternately suffering from a disease of the imagination or no disease at all. Depending on what degree of self-awareness the physicians attributed to her, she was then viewed as pitiable and incompetent or as foolish and histrionic. Either way her sense of self was challenged and she perceived rightly when she thought that some of the housestaff physicians were laughing at her.

Mrs. D was seen by three psychiatrists in the next two days. Dr. Adams, a female psychiatrist with a psychodynamic orientation, arrived to find Mrs. D crying in her hospital room. When she introduced herself as a psychiatrist, the patient angrily threw her lunch tray across the room and the interview ended abruptly. The psychiatrist offered to return at a more opportune time, spoke briefly with the nurses and wrote a *prn* (if needed) order for sedation. In her notes, Dr. Adams described the patient as hostile and uncooperative, "a personality disorder, hysteric -- maybe borderline."[3] She noted from the medical chart that Mrs. D, had recently lost her mother and speculated that she had difficulty relating to women due to her preexisting psychodynamic conflicts.

Dr. Brown, the usual consultant, saw Mrs. D the following day and found her much calmer and apologetic about her outburst. She cried easily and stated that the doctors were not taking her seriously and did not seem to know what was wrong. She had no need for a psychiatrist but was willing to provide additional personal history.

Mrs. D had been married for just two months. Her mother, with whom she had a very conflicted relationship, died after a protracted illness two weeks prior to Mrs. D's stroke. At the time of her mother's death, Mrs. D was out of the country, on her honeymoon. She cut her trip short to come back for the funeral. She recalled feeling alternately numb with grief and angry with her mother for dying while she was away.

Dr. Brown reassured Mrs. D that her problems were understandable given the losses she had experienced. He completed a consultation report with diagnoses of conversion disorder and grief reaction. By now the neurologists were anxious to discharge Mrs. D to follow her recovery as an outpatient. If any active treatment was needed it was clearly psychiatric and Dr. Brown referred her to Dr. Clark, a psychiatrist with a special interest in behavioral medicine.

Dr. Clark interviewed Mrs. D in her hospital room and found her reluctant to speak to yet another psychiatrist. When he focused on Mrs. D's physical symptoms, she became more engaged in conversation and gradually spoke about the impact of her recent life events. She reported many symptoms of grief or depression: little appetite for several weeks with a weight loss of 4 kilograms; waking early in the morning and unable to fall back asleep; crying frequently; feeling hopeless with occasional thoughts of suicide. Dr. Clark discussed his diagnoses with Mrs. D, describing conversion disorder and depression as bodily reactions to stress. He offered to treat her as an outpatient with medication and "stress-reduction" to aid her physical rehabilitation. Mrs. D readily acknowledged that she was under a lot of stress but declined the offer of psychiatric help.

Several days later Mrs. D contacted Dr. Clark and asked if she could come in "to talk." She stated: "I was wrong. Maybe I do have a psychiatric problem. My mother's death is really getting to me. I have so much guilt because I was away when she died. And it's very hard being married. My husband cares but he doesn't understand. I have so many bad thoughts I can't get out of my head." Dr. Clark interpreted her distressing thoughts as symptoms of grief complicated by a major depression. He contracted with her for a course of antidepressant medication and weekly sessions of psychotherapy to help her "get over this difficult time."

This case history illustrates the way in which the division of labor between neurologist and psychiatrist marks a division between real disease and imaginary illness. Real disease sustains the interest of biomedical practitioners. Patients receive extended evaluation and care and are made to feel they are the possessor of something serious, worthy of close attention, even having a kind of intrinsic value as an object of medical study. Patients whose illness is identified as imaginary receive the contrary message. They have no disease; their illness is illegitimate, their condition not serious. Once this is established, the disinterest of the biomedical practitioner casts doubt on the patient's right to the sick role. Medical disinterest becomes social ambiguity as family and others can no longer tell to what extent the sick person is to be relieved of social obligation and nursed back to health.

Mrs. D felt she had a physical problem that deserved serious physical treatment. She was well aware of the negative connotations that psychiatric diagnosis and treatment would have within her family and at work.

It would just confirm that her problems were not real or were due to her own weakness -- mental or moral. When it became clear that the neurologists were through with her she decided to contact the psychiatrist to obtain the only help that was being offered to her, despite its cost in social stigma.

Gender, Blame and Self-Control

Conversion symptoms are often held to be associated with an hysterical or histrionic personality, although the evidence for this has been challenged.[4] The characteristics of the hysteric personality are said to include: vague, diffuse emotion-laden memories; use of intuition to solve problems; high suggestibility and distractibility; attraction to romance and fantasy; sexualized and romanticized relationships; general lack of interest in objective fact; and, a direct and active engagement with the human world. Although hysterics appear emotional and expressive their emotions are said to be superficial or shallow (Winstead 1984). As Chodoff (1982) has noted, the characteristics of the hysterical personality read like a hostile caricature of the feminine in Western society.

Hysteria and the feminine are closely related in Western thought. There is a traditional view of women as dominated by passion, ruled by their bodies, and essentially irrational (Fortenbaugh 1975; Phillips 1984; Spelman 1983).[5] This is commonly linked to woman's reproductive capacity (Lange 1983). In classical theories of reproduction, women provided the matter for the fetus, generally as blood, while men transmitted form, sometimes embodied in an already perfect homunculus contained within the semen. Hysteria owes its name to the causal theory of antiquity which implicated the womb (Gr. hystera, cf. Veith 1965). The uterus was a sort of "wild animal" that moved within the body in response to sexual frustration or other privations. Floating or pressing upward against the stomach, liver or diaphragm, the uterus gave rise to the myriad symptoms of hysteria.

Women tend to be diagnosed with hysterical neurosis or conversion symptoms two to seven times more frequently than men (Winstead 1984:78). Clinically, men with conversion symptoms are more likely to be thought to be seeking compensation while women are more likely to be seen as suffering from an emotional disorder. Thus, men's behavior is viewed as rationally motivated -- the pursuit of money -- albeit through maladaptive means, while hysterical women's behavior is frankly irrational. When a portrait of hysteria was presented to a sample of mental health professionals and they were told the subject was female, three quarters diagnosed hysteria; when told the subject of the case history was male, the diagnosis was split between hysteria and antisocial personality disorder (Warner 1978). The antisocial person is perceived as acting voluntarily but with bad intent.

These findings may reflect the tendency to view men as responsible χ for their actions while women are more likely to be seen as "out of control." Self-control implies rationally guided behavior. The claim to be out of control carries far more stigma for men in North American culture. While an "out of control" woman brings to mind someone weeping, an "out of control" man is likely to be imagined as aggressive or violent. χ Anger is the only common emotion that is more frequently associated with men than women (Shields 1987). Men are expected to maintain a high level of self-control to ward off socially dangerous impulses.

When symptoms are perceived to arise from a desire to avoid social responsibilities, men and women are judged more equally culpable. But there remains a tendency to view men as more often seeking compensation or avoiding social responsibilities while women are simply overwhelmed by emotions they are unable to control. Women are thus extended the sick role through the diagnosis of hysteria but at the cost of promoting the view that they are weak-willed, irrational, and subject to excessive passion and emotion.

There is, thus, a split between two types of psychological causation: rational and irrational, reasoning and emotion, motivated and capricious, conscious and unconscious. When willful or conscious action gives rise to symptoms, the person tends to be viewed as responsible for his sickness though mentally competent. He is mentally strong but morally weak. In contrast, unconscious or inadvertent action gives rise to symptoms for which the person is not responsible. He is not bad but may well be perceived as mentally weak. His conscious will is not strong enough to master the irrational passions. Although moral and mental strength go together in traditional ethical thought, in the classification of psychological disease they are distinct and often contrary.

Psychosomatic Explanation as a Response to the Irrational

The psychiatrists who saw Mrs. D brought three different conceptual models to bear on her situation: biological, psychodynamic and behavioral. Each practitioner used an amalgam of these approaches although often favoring one mode of explanation. However, all three models present the same dilemma over the boundaries of rational control and responsibility for doctor and patient -- as a brief discussion of each will show.

From the perspective of biological psychiatry, Mrs. D had suffered a stroke resulting in an organic vulnerability. Her subsequent conversion symptoms could be understood as a consequence of her diminished capacity for coping due to brain damage (Merskey 1982:97ff). Alternatively, her conversion symptoms could be seen as part of a broader syndrome of hysteria associated with biologically determined personality traits (Merskey 1982:191ff). In either case, conversion symptoms are seen as a reversion to a primitive mode of nervous functioning (Jones 1980). This leads to a pathological inhibition of sensorimotor pathways or a dissociation of the personality into subordinate parts due to a lack of higher integrative processes (Roy 1982).

The explanations of biological psychiatry apparently eliminate the dichotomy between mind and body; both neurological disease and psychological disorder reflect the activity of the central nervous system (Cobb 1957). But the contrast between mind and body is preserved in the mapping of reason and emotion onto brain structures. Higher mental functions like thought and imagination are localized in the cerebral cortex while bodily processes of arousal, motivation and emotion, are located in lower, more primitive brain structures (Averill 1974). The limbic system, a group of structures common to mammals and situated between the brainstem and the cerebral cortex, is held to be the seat of emotion (Papez 1937). The limbic system constitutes a "visceral brain" whose functioning "eludes the grasp of intellect because its animalistic and primitive structure makes it impossible to communicate in verbal terms" (MacLean 1949:348). The anatomical hierarchy of the brain is said to parallel our evolutionary ascent from mammals driven by passion and aggression to civilized beings capable of rational thought and reflection.

The mature person has the primitive brain firmly under control. Psychosomatic symptoms may be attributed to abnormal activity of the limbic system, which gives rise to "infantile" expressions of distress, that is, somatic symptoms in place of the verbal expression of abstract symbols (MacLean 1949; Ruesch 1948). Symptoms that are held to arise from lower brain centers (or the release of lower centers from higher control) thus reveal that patients' rational, adult control of the body is compromised. The organic explanations of biological psychiatry offer the sick role to the person willing to objectify his own mind as a disordered brain, acknowledge an impairment in social competence, and accept a diminished sense of self.

Due largely to the dramatic success of antidepressant medication, many psychodynamically and socially oriented psychiatrists have come to view major depression as a biological disorder. Mrs. D was told she was suffering from a physical depression, a disease of the body that affected her mood, energy level, sleep and appetite. This objectified her distress and shifted it away from her interpersonal concerns. But it left her with no sense of what action to take beyond following the doctor's orders for medication. Biological explanations for emotional distress tend to move it to an impersonal causal realm. This decreases the pressure on the person and his social network to identify and change problematic relationships. Others are free to adopt a solicitous stance to the suffering individual because the biological explanation eliminates the dimension of social protest implicit in psychological distress.

The psychodynamic approach views conversion symptoms as the result of conflicting desires that are not directly accessible to consciousness. In Mrs. D's case this was conceptualized by Dr. Adams as conflict between her need for her mother's nurturance and anger at her mother for abandoning her by dying. As Reiser (1985:33) puts it:

Since the patient is not and cannot be aware of the nature and extent of the opposing motives, the conflict is not amenable to resolution by rational cognitive processes. Unresolved and actively barred from conscious awareness, the incompatible dispositions are nonetheless considered to remain active, that is to continue to press for expression. Under these circumstances, the neurotic symptom, behavior or character trait is thought to develop a *symbolic representational compromise*, which simultaneously expresses both of the opposing motive forces but in a form that *escapes conscious recognition*. The major hysterical conversion symptom serves as the prototypical example: hysterical paralysis of the arm may simultaneously represent an unrecognized impulse to use the arm for murderous intent and "appropriate biblical" punishment for it.

The great appeal of the psychoanalytic perspective is that it allows the person the possibility of self-mastery through understanding. Psychodynamic theory relocates the rational/irrational split within the patient's mind in the topographical division between conscious and unconscious. Psychoanalytic treatment aims at the heroic victory of rational consciousness over the self-destructive irrationality of the unconscious. The patient can be accorded the normative sick role and join with the physician against the irrational forces which are now perceived to occupy not just the body but a portion of the mind as well. But this distancing from objectionable aspects of the self is never completely successful. Thus, psychoanalysts often speak of patients' "unwitting complicity and participation in the development and generation of the stressful situation" (Reiser 1985:181). In its studied avoidance of the social context of its own practice, psychoanalysis exaggerates the importance of patients' idiosyncratic inner motives. Dr. Adams viewed Mrs. D's lunch tray throwing as a display of her psychodynamics rather than as an expression of her anger at being ridiculed and having to face the embarrassment of a psychiatric evaluation.

Psychoanalysis emphasizes psychological knowledge as the basis for self-mastery. It holds that external reality is largely beyond our control and proposes a stoic ethical response to the origins of distress (Gellner 1985). We can reduce suffering by mastering the passions that reside within, making our desires congruent with reality. As Gellner points out, however, the stoic assumption that willful action can control the inner world is exaggerated. In its vision of man held hostage by his instincts, psychoanalysis goes beyond stoicism. But, while acknowledging the power of the irrational unconscious, psychodynamics promotes a new rational order in which the emotions are controlled not by suppression but through verbal description and scientific explanation.

Psychodynamic explanations are based on hydraulic metaphors that reveal their origins in the neurology of the late nineteenth century (Sulloway 1979). Contemporary psychoanalysis has moved away from such overtly mechanistic accounts and tends to conceive of its work as hermeneutic -- aimed at interpretive understanding rather than scientific explanation. However, patients' distress is still interpreted in language that emphasizes the inexorable press of biological drives or the need for psychic energy to find an outlet -- a sort of hybrid "biohermeneutics" (Gellner 1985). Further, according to psychoanalytic theory the clinician is in a privileged position to discover the true origins of the symptom: "[the] ideational content of pathogenic psychological conflict is usually not present in or readily accessible to the patient's consciousness and the patient may be quite unaware of being in a situation of special personal or psychosocial stress" (Reiser 1985:180). The psychoanalytic observer is placed by theory in an unassailable position as the broker of personal significance for the patient and his monopoly on scientific rationality is complete.

Behavioral medicine, a third contemporary approach to psychosomatics, is based on the learning theories of academic psychology (Schwartz 1982). Conceptualized in these terms, Mrs. D's conversion symptoms were based on learning that the symptoms of a stroke elicited care-giving from others. The reinforcement of symptomatic behavior could occur covertly, without Mrs. D's awareness. Learning theory tends to be less pejorative than psychodynamics, suggesting that maladaptive behaviors are acquired through normal psychological processes and maintained by environmental contingencies beyond the individual's control.[6]

Behavioral medicine has adopted "stress" as an organizing metaphor for its discourse on health and illness. Stress stands for the inexorable demands of the environment that can, at times, exceed the human capacity for mastery. Stress has the great clinical advantage that its moral implications are hazy. Practitioners can focus on stress "management" without emphasizing patients' responsibility for their illness.

But people differ in their capacity to master the environment and some may be judged more susceptible to stress than others. Further, individuals choose to subject themselves to stress in the pursuit of goals that are themselves judged morally upright or questionable. Stress is determined by lifestyle which, in turn, can be viewed as a result of economic necessity, character or moral choice. Since health care practitioners tend to view the social origins of stress as implacable givens, the focus is on their adjustment through lifestyle change and psychological coping (Young 1982). The failure to reduce or cope with stress then reflects no limitation in the physician's skill but rather the irrationality or weakness, psychological and moral, of the patient who is unable to curtail unhealthy habits. Relating disease to stress may thus be a step in shifting responsibility for disease to the patient. Examples of practitioners viewing patients as responsible for causing their illness are far more common than the rhetoric of behavioral medicine would suggest.

Psychosomatic explanation, whether it takes the form of psychophysiology, intrapsychic conflict or stress, occupies a morally charged realm. Simply put, there are things we are responsible for and things we are not. For the most part, these correspond to things we do and things that happen to us. And "real" sickness in Western culture is an exemplar of the kind of misfortune that just happens to us. Consequently, wherever responsibility is imputed, a person cannot be "really" sick until an autonomous (biological) process takes hold. Psychogenic explanation is attractive to physicians because it shifts responsibility for the unexplained and uncontrollable to the patient. If, as the psychosomatic perspective suggests, the individual has voluntary control over illness, then he must in some way be morally responsible for his condition. In this way, the psychogenic approach of psychosomatics reintroduces the value of rational self-control. While the notion of personal responsibility is stigmatizing and confusing to patients who experience their problems as happening to them, it remains attractive to some because it holds out the hope of personal empowerment through self-control.

CAN DUALISM BE TRANSCENDED?

> "It is humiliating to have to appear like an empty tube which is simply inflated by a mind" Wittgenstein, 1931 (1980:11).

> "The human body is the best picture of the human soul" Wittgenstein, c. 1947 (1958:II, iv).

There is a continuous metaphysical tradition in Western thought, traceable from the pre-Socratic philosophers down through the medieval alchemists, expressed by such metaphoric contrasts as active, hot, male and right on the one hand versus passive, cold, female, left on the other (Needham 1985). Paralleling this are contrasts between reason and passion, thought and emotion, free will and compulsion, matter and spirit, mortality and immortality, and so on. These contrasts reflect an underlying set of values that inform the concepts of mind and body, playing one against the other in an unequal contest. Mind is generally UP, RATIONAL, ACTIVE, VOLUNTARY, CONTROLLED, STRONG, MASCULINE, FORM (IMMATERIAL), IMMORTAL, while body is DOWN, IRRATIONAL, PASSIVE, INVOLUNTARY, UNCONTROLLED, WEAK, FEMININE, MATTER, MORTAL (Lakoff and Johnson 1980). The precise location of these dualities varies with different metaphysical systems but it is striking how similar values persist. Thus, for Aristotle the body may be a part of the soul, but the soul is then divided into upper and lower halves, with the lower part more material, animal, emotional and bodily while the upper is immaterial, divine, pure thought (Fortenbaugh 1975).[7]

Table II collects some of the contrasts that parallel mind and body in traditional philosophy. These polarities are hidden behind the colloquial use of "mind" and "body" and also condition their implicit use in the medical classification of distress. The mind in health is quick, active, powerful, and pure. Used in metaphor, it stands for what is essentially human -- the domain of intelligence, responsibility, freedom and the spiritual or sacred. The body, when its essential nature is being described, tends to be associated with the opposite qualities: passive, weak, impure and coarse. The body stands for the animal, dumb, blind, enslaved and profane; it is ultimately governed, not by human will, but by the inexorable rules of the material world. The body is healthy just insofar as it transcends its own nature -- being experienced as like the mind -- limitless, powerful, invulnerable. Sickness inverts the healthy order of things. Sickness of the mind makes it like the body: irrational, slow, passive, involuntary, out of control. But sickness of the body reveals its essential nature -- fragile, mortal, clumsily navigating the material world, a constant reminder of the limits of reason.

Social Origins of Dualism

Duality and binary opposition underlie more complex symbolic classifications in many cultures (Needham 1979). The universality of dual classification may reflect hard-wired aspects of the nervous system that underlie the bipolar structure of motivation (approach/avoidance: Schneirla 1959) and emotion (Russell 1979). Asymmetries in the organization of the

TABLE II.

Dual symbolic classification in Western metaphysics

mind	body
spirit	soul
active	passive
form	matter
rational	irrational
reason	emotion
free	determined
voluntary	involuntary
master	slave
adult	child
male	female
immortal	mortal
right	left
culture	nature

Sources: Corballis (1980), Fortenbaugh (1975), Hertz (1973), Keller (1985), Lloyd, (1964, 1973), Needham (1979).

cerebral hemispheres might also contribute to pancultural commonalitites in dual symbolic classifications (TenHouten 1985). But these universal dualisms do not dictate the specific oppositions of mind and body found in Western culture.

Durkheim and Mauss (1903/1963) argued that complex categorizations arise out of forms of social organization. The unity of classified knowledge reflects the unity of social structure. The categories and qualities of experience can therefore be as varied as the forms of social life (Bloor 1982). But Durkheim then presented himself with this puzzle: despite variations in social structure, mind-body duality appears to have wide currency (Durkheim 1973). Duality must then be due to some very general aspect of social structure and Durkheim located it in the inevitable tension between the collective and the individual. The individual's life is rooted in his own body which comes into conflict with the demands of society. "Because society surpasses us, it obliges us to surpass ourselves; and to surpass itself, a being must, to some degree, depart from its nature – a departure that does not take place without causing more or less painful tensions" (Durkheim 1973:163). The social evokes a sense of mystery and awe because it rules the physical life of the individual and yet it remains invisible. For Durkheim, the intuition of one's embeddedness in the social organism is the source of the sense of the sacred. Morality resides in subordinating the profane desires of the body to the sacred order of society. Mental representations are created by the forms of social life and so the mind itself carries the sacred quality of the social. Mind and body thus rest on sacred and profane which in turn are revealed to be the collective and the individual.

Durkheim overstated the case for the universality of the categories of mind and body, but variations in social structure could account for the specific metaphors of mind and body within a culture -- not only because commonsense models of mind are based on analogies from forms of social life, but also because the intrapsychic world is a social creation. The "existential givens" of psychological theory are the result of specific child-rearing practices, family configuration and social structure. Where human relationships are substantially different, the duality of existence will be altered as well. The nature of mind, as the interior world or agency of the person, and body, as the medium of sensation and action, would then depend on the way in which social structure shapes the development of the sense of self.

For example, Hsu (1972) describes *jen*, the Confucian Chinese concept of personhood in which the sense of self does not end at the boundary of the skin but extends into the family and the intimate social circle. Social responsibility is not something separate from the self -- an onerous burden imposed on the ethically conscious but isolated individual -- it is part of the person's will. The basic unit of identity is the family not the individual and the person is conceived of always in transaction with others. The body is thus, not something whose desires are in constant conflict with the social order but the agent of the family. A domesticated body is the source of the family's economic wealth and moral integrity. The body belongs not only to the individual but to society and bodily illness therefore reflects disharmony in the social order (Wu 1982). Illness is not construed in terms of the metaphors of psychology with its emphasis on activities of the mind's inner world and the rhetoric of psychological motives. The proper treatment for all distress is not psychological but somatic and socio-moral.

Another example is provided by Doi (1973) who describes Japanese child-rearing where the needs of the infant are constantly anticipated by the mother. The mother's indulgent love is not directly contingent on the child's willful striving for social approval. Love comes from belonging more than striving, from fitting in rather than controlling. Accordingly, although inside and outside are clearly demarcated, the sense of opposition between self and other is muted (Smith 1983). The person is socialized to develop an acute awareness of the public presentation of self. The person is an actor on the social stage who views himself from the perspective of the audience (Takano 1985). This is clearly evident in the narrative voice of the modern Japanese novel where the author observes his body as one social object among others. Society is not a collection of individual minds but, in an inversion of Durkheim's opposition, an over-arching body to which the individual belongs.

For the Japanese, the inner sense of self, *jibun*, and the exercise of autonomy in interpersonal relationships are difficult achievements, much less valued than in the West (DeVos 1985). The moral value of the self is expressed through social connectedness and endurance rather than through mastery or control of the physical and social environment. This supports a sociosomatic theory of the origins of distress, in which the self endures bodily suffering as a consequence of inescapable social condi-tions. Japanese sociosomatics emphasizes inborn constitution and the stress of fulfilling social roles as causes of disease. Morally upright behavior may lead to illness when the person overextends himself -- a common mode of explanation in contemporary Japanese biomedicine (Lock 1987). Although similar to Western stress or occupational medi-cine models, this contrasts with Western psychosomatics, in which the intrapsychic self mediates the body's suffering and the person is morally culpable not only because of how he acts toward others but because of how he acts toward himself.[8]

The Western concept of the person is of a rational agent, which occu-pies a space within the body, which itself dwells within the social world. The self has goals that are distinct from, and in many cases conflict with, the goals of others who occupy the social world "outside." The value of the person lies in his strength or will which is defined always in opposi-tion to the other -- whether that other be society, nature or the body itself. The person is identical with that rational agency that establishes its unique worth by promoting its own goals over those of others. The potential divisiveness of this individualism and rational self-interest is held in check by appeals to moral obligation. The social face of individu-alism is the rational acceptance of responsibility.[9]

The emphasis on rationality and responsibility means that the person cannot be identical with the body and its irresponsible passions and irra-tional concerns. Instead the person is held to be identical with an imma-terial mind that gives rise to a personality through psychological pro-cesses. People can rationalize the vagaries of their actions by giving a psychological account of conflicting motivations. Some psychological processes are owned by the self as strengths (e.g. coping) while others are distanced from as weaknesses. The cause of much human suffering is held to be psychological and hence, potentially under personal control.

Body, self, and society form a system whose relationships define the separate terms. The resulting phenomenological distinction is between events under personal control, events under social control and events under bodily or physiological control. The self is what society holds accountable for the body. The body is what the person holds to account for his limitations when answering to society. When political power permits, the person may hold society to account for the body. Thus, bodily illness may be interpreted as a result of social inequity or political misfortune (Kleinman and Kleinman 1985). The power struggle between self and other over responsibility for the body is played out both at the level of the physician-patient interaction and in the relationship of the health care system to other social institutions. Each wishes to attribute to the other all that is dis-valued -- weakness, sickness, irrationality -- and keep for itself what is good. The psychologization of distress is a move in this struggle over the distribution of power, responsibility and blame.

Biomedicine and the Moral order

Biomedical diagnosis aims to uncover the causes of disease. Lay thought, however, is more concerned with moral and juridical questions of responsibility and blame. The commonsense categories of misfortune divide broadly in those caused by human agency and those that result from impersonal accident. No matter how limited and technical they conceive their diagnostic task to be, physicians can hardly avoid employing these commonsense categories in their thinking about sickness.

Causality, responsibility and blame are distinct concepts (Shaver and Drown 1986). Attributions of responsibility are made more readily than attributions of cause and people may hold themselves or others responsible for events they clearly did not cause. Responsibility is a matter of degree, depending not only on causality, but also on awareness of the consequences of action; intention or voluntariness of action; absence of external coercion; and, appreciation of the moral wrongfulness of action. For example, the child who pretends to have a stomach ache is held less responsible than the adult who fakes illness to seek monetary compensation because of presumed differences in their social awareness. Finally, blame is the attribution made after an individual's justifications for an action with a negative outcome have been rejected.

Blame is a particular form of moral behavior that censures the individual for negative voluntary actions. From a rationalist perspective, volition always involves reasons for action. But, because so many negative actions or outcomes are not actually intended, blame exaggerates the role of consciousness and rationality in daily life. Blaming people for their ill fortune moralizes events that might otherwise be viewed as the result of external malign forces or simply chance. Both chance and nonhuman evil are threats to the notion of a "just" and morally ordered world. Thus, one function of blame is to support and enlarge a coherent moral order centered on rational human agency. As Bernard Williams puts it:

The institution of blame is best seen as involving a fiction, by which we treat the agent as one for whom the relevant ethical considerations are reasons. The "ought to have" of blame can be seen as an extension into the unwilling of the "ought to have" we may offer, in advice, to those whose ends we share. This fiction has various functions. One is that if we treat the agent as someone who gives weight to ethical reasons, this may help to make him into such a person (Williams 1985:193).

Attributing blame then, is also attributing a sort of moral personhood. The individual who can be blamed is presumed to have been capable of acting otherwise. Blame is withheld when people act under coercion, with unavoidably incomplete knowledge or mistaken premises, or with diminished capacity to control their behavior or understand its consequences. In every case, exemption from blame is linked to a reduction in the person's moral agency.

The effort to diminish the moral blame attached to deviant behavior has led to labelling alcoholism, drug abuse, criminal behavior, and other social problems as forms of sickness. But sickness that affects behavior implies a diminished capacity of the person to govern his action. The person then escapes moral blame, not primarily because his actions have no moral import, but because he himself has a diminished capacity for moral behavior. In this way, the medicalization of deviance reduces the person's "social being."

The moral dimension of medicine, then, is not something imposed by doctors on patients, it arises from the cultural concept of the person. Given the current Western concept of the person, some form of mind-body dualism is inescapable. Medical theory offers a subtly articulated expression of the person's alienation from the body in Western society, but this alienation is found, as well, in every sphere of economic and political life (cf. Turner 1984). Medical training certainly has some deforming effects on physicians, contributing to a sense of elitism, entitlement and distancing from feeling. Medical students are treated as disembodied intellects who can absorb endless amounts of detail with little attention to their own emotional and physical needs. General hospitals, with their concentration of suffering, make exceptional demands on practitioners and students. The rationalization of medical care and the preoccupation with technology further widen the gulf between doctor and patient. Medical practice provides important opportunities to amplify the division of person and body by focusing attention on diseased bodies at the expense of caring for people. In the technical distancing from the body and from interpersonal relationships, the physician loses the opportunity to participate consciously in the construction of social and moral value.

Some notion of accident versus agency is undoubtedly a cultural universal. What is specific to a culture is just where this distinction is drawn and how each category of action fits with other aspects of the social system. Western culture extends the realm of personal agency by psychologizing experience. An extreme example is given by the early psychoanalytic writer George Groddeck (1949) who viewed all disease as the result of the workings of the psyche. In this view, a person is always responsible for his own misfortune. Currently, a similar idealism is found in the work of some holistic health practitioners who grant limitless power to the imagination. In subtler form, the same heroic view of the individual mind can be found in the optimism of workers who advocate the use of mental imagery to control or cure cancer (e.g. Simonton, Matthews-Simonton and Creighton 1980). This holism tends to be not a transcendence of mind-body duality but a negation of one pole -- the body is made subordinate to the mind. Susan Sontag's (1977) polemic on the dangers of psychologizing cancer illustrates the damaging effects of such "holism." Clinicians and laypersons convinced that cancer is caused by repressed anger alienate the sufferer from her own distressed body. Thus, the moral problem of sickness does not vanish either when biomedical physicians ignore the person in favor of the physiology of the body, or when holistic practitioners ignore the body by retreating into a world of fanciful imagery.

The transcendence of mind-body dualism that *is* possible for Western medicine right now, involves a shift from a preoccupation with causes to an emphasis on care (Cassell 1982). Caring begins with accepting the phenomenal reality of the patient's suffering, including its moral significance to the patient and others. Accepting the patient as a person leads to a willingness to explore the personal meanings of distress beyond the causal theories of biomedicine. The physician who is able to remain agnostic about the cause of obscure symptoms will face opposition from the need of patients, families and social institutions to attribute responsibility. The physician can try to educate the public that responsibility for cause is distinct from responsibility to try to get well, and that both should be distinguished from moral blame. But biological, psychological and social attributions of cause and responsibility must be seen as therapeutic measures that express values rather than as morally neutral descriptions of objective reality (Kirmayer 1986, 1987). The choice of explanations in medicine is always a choice of values and clinical practice, no less than public debate about medical care, must be concerned with the moral implications of diagnosis and treatment.

CONCLUSION

I have tried to show some ways in which the metaphors of mind and body contribute to the alienation of the person and the opposition of doctor and patient in biomedicine. Mind-body dualism is so basic to Western culture that holistic or psychosomatic medical approaches are assimilated to it rather than resulting in any reform of practice. Distress is dichotomized into physical and mental, real and imaginary, accident and moral choice. The duality of mind and body expresses a tension between the unlimited world of thought and the finitude of bodily life. It provides a metaphoric basis for thinking about social responsibility and individual will. It conditions the emotional response to pain and suffering, leading us to view the stoic as mentally sound and morally upright.

The passivity of matter makes the body a fit object for technical control. Issues of moral choice and value are set aside and patients are enjoined to act with the doctor as scientists toward their own bodies. But sickness has a corrosive effect on rationality. Personal and social distress, along with those mysteries of the body that have not yet yielded to biological explanation, are challenges to the rational order. The threat that unexplained or uncontrolled sickness presents to the authority of biomedicine is neutralized by making the patient accountable for the illness. Patients are then either rational but morally suspect in choosing to be sick or irrational and thus morally blameless but mentally incompetent. From this marginal state, half-sick/half-bad, patients have various options: they may reject biomedicine for some heterodox medical system consonant with their cultural beliefs and economic resources; they may "doctor-shop," seeking to re-enter the biomedical health care system through another practitioner; or, they may be attracted to a holistic or psychosomatic medicine designed to "rescue" marginal patients.

Psychosomatic medicine, despite its aspirations toward wholeness, reproduces the same Cartesian split. Attributing symptoms to psychological causes attaches blame to patients or diminishes trust in their rationality. Psychological insight, when applied with the distant "objectivity" of the scientist, can reproduce the same hierarchy of values as the organic explanations of biomedicine. For those who accept the value of relatedness, receptivity and responsiveness rather than power, domination and control, the exploration of alternate epistemologies and conceptions of the person is a pressing concern.[10]

Without consistent attention to the experience of illness and the sociomoral dimensions of sickness, the "biopsychosocial approach" of contemporary medical education will become just another technique for rationalizing the patient as a system of medical facts. Personality and social stress will be variables duly noted and entered into the equation of the patient's distress, while disease remains the one solid fact about the person before which emotional reactions and personal values fade to insignificance. A more substantive change in medical training would involve learning to work with the emotions and relatedness. To achieve this we need not additional subjects in the medical school curriculum, but significant changes in the competitive, individualistic, rationalistic atmosphere of the classroom and the rigid hierarchy of apprenticeship, in which the dominance of student by teacher reproduces the dominance of patient by physician.

Realizing the metaphoric basis of our world view enlarges the domain of moral choice. This does not mean rejecting the rationality of scientific empiricism in a "romantic rebellion" (Shweder 1984), but it frees us to give serious attention to the role of feeling and value in the practice of medicine. In a profound and often beautiful book, Jacob Needleman (1985) explores the physician's calling and the importance of the emotions in the moral problems of medicine. But, as Needleman emphasizes, it is the emotions of the physician, as much as those of the patient, that must be understood and reckoned with. In the confrontation with mortality, the practice of medicine offers a unique opportunity for a true marriage of reason and feeling. This will occur only when both healer and sufferer are open to bodily-felt meaning and the social context of sickness and respond, not just with a flurry of technical activity, but with a human relatedness that nurtures the seeds of contemplation and compassion.

ACKNOWLEDGEMENTS

I would like to thank Barbara Hayton, David Howes, Margaret Lock, Joel Paris, Jim Robbins, and Margaret Schibuk for stimulating discussions and helpful comments on earlier versions of this paper.

NOTES

1. The vignettes and case examples in this paper are drawn from my experience in general hospitals as a consultation-liaison psychiatrist.
2. Psychosomatics, as a named body of knowledge and clinical practice, is a phenomenon of the twentieth century. According to Margetts (1950), the word "psychosomatic," coined by J.C.A. Heinroth in 1818, was used infrequently in medical literature prior to 1935. In the nineteenth century, neurosis, nervousness and neurasthenia were viewed as physical conditions of the nervous system (Stainbrook 1952). Contemporary psychosomatic medicine,

intended as a corrective to the segmentation of medical care, remains compartmentalized as a subspeciality of psychiatry and health psychology practiced in general hospitals. For the history of psychosomatic medicine and useful bibliographies, see also: Ackernecht (1982), Fischer-Homberger (1979) and Lipowski (1984).

3. "Personality disorder" refers to a pattern of behavior or character that is socially deviant. Often personality disorders are more troubling for others than for the patient since they form part of a more or less stable adaptation. Borderline personality disorder is a particularly severe type marked by an unstable sense of personal identity, polydrug abuse, promiscuous sexuality, predominantly depressed and angry affect, repeated suicide attempts and tumultuous interpersonal relationships (American Psychiatric Association 1980). "Borderline" is often used as a pejorative label in psychiatry to describe patients who form immediate intense and angry attachments to others, including the psychiatrist or health care provider. The term "borderline" derives from older psychodynamic theory which held that these patients' problems stood on the border between neurosis and psychosis. This sort of continuum theory of mental disorder is less popular now and borderline disorder is thought to be related to affective disorders like major depression.

4. In current psychiatric thought, hysteria has been "split asunder" to yield a psychiatric syndrome of multiple somatic symptoms of unknown etiology (Briquet's syndrome or somatization disorder), a personality style (histrionic), and isolated conversion symptoms (Hyler and Spitzer 1977). Despite the effort to distinguish these three uses of the term hysteria there is still a tendency to confuse them (cf., Merskey 1979).

5. Despite significant changes in sex roles, the classical notion that women are emotional while men are rational survives today. Asked to name a "most emotional" person, both men and women tend to identify a woman relative or close friend (Shields 1987:242). This emotionality is generally viewed in a negative light and extends to the evaluation of other aspects of sex roles. In a review of fifty North American child-rearing manuals published since 1915, Shields and Koster (cited in Shields 1987:246) found that, across all eras, fathers were viewed primarily as disciplinarians, "controlled and controlling." Fathers were noted to be less emotional than mothers and, hence, better able to be objective about their children's behavior. In contrast, "mothers were seen as having the tendency to overreact emotionally which could have deleterious effects on both daughters and sons . . . A mother's feminine tendency to be emotional is construed as a threat to the child's achievement of rational maturity, especially if the child is a male." Women's emotion is identified with the irrational, primitive and immature. It is the addition of rationality that makes emotions less shallow or superficial. Because emotional expression in men is tempered by rationality, male emotion is viewed as a hard-won accomplishment with more positive connotations of interpersonal sensitivity

or warmth.

6. The strict behaviorism that once denied the utility of the concept of mind has given way to cognitive psychology, with its dominant metaphor of the mind as an information processing machine (Gentner 1985). Cognitive psychotherapy is founded on the concept of the person as rational agent. Physical symptoms, like other maladaptive behaviour, may be attributed to irrational thinking or reasoning from mistaken premises (Turk, Meichenbaum and Genest 1983). Patients are taught to observe and record their own thoughts in a ledger, noting their uplifting or demoralizing quality. Practitioners then attempt to modify patients' coping strategies by correcting lapses in logic or false premises and substituting positive, encouraging thoughts for negative or self-punitive ones. Mind is thus reified as a written list of positive and negative statements to which accounting principles can be applied.

7. The arrangement of the body itself along a continuum from mortal, base and impure to immortal, refined, and pure, is reflected in biomedicine in the status accorded different medical specialities. Examining *Who's Who in Britain*, Hudson, (1972:70) found that medical "specialists from English private schools are more likely than others to achieve eminence by working on living bodies as opposed to dead bodies; on the head as opposed to the lower trunk; on the surface of the body, as opposed to its insides; and on male bodies as opposed to female."

8. In conversation with child psychiatry residents at Kyoto University in May 1984, I found the residents tended to explain psychiatric problems in terms of the interaction of psychophysical constitution and social stress, rather than in terms of the intrapsychic forces or family dynamics that dominate the case discussions of North American psychiatric residents.

9. The moral and juridical basis of the Western concept of the person was emphasized by Marcel Mauss (1979). For recent philosophical and anthropological reconsiderations of Mauss see: Carrithers, Collins and Lukes (1985). More psychological approaches to the concept of the person and the sense of self can be found in edited volumes by Gergen and Davis (1985) and Marsella, DeVos and Hsu (1985).

10. Provocative accounts of the masculine biases of our current epistemologies can be found in Harding and Hintikka (1983) and Keller (1985) which also offer hints of a more feminine approach to knowledge. This work is complemented by psychological accounts of woman's moral development (Gilligan 1982) and "ways of knowing" (Belenky, Clinchy, Goldberger and Tarule, 1986). See also the novel by philosopher Rebecca Goldstein (1983) in which male rationality is contrasted with women's attention to the bodily-felt significance of relationships and events which yields a "mattering map" that represents reality through embodiment.

REFERENCES

Ackernecht, E.H.
  1982   The History of Psychosomatic Medicine. Psychological Medicine 12:17-24.
American Psychiatric Association
  1980   Diagnostic and Statistical Manual, Edition III. Washington: American Psychiatric Association.
Averill, J.
  1974   An Analysis of Psychophysiological Symbolism and its Influence on Theories of Emotion. Journal for the Theory of Social Behavior 4:147-190.
Bateson, G.
  1979   Mind and Nature: A Necessary Unity. New York: Dutton.
Beckman, H.B. and Frankel, R.M.
  1984   The Effect of Physician Behavior on the Collection of Data. Annals of Internal Medicine 101:692-696.
Belenky, M.F., Clinchy, B.M., Goldberger, N.R. and Tarule, J.M.
  1986   Women's Ways of Knowing. New York: Basic Books.
Blackwell, B.
  1967   Upper Middle Class Adult Expectations about Entering the Sick Role for Physical and Psychiatric Dysfunctions. Journal of Health and Social Behavior 8:83-95.
Bloor, D.
  1982   Durkheim and Mauss Revisited: Classification and the Sociology of Knowledge. Studies of the History and Philosophy of Science 13:267-297.
Carrithers, M., Collins, S. and Lukes, S. (eds.)
  1985   The Category of the Person. Cambridge: Cambridge University Press.
Cassell, E.J.
  1982   The Nature of Suffering and the Goals of Medicine. New England Journal of Medicine 306:634-645.
Chodoff, P.
  1982   Hysteria and Women. American Journal of Psychiatry 139:545-551.
Cobb, S.
  1957   Monism and Psychosomatic Medicine. Psychosomatic Medicine 19:177-78.
Corballis, M.
  1980   Laterality and Myth. American Psychologist 35:284-295.
Dearing, B.
  1980   Aequanimitas Revisited: Personal and Professional Styles of Physicians. Man and Medicine 5:139-148.

DeVos, G.
    1985   Dimensions of the Self in Japanese Culture. *In* A.J. Marsella, G.
          DeVos and F.L.K. Hsu (eds.), Culture and Self. Pp. 141-184. New
          York: Tavistock Publications.
Doi, T.
    1973   The Anatomy of Dependence. Tokyo: Koadansha International.
Durkheim, E.
    1973   The Dualism of Human Nature and its Social Consequences. *In* R.N.
          Bellah (ed.), Emile Durkheim on Morality and Society. Pp. 149-163.
          Chicago: University of Chicago Press.
Durkheim, E. and Mauss, M.
    1963   Primitive Classification. Chicago: University of Chicago Press.
Engel, G.L.
    1970   Conversion Symptoms. *In* C.M. MacBryde and R.S. Blacklow (eds.),
          Signs and Symptoms: Applied Pathologic Physiology and Clinical Inter-
          pretation (5th edition). Pp. 650-668. Philadelphia: Lippincott.
    1977   The Need for a New Medical Model: A Challenge for Biomedicine.
          Science 196:129-136.
Eisenberg, L.
    1977   Disease and Illness: Distinctions Between Professional and Popular
          Ideas of Sickness. Culture, Medicine and Psychiatry 1:9-24.
Fischer-Homberger, E.
    1979   On the Medical History of the Doctrine of Imagination. Psychological
          Medicine 9:619-628.
Fortenbaugh, W.W.
    1975   Aristotle on Emotion. London: Duckworth.
Gellner, E.
    1985   The Psychoanalytic Movement. London: Paladin Books.
Gentner, D.
    1985   The Evolution of Mental Metaphors in Psychology: A 90-year Retro-
          spective. American Psychologist 40:181-192.
Gergen, K.J. and Davis, K.E. (eds.)
    1985   The Social Construction of the Person. New York: Springer-Verlag.
Gilligan, C.
    1982   In a Different Voice. Cambridge: Harvard University Press.
Goldstein, R.
    1983   The Mind-Body Problem: A Novel. New York: Random House.
Greenhill, M.H.
    1977   The Development of Liaison Programs. *In* G. Usdin (ed.), Psychiatric
          Medicine. Pp. 115-191. New York: Brunner/Mazel.
Groddeck, G.
    1949   The Book of the It. New York: Random House.

Hahn, R.A.
  1985a A World of Internal Medicine. *In* R.A. Hahn and A.D. Gaines (eds.),
    Physicians of Western Medicine. Pp. 51-111. Dordrecht, Holland: D.
    Reidel.
  1985b Between Two Worlds: Physicians as Patients. Medical Anthropology
    Quarterly 16:87-98.
Hankins, T.L.
  1985 Science and the Enlightenment. Cambridge: Cambridge University
    Press.
Harding, S. and Hintikka, M.B. (eds.)
  1983 Discovering Reality. Dordrecht, Holland: D. Reidel.
Harré, R.
  1984 Social Elements as Mind. British Journal of Medical Psychology
    57:127-135.
Hertz, R.
  1973 The Pre-Eminence of the Right Hand: A Study in Religious Polarity. *In*
    R. Needham (ed.), Right and Left: Essays on Dual Symbolic Classifica-
    tion. Pp. 3-31. Chicago: University of Chicago Press.
Hesse, M.
  1978 Theory and Value in the Social Sciences. *In* C. Hookway and P. Pettit
    (eds.), Action and Interpretation. Pp. 1-16. Cambridge: Cambridge
    University Press.
Hsu, F.L.K.
  1971 Psychological Homeostasis and *Jen*: Concepts for Advancing Psycholog-
    ical Anthropology. American Anthropologist 73:23-41.
Hudson, L.
  1972 The Cult of the Fact. New York: Harper & Row.
Hyler, S. and Spitzer, R.
  1978 Hysteria Split Asunder. American Journal of Psychiatry 135:1500-1504.
Jones, M.M.
  1980 Conversion Reaction: Anachronism or Evolutionary Form? A Review
    of the Neurological, Behavioral and Psychoanalytic Literature. Psycho-
    logical Bulletin 87:427-441.
Keller, E.F.
  1985 Reflections on Gender and Science. New Haven: Yale University
    Press.
Kirmayer, L.J.
  1986 Somatization and the Social Construction of Illness Experience. *In* S.
    McHugh and T.M. Vallis (eds.), Illness Behavior: A Multidisciplinary
    Model. Pp. 111-133. New York: Plenum Press.
  1987 Languages of Suffering and Healing: Alexithymia as a Social and Cul-
    tural Process. Transcultural Psychiatric Research Review 24:120-136.

Kleinman, A. and Kleinman, J.
   1985   Somatization: The Interconnections Among Culture, Depressive Experi-
          ences, and the Meanings of Pain. A Study in Chinese Society. *In* A.
          Kleinman and B. Good (eds.), Culture and Depression. Pp. 429-490.
          Berkeley: University of California Press.
Lakoff, G. and Johnson, M.
   1980   Metaphors We Live By. Chicago: University of Chicago Press.
Lange, L.
   1983   Woman is Not a Rational Animal: On Aristotle's Biology of Reproduc-
          tion. *In* S. Harding and M.B. Hintikka (eds.), Discovering Reality. Pp.
          1-16. Dordrecht, Holland: D. Reidel.
Lipowski, Z.J.
   1984   What Does the Word "Psychosomatic" Really Mean? A Historical and
          Semantic Inquiry. Psychosomatic Medicine 46:153-171.
Lloyd, G.
   1964   The Hot and the Cold, the Dry and the Wet in Greek Philosophy. Jour-
          nal of Hellenic Studies 84:92-106.
   1973   Right and Left in Greek Philosophy. *In* R. Needham (ed.), Right and
          Left: Essays on Dual Symbolic Classification. Pp. 167-186. Chicago:
          University of Chicago Press.
Lock, M.
   1987   Protests of a Good Wife and a Wise Mother: The Medicalization of Dis-
          tress in modern Japan. *In* E. Norbeck and M. Lock (eds.), Health, Ill-
          ness and Medical Care in Japan: Continuity and Change. Pp. 130-157.
          Honolulu: University of Hawaii Press.
MacLean, P.D.
   1949   Psychosomatic Disease and the "Visceral Brain." Recent Developments
          Bearing on the Papez theory of Emotion. Psychosomatic Medicine
          11:338-353.
Margetts, E.L.
   1950   The Early History of the Word "Psychosomatic." Canadian Medical
          Association Journal 63:402-404.
Marsella, A.J., DeVos, G. and Hsu, F.L.K. (eds.)
   1985   Culture and Self. New York: Tavistock Publications.
Mauss, M.
   1979   Sociology and Psychology. London: Routledge & Kegan Paul.
Mechanic, D.
   1972   Social Psychological Factors Affecting the Presentation of Bodily Com-
          plaints. New England Journal of Medicine 286:1132-39.
Merskey, H.
   1982   The Analysis of Hysteria. London: Balliere Tindall.
Mishler, E.G.
   1985   The Discourse of Medicine. New Jersey: Ablex Publishing Company.

Needham, R.

1979   Symbolic Classification. Santa Monica: Goodyear Publishing Company.

1985   Exemplars. Berkeley: University of California Press.

Needleman, J.

1985   The Way of the Physician. New York: W.W. Norton.

Osherson, S. and AmaraSingham, L.

1981   The Machine Metaphor in Medicine. In E. Mishler, G., Mishler, L. AmaraSignham, et al. Social Contexts of Health, Illness and Patient Care. Pp. 218-249. New York: Cambridge University Press.

Osler, W.

1928   Aequanimitas, 2nd edition. London: H.K. Lewis.

Papez, J.W.

1937   A Proposed Mechanism of Emotion. Archives of Neurology 38:725-743.

Parsons, T.

1951   The Social System. Glencoe, Ill: The Free Press.

Perelman, C.

1982   The Realm of Rhetoric. Notre Dame, Indiana: University of Notre Dame Press.

Phillips, J.A.

1984   Eve: The History of an Idea. San Francisco: Harper & Row.

Reiser, M.F.

1985   Mind, Brain and Body. New York: Basic Books.

Roy, A. (ed.)

1982   Hysteria. New York: John Wiley & Sons.

Ruesch, J.

1948   The Infantile Personality: The Core Problem of Psychosomatic Medicine. Psychosomatic Medicine 10:134-144.

Russell, J.A.

1979   Affective Space is Bipolar. Journal of Personality and Social Psychology 37:345-356.

Schneirla, T.C.

1959   An Evolutionary and Developmental Theory of Biphasic Processes Underlying Approach and Withdrawal. In M.R. Jones (ed.), Nebraska Symposium on Motivation (Vol. 7). Pp. 1-42. Lincoln: University of Nebraska Press.

Schwartz, G.E.

1982   Testing the Biopsychosocial Model: The Ultimate Challenge Facing Behavioral Medicine. Journal of Consulting and Clinical Psychology 50:795-796.

Segall, A.

1976   The Sick Role Concept: Understanding Illness Behavior. Journal of Health and Social Behavior 17:163-170.

Shaver, K.G. and Drown, D.
    1986   On Causality, Responsibility and Self-Blame: A Theoretical Note. Jour-
           nal of Personality and Social Psychology 50:697-702.
Shields, S.A.
    1987   Women, Men and the Dilemma of Emotion. *In* P. Shaver and C. Hen-
           drick (eds.), Review of Personality and Social Psychology 7:229-250.
Shweder, R.A.
    1984   Anthropology's Romantic Rebellion Against the Enlightenment, or
           There's More to Thinking than Reason and Evidence. *In* R.A. Shweder
           and R.A. LeVine (eds.), Culture Theory: Essays on Mind, Self and
           Emotion. Pp. 27-66. Cambridge: Cambridge University Press.
Simonton, O.C., Matthews-Simonton, S. and Creighton, J.L.
    1980   Getting Well Again. New York: Bantam Books.
Smith, R.
    1983   Japanese Society. New York: Cambridge University Press.
Sontag, S.
    1977   Illness as Metaphor. New York: Random House.
Spelman, E.V.
    1983   Aristotle and the Politicization of the Soul. *In* S. Harding and M.B.
           Hintikka (eds.), Discovering Reality. Pp. 17-30. Dordrecht, Holland:
           D. Reidel.
Spicker, S. (ed.)
    1970   The Philosophy of the Body: Rejections of Cartesian Dualism. Chicago:
           Quadrangle Books.
Stainbrook, E.
    1952   Psychosomatic Medicine in the 19th Century. Psychosomatic Medicine
           14:211-226.
Sullivan, M.
    1986   In What Sense is Contemporary Medicine Dualistic? Culture, Medicine
           and Psychiatry 10:331-350.
Sulloway, F.J.
    1979   Freud: Biologist of the Mind. New York: Basic Books.
Takano, R.
    1977   Anthropophobia and Japanese Performance. Psychiatry 40:259-269.
Taussig, M.T.
    1980   Reification and the Consciousness of the Patient. Social Science and
           Medicine B 14:3-13.
Taylor, C.
    1982   Rationality. *In* M. Hollis and S. Lukes (eds.), Rationality and Relativ-
           ism. Pp. 87-105. Cambridge, Mass.: MIT Press.
TenHouten, W.
    1985   Cerebral-lateralization Theory and the Sociology of Knowledge. *In* D.F.
           Benson and E. Zaidel (eds.), The Dual Brain: Hemispheric

Specialization in Humans. Pp. 341-358. New York: The Guilford Press.

Tumulty, P.A.
1970   What is a Clinician and What Does He Do? New England Journal of Medicine 283:20-24.

Turk, D.C., Meichenbaum, D. and Genest, M.
1983   Pain and Behavioral Medicine: A Cognitive-Behavioral Perspective. New York: Guilford Press.

Turner, Bryan
1984   The Body and Society. Oxford: Basil Blackwell.

Twaddle, A.C.
1969   Health Decisions and Sick Role Variations: An Exploration. Journal of Health and Social Behavior 10:105-114.

Veith, I.
1965   Hysteria: The History of a Disease. Chicago: University of Chicago Press.

Warner, R.
1978   The Diagnosis of Antisocial and Hysterical Personality Disorders: An Example of Sex Bias. Journal of Nervous and Mental Disease 166:839-845.

Williams, B.
1985   Ethics and the Limits of Philosophy. Cambridge: Harvard University Press.

Winstead, B.A.
1984   Hysteria. In C.S. Widom (ed.), Sex Roles and Psychopathology. Pp. 73-100. New York: Plenum Press.

Wittgenstein, L.
1958   Philosophical Investigations. New York: Macmillan.
1980   Culture and Value. Chicago: Chicago University Press.

World Health Organization
1964   Psychosomatic Disorders (Technical Report Series No. 275). Geneva: World Health Organization.

Wu, D.
1982   Psychotherapy and Emotion in Traditional Chinese Medicine. In A. Marsella and G. White (eds.), Cultural Conceptions of Mental Health and Therapy. Pp. 285-302. Dordrecht, Holland: D. Reidel.

Young, A.
1982   The Anthropologies of Illness and Sickness. Annual Review of Anthropology 11:257-285.

CECIL G. HELMAN

PSYCHE, SOMA, AND SOCIETY: THE SOCIAL CONSTRUCTION OF
PSYCHOSOMATIC DISORDERS

INTRODUCTION

Since World War Two, there has been increasing interest in the concept
of "psychosomatic" disorders. A new field of study -- psychosomatic
medicine -- has been developed, with the aim of understanding ill-health
from a more holistic perspective; its purpose, according to Lipowski
(1968), is "to study, and to formulate explanatory hypotheses about, the
relationships between biological, psychological, and social phenomena as
they pertain to person." As a result of this approach, a wide range of
conditions have been described, all of which have some psychosomatic
component (Knapp 1980). However, little research has been done on
how -- and why -- the diagnosis of "psychosomatic" disorder is negotiated
between clinicians and patients, and on the lay explanatory models (See
Kleinman 1980:105) used by patients with these conditions. In particular,
it is important to understand how these patients make sense of their
physiological experiences, of the diagnostic label of "psychosomatic,"
and of the "stress," "emotions" or "tension" said by clinicians to cause
or exacerbate their disorders. The pilot-study described below attempts
to shed light on these problems, in the case of certain gastrointestinal
and respiratory conditions.

THE CATEGORY OF "PSYCHOSOMATIC"

Despite decades of research, psychosomatic disorders remain, to some
extent, an anomalous category within the biomedical model. As numerous authors (Cassell 1976; Eisenberg 1977; Engel 1977; Kleinman et al.
1978) have pointed out, contemporary biomedicine is characterized by a
mind-body dualism, the reduction of ill-health to physicochemical terms,
and an emphasis on biological (rather than social or psychological) information in reaching a diagnosis. As a result psychosomatic disorders are
often difficult to diagnose, or to confine within the biomedical model --
especially as they are often "illness without disease," where emotional or
behavioral changes occur in the absense of any identifiable organic
abnormality. This group, which Minuchin et al. (1978:29) term *secondary*
psychosomatic disorders, should be distinguished from *primary* disorders

95

M. Lock and D. R. Gordon (eds.), Biomedicine Examined, 95–122.
© *1988 by Kluwer Academic Publishers.*

where an identifiable physiological dysfunction is already present, but is exacerbated by psychological factors. But even in primary disorders -- which Engel (1975:657) terms "somatopsychic-psychosomatic" -- the relationship of the organic abnormality to the patient's symptoms and signs is often tenuous and unpredictable. In both groups the clinical picture is frequently time- or context-specific, and diagnosis depends on knowing why a particular individual got a particular symptom, at a particular time. In many cases these contexts are social, psychological or environmental, and this information may be inaccessible to some clinicians, especially those with a bias towards biological explanations of ill-health.

A further difficulty, from the biomedical point of view, is that many psychosomatic disorders have a chronic, relapsing, and unpredictable course (e.g., Drossman 1977); where social or psychological factors play an important role in exacerbations, this course is less controllable by clinicians reliant on chemotherapy or a "technological fix." Another problem with these disorders is the difficulty in explaining, of predicting "organ choice" -- i.e., why a particular organ or psychological system is involved in a particular individual -- by using a strict biomedical paradigm. This is because each occurrence of the disorder may only be explained by reference to the unique biological, social and psychological aspects of that patient's life -- and not by the characteristics of a particular disease entity.

Lipowsky (1968) has also pointed out that the very term "psychosomatic" "connotes an assumption that there exist 2 classes of phenomena, i.e., psychic (mental) and somatic, which require separate methods of observation and distinct languages for their description." The term imposes, therefore, both a semantic and a methodological dualism on the study of ill-health. Engel (1967) pointed out the difficulties, given this assumed dualism, in reconciling the paradigms used to explain phenomena in the biological and the psychological domains. He noted the difficulties in establishing relationships between these two frames of reference, since the principles used in establishing relationships *within* the psychological frame are different from those needed to establish relationships within the somatic frame of reference. For example, behavior and mental activity are subdivided into subcategories such as affects, object relations, ego defences, and cognitive functions, and these abstractions do not constitute discrete, measurable functions in the same sense as, say, the components of gastric secretions. Also, each subcategory is always part of others; e.g., effects always involve drives, ego defences, cognitive functions, and so on. Psychic activity, therefore, is a "configurational unit," while somatic activity involves "multiple interrelated unitary functions more or less susceptible to identification and measurement." To overcome the seeming incompatibility of paradigms, both

Lipowski (1968) and Engel (1977) proposed a more holistic model that would integrate the biological, psychological, interpersonal, and social aspects of the patient's life. Engel, in particular, has stressed the inter-relationships -- especially positive and negative feedback loops -- between these different domains. Other writers, however, have tended to view psychosomatic disorders along one of three etiological axes -- as originating in the "psyche," "soma" or "society."

Psyche: *The Psychogenicity Hypothesis*

Early writers on psychosomatic disorders explained both their pathogenesis and "organ choice" by "psychogenicity" -- that is, a linear, causal relationship between certain psychological factors -- such as personality type (or trait), intrapsychic conflicts, defensive patterns, or dysphoric affects -- and specific somatic symptoms or structural changes. This approach implied a dichotomy, as well as etiological link, between "inner" (psyche, emotions, conflicts) and "outer" (somatic signs, symptoms, and behavioral changes) realities. In a sense, psychic factors were seen as "pathogens" acting upon, or expressed in, the material body in a patterned way. This formulation followed Freud and Breuer's (1966) earlier theory of conversion hysteria, as the symbolic somatic expression of intrapsychic conflicts. Weiss and English (1942) postulated a subjective "organ language," whereby selection of affected organs was based on the symbolic meaning of that organ for the individual: diarrhea and vomiting were attempts by the patient "to rid himself of a guilty feeling or of the thoughts that produce it," while asthma was a form of "symbolic crying." Sontag (1948) suggested that the individual developed a conditioned response in childhood which later determined "the expression of an anxiety through certain physiological channels." Other writers focussed less on specific emotions, or conflicts, and described instead dysfunctional "personality types" or "traits." According to Dunbar (1948), each psychosomatic disorder was associated with a specific "personality profile" which contributed towards its development. In this context, the notion of pathogenic personality types can be seen as cultural constructs, based on normative models of social behavior. For example, Groen (1948) described the "character structure" of ulcerative colitis patients as: "They are fearful, and when in imminent danger often overtly cowardly." More recently, the emphasis has shifted from an emphasis on the pathogenic "personality type" to specific traits, or clusters of traits. Gildea (1968) reviewed the "personality malfunctions" of psychosomatic disorders, reported by various authors; his study attempted to link psychological "maladjustments" to specific traits such as "subnormal assertiveness"

("Rarely able to verbalize feelings even when aware of them. Just takes it from boss, wife, friends, etc.") with specific disorders, such as hypertension, peptic ulceration and rheumatoid arthritis. Knapp (1975) described the "consistent characterological features" of asthmatics, including "unusually strong passive and dependent personality traits" reflecting needs to maintain gratification and support from key persons. Cheren and Knapp (1980) have also described the "typical personality traits" or "character disturbances" of patients with ulcerative colitis, Crohn's disease, irritable bowel syndrome, and bronchial asthma; e.g., those with Crohn's disease had "compulsive or paranoid traits" with excessive dependency, compliance, and "explosive manipulation."

While the linear, psychogenic model has gradually given way to more complex, multi-causal or systems models (e.g., Minuchin et al. 1978), the contributory role of intrapsychic conflicts or personality traits in psychosomatic disorders remains part of the biomedical model -- and of lay explanatory models. This is despite Lipowski's (1968) review of the literature, which casts doubt on the existence of disease-specific personalities or psychodynamic constellations in these disorders.

Soma: Physical Weakness or Vulnerability

Parallel with theories of psychogenicity, several researchers postulated an inherited or acquired physical "weakness" -- either local or generalized -- which determined the choice of "target organ," especially in the presence of pathogenic personality traits or intrapsychic conflicts. Bauer (1942) suggested that there was "inherited organ inferiority" which led to specific psychosomatic disorders, but he largely ignored the psychodynamic aspects, while Hendrick (1948) suggested "physical infantilism" -- an "immaturity in the homeostatic processes in an organ system" -- as the reason for "organ choice." Sontag (1948) and other "constitutionalists" argued for more general physiologic predispositions to these conditions, such as inherited disorders of enzyme systems or autonomic nervous system functions. Alexander and his colleagues (1968) have tried to explain organ choice or psychosomatic specificity by a multi-causal model, involving (1) the individual's "characteristic psychodynamic conflict pattern" present from childhood (such as "frustrated urges for accomplishment" in ulcerative colitis patients), (2) a specific "onset situation," which involved activation of this conflict pattern, and (3) a physical, constitutional "Factor X," defined as a "specific organ vulnerability" or weakness. Alexander suggested that some organic diseases have not only a specific pathophysiology, but also a specific psychopathology. This implies either psychogenesis, or that physiological phenomena are the expression of "certain basic qualities of the organism which manifest themselves on the somatic side as an organic predisposition." However, this hypothesis assumes a consistency of conflict patterns, onset situation, and constitutional predisposition. If "characterological patterns or emotional constellations" are consistent phenomena, then one would expect a particular "target organ" to be consistent over an individual's life. While some patterns -- such as ulcerative colitis -- seem more consistent, others vary greatly. Vaillant (1968), in his 30-year prospective study of 95 men, found that there were no consistent target organs over the years in those who developed psychosomatic disorders.

Engel (1975:667) has suggested that physical factors -- particularly in the "somatopsychic-psychosomatic group" -- may in their turn influence the development of specific psychological characteristics. The relationship of physical to psychological symptoms was also examined by Grace and Graham (1952) and Grace et al. (1962). They related specific psychological "attitudes" to certain physiological symptoms in psychosomatic disorders; for example, diarrhea occurred "when an individual wanted to be done with a situation or to have it over with, or to get rid of something or somebody." However, their research did not indicate the nature of the link between attitudes and symptoms.

Society: The Role of Social Categories

An alternative to the mind/body dualism outlined above has been the study of the categories through which physical and psychological states are cognized, labelled, ordered, and acted upon. In this view, "mind," "emotions," "personality," "body" and "organs" are all cultural categories, which are socially derived. They result from, and are maintained by, personal experiences, interpersonal interactions, and institutionalized role structures (Katon et al. 1982). Douglas (1973) has pointed out the social dimensions of the human body, whereby experiences of the body are modified by the social categories through which it is known, and the categories of social experience and bodily experience enhance one another. In every society, according to Douglas, there is pressure to create consonance between the perceptions of social and physiological experience. Littlewood (1984) points out the reciprocity of this relationship, whereby natural symbols derived from human biology help pattern or reinforce social categories, as in the binary division of the social and natural worlds into "male" and "female" categories. Taussig (1980) has also noted how the body is a condensation of social categories -- a "cornucopia of highly charged symbols -- fluids, scents, tissues, different surfaces, movements, feelings, cycles of changes constituting birth, growing old, sleeping and waking" -- all of which are linked to different forms of social relations. The social and cultural worlds, therefore, provide the categories through which both bodily and psychological experiences are perceived, and interpreted (Fisher 1968; Mechanic 1972). Similarly, the internal perception of bodily disease, or of abnormal symptoms, is in part learnt from social interactions, and depends on "external cues, interpersonal communications, and the information at hand about the situation" (Barsky and Klerman 1983). These also influence whether, and how, emotional problems are expressed via somatic complaints (Katon et al. 1982), and vice versa.

To some extent, both emotions and personality are cultural constructs. Kleinman (1980:147) points out that while affects occur as universal psychobiological states, cultural beliefs and values determine how they are cognized -- before they take the form of "perceived, felt, labelled, and valuated experiences recognized as emotions." In his study of the Australian Pintupi, Myers (1979) suggested that emotions be seen as an ideology, as models of how one *should* feel and behave in relation to others. The Pintupi view of the emotions provides individuals with a moral system, an internal representation of the normative order, with some emotions more socially acceptable that others. In that sense, emotions recognized and interpreted through socially-derived categories represent different types of social relationships, and different types of relationship of the self to the moral order. Engel (1960), for example, noted how some (socially) "bad" emotions -- such as rage, fear, envy, greed or disgust -- are reified by some individuals into external agents that somehow "cause" them to feel ill or unhappy. The perception of emotions, therefore, may involve the cognitive distinction between socially "good" and socially "bad" emotions, with the latter -- like diseased organs (Cassell 1976) -- reified, and thus separated from the concept of "self." This reification may be a function of the construction of illness realities in the encounters between clinician and patient (Good and Good 1981). On this basis I would hypothesize that the diagnosis, in psychosomatic medicine, of the pathogenic social personality trait or emotion, may be a way of shifting responsibility for etiology, exacerbations or therapeutic failure from the clinician to the patient -- or to reified parts of the patient's "self." These socially "weak" parts of the self can be seen as analogous to the physical weaknesses described earlier, such as Alexander's constitutional "Factor X."

In psychosomatic disorders, as in other conditions, it is important to understand the cognitive categories through which patients (and clinicians) are interpreting the illness, and also how these categories are derived from the patient's social milieu. There is evidence that once a "cognitive set" for the interpretation of psychological or bodily states has been established, it may adversely affect both physiological responses and mood states. Wright and Beck (1983), for example, have described the "self-defeating cognitions and behavior" of depressed patients, while Barsky and Klerman (1983) point out how even physiological responses such as heart rate, breathing, or pain response may all be affected by "the thoughts a person has about his or her physical state and the ideas supplied by others." The research of family therapists, such as Minuchin et al. (1978) sheds light on the social origins of these ideas, especially the role of family members in the development, and maintenance, of the symptoms and cognitions of the psychosomatic child. They point out that "the psychological unit is not the individual. It is the individual in his significant social contexts." One dimension of this social context is described in this study -- the explanatory models and self-perception of patients with chronic psychosomatic disorders.

METHODS

The study sample consisted of 42 adult patients, diagnosed by their clinicians as having chronic disorders with a definite "psychosomatic component" -- either primary or secondary, as described above. Half the sample (21 patients) had a psychosomatic disorder of the *respiratory* system (in this case, bronchial asthma), while the other half had disorders of the *gastrointestinal* tract (ulcerative colitis, irritable bowel syndrome, functional vomiting, and Crohn's colitis). A further 16 patients, approached to take part in the study, either refused or failed to turn up to interviews. The patients were attending the Departments of Medicine at the Cambridge City and Mount Auburn Hospitals in Cambridge, Massachusetts, and a private family practice in that same city. The interviews were semi-structured, and were based on a detailed and standardized questionnaire. They usually lasted 1 -- 1 1/2 hours. The interviews were designed to collect three types of data: (1) social and demographic details of the patient sample, (2) patients' explanatory models of their disorder, particularly of its "psychosomatic" component, and (3) patients' perceptions of the role of their social relationships in causing, maintaining, or alleviating their disorders. The aim was to discover common themes between, and within, each diagnostic category -- especially the ways that patients made sense of their physiological experiences, of the diagnostic label of "psychosomatic," and of their interactions with clinicians over the year.

TABLE I

The patient sample

	Combined sample (N = 42)	Gastrointestinal group (N = 21)	Respiratory group (N = 21)
Sex			
Male	10	2	8
Female	32	19	13
Age			
18-39 years	30	13	17
40-59 years	5	4	1
60+ years	7	4	3
Mean age			
Years	39.7	41.9[a]	37.5[b]
Marital Status			
Married	13	9	4
Single	14	4	10
Divorced/Separated	9	6	3
Widowed/Other	6	2	4
Education			
College graduate	19	6	13
Partial College or high school graduate	17	10	7
Partial high school or less	6	5	1
Ethnicity			
Caucasian	31	13	18
Black	6	4	2
Hispanic	4	4	-
Indian	1	-	1
Religion			
Roman Catholic	20	10	10
Protestant	11	7	4
Jewish	5	1	4
Other	6	3	3
Mean years since diagnosis			
Years		20.9[c]	8.1[d]

[a] s.d. 16.2; s.e. 3.5   [b] s.d. 13.6; s.e. 3.0   [c] s.d. 16.2; s.e. 3.5   [d] s.d. 10.7; s.e. 2.3

Gastrointestinal group: *diagnostic categories*

Ulcerative colitis	12
Irritable bowel syndrome	6
Functional vomiting	2
Crohn's colitis	1

RESULTS

Details of the patient sample, and of their diagnoses, are shown in Table I. The majority of the sample had been diagnosed as having the psychosomatic disorder many years previously -- the mean number of years since diagnosis was 20.9 years in the asthmatics, and 8.1 years in the gastrointestinal group. In most cases this period was characterized by chronic, relapsing, and unpredictable ill-health. It also corresponded to exposure to biomedical models of psychosomatic disorders, especially of their etiology and treatment.

Although this pilot-study was based on a small, and not necessarily representative sample, it was possible -- in analyzing the transcribed interviews -- to isolate common themes within each diagnostic category. Despite the heterogeneous educational and ethnic composition of each diagnostic group, a majority of its members seemed to share a similar explanatory image -- an image based on physiological experiences, and one which condensed the physical, psychological and social aspects of their condition. These common images will be described in more detail below.

Explanatory Models of Psychosomatic Disorders

The range of patients' explanatory models of the etiology of their condition is shown in Table II. These models show some resemblance to the biomedical explanations of psychosomatic disorders described above. Like them, they include notions of the role of pathogenic emotions or personality, physical weakness (inherited or acquired), and social factors, in their development or exacerbation. Most of the patients' models involved *multi-causal* explanations for their disorder, involved *multi-causal* explanations for their disorder, with an average of 4.9 etiologies per asthmatic patients, and 3.3 for the gastrointestinal group. These models constitute a lay "biophychosocial" model (see Engel 1977), which links together the physical, psychological, and social aspects of the patient's life. For example:

(54 year old woman with asthma) I guess I was having the kids when it began. That might be one major catastrophe in my life. Also I was separating from my husband at the time. I used to get upset. I used to get an attack when my father went back to Maine. He lived in Maine, he worked there. I think it was all the factors together, plus the fact that I had a series of bronchitis when I was a child. I was a croupy child. My father was bronchial, my mother was bronchial when we were kids. Also, I used to live on coffee and cigarettes at the time -- 30 cigarettes a day, maybe 40 --

TABLE II
Explanatory models of etiology

Etiology	Respiratory group (N = 21) No. (%)	Gastrointestinal group (N = 21) No. (%)	Total (N = 42) No. (%)
1. Emotion and stress	21 (100)	19 (91)	40 (95)
2. Personality type	14 (67)	18 (86)	32 (76)
3. Hereditary predisposition	14 (67)	7 (33)[a]	21 (50)
4. Allergies	15 (71)	-	15 (36)
5. Infection	9 (43)	2 (10)[b]	11 (26)
6. Constitutional weakness (local or generalized)	4 (19)	10 (48)[c]	14 (33)
7. Weather	11 (52)	-	11 (26)
8. Smoking	7 (33)	-	7 (17)
9. Dust or sawdust	3 (14)	-	3 (7)
10. Exercise	2 (10)	-	2 (5)
11. Divine punishment	1 (5)	-	1 (2)
12. Ethnicity	1 (5)	-	1 (2)
13. Gender	-	1 (5)	1 (2)
14. Medication	-	2 (10)	2 (2)
15. Eating habits	-	5 (24)	5 (12)
16. Bowel training	-	1 (5)	1 (2)
17. Swallowing ocean water	1 (5)	-	1 (2)
18. Auto-immune	-	2 (10)	2 (5)
Total	103	69	172

[a] p < 0.04.   [b] p < 0.02.   [c] p < 0.05.

Mean number of etiologies per patient: Respiratory group   4.9
                                          Gastrointestinal group 3.3

A large majority of the patients (40) saw "emotion," "stress" or their own "personality" as contributing towards their ill-health. Half (21) ascribed it to heredity, and a third (14) to a local or generalized "weakness." Overall, etiologies over which the individual had *no* control -- such as heredity, constitutional weakness, weather, allergies, or gender -- were more common in the asthmatic group. This difference may be related to gender, since the majority were female (19); when selected illness etiologies were compared between females in each group some differences persisted, though with the small size of the sample these were not statistically significant. Like the biomedical models, patients viewed the origin of their ill-health along the three etiological axes -- "psyche," "soma," and "society."

Psyche

While the majority of the sample saw emotional state as a causative factor, they varied in defining what emotions were defined as negative or pathogenic -- though there were no significant differences between the two diagnostic groups. Often two or three emotions were defined as negative by each patient: the asthmatics identified "tension" (15 patients), anger (6), "stress" (3), uncertainty (3), upset (7), unhappiness (2) and guilt (1), while the GI identified "tension" (18), anger (10), frustration (4), "stress" (8), and uncertainty (2). In 31 of the interviews (twelve asthmatics and nineteen GI group), unpleasant or negative emotions were spoken of as an "it," separate in some way from the concept of "self." These reified emotions were described as pathogens, attacking vulnerable organs or systems within the body. This image may be derived from contemporary US popular culture (e.g., "Suddenly, without any warning, she was gripped by an attack of raw, stomach-wrenching terror" -- Houck 1983:30). There were significant differences -- to be described below -- in how most patients in each diagnostic group described the effect of these emotions on their health.

As with emotions, a majority (40) of the patients saw their own "personality" as contributing towards the origin, exacerbation, or chronicity of their condition. In that sense, personality type or trait was described as separate, to a variable degree, from the "self" and, as such, only partially under its control. Personality was conceived of as either congenital or inherited (familial). Eleven asthmatics and eleven in the GI group described themselves as too "sensitive," "nervous," "tense," or "vulnerable." Twelve GI patients (and three asthmatics) described themselves as people who "held too much in," while three of the GI group described themselves as being "obsessive," "perfectionist" or "anal" personalities. In some cases patients saw themselves as having several of these pathogenic personality traits.

Soma

Half the sample (21) ascribed their condition to a hereditary physical "weakness" or predisposition (see Table II), while others saw this weakness as being constitutional (14) or acquired (25). Examples of these three, overlapping, types of "weakness" are:

(31-year old woman with asthma) All my family have it (asthma). My aunt, mother, two sisters, a brother, three nephews, a baby niece. I don't know if it's hereditary. Some doctors say no, but I figure yes -- how come a baby be born with a sickness? Maybe from the father, no?

(24-year old woman with asthma) Everyone has a weakness. It's my weakness. It's the way your body reacts to negative stimuli – like allergies --

(67-year old woman with ulcerative colitis) If you're upset about something, your stomach is the weakest part. Like when my sister was at the hospital and she passed away, and the doctor called to tell you she died, and then I had to fly up to the bathroom.

(64-year old woman with ulcerative colitis) A hereditary tendency. It's a problem in our family. When we're upset colitis -- not ulcerative -- would result. My mother, my daughter, my uncle in Ireland, have it if they're stressed. Also, in November I was awaiting a hip replacement, and I was given Naprosyn 500 mg two or three times a day, plus aspirin. Within one week I started with colitis, bleeding. Yes, it was the over-strong medication – a physical cause.

As illustrated in Table II, physical "weakness" can be acquired by exposure to a number of external influences, including drugs, infection, allergens, tobacco smoke, or weather conditions. In addition, nineteen patients (fifteen asthmatics and four in the GI group) spoke of their diseased organs (lungs, bronchi, stomach, colon) as being in some way autonomous, and independent of the individual's control. This reification of diseased organs was also noted by Cassell (1976). He suggested that this was a method of distancing a disease, or diseased part, and thus enabling the victim to see it as "non-self": no longer part of the intact body image -- but rather something intrusive on it. In this study, "autonomous" organs were understood as "attacking" the self, sometimes in association with pathogenic emotions or personality traits, or external physical or social influences. For example:

(54-year old woman with asthma) The bronchial tubes close, tighten up on me, so no air goes into the lungs. When I feel nervous, I feel them tightening up.

(30-year old woman with irritable bowel syndrome) (To help the condition) Trying to monitor what's happening. Being in touch with what's inside before the colon knows. Using relaxations. Talking to the various organs. It helps.

(32-year old woman with asthma) When I'm under acute stress, I can feel myself tightening up. If people are not aware of how their emotions can damage their health, it can be detrimental. If you're unhappy, angry, it can cause them to tense up, cause their resistence to be low so that they're open to infection. With asthma it causes them to tighten up their bronchi. If my son is going through a bad time, the stress can cause my bronchi to spasm.

(60-year old woman with asthma) Bad relationships – they can make it worse -- anxiety, upset. If something happens, it may make an attack come on. A row can make me tighten up. You get anxiety, and that gives you an attack.

As with reified emotions or personality, these organs are to some degree separate from the concept of self; in some cases they are "public" organs, directly responsive to interactions with other people, as mentioned in this last example. The experiences of bronchospasm and feeling tense -- "tight" bronchi and feeling "uptight" -- were fused in a number of cases. For example:

(24-year old woman with asthma) When I get uptight with my husband, which has been happening a lot lately, I feel a tightness in my chest and I know I'm going to have an attack.

Overall, these explanatory models saw the psychosomatic disorder as arising in part from a physical "weakness," or from an autonomous organ. In both cases, therefore, its etiology was largely out of the patient's control.

Society

The third etiological axis -- the perceived role of social relationships in the disorder -- is illustrated in Table III. A majority of the sample (thirty-nine patients) believed that "bad" relationships could adversely affect their condition, or explain its etiology. Similarly, thirty-one patients believed that "good" relationships could affect their health in a positive way, while twenty-eight believed that these good relationships could actually cause, maintain, or restore good health. A large number of GI patients also believed that their relationships with others would have been different if they had not been ill. There was wide variation, however, in how both "good" and "bad" relationships were defined. A comparison of replies by male and female patients showed some differences, though these were not statistically significant.

In terms of the social "onset situation" (see Alexander et al. 1968), a majority of the asthmatic group (17 patients) saw the "stress" that precipitated their attacks as originating in their pre-existing social networks -- such as conflict with spouse, children, relatives, friends, or neighbors. In the GI group, these social etiologies were identified by eleven patients, while five mentioned "pressure" at work.

Analysis of the questionnaires suggested that patients' explanatory models of psychosomatic disorders are, to a lesser or greater extent, *socially constructed*. The majority of the sample (thirty-nine patients) indicated that they had *learnt* about their condition -- particularly the etiological roles of stress, emotion, and personality -- from other people, especially from clinicians. This supports Good and Good's contention (1981) that illness realities are socially constructed, particularly in the interactions between doctors and patients. They also incorporate, however, patients' personal experiences, the experience of friends or family, inherited folklore, and information gathered from printed material or

TABLE III

Health and social relationships

Etiology	Respiratory group (N = 21)		Gastrointestinal group (N = 21)		Total (N = 42)	
	No. (%)		No. (%)		No. (%)	
1. Bad relationships can influence one's health	20	(95)	19	(91) n.s.	39	(93)
2. Relationships can cause illness	20	(95)	19	(91) n.s.	39	(93)
3. Good relationships can influence one's health	16	(76)	15	(71) n.s.	31	(74)
4. Relationships can cause, maintain or restore health	11	(52)	17	(81) a.	28	(67)
5. One's health affects one's relationships	9	(43)	10	(48) n.s.	19	(45)
6. One's relationships would be different if one were healthy	1	(5)	6	(29) b.	7	(17)
7. Have had bad relationships with family members	13	(62)	15	(71) n.s.	28	(69)
8. Have had bad relationships with non-relatives	10	(48)	11	(52) n.s.	21	(50)

a = $p < 0.05$, b = $p < 0.05$

television. In the case of psychosomatic disorders, the social construction involves the "psychologization" of the condition -- redefining physical symptoms as emotional or psychological in origin. This implies, also, a shift of responsibility from clinician to patient -- especially to their personality, emotional makeup, lifestyle, or social relationships. In some cases the shift was to the patients' physical predisposition, or "genetic makeup." This social construction of psychosomatic disorders is illustrated in the following four vignettes:

(25-year old medical student with ulcerative colitis) Other people were saying there was something wrong with me psychologically. If I'd had appendicitis or a cough, I would have been spared this. Some of these friends were doctors, others not. It's often associated with a psychological component. I searched very hard and for a reason -- Why me? Everyone told me it must be psychological, there must be a large psychological component -- it's in the medical textbooks. Our society associates the bowel and stomach with nervousness -- it's more sensitive to tension.

(33-year old man with ulcerative colitis) It's a kind of strange thing -- my parents left about one week before it happened. They left for Florida. I didn't know if that was the cause. I read it in magazines. My girl friend -- she's a social worker -- she's seen it in books. I've heard it's stress. Maybe because of my background, my Armenian background, my Far Eastern background I've heard it I have a lot of stress. They keep asking me if I have a lot of stress in life, but I don't think I have a lot of stress.

(29-year old male sociology professor) I had been in analysis and my sense is that my background would predispose me to get an ulcerative colitis-like illness, but there was no reason for me to get one. I'm very perfectionist, oriented towards control of situations -- the uncontrollable -- and this predisposes me to a colitis-type of disease. But I have no idea why I got it. It came after a period of my life when I was at my healthiest. It's as a result of psychoanalysis, and also from having the illness. I didn't know this before last April. I learnt it from the medical doctors and from the shrink. It was new knowledge, the connection of these types of colitis symptoms and those psychological symptoms.

(34-year old woman with ulcerative colitis) They tell me it's a lot of genetic tendencies to put psychological stress into my body. Until a year ago the doctors were telling me I was crazy. It was my own fault because of what I did. Some said it's because of how you ate. Some said because I was too sensitive. Others said it was because of exercise. Others said because I wasn't taking enough Azulfidine. What I heard from all the doctors was that it was my fault, and if only you did what they said, everything would be OK. Caroline, my (present) doctor, she asked me what psychosomatic meant. She said "It's not what you think -- people think psychosomatic means that it's your fault" -- but she says that there's a genetic tendency to put stress in your body, and that's a weakness you have. Before Caroline, they said it was all my fault, and that if only I did things differently everything would be OK.

Another social aspect of these conditions were the patients' beliefs about whether they had *control* over their social relationships -- particularly as to whether the relationship had positive or negative effects on their health. Half the sample (eight asthmatics, thirteen GI patients) felt that they *did* have control over the types of relationships that they had with other people. Eight (80%) of the men and thirteen (41%) of the women thought that they did have control. When the two groups of female patients were compared, slightly more GI patients (62%) had control, compared with 42% of the asthmatics.

COMPARISON OF THE TWO DIAGNOSTIC GROUPS

As illustrated in Table I, the patient sample had a heterogeneous educational, ethnic and religious background. In addition, the GI group consisted of patients with four different diagnosis -- ulcerative colitis, irritable bowel syndrome, functional vomiting, and Crohn's colitis. Despite this heterogeneity, it was possible in analyzing the interviews of the two diagnostic groups (respiratory and gastrointestinal) to isolate common themes within each group. These themes are illustrated in Table IV. I argue that the majority of patients in each group, whatever their background, use a common explanatory *image* – an emic illness category which condenses the physical, psychological and social aspects of their disorder. In each case this image is a natural symbol – a socially-derived category through which a natural physiological process is perceived and understood (see Douglas 1973:93). The natural symbols here are the process of *respiration* and *digestion*, and the associated symptoms of bronchospasm, diarrhea, or vomiting. Like "heart distress" in Iran (Good 1977), this physiological image "draws together a network of symbols, situations, motives, feelings, and stresses" which are rooted in patients' daily lives and experiences. I suggest that chronic disorders of the respiratory or gastrointestinal tract lead to a pre-occupation with these physiological processes, and that the illness realities constructed between clinician and patient (see Good and Good 1981) link these natural symbols to the wider world of social and moral values. As well as explaining a physical process, the symbol organizes both social and emotional experiences, and helps define certain emotions, thoughts, personality traits, and parts of the body as either "self" or "non-self." Defining some of these as "non-self" can bring the patient's self-image closer to the normative order of contemporary life -- to social values of independence, fitness, youthfulness, contentment, social success, and control over bodily functions and emotions.

The differences in physiological imagery utilized by a majority of patients in each diagnostic category, is illustrated in Table IV. In *respiration*, the individual has only partial control over the physiological process, in that he can neither "choose" whether to breathe, nor what he breathes in -- whether this is dust, odors, or certain allergens. Respiration is a visible physiological process, with immediate response to environmental influences -- e.g., by coughing, sighing, speech, or hyperventilation. Expiratory sound, such as speech or crying, is patterned by culture and is part of all social relationships. Bronchial asthma involves a disturbance of this process, often with symptoms of wheezing, cough, or shortness of breath. By contrast, *digestion* is a slower and more invisible physiological process, and the conversion of food to feces can take up to 24-hours. Unlike respiration, the individual has some control over what, and when to eat, and the time and place of excretion. All food eaten must eventually be discharged from the system as either feces or vomitus. The preparation and ingestion of food, and the disposal of waste products, are all closely patterned by culture, and are also linked to the maintenance of social relationships. In the gastrointestinal conditions in this group (see Table I), a disturbance of function can involve diarrhea, bloody stools, abdominal pain, and vomiting.

TABLE IV
Comparison of the two diagnostic groups

	Respiratory group (N = 21)	Gastrointestinal group (N = 21)
NATURAL SYMBOL	*Respiration* Visible, continuous physiological process. No control over what is inhaled. Immediate, visible response to social or environmental stimuli. Expiratory sound, such as speech or crying, patterned by culture.	*Digestion/excretion* Hidden physiological process. Control over food ingestion, and time and place of excretion. Long delay between input and output processes. No immediate, visible response to social or environmental stimuli. Processes of eating and excretion, and disposal of waste products, patterned by culture.
PSYCHE Emotions	Negative emotions, usually linked to outside events, attack (weak) lungs or bronchi, causing them to tighten up and obstruct breathing (12 patients) Negative emotions described as an "it" (12 patients)	Negative emotions, linked to outside events, accumulate within the self, and are expelled via (weak) stomach or bowels as anger, "stress," feces, or vomitus. (14 patients) Negative emotions described as an "it" (19 patients)
Personality type or trait	Too sensitive, nervous or vulnerable to outside events (11 patients) Hold too much in (3 patients)	Too sensitive, nervous, or tense (11 patients) Hold too much in (12 patients) Obsessive, perfectionist, or anal (3 patients)
SOMA Body organs or systems	1. Independent of patient's control. Attack patient's breathing in response to outside events or to negative emotions responding to those events. (15 patients) 2. Weakened by hereditary, constitutional, or acquired factors -- or combination these (Hereditary: 14 patients) (Constitutional: 4 patients) (Acquired: 16 patients)	1. Independent of patient's control Accumulate negative emotions, or respond to outside events by producing vomiting or diarrhea. (4 patients) 2. Weakened by hereditary, constitutional, or acquired factors -- or combination these (Hereditary: 7 patients) (Constitutional: 10 patients) (Acquired: 9 patients)
SOCIETY Social relationships	1. Can cause illness (20 patients) 2. Can cause, maintain, or restore good health (11 patients)	1. Can cause illness (19 patients) 2. Can cause, maintain, or restore good health (17 patients)

Table IV (continued)

	Respiratory group (N = 21)	Gastrointestinal group (N = 21)
	3. Patient has control over their relationships (8 patients)	3. Patient has control over their relationships (13 patients)
PATHO-GENESIS	External social or environmental influences attack (weakened) lungs directly, or in association with negative emotions and/or autonomous organs. Pathogenesis may be aided by personality type (18 patients)	External social influences taken into the self, transformed into negative emotions, and allowed to accumulate; must then be expelled via (weakened) bowels or stomach in form of anger, stress, feces, or vomitus. Pathogenesis may be aided by personality type (14 patients)

Social stresses or environmental factors / 'Bad' / Weak Organ / Emotions

Social stresses / 'Bad' Emotions / Weak Organ / Feces/Vomitus

Direction of pathogenesis	From without	From within
Reaction to external influences	Immediate	Delayed
Therapeutic strategy	"Unload" or express negative emotions to other people (2 patients) Avoid, or withdraw from, external social or environmental influences (10 patients) Get control over emotions (2 patients) Take medication (20 patients)	"Unload" or express negative emotions to other people (14 patients) Relax (5 patients) Take medication (17 patients)

1. The Gastrointestinal Group

Fourteen patients in this group described a process of taking social stresses into the "self," transforming them into negative emotions -- such as anger, tension, hostility, fear, or stress -- and failing to release them quickly. These (reified) emotions accumulate within the self, become more dangerous, and must be expelled to the outside as feces, vomitus, anger or "stress." In some cases, these two types of socially-unacceptable waste products are used interchangeably, with "stress" taken in and "digested" eventually to either anger or feces. The escape of these negative emotions from the self was facilitated by an acquired, hereditary, or constitutional physical "weakness" in the stomach or bowel. In four cases these organs acted, to some extent, independent of their owner's control. Personality, too, played a part in "taking" or "holding" too much in. Twelve patients described themselves as someone who "held too much in," while 11 saw themselves as too sensitive, nervous or tense. Three patients, all college graduates, described themselves as either too "obsessive," "perfectionist," or "anal."

Examples of this "digestion/excretion" metaphor for negative emotions are:

(64-year old woman with ulcerative colitis) Stress -- physically, or in any other way, can react different with different people. That's how it reacts with our family. We keep it internal. My mother was dealing with my father who had an alcohol problem. She was keeping it all inside. This Irish stoicism is for the birds -- our grief has to come out somewhere.

(30-year old woman with irritable bowel syndrome) I tend to hold lots of things inside. I don't express emotion freely. Anger, tension, hostility, fear, any kind of upset -- I think of them as being crammed into my colon.

(30-year old woman with irritable bowel syndrome) Maybe not letting anger out. Even today if I'm angry with someone, I can get an attack. But I can't let my anger out. When I get angry I get pretty violent, I'd really like to smash someone. When you're angry and you can't let it out, it causes tension, and that's connected to your nerve, and it's a way of releasing itself -- instead of angry words, it come out like this.

(29-year old man with ulcerative colitis) People who have a short fuse are unlikely to be affected by it. You would just blow up and get all the anger out. Someone who tears himself up [gets it]. It's not expressing anger, bottling anger up.

(31-year old woman with functional vomiting) Things build up. You lets it all build up till it comes out one way or another, but if you let it out, you let it out. You don't realise you're doing it. You have to find a way of letting it out. Sometimes you can't control your nerves to your stomach. If you hold it in, like with your neighbors -- if you dwell on them -- it can make you sick.

(34-year old woman with ulcerative colitis) Being the oldest child I had five ador-
ing adults, then two of them died and my brother came when I was two years old.
It shook my world upside down. I was raped when I was three, and I stored things
inside, inside myself. I put negative feelings inside myself, rather than put them
outside myself. Doctors often say anger gets stored in the colon. Someone who's
read my chart says my colon's weak. Stress goes to the weakest organ. I let it get
to me, and eat me away. Once something gets inside of me it just bounces around
inside of me, until I can get rid of it. If I can catch anger while it's fresh and
pound something, it'll get out of me -- or someone will help me get it out.

A large majority (19 patients) of the GI group believed that certain
types of relationships with others could cause illness, while 17 felt that
some relationships could cause, maintain or restore good health. The
predominant image (14 patients) in this group was of the therapeutic
value of expressing or "unloading" the negative emotions to another person. For example:

(33-year old man with ulcerative colitis) If you have people you can confide in --
can unload on -- so that you're not carrying emotional stress along -- someone
you're concerned about. It can't effect a cure, but it could prevent some diseases.

(36-year old woman with irritable bowel syndrome) A good relationship can make
you stay healthy, because you can ventilate a lot of stress, and enjoy them because
you can get perspective.

By contrast to this group, a bad relationship was particularly one
where one could not "unload" the accumulated stress. When stress or
tension did build up in the system, it could attack the organ already
weakened by hereditary (7 patients), constitutional (10) or acquired (9)
factors. In each case, the direction of the attack was *from within* -- from
inside the self, towards the world outside. "Good" emotions were less
clearly defined than bad ones; 16 patients in the GI group described these
in a more diffuse way -- as serenity, calmness, peacefulness and relax-
ation -- and more of these seemed linked to specific types of social relationships.

In the 14 patients (3 men, 11 women) therefore, who utilized the metaphor of digestion/excretion, this image linked together negative emotions, concepts of personality and body image, unpleasant physical symptoms, and notions about relationships with other people.

2. The Respiratory Group

The explanatory models used by most patients in this group differed significantly from those of the GI group (Table IV). The predominant
image among the asthmatics can be summarized as: (1) an excessive
permeability to outside influences, both social and physical; (2) a

diminished control over one's physical or emotional responses to those influences; (3) only limited control over some emotional states, or parts of the body; (4) an immediate and dramatic response to outer influences, and/or emotional states; and (5) the patient's vulnerability to these influences is increased either by a physical weakness, or by certain personality factors.

The pathogenesis of asthma was seen as an attack *from without* on the respiratory system, and on the "self." This attack, triggered by outside influences, was often aided by pathogenic, uncontrollable emotions, or by semi-autonomous parts of the body. For 16 patients, the "onset situation" (see Alexander et al. 1968) was a change in the physical environment -- such as exposure to dust, pollen, grass, or smoke -- while 17 identified a sudden increase in interpersonal tension -- such as an argument with a spouse or child -- or other sources of "stress." In most cases the attack was immediate, though in a few cases the stress took longer to cause the attack. The permeability to outside influences is illustrated in these examples:

(24-year old woman) My lungs are filled with fluid. The chest is congested with phlegm. It constricts the passages and restricts oxygen getting to the lungs. Maybe there's an outside amount of fluid absorbed into my skin. If it's damp and rainy, if there's general wetness, humidity in the air, the water is absorbed into the system.

(73-year old woman) If you have it already, from an infection, emotional things can make it worse. Even talking on the telephone. Like problems with my three children. One is divorced, and she's had a lot of emotional upsets. Even though she's 46 she phones me from New York, and that brings on an attack.

All 21 patients felt that emotional factors were linked with the origin, or exacerbation, of their asthma. Twelve of them spoke of their emotions as separate from "self," and sometimes acting on their own accord, as illustrated in some of the earlier examples ("The anxiety of not having my medication with me causes anxiety, and that brings on an attack"). These reified, pathogenic emotions often acted in concert with external stressors, to cause an attack to come on. As with the GI group, personality was seen as a potentially pathogenic factor. Eleven patients described themselves as being too sensitive, nervous, or vulnerable to outside stresses, so that they "took too much in." Only three "held too much in." For example:

(62-year old woman) People who are more sensitive, who take more effect from everyday living get it. Some people can have a fight and it runs off from them like water off a duck's back, but it would affect me.

(72-year old woman) A nervous person gets asthma. All through my life I never thought I was a nervous person, but I must have been. Behind it all there must have been a case of nerves.

These personality or emotional factors were particularly potent in the presence of a physical "weakness" in the lungs, or in the body generally. Fourteen patients saw this predisposition as being hereditary, four as constitutional, and 16 as acquired. A total of 15 patients described their lungs or bronchi as largely independent of their control. For example:

(18-year old woman) The chest tightens up. It hurts when you breath. I'm struggling to get a breath in. They're closing in. Not expanding properly. Hardening up.

As well as using prescribed medication, the commonest therapeutic strategy (10 patients) was withdrawal, or avoidance of the stressful situation or physical environment. Only 2 believed asthma would be relieved by "unloading" stress, or sharing it with other people. Overall, the respiratory group gave the impression of having less *control* over external stressors, the social or physical environment, emotional states, and parts of the body. Since 19 of them were female, this may be partly explained by gender (even though their explanatory image differs from that of females in the GI group), and this should be tested further in a larger study.

Like the GI group, a majority of the patients in the respiratory group reified both negative emotions and/or malfunctioning body parts, separated these from the concept of "self" to a variable degree, and saw them as agents of an "asthmatic attack" -- sometimes in association with external influences. Three of the patients described this attack in anthropomorphic terms - "as though someone was choking me," "like someone's running a knife through it (the chest)," and "my chest is so tight like someone is sitting on it." This image of personal vulnerability, weak or autonomous organs, attack from outside, and immediate physiological response, is different from that used by the GI group. My hypothesis is that part of this difference lies in the differing physiological experiences of the two groups.

DISCUSSION

This pilot-study -- although based on a small sample -- illustrates the importance of eliciting patients' explanatory models of their ill-health. This is particularly true with chronic disorders, where long exposure to biomedical models may influence the answers to questions such as "Why am I not getting better?" Clinicians, too, ask these questions of patients with chronic, relapsing, disorders. As noted above, psychosomatic disorders are an anomalous category within biomedicine; conditions such as bronchial asthma (Knapp 1975), ulcerative colitis (Engel 1975:675), irritable bowel syndrome (Drossman et al. 1977), functional vomiting (Engel 1975:660) and Crohn's colitis (Cheren and Knapp 1980: 1866) are all characterized by a chronic, relapsing, and unpredictable course, often with a poor prognosis. As such, they are less controllable by conventional biomedical treatments. It is suggested that clinicians respond to the uncontrollable aspect by increasingly "psychologizing" the condition – i.e., by shifting the responsibility for etiology or flare-ups to patients' emotions, personality, or early psychological experiences. This shift matches the contemporary emphasis on social values of fitness, contentment, youthfulness, autonomy, individualism, self-control, and responsibility for one's own health and lifestyle. Clinical encounters between doctors and patients take place within this matrix of social values, and may help reinforce these in the minds of the participants.

This study revealed important agreements between lay and biomedical explanatory models of psychosomatic disorders -- particularly on the etiological role of personality, emotions, "weak" body parts, and social relationships. In this respect, "personality" can be seen as a "map" of ideal social values or, as Riesman (1983) suggests, "the expression in a given social context, of one's sense of who one is"; it is "a relational process, a process that relates a person to his situation as he sees it." Riesman suggests further that one should understand "how people perceive themselves to be located in the social contexts they live in, what factors influence this perception of themselves, and how the choice and performance of acts from the cultural repertoire express the sense of self." In the case of the illness realities negotiated between clinician and patient, I would suggest that there is pressure to achieve consonance between predominant social values (as expressed in biomedical categories), and patients' self-perception. In the case of psychosomatic disorders, patients *learn* to explain their chronic ill-health by defining themselves as too "obsessive," "perfectionist," and "anal," or "sensitive," or as people who "hold too much in"; their personalities, emotions, lifestyle, or physical weaknesses are blamed for their failure to conform to the ideal social values listed above.

My hypothesis is that, in response to "psychologization" and the stigma that this implies, patients *reify* concepts of pathogenic (or "weak") personality, emotions, and bodily parts, and separate these from the idealized concept of "self" (cf. Cassell 1976). This shifts responsibility onto these reified entities, which become instead part of the outside world, a more public interface between the self and the environment. In that sense, blame for the loss of fitness, autonomy, contentment, self-control, etc. is shifted to emotions or body parts that are under the control (to a variable degree) of outside forces. Parts of the body, and the personality, are seen as "non-self," either part of other people or of the natural environment. This is particularly true of socially "bad" emotions -- such as anger, hostility, frustration, and tension -- and parts of the body that are "weak" or vulnerable.

The emphasis on heredity as an etiological factor by half the sample (14 asthmatics, 7 GI patients) can also be seen as shifting responsibility for the "weakness" to previous generations. Among the asthmatics, especially, the condition was also blamed on factors outside the individual's control -- such as the weather, allergies, or infection, or the quality of one's social relationships. By these shifts of responsibility, therefore, patients were able to maintain their idealized concept of the healthy, autonomous self, reduce anxiety and uncertainty, and decrease the social stigma of the disorder. It should be noted that the stigma of a "nervous" condition is reinforced by diagnostic labels such as *psychosomatic*, a *nervous* stomach, or an *irritable* bowel.

A second hypothesis to emerge from this pilot-study is that *physiological* experiences may, to some extent, structure patients' explanatory models of their condition. This was particularly clear in the 21 patients in the gastrointestinal group. Despite the heterogeneity of each diagnostic category, the majority of patients shared a common explanatory image -- a natural symbol of either respiration, or digestion/excretion. This image explains, and links together, physiological, emotional, and social experiences, and also links these to wider, contemporary social values (cf., Good 1977). For the individual patient, therefore, the physical symptoms condense a range of associated personal meanings, experience, memories, and expectations. It also helps structure experiences of time ("old anger"), space (environments that trigger asthma), emotions ("get all the anger out"), body parts ("the bronchial tubes close, tighten up on me"), social relationships ("a row can make me tighten up"), personality ("certain personalities, I guess anal, get it"), and physical vulnerability ("stress goes to weakest organ"). This influence of physiological imagery may explain the specific "attitudes" in psychosomatic patients, noted by Grace and Graham (1952) and Grace et al. (1962). Also, the use of somatic imagery in describing emotions by this sample, tends to contradict Leff's point (1981) that this was only a feature of earlier periods in history.

How patients perceive and interpret their symptoms is obviously of clinical importance. Not only may their "cognitive set" affect both their emotional and physiological state, but it may affect how they perceive, and are perceived by, members of their families (see Minuchin et al. 1978). In addition, since at least two of the disorders examined in this study have been found to be amenable to psychotherapy -- i.e. ulcerative colitis (Karush et al. 1977) and irritable bowel syndrome (Svedlund et al. 1983) -- it is important for therapists to understand patients' self-image, and perceptions of emotions and somatic symptoms. This small study was designed as a preliminary examination of these phenomena, and it is hoped that the initial findings of the study will be tested in a larger patient sample in the future.

ACKNOWLEDGEMENTS

This study was carried out while I was a Visiting Fellow in Social Medicine and Health Policy, and Visiting Scholar in the Division of Primary Care, at Harvard Medical School. I wish to acknowledge the help and encouragement of Dr. Leon Eisenberg, Dr. Arthur Kleinman, Dr. Mary-Jo Good, and Dr. Maurice Eisenbruch of Harvard Medical School; Dr. Robert S. Lawrence of the Department of Medicine, Cambridge City Hospital, and the medical residents and attending physicians in his department; Dr. Charles J. Hatem of Mount Auburn Hospital; Dr. Paul E. Lesser of Cambridge Hospital; and Dr. Stanley E. Sagov. This article first appeared in *Culture, Medicine and Psychiatry* 9 (1985), 1-26.

REFERENCES

Alexander, F., French, T.M. and G.H. Pollock (eds.)
    1968   Psychosomatic Specificity. Vol. 1. Chicago: University of Chicago Press.
Barsky, A.J. and G.L. Klerman
    1983   Overview: Hypochondriasis, Bodily Complaints, and Somatic Styles. American Journal of Psychiatry 40:273-283.
Bauer, J.
    1942   Constitution and Disease. New York: Grune and Stratton.
Cassell, E.J.
    1976   The Healer's Art: A New Approach to the Doctor-Patient Relationship. New York: Lippincott.
    1976   Disease as an "It": Concepts of Disease Revealed by Patients' Presentation of Symptoms. Social Science and Medicine 10:143-146.
Cheren, S. and P.H. Knapp
    1980   Gastrointestinal Disorders. *In* M.I. Kaplan, A.M. Freedman and B.J. Sadock (eds.), Comprehensive Textbook of Psychiatry, Vol. 2. Third Edition. Pp. 1862-1972. Baltimore: Williams and Wilkins.
Douglas, M.
    1973   Natural Symbols. Harmondsworth: Penguin Books.
Drossman, D.A., Powell, D.W. and J.Y. Sessions
    1977   The Irritable Bowel Syndrome. Gastroenterology 73(4):811-822.

Dunbar, F.
    1948   Introduction. *In* F. Dunbar (ed.), Synopsis of Psychosomatic Diagnosis
            and Treatment. Pp. 13-27. St. Louis: C.V. Mosby and Company.
Eisenberg, L.
    1977   Disease and Illness: Distinctions between Professional and Popular
            Ideas of Sickness. Culture, Medicine and Psychiatry 1:9-23.
Engel, G.L.
    1960   A Unified Concept of Health Disease. Perspectives on Biology and
            Medicine 3:459-485.
    1967   The Concept of Psychosomatic Disorder. Journal of Psychosomatic
            Research 11:3-9.
    1975   Psychological Aspects of Gastrointestinal Disorders. *In* M.F. Reiser
            (ed.), American Handbook of Psychiatry, Vol. 4, 2nd Edition. Pp.
            653-692. New York: Basic Books.
    1977   The Need for a New Medical Model: A Challenge for Biomedicine.
            Science 196:129-136.
Fisher, S.
    1968   Body Image. *In* Polhemus, T. (ed.), Social Aspects of the Human Body.
            Pp. 154-173. Harmondsworth: Penguin.
Freud, S. and J. Breuer
    1966   Studies on Hysteria. Trans. J. Strachey. New York: Avon.
Gildea, E.F.
    1968   Special Features of Personality which are Common to Certain Psychoso-
            matic Disorders. Psychosomatic Medicine 11:273-281.
Good, B.J.
    1977   The Heart of What's the Matter: The Semantics of Illness in Iran. Cul-
            ture, Medicine and Psychiatry 1:25-58.
Good, B.J. and M-J.D.V. Good
    1981   The Semantics of Medical Discourse. *In* Mendelsohn, E. and Y.
            Elkana (eds.), Sociology of the Sciences. Vol. 5. Dordrecht, Holland:
            D. Reidel Publ. Co.
Grace, W.J. and D.T. Graham
    1952   Relationship of Specific Attitudes and Emotions to Certain Bodily Dis-
            eases. Psychosomatic Medicine 14(4):243-251.
Graham, D.T. et al.
    1962   Specific Attitudes in Initial Interviews with Patients Having Different
            "Psychosomatic" Diseases. Psychosomatic Medicine 24:257-266.
Groen, J.
    1948   Psychogenesis and Psychotherapy of Ulcerative Colitis. *In* Dunbar, F.
            (ed.), Synopsis of Psychosomatic Diagnosis and Treatment. Pp.
            108-114. St. Louis: C.V. Mosby and Company.
Hendrick, I.
    1948   Homeostasis. *In* Dunbar, F. (ed.), Synopsis of Psychosomatic Diagnosis

and Treatment. Pp. 55-56. St. Louis: C.V. Mosby and Company.

Houck, C.
  1983 Anxiety Attacks That Trigger Terror. Woman's Day 13 (September):
        30-35.

Karush, A. et al.
  1977 Psychotherapy in Ulcerative Colitis. Philadelphia: W.B. Saunders.

Katon, W., Kleinman, A. and G. Rosen
  1982 Depression and Somatization: A Review. American Journal of Medi-
        cine 72:241-247.

Kleinman, A., Eisenberg, L. and B. Good
  1978 Culture, Illness and Care: Clinical Lessons from Anthropologic and
        Cross-Cultural Research. Annals of Internal Medicine 88:251-258.

Kleinman, A.
  1980 Patients and Healers in the Context of Culture. Berkeley: University of
        California Press.

Knapp, P.H.
  1975 Psychosomatic Aspects of Bronchial Asthma. In M.F. Reiser (ed.),
        American Handbook of Psychiatry, Vol. 4, 2nd Edition. Pp. 693-707.
        New York: Basic Books.
  1980 Current Theoretical Concepts in Psychosomatic Medicine. In H.I.
        Kaplan, A.M. Freedman, and B.J. Sadock (eds.), Comprehensive Text-
        book of Psychiatry, Vol. 2, 3rd Edition. Pp. 1853-1862. Baltimore: Wil-
        liams and Wilkins.

Leff, J.
  1981 Psychiatry Around the Globe: A Transcultural View. New York: Marcel
        Dekker.

Lipowski, Z.J.
  1968 Review of Consultation Psychiatry and Psychosomatic Medicine. Psy-
        chosomatic Medicine 11:273-281.

Littlewood, R.
  1984 The Individual Articulation of Shared Symbols. Journal of Operational
        Psychiatry (in press).

Mechanic, D.
  1972 Social Psychologic Factors Affecting the Presentation of Bodily Com-
        plaints. New England Journal of Medicine 286:1132-1139.

Minuchin, S., Rosman, B.L. and L. Baker
  1978 Psychosomatic Families. Cambridge: Harvard University Press.

Myers, F.R.
  1979 Emotions and the Self: A Theory of Personhood and Political Order
        among Pintupi Aborigines. Ethos 7:342-370.

Riesman, P.
  1983 On the Irrelevance of Child Rearing Practices for the Formation of
        Personality. Culture, Medicine and Psychiatry 7:103-129.

Sontag, L.W.
    1948 Determinants of Predisposition to Psychosomatic Dysfunction and Disease: Problem of Proneness to Psychosomatic Disorder. *In* F. Dunbar (ed.), Synopsis of Psychosomatic Diagnosis and Treatment. Pp. 38-66. St. Louis: C.V. Mosby and Company.
Svedlund, J. et al.
    1983 Controlled Study of Psychotherapy in Irritable Bowel Syndrome. Lancet 2 (8350):589-592.
Taussig, M.T.
    1980 Reification and the Consciousness of the Patient. Social Science and Medicine 14B:3-13.
Vaillant, G.E.
    1978 Natural History of Male Psychological Health: What Kinds of Men Do Not Get Psychosomatic Illness. Psychosomatic Medicine 40:420-431.
Weiss, E. and O.S. English
    1942 Psychosomatic Medicine. Pp. 10-11. Philadelphia: W.B. Saunders.
Wright, J.H. and A.T. Beck
    1983 Cognitive Therapy of Depression: Theory and Practice. Hospital and Community Psychiatry 34(12):1119-1127.

PART III

REPRODUCING MEDICAL PERCEPTION AND PRACTICE

JOSEPH W. LELLA AND DOROTHY PAWLUCH

MEDICAL STUDENTS AND THE CADAVER IN SOCIAL AND CULTURAL CONTEXT

WHAT GOES ON IN THE LAB?

If, on a bright September morning, you were to walk into the anatomy laboratory of a typical North American faculty of medicine, one of the first things to strike you might be the smell of formaldehyde -- next the air of calm, studious concentration of white coated figures, men and women in teams of four, bent over their gleaming stainless steel tanks. Moving closer to one of the groups, and glancing over a medical student's shoulder, you might be horrified to see that the young man or woman was contemplating a cadaver, perhaps its face and midriff draped, a lab manual or textbook nearby opened for easy consultation to help in identifying elements of a neatly dissected portion of human anatomy.

Some commentators upon such scenes have been impressed with the calm and the studiousness, and have reported that students experience such work as anatomy teachers intend, as a matter-of-fact way of learning about the structures of the human body (McGuire 1966; Becker et al. 1961). Others have portrayed more stormy emotions boiling beneath the surface. Students are said to have been disturbed by their work. To defend themselves, they depersonalize the cadaver, suppress personal feelings and stress the objective-scientific benefits to be gained by their dissection. This reaction is seen as the first step in a long process by which students learn to subvert personal, emotional and humanistic concerns for objective, scientific professional goals (Gregg 1957; Lief and Fox 1963; Reiser and Schroder 1980; Coombs 1978; Coombs and St. John 1979).

Some of these authors believe that the cadaver experience can and should be used to facilitate the integration rather than separation of personal and professional development. Such integration would promote the linking of scientific and humanistic concerns both in patient care and in all aspects of medical practice. In short, they seem to argue, if students could learn to treat their own reactions to the cadaver with openness and respect for both science and the self, they might continue to do the same for the concerns of patients (Reiser and Schroder 1980; Coombs 1978).

It was easy for a behavioral scientist teaching in a faculty of medicine to agree with such views, however Lella was curious to learn more about the content of students' experiences. Were they indeed disturbed by

M. Lock and D. R. Gordon (eds.), Biomedicine Examined, 125–153.
© 1988 by Kluwer Academic Publishers.

dissection, and if so, in what ways? Why? What, if anything, could be done about it?

It was with the aim of taking at least some steps toward answering these questions that the work reported in this paper was undertaken. The first section discusses students' own accounts of their "cadaver experience." It briefly describes the sources of these accounts, then outlines: (a) students' physical and emotional reactions; (b) more abstract concerns, issues and dilemmas; and, (c) the coping mechanisms they reported using. This section then presents some social, psychological and cultural hypotheses to account for students' reactions.

The next section notes that anatomy teaching from the sixteenth until the early nineteenth centuries explicitly addressed such reactions through integration with "philosophy and feeling." As background for thinking about whether such an integration might be tried again, this section also attempts to explain how and why it largely disappeared from the teaching of anatomy and cadaver dissection in the late nineteenth and early twentieth centuries.

Our final section examines the current social and cultural context of the teaching of anatomy and argues that the time is ripe for a return to more integrated teaching. This section briefly looks at some current attempts to do just that and uses these and our prior analyses to suggest the broad outlines of what might be successful pedagogy.

For centuries, gross anatomy, taught through cadaver dissection, was perhaps the most basic of the medical sciences. Though it has been somewhat eclipsed recently, it retains an important place in the contemporary curriculum (Blake 1980). We contend that understanding its evolving symbolic, social and psychological functions is crucial to a full understanding of the changing shape and outcomes of medical education itself.

Before beginning, we warn the reader that our paper really just suggests what might be accomplished with further research. If it can stimulate new work in the directions indicated here, its authors' goal will have been achieved.

A STUDY OF STUDENTS' RESPONSES TO THE CADAVER EXPERIENCE

Several years ago we asked medical students at McGill University in December of their first year to write an essay on one of a number of themes present in a currently available textbook on interviewing patients (Reiser and Schroder 1980). The first chapter, "Becoming a Doctor," contains a descriptive-analytic discussion of the issues first year medical students typically confront. The second, "The Interview Process," discusses and analyzes the medical interview noting the various ways through which information is communicated and interviewing can be facilitated as an interpersonal process. During the course of a two-hour examination of four essay questions, students were required to answer either question 1 or 2 (Figure 1). Question 2 was included to give students a chance to discuss a less intimate, more "objective" topic if they so desired. They were told that they would receive full credit for this answer if they made a serious effort. They had no idea that their work would be used for research purposes.[1]

1. In Chapter 2 of *Patient Interviewing*, entitled "Becoming a Doctor," Reiser discusses:

    1)   a number of important experiences which medical students undergo in their early years of medical studies;

    2)   the implications which these may have for the student's professional development, especially his/her relations with patients. Choose an experience noted by Reiser which seems to echo one of your own.

DISCUSS:

    (a)   similarities and differences between your experience and that cited by the author;
    (b)   similarities and differences (from Reiser's descriptions) in your own reactions, and the conclusions which you have drawn from them concerning your own development as a physician, especially in how you feel and you will relate to patients.

2. In Chapter 6 of *Patient Interviewing*, Reiser and Schroder discuss the "Interview Process" and outline 12 "process tracks," i.e. ways through which information is communicated in the interview, though not necessarily in explicit verbal form.

    (a)   Think about a particular conversation in which you have recently participated with one other person.
    (b)   Describe the setting and explicit purpose of the conversation briefly.
    (c)   Give a clear explanation of what Reiser and Schroder seem to mean by *each* of *at least four process tracks*, and
    (d)   show how aspects of your conversation illustrated these tracks.

Fig. 1. Essay Questions

Table I demonstrates that although only 11.8% of the reading material dealt with the cadaver experience, 37.5% (60) of our class of 160 students chose to write about it. These proportions confirm the impressions of others that the cadaver experience has a significant impact. The rich content of students' essays corroborates this.

TABLE I

Difference between space devoted to topic in
textbook and student essays

Topic	No. of Centimetres of Text	%	No. of Student Essays	%
Cadaver	107	11.8	60	37.5
Other Experiences	314	34.6	66	41.3
Interview and Process Tracks	487	53.6	34	21.3

Initial Reactions

Whether or not they had prior experience with death and dissection, many students approached their first anatomy lab with apprehension. Some students had lived with this apprehension from the moment they began contemplating a medical career or received their medical school acceptance letters; others found themselves getting anxious a few days or moments before the lab. Many attempted to "psyche themselves up" for the experience by mentally rehearsing it, hoping thereby to dull its impact. Others tried to assume what one student labeled a "defensive" or professional stance and concentrated on holding their emotions "in check" before entering the lab.

The sight of the lab and the cadavers produced a myriad of emotional or "gut level" responses, not all of which were negative. Many students did feel horrified and traumatized. Among them, perhaps surprisingly, were some who had previously dissected animals, or had witnessed the death of a relative or friend, or had worked in a hospital setting.[2]

I was angry with myself. I expected that after visiting slaughterhouses and seeing all kinds of sick and dead animals, a human corpse would not bother me. But it did.[3]

Emotional responses were occasionally accompanied by physical reactions including nausea and dizziness. Though many described the urge to run out of the laboratory none admitted to doing so. They did admit to having difficulty with meals for the next few days and one student described a delayed physical reaction:

I went home that day and vomited my gut out. [Then] I got drunk to forget.

Other students were surprised to find that they had experienced more positive reactions over-riding the repulsion, including fascination and curiosity. One student wrote that she was "in awe of the cadaver." These sentiments were augmented as the dissections began and as the students began to discover the intricacy and complexity of the human body.

Whatever their initial responses, all our respondents eventually were able to function more or less comfortably in the lab. However, several mentioned the constant vigil required to keep their emotions in check. Even those who were not bothered initially by the sight of the cadaver and those who believed they had resolved their fears and anxieties, discovered how easily a dream, a glimpse of the cadaver's face, the sight of a particular prosection at a particular angle or even some apparently unrelated incident outside of the lab could evoke an emotional response.

One exercise that disturbed even the most detached and controlled of the respondents required students to saw their cadavers across the mid-section, stand the lower section up on its waist and then saw through the genitalia. (There were several vivid descriptions of the procedure in the students' accounts.) Some students were simply not able to complete the task and left the lab until the job was done. In several cases there appeared to be an undercurrent of sexually related anxiety. One group had difficulty performing the procedure on the same-sexed cadaver and, in fact, traded cadavers:

[We found this] part of the anatomy lab a particularly emasculating manoeuvre – so much so, we asked a female colleague to do it for us.

Values and Feelings

As difficult as these types of emotions were to deal with, students also found themselves having to contend with more complex feelings and with the thorny moral, normative and metaphysical issues that the cadaver experience raised for them. Four main themes emerged in students' discussions of these concerns:

(1) *Guilt over dissecting a human body and violating a cultural taboo.*
Some respondents stressed the first incision rather than the sight of the
cadaver as the single most difficult moment of their experience. They
referred not to "dissection" but to "gross mutilation," "desecration,"
"defiling" and "butchering" of their cadavers. One student was particularly insightful and articulate about the source of his guilt:

According to Judaic tradition, dead people are supposed to be buried without
being displayed. No religious Jew will donate any part of his or her body to science. But I don't think it was the religious aspect particularly that bothered me. . .
I think it was more disturbing that I was invading, almost violating this body. It
seemed to go against some implicit primal precept. There was a feeling of tres-
passing a border that should not be crossed.

Another student, in a similar vein, stated that she felt as though she were
committing "a fundamental sin."

(2) *The anonymous invasion of privacy or personal space.* Some students
described the difficulty they had reconciling the seemingly intimate nature
of the act (the dissection) with its context:

. . . here I was a complete stranger to her and those who knew her and without
explanation to her, her friends or her relations, I was doing this to her.

(3) *Reflections of life, death and the human condition.* Many students
admitted that it was the first time they had confronted their feelings
about these issues so directly. One wrote that the anatomy lab had
deprived her of the luxury of treating death like a "never-to-be encountered stranger."
    But as disquieting as such thoughts sometimes were, some students
also found them liberating. They appreciated the opportunity the cadaver
experience provided to penetrate the mystique of death and to explore
their feelings on the subject:

I felt . . . it was an escape from the over-protectiveness of a society obsessed with
cleanliness, where a group's standard of living is directly proportional to the extent
to which it can dissociate itself from its own refuse, including its sick, dying and
dead. All generations before me, and even my rural friends, have lived in contact
with disease and death. I hadn't even seen a dead cat. I felt the urban ignorance
and terror of my physical essence to be an obstacle to self-knowledge which had to
be removed.

The exercise took on special significance given the suffering and death
they knew they were likely to encounter as physicians.
    These reflections were not merely abstractly philosophical. Students

also wondered about the life and death of the person who had been the cadaver; about relatives and friends, and the self. What kind of person had the cadaver lying before them been? What kind of life had she led? Did she have a family? Had she been happy? Had she found her life meaningful and worthwhile?; under what circumstances had she died?; how had she experienced her death, and so on.

Others, particularly those students who were concurrently facing illness and death within their own families or circle of acquaintances, found themselves dwelling on and anticipating the death of those close to them. But more asked themselves about their own mortality, not only about its inevitability, but about its meaning. They voiced fear and anxiety about the unknown.

In some instances, these reflections led to a re-assessment and shifting of values, best expressed in the following statement:

My priorities have changed. I attach increased importance to friendships. I grab opportunities as they present themselves, I have a heightened appreciativeness for life, people, nature — overall, a more "live for today" and "appreciate being alive" type of attitude.

Other students reaffirmed their religious beliefs:

As a Christian . . . I was taught to believe that death was an enemy, the last enemy and that death was defeated when Christ came again. There is hope and life beyond death.

(4) *Positive and negative changes in self-image*. Students were keenly aware of the status change the cadaver experience represented. They perceived it as a threshold crossed, a symbolic rite of passage into the hallowed realm of medicine. Some described the experience as the first that made them feel they were "actually in medicine." Others referred to an "over-riding sense of privilege," and "becoming one of the chosen few." Indeed, in one case the sense of privileged status was sufficiently strong to produce an unexpected zeal for the anatomy lab. But being set apart in this manner was not a positive experience for all. For some, the separation was, at least initially, lonely and even stigmatizing since it was based on their "tabooed" experience with the dead:

. . . I suddenly became aware of how different I was from those around me. Just twenty minutes ago I had my hands in a human corpse!

This type of response could also be provoked by the reactions of friends outside of medicine:

My friend reacted with incredulity, horror and disgust [when I described the lab]. I felt him pulling back from me somewhat. He kept glancing at my hands. In his eyes, I had done something terrible.

Coping Mechanisms

Among the main coping strategies that students discussed were objectification and rationalizations. On the basis of the literature review, we were not surprised to find that the invariable and most basic coping strategy was the objectification of the cadaver. Students reaffirmed that the cadaver became "a thing to be prodded, dissected and studied" or a "mere mass of tissues and organs." They marvelled at how quickly they were able to assume a cool and detached stance to the point where they could even study a prosection while munching on a sandwich.

They showed awareness of some of the structural features of the setting that facilitated objectification For instance, several students stressed the context of the lab, the odour of formaldehyde, the pale colour and "rubbery," "waxy" or "plastic" texture of the cadaver's skin, the absence of blood -- all of which rendered the cadaver "something far removed from human." One student wrote:

I spent an inordinate amount of time looking into his lifeless eyes, taking him in and convincing myself that there was nothing here one could call "human."

Others commented on the pressure to absorb vast amounts of information before up-coming exams.[3] "We were all too concerned with passing anatomy," admitted one student, "to bother for very long with our hang-ups about the cadaver." Thus it became their "textbook."

Many felt that the head and face were powerful symbols of the cadaver's humanity. One student put it this way: "There's something about the head that gives life to the dead." Thus they made every effort to keep the cadaver's face covered:

Sometimes the cloth slipped off his head. I always replaced it immediately. Only with the face covered did the cadaver somehow cease to be "a dead person" and become instead something to study.

Others developed exaggerated ways of treating the cadaver as an object. For instance, one student found himself using his cadaver as a prop for his tool box. Upon reflection, he stated that it was as if, by reducing the cadaver in a physical sense to a mere object, he could reduce its emotional impact.

Students rationalized the objectification in their own minds in several ways. For some, the ability to assume an objective and detached stance was an essential and requisite component of learning and medical practice. "It is absolutely crucial," one student argued, "for physicians to have control over their feelings if they're going to be at all effective." Others emphasized that objectification was necessary if physicians were to retain their hold on sanity in the face of all the suffering and death they would come up against over their careers. If physicians had to deal with their own emotional crises every time they lost a patient, a medical career would simply not be possible.

In one sense, the ability to objectify and become emotionally detached was a source of gratification and pride for the students. They were learning "how to be doctors." But much more prevalent was the concern, anxiety and doubt the objectification raised for students, about their ability to relate to others and particularly to patients on an emotional level. Indeed, this concern was the single most striking theme in the data and comprised the most articulate and personal of all the responses. Perhaps this was because our course and the text were concerned to sensitize students to such issues. Their responses, however, were far from merely giving the teacher what he or she expects:

It worried me that the cadaver could so quickly cease to bother me at all. Deep down, I was disappointed with myself for "losing touch" with my feelings so easily. Would I end up just as easily setting aside the compassion I might feel towards my patients as well.

I recognize the callousness working with the cadavers foments, and I worry about it being transferred towards living humans. I am concerned about my development as a physician and the way I'll relate to my patients. If it is so easy to suppress my feelings while dissecting the cadaver, perhaps there will be the temptation to do the same when dealing with patients.

I have found myself not only cutting off my feelings in the lab, but out of the lab as well. In quest of being truly objective I remember ruthlessly quizzing my girlfriend as to the exact nature and extent of a lower abdominal pain she was experiencing. So absorbed was I in attempting to diagnose her ailment, I had blinded myself as to how much pain she actually felt. I was quite surprised when she soon started to cry and asked me to stop questioning her.

Another student, who had no difficulty with the cadaver initially, found himself well into the term, reacting emotionally to the sight of a particular prosection. Rather than feeling disappointed with himself he was jubilant:

I happily realized I hadn't lost that human and sensitive part of me. I felt a great sense of relief at discovering I still had some feelings.

Students coped not only by objectifying the cadaver but also by rationalizing their dissections. Many came to consider dissection as an indispensible learning experience, one which imparted knowledge that would simply not be accessible through other means. Connected to this rationalization was the idea that the knowledge was of a vital nature – that it could be applied directly towards the alleviation of suffering and the sustenance of life. The cadaver was perceived as a tool with the unique power to instruct in the preservation of life.

Other students stressed the value and the promise of future good to be derived from the dissection experience with reference to the cadaver itself. For these students the dissection was perceived as the cadaver's last and perhaps most noble act, a living and valuable legacy left behind.

Both rationalizations were accompanied by a strong sense of duty and responsibility to learn as much as possible and to use one's skills and talents to the best of one's ability. In other words, for many students the dissection was justified by an ardent commitment to the practice of medicine:

I will always consider it my responsibility . . . to repay them [the cadavers] by doing my best in using the knowledge I've acquired from them in the future.

Another rationalization involved the attribution of awareness, understanding, and complicity to the cadaver:

I felt the woman's presence and her compassion for me . . . It is my feeling that whatever she is, she accepts and understands what we must do.

Finally, there was a type of rationalization by those whose greatest anxieties were connected to the violation of their cadavers. For many of these students the guilt they experienced as a result of this violation was alleviated or at least mitigated by efforts to show the cadaver "due respect." Denied the possibility of respecting the physical integrity of their cadaver, they sought other means of expressing their respect. In some cases these means entailed verbal references and general demeanor in the lab: students would refrain from joking about or nick-naming their cadavers, and would disapprove privately or perhaps chastize openly colleagues they felt were being disrespectful in their reference to and/or treatment of the cadaver.

For other students it was possible to show respect through the method of dissection:

. . . it means keeping the body as intact as possible, proceeding in an unhurried manner, being as careful as possible and above all, not wasting the opportunity.

Students did not describe joking about or nick-naming their cadavers although they described "others" doing these things. Perhaps it is too much to ask of students to have them tell these tales on themselves. Others have reported that such behavior is often used to lighten what may be a tense and anxiety-laden situation (Robinson 1957; Reiser and Schroder 1980:25).

Despite the insistence by some observers on the emotional neutrality of the dissecting experience, this analysis should come as no great surprise. In our daily relations with the living, we all respond to the symbolism of bodyparts, their size, constitution and color, their physical gestures, facial and other bodily expressions, signs of age and weakness, or youth and vigour, and to the extent of their dress or undress (Polhemus 1978; Kern 1975). It would seem likely that such individual characteristics, habitually responded to in the living, would provoke some emotional and/or cognitive response when students perceive them in the dead, especially upon an initial encounter.

Further, explicit, shared beliefs and/or rituals are invoked in most human societies when people encounter the dead. Such beliefs and rituals seem to have been tailored by an innate wisdom to allow human beings to support one another in a shared meaning of life, to allow life to go on, despite our confrontation with its extinction (Lifton and Olson 1974). Philippe Aries notes:

The necessity of organizing work and maintaining order and morality in order to have a peaceful life in common [has] led society to protect itself from the violent and unpredictable forces of nature . . . the ecstacy of love and the agony of death. . . . Death has been imprisoned in ceremony, transformed into spectacle . . . not a solitary adventure but . . . a public phenomenon involving the whole community (Aries 1981:604).

In the absence of explicitly recognized death ritual, a "public phenomenon involving the whole community," and of beliefs, explicitly invoked to support them, students either supplied their own or identified with those briefly stated by the professor at the outset of the course -- that the exercise was an important learning experience which would aid them to become knowledgeable and skilled physicians. Cadavers were to be treated with respect because they once were human and had been donated to help students achieve these goals. It was not always that such slight attention was given to feeling and philosophy in the teaching of anatomy through dissection. We turn to this topic in the next section.

DISSECTION, FEELING AND PHILOSOPHY FROM THE 16TH TO MID-20TH CENTURIES

Explicit, shared and sometimes public answers to ultimate questions about life and death were more fully integrated into the teaching of human anatomy and dissection until well into the 19th century. As the middle ages gradually faded into the renaissance, death and the cadaver came to be seen more and more as natural phenomena. Dissection for scientific and teaching purposes became more accepted, yet it was not seen merely as an investigative, scientific act (Illich 1975:133-134). Vesalius' pioneering mid-16th century text, the *De Humani Corporis Fabrica*, describes the results of dissection but also includes mediative references to life and the human condition. For example, his illustration, "A Delineation From the Side of the Bones of the Human Body," shows "the skeletal Hamlet soliloquizing beside the tomb upon some poor Yorick . . . " On the tomb we see, in Latin, "Genius lives on, all else is mortal" (Saunders and O'Malley 1950:86; see also Singer 1946). But beyond this, as Edelstein convincingly argues, Vesalius' anatomy was a conscious expression of renaissance humanism's identification with man himself rather than with God as man's own destiny (Edelstein 1943:547-561).

In the middle of the 17th century, Rembrandt's "Anatomy Lesson" memorialized the passion of some for teaching self-knowledge and knowledge of God through dissection in public "anatomies."

The subject of the picture . . . is a pictorial adaptation of a type of scene which is well attested in the history of anatomy. At Leiden, when Pieter Paaw was about to perform one of his winter anatomies, crowds packed the stalls of the anatomy-theatre, and soon forgot the cold as Paaw kept them constantly astonished by revealing "the ingenious works of God in the human body" and by expounding "the proper office of every part" (Schupbach 1982:25 and *passim;* see also Lassek 1958:108).

In England, in 1697, John Browne, Surgeon to the King and Senior Surgeon to St. Thomas' Hospital, prefaces his anatomy textbook *Myographia Nova*, with the following words to his "Kind Reader":

There is nothing affords greater Light into the Mysterious Recesses of the Supreme Architect, than the True Knowledge and Understanding of the Frame and Admirable Structure of the Humane Body. Man being made as a Stately Pile finely built up and curiously wrought into variety of parts, wonderfully put together, in due order, frame and symmetry . . .

Whoever therefore hath been much concersant in dissecting of Bodies, cannot deny, but that Anatomy is well worth his Care and Study, it very much conducing towards the knowing of our Wise Creator, of his Hands; by which are shewn his unspeakable Wisdom and Power in thus forming Man with that Harmony of similar and dissimilar Parts (Browne 1697).

As late as 1817, John Abernethy, Anatomical Lecturer to the Royal College of Surgeons in England, interpreted for the college his views and those of the celebrated anatomist and surgeon, John Hunter, saying that they strove to "elevate the thoughts of the student from the contemplation of Nature to Nature's God" (Abernethy 1830:vii-viii).

Perhaps some of these expressions did not invoke *shared* meanings. Perhaps they were purely "ceremonial." They were, it seems, anchored in the philosophical views of their time -- linked to basic concepts of the place of man and his body in the universe (Temkin 1977). And they did provide some point of reference for human feeling, some authoritative indication that anxiety or other deep emotions might be calmed, some hint that questions about life and death might be answered.

It was well into the 19th century before this *explicit linking* of *description, dissection and meditation* largely disappeared. Increased understanding of the details of human anatomy had accumulated throughout the 17th and 18th centuries (Singer 1957; Puschmann 1966). Against this background, physicians and biological scientists began to think of life and of bodily processes somewhat differently -- less as qualitative entities, more as observable phenomena. Diseases were now seen not so much as states of being or essences to be grasped verbally or conceptually, but more as visible lesions or disturbances in tissues and cells and their relations (Foucault 1963; Armstrong 1984). This was helped both by increased precision in the techniques of physical examination (e.g. percussion and auscultation using the stethoscope), and by advances in postmortem, pathological investigation. Michel Foucault's *Birth of the Clinic* (1963) analyzes these developments and their institutional and cultural supports in some depth.

Later in the 19th century, the discovery of anaesthesia and the development of aseptic operating techniques allowed surgeons more time to see, to saw and to sew, lessening the pain and iatrogenic consequences of their work. Some of the causes of disease could be understood and eliminated in dramatic fashion. The discovery and use of the small-pox vaccine, for example, seemed to have an important and positive impact on mortality rates.

Through these and other achievements, medicine became more and more capable of striking a credible claim to improving human health at individual and social levels. It found a broader clientele more willing to trust in its ministrations. It also found the modern state more willing and able to grant it a monopoly on practice and on training for it. Under the impetus of state supervisory bodies, medical training gradually abandoned the loosely organized methods of individual apprenticeship, university study, hospital training, or various combinations of these, and moved towards more tightly defined curricula. These combined university study of basic sciences with clinical training in university hospitals, staffed by clinician-scientists (Puschmann 1966; Billroth 1924; O'Malley 1970). In short, curricula were now oriented toward producing physicians who could be seen as justifying the state's trust, as able to practice scientifically legitimate medicine, medicine that could make a difference. (Whether and to what extent it did in actuality make a difference, is, of course, another story [McKeown 1979].)

Anatomy and dissection were assumed to be crucial to the training of a physician, especially since disease had come to be defined as located and treatable in the tangible body. Anatomy now was no longer taught through mere demonstration, that is by means of dissection and exposition by the learned professor, but through students doing it themselves. In many places, pathological anatomy, learned by performing autopsies, came to be required of students so that they could see the causes, form and consequences of particular diseases (Blake 1980; Maulitz 1980).

By the beginning of the 20th century, anatomy and its associated sciences were charged with teaching the results but also the method of science. *Unbiased observation, and experimentation were seen as the basis of medicine and these were thought to preclude the simultaneous presence and expression of faith, of philosophy, of emotion, and of feeling.* As Sir William Osler noted in 1903:

One and all of you will have to face the ordeal of every student of this generation who sooner or later tries to mix the waters of science with the oil of faith. You can have a great deal of both, if you only keep them separate. The wrong comes from the attempt at mixture (Osler 1903:26).

Medical education and scientists at this time felt that they had won a difficult victory over remnants of a medieval vagary and mysticism which made science impossible. Abraham Flexner contrasted medieval and modern medical science as follows:

Medieval medicine . . . starts with a . . . pre-supposition, a notion, a metaphysical principle, and purports thence to deduce its procedure . . . modern medicine strives to be honestly and modestly inductive, consulting the situation for relevant facts, and cautiously drawing provisional conclusions, subject to revision whenever the issue of experience suggests modification . . .

[Since the 18th century] . . . the scientific viewpoint [has] made its way by slow stages . . . in consequence [however] the human body is now viewed as an item in the universe of matter and life, without recourse to essences and principles (Flexner 1912:5-6).

Of course, scientifically influential and informed debates about the nature of life and other issues dealing with "essences and principles" continued well into the 19th century and beyond (Temkin 1977; Pagel 1945). But it seems that these gradually became accepted as the unexamined and indeed hardly conscious core of the scientific enterprise. (Gerald Holton's discussion (1984) of the philosophic underpinnings of modern physics is *a propos*). In any event it seems probable that by the early 20th century, they were not explicitly stated and influential in the day-to-day work of undergraduate medical education.

Other factors militated against an explicit integration of philosophy and feeling with the teaching of anatomy at this time. The 19th century had seen vigorous religious opposition to some scientific tenets. Its condemnation of Darwinian evolutionism is perhaps *the* case in point. Thus, scientists and believers often found themselves in opposing camps. Those who wanted to keep one foot in both often kept "the oil of faith and the water of science" in separate jars.

This century too has seen physicians wage a long, hard struggle against other practitioners who claimed equally truthful and comprehensive systems of health and healing, among them: "homeopaths, magnetizers, and phrenologists" (Pagel 1945:2-3; Starr 1982).

Further, there was no longer even the semblance of unity in the west on a broad range of moral and philosophical questions. The "universal" church had long split into competing sects and denominations. Academic philosophy harbored many schools of thought on the ultimate meaning of life and death, and the afterlife. He who would refer to such things while simultaneously pursuing science and scientific thinking would be introducing chaos. Flexner (1910) thus praised the "laboratory" for the medical student as "a wholesome discipline [since] it banishes from the mind metaphysical principles such as vital force depression, etc . . . His actual contact with the facts puts him squarely on his feet and cures him once and for all of mystical vagaries."

One further complex of factors that may have influenced the elimination of explicit reference to philosophy and human emotion from the anatomy laboratory was the evolution of feelings, attitudes, beliefs and practices regarding death itself. After the middle ages, in the west, social life became increasingly shaped by the demands of a rationalized, industrial economy. Religious institutions and beliefs became fragmented, reflecting somewhat independent and competing economic, national and class social units. Individuals and families became more mobile and less tied to local communities, and thus death ritual, belief and practice belonged less to the total community and became more family and individual-centered, more of a "solitary" adventure. Aries notes that, as a consequence, in the late 18th through early 20th centuries, feelings regarding death were more difficult to repress, more labile, more linked to erotic feelings, and indeed, more threatening. In his data on this period -- letters, diaries, novels, drama and other cultural products -- Aries observes anxieties over pre-mature burial, a fascination with the cadaver, and a linking of erotic and death themes including widespread fantasies about and fears of necrophilia.

As the 19th century progressed, Aries finds an emergence of romantic attitudes toward one's own death and that of family members. He interprets these as exaggerated response-defenses against the fears and fantasies noted above. Thus, stress on a relatively new belief in the immortality of the conscious soul, and on loving reunions in the "great beyond" served to channel newly liberated and dangerous emotions. The same function was served (even among the atheistic) by lush funerals designed to heighten emotions, funeral statuary, and long and heart-rending deathbed partings.

It is no wonder, in this context, that the newly exuberant biological sciences, including anatomy, and their medical allies, no longer included explicit reference to the emotions, to philosophy and to the meaning of death in their teaching. Even if one rejects the hypothesis that science and medicine and their ideological stances were in part defenses against such feelings, one can see how their expression could hardly have been tolerated within scientific and medical institutions.

In 1865, Claude Bernard, the eminent physiologist, lauded the difference between the scientist and the ordinary man, showing how he could never permit himself the feelings of the layman while playing the role of a scientist:

A physiologist is not a man of fashion, he is a man of science, absorbed by the scientific idea which he pursues; he no longer hears the cry of animals, he no longer sees the blood that flows, he sees only his idea and perceives only organisms concealing problems which he intends to solve. Similarly, no surgeon is stopped by the most moving cries and sobs, because he sees only his idea and the purpose of his operation. Similarly again, no anatomist feels himself in a horrible slaughterhouse; under the influence of a scientific idea he delightedly follows a nervous filament through stinking livid flesh, which to any other man would be an object of disgust and horror (Bernard 1949:103).

With a few recent exceptions (to be discussed below), relatively little has changed since Osler and Flexner's time. Beyond a few introductory remarks describing the source of cadavers and their disposal, the history of anatomy, and the respect due this "human material" (e.g. Tobin and Jacobs 1981), little explicit attention is paid by teachers to the possibility that such work might arouse feelings and thoughts in students other than those related to objective learning. Anatomy textbooks contain no acknowledgement that the body to be cut, handled and described is other than an "object" of study. Texts present drawings and photographs – atlases of human anatomy could just as easily be terrestrial geographies were not the lands in question more familiar (Hobart 1984; Sauerland 1984; Christensen 1978; Laurenson 1968).

Recent general discussions of the teaching of anatomy reveal little or no recognition of the students' emotional and value-based responses to the experience (WHO 1964; Blunt 1976). Blunt recognizes "affective learning" as important to medical education. He outlines ways of using group and interpersonal process as learning tools, and yet explicitly rejects teaching oriented toward "outcomes that involve feeling or emotion . . . receiving, responding, valuing or developing a philosophy" (Blunt 1976:4). The author notes, "Outcomes in the affective domain are not normally of great importance in a course in topographical anatomy" (Blunt 1976:4).

The stated goals of contemporary anatomy in the medical curriculum, then, seem similar to those of the late 19th and early 20th centuries: first, explicit knowledge of (ability to name, visualize, and manually locate) the structures of the human body and their inter-relations; second, ability to relate knowledge of the functioning and malfunctioning of the body (derived from other scientific and clinical disciplines) to this knowledge; third, ability to relate and locate the problems of particular patients in reference to the above knowledge in an "objective" fashion (Blake 1980).

It would seem that philosophy and feelings were eliminated from the teaching of anatomy in part because they seemed incompatible with these goals. But, we would submit, they were seen as incompatible with these goals because of the evolving social and cultural context of modern medicine.

THE CURRENT CONTEXT

It is in reference to this supposed incompatibility, too, that some of our students interpreted anatomy and dissection as a necessary "initiation" into medical objectivity and detachment, as an experience which marked their separation from the profane world, and endowed them with a special, medical "mantle" or status, and which impressed them with the need for care, respect and reverence for their task. Several analysts have interpreted the experience in a similar way (Lief and Fox 1963; Coombs 1978; Reiser and Schroder 1980). Perhaps our students and these analysts have touched upon some of the real reasons (aside from departmental power) for the survival of dissection. Perhaps they perceive a purpose somehow intended by medical educators. Dissection does persist despite influential calls for a radical reduction in time devoted to it (Blake 1980) and despite the urging of some for radical changes in its pedagogy (Lloyd 1970; Halasz 1972; Bernard 1972; Neame 1984). Perhaps anatomy, taught through dissection, with little explicit reference to what we have called "feeling and philosophy" *is* an important initiation rite. Perhaps the cutting of preserved human flesh, the resulting spiritual, cognitive, and emotional disturbances, their private resolution and students' concomitant actual and symbolic identification with medicine and its norms are crucial reasons why dissection and its pedagogy remain as they are (Morinis 1985). Perhaps explicitly reintegrating feelings and philosophy into the teaching of anatomy and dissection would be dangerous tinkering. Perhaps the divorce is necessary.

If it is, an increasing number of students and medical educators are going in the wrong direction. In a recent article, June Penney, an anatomist at Dalhousie University, after describing quantitative findings analogous to our own concerning student experiences, reported that 64% of first year anatomy students at her school found themselves "inadequately prepared emotionally by faculty" for cadaver dissection; 53% wished to have more emotional preparation in the form of discussions on death and the "sharing of fears." Professor Penney recommends that an orientation program be provided to students before entering the anatomy lab and that while they are dissecting they should be formed into small groups to be led by faculty which should encourage the expression of reactions and emotions. Since dissection is the students' first contact with death, Penney feels that the program should be identified as part of the "death and dying" component of the curriculum (Penney 1985).

Several other medical educators, some associated with anatomists, have made the same or similar recommendations, and some have described on-going programs. Several have been reported in the literature (Blackwell et al. 1979; Marks and Bertman 1980; Terry 1985), while others are known to the authors through informal communication. Those which have been evaluated have used students' ratings. Small discussion groups seem to be highly valued by a majority of students in the programs studied but a significant minority express negative opinions. Other techniques do not fare as well as discussion groups, but are still valued by some.

One well might wonder whether these programs are going against or with the grain, and whether their example should be heeded. We believe it should. Since Osler and Flexner, medicine has become even more secure. Current battles between medicine and non-scientifically based alternatives are minor skirmishes. Conflicts between science and religion are being fought on the periphery of society. The churches are, if anything, among the more ardent supporters of modern medicine. Liberal theologians have long since learned to live with and or explain away "apparent" discrepancies between religious and scientific thinking, and fundamentalists seem to feel comfortable keeping both in separate categories.

For its side, science has become less absolutist. Although these views are hardly medical orthodoxy, in recent years, historians, anthropologists, sociologists of knowledge and some medical colleagues have gained a respectable hearing for the view that medical knowledge and practice, despite their grounding in observation and experimentation are historically, socially, and culturally conditioned. The gap between medical science (indeed science generally) and other forms of cultural expression seen by 19th and early 20th century physicians and scientists is no longer viewed by some as so great (Lock and Lella 1986; Foucault 1963; Hahn and Gaines 1985; Lock 1980; Wright and Treacher 1982; Freidson 1973; Mulkay 1979).

In clinical medical practice, a trend toward a revalidation of the emotional *vis à vis* the physiological, can also be seen. Armstrong documents how, from the late 1930's, a gradual change in the content of medical reasoning became evident in clinical textbooks and other work, through the influence of modern psychiatry (and its social-psychological ally). The concept of mental hygiene has been accepted; the psychiatric examination is acknowledged as important and disease is now perceived by some, at least, to be located not solely in the tangible body, but also at least partially in the patient's less tangible psychology, and social relations. As Armstrong notes, it has become necessary once again to pay close attention to the patient's account of the problem (1984:739).

Medical sociology and social epidemiology have emerged, and with them techniques through which to elicit patients' views, social psychological influences upon them, and their influence on the emergence of disease and illness. Thus, the concept of multi-level-causality has come into its own. It is no longer unchallenged medical orthodoxy to see disease merely in the tangible body (Berkman 1980; Engel 1980; Becker 1974).

Thus, the physician as recording instrument has come back into focus not only in his five senses and cognition, but also in his emotions and values. If the patient's account of the disease is crucial, it is now important for the physician to be able to understand and control his interactions with the patient in order to be able to elicit an uninhibited account; to interact in ways which influence behavior, to work toward mutually agreed upon versions of health; and to maximize the benefit of these for the patient (Smith 1984; Rosenberg 1984; Gorlin and Zucker 1983). The physician can no longer treat the patient's body, out there, as a mere object of the scientific recording mind, in here. The physician and his emotions, his philosophy, insofar as these influence his relations with patients, are just as much a part of his doctoring as are his cognitive skills, his knowledge, his ability to be objective (Helman 1985; Good and Good 1980).

Finally, the basic values behind quantitatively measurable care and cure are also being questioned. Patients' rights movements, and their legal and philosophical allies in medical law and ethics, have succeeded in redefining the goals of medicine. These now include not only measurable cure, or length of life, but also helping the patient achieve his own goals which may differ from those of his doctor. Current insistence upon informed consent as the backbone of medical ethics and law reflects this concern (Jonsen et al. 1982).

Medical students, consequently, are being taught an awareness of the impact of their feelings, their thoughts, and values on patients and on treatment. Information and skills linked to this new awareness are being taught in behavioral science, psychiatry, humanities and ethics courses, and appear in national licensing and board qualifying examinations in a number of medical specialties (Menken and Sheps 1984; American Board of Internal Medicine 1979; Council on Medical Education 1982).

Medical schools have developed educational programs in "death and dying" in response to a changing social context. Aries argues that in the post-Victorian era, the continuing fragmentation of society, the further attenuation of community-based ritual, the rise of medicine as technical champion in a war against death and evil, all militated against the explicit social recognition of death, and for the suppression of thoughts and feelings concerning it. Death has become a private matter. There are no agreed upon definitions of the meaning and significance of life against which to measure and master the force of our feelings about death. These are individual, private and often repressed.

Recently, however, some social scientific, psychiatric and other medical circles have reacted against this. Research and writing from these sources point to the psychological devastation arising from repression (Becker 1973; Shneidman 1984). We are now urged to acknowledge and come to terms with feelings about death and dying according to our individual preferences and beliefs. Health professionals, including physicians, are now being taught how to promote "good death," within and even outside of special hospices for the dying. Professionals are being taught to acknowledge and develop their own emotional and intellectual stances toward death so that they can recognize its influence on their work with others (Dickinson 1981; Field 1984).

All of this is recent, and by no means uniformly distributed throughout medicine. It is also, by no means, completely non-controversial. The view which we have described as 19th century in origin is still an important one. The newer view, however, is also influential and the pedagogy of anatomy and students' reactions to it should be considered against its background (as Penny and others have done).

It can be argued, then, that completely separating the teaching of dissection from explicit reference to human emotions and values can no longer be justified. Indeed a strong case can be made for re-establishing a link. Many students have expressed a desire for and seem to feel that they could handle it while maintaining their objectivity. Such a link would explicitly recognize the significant (perhaps ritual) experience that dissection is and expose that experience to "reflection" as a legitimate and integral part of medical education. Such reflection would powerfully re-enforce messages being sent in courses in behavioral science, in communication skills, and in human values and ethics courses: that science and objectivity need not be sacrificed if subjectivity and sensitivity are also promoted in teaching the work of doctoring.

The students' responses to dissection, and the above historical and cultural analyses hold important hints for those who would try to aid such reflection. First, it seems obvious that in our modern context, one cannot teach *one* set of values, attitudes and feelings toward death and the cadaver experience. Our society is fragmented, and we must respect the pluralism that exists. We can no longer officially, and as part of the curriculum, point to the cadaver as a way to God.

We can, however, on the cognitive level, teach the history of attitudes and values regarding death and of practices dealing with dissection and the teaching of dissection. Sketching out the context somewhat as we have done above can give students a sense of the relativity of these things -- some notion that their feelings or values have a history and can be examined in its light for further personal growth.

At a level closer to personal feelings and values, we can expose students to literature, novels, published personal diaries, analyses of medical students' and physicians' reactions to dissection, autopsies and similar experiences and give them the opportunity to reflect upon their own. In the experience described in the first section of this paper, it is clear that students saw our examination questions as opportunities to review their experiences, to compare and contrast them with those outlined in our textbook, to interpret them in ways which would allow them to live with their experiences and their coping mechanisms. In different ways they seemed to seek reassurance about the normalcy of their spontaneous emotional reactions. Reading about reactions similar to their own seemed to yield this. Students also expressed and attempted to justify their "violations" of deeply felt social values, privacy, and the sacredness of the dead. Although our text did not treat these matters in great detail, it and our exam question fostered and legitimated reflection on feelings and their possible meaning.

We should also make discussion groups available to students for supportive airing, comparing and reflecting upon emotions, thoughts and experiences. Participation, we feel, should not be required since such matters in our society are considered intensely private affairs. As we have seen, however, many medical students have indicated a willingness, and indeed a need, to express and discuss such feelings.

CONCLUSION

Cadaver dissection is not the last of the educational experiences in which medical students will face disturbing personal reactions. They will have to probe healthy and diseased breasts, vaginas, penises, scrotums and rectums. They will cut into living human flesh, assist at abortions and at the death both of babies and of octogenarians. They will have to decide how hard to work to save, sustain, or terminate lives. In sum, they will be introduced to activities which most members of our society do not share, and which in some ways violate deeply held social values.

Perhaps the most important goal which students' essays and our analysis suggests for educational activities is introduction of a means of reflection on the cadaver and similar experiences -- making accessible a process to help students cope in a way that allows them to feel enhanced rather than diminished as human beings and physicians.

Finally, while helping students cope with their own subjectivity, such programs would also model and promote attention to the subjective dimensions of patients' lives. By reflecting on dissection, considered among the most purely bio-medical and objective of medical school experiences, students might begin to learn how to integrate the objective-scientific with the subjective-humanistic to the benefit of their overall education.

ACKNOWLEDGEMENTS

This paper was made possible by a grant from Formation des Chercheurs et d'Action Concertée, F.C.A.C. (Québec).

NOTES

1. This only occurred to us after reading their responses and appreciating their richness. In talking with the class after reading the essays, we asked them to

approve our plans to analyze their essays. We told them that any use would be anonymous and they unanimously gave their permission.

2. Lief and Fox (1963:14) assume that these prior experiences prepare medical students for work in the anatomy lab.

   Penney (1985) has the data to determine whether prior experience correlates with degree of apprehension but has not looked at the link. These students seemed particularly disconcerted and even angry to find themselves reacting emotionally.

3. Lief and Fox (1963:15) argue that the anxiety connected to simply making it through medical school and more specifically passing exams becomes a "psychic counter-irritant" under which anxieties about death and dissection are submerged.

REFERENCES

Abernethy, J.
   1830   The Surgical and Physiological Works of John Abernethy. Vol. 1. London: Longman, Rees, Orme, Brown and Green.
   1814   An Enquiry into the Probability and Rationality of Mr. Hunter's Theory of Life; Being the Subject of the First Two Anatomical Lectures Delivered Before the Royal College of Surgeons. London: Longman, Hurst, Rees, Orme, Brown, Paternoster-Row.
American Board of Internal Medicine
   1979   Competence in Internal Medicine. Annals of Internal Medicine 90:402-411.
Aries, P.
   1981   The Hour of our Death. New York: Alfred A. Knopf.
Armstrong, D.
   1984   The Patient's View. Social Science & Medicine 18:737-744.
Becker, E.
   1973   The Denial of Death. New York: Free Press.
Becker, H.S., B. Geer, E.C. Hughes and A.L. Strauss
   1961   Boys in White: Student Culture in a Medical School. Chicago: University of Chicago Press.
Becker, M. (ed.)
   1974   The Health Belief Model and Personal Health Behavior. Health Education Monographs 2:324-473.
Berkman, L.F.
   1980   Physical Health and the Social Environment: A Social Epidemiological Perspective. In L. Eisenberg and A. Kleinman (eds.), The Relevance of Social Science for Medicine. Pp. 51-75. Boston: D. Reidel Publishing Co.

Bernard, C.
  1949 An Introduction to the Study of Experimental Medicine. New York:
        Henry Schuman (Translated by H.C. Greene and with an introduction
        by L. Henderson from the French published in 1865).
Bernard, G.R.
  1972 Prosection Demonstrations as Substitutes for the Conventional Human
        Gross Anatomy Laboratory. Journal of Medical Education 47:724-728.
Billroth, T.
  1924 The Medical Sciences in the German Universities: A Study in the His-
        tory of Civilization. New York: The Macmillan Company.
Blackwell, B.A.E., F. Rodin, F. Nagy and R.D. Reece
  1979 Humanizing the Student-Cadaver Encounter. General Hospital Psychia-
        try 1:315-321.
Blake, J.B.
  1980 Anatomy. In R.L. Numbers (ed.), The Education of American Physi-
        cians: Historical Essays. Pp. 29-47. Berkeley: University of California
        Press.
Blunt, M.J.
  1976 A New Approach to Teaching and Learning Anatomy: Objective and
        Learning Activities. London: Butterworth's.
Brown, G.W. and T. Harris
  1978 Social Origins of Depression: A Study of Psychiatric Disorder in
        Women. London: Tavistock Publications.
Browne, J.
  1697 Myographia Nova. London: Tho. Milbourn.
Christensen, J.B.
  1978 Synopsis of Gross Anatomy. (3rd ed.) New York: Harper and Row.
Coombs, R.H.
  1978 Mastering Medicine: Professional Socialization in Medical School. New
        York: The Free Press.
Coombs, R.H. and J. St. John
  1979 Making it in Medical School. New York: Spectrum Publications.
Council on Medical Education, A.M.A.
  1982 Future Directions for Medical Education (adopted June 15, 1982 by the
        American Medical Association House of Delegates). Journal of the
        American Medical Association 248:3225.
Dickinson, G.E.
  1981 Death Education in U.S. Medical Schools: 1975-80. Journal of Medical
        Education 56:111-114.
Edelstein, L.
  1943 Andreas Vesalius, The Humanist. Bulletin of the History of Medicine
        14:547-561.

Engel, G.
  1980 The Clinical Application of the Biopsychosocial Model. American Journal of Psychiatry 137:535-544.
Field, D.
  1984 Formal Instruction in United Kingdom Schools about Death and Dying. Medical Education 18:429-434.
Flexner, A.
  1910 Medical Education in the United States and Canada. Bulletin No. 4. New York: The Carnegie Foundation for the Advancement of Teaching.
  1912 Medical Education in Europe. Bulletin No. 6. New York: The Carnegie Foundation for the Advancement of Teaching.
  1925 Medical Education: A Comparative Study. New York: The Macmillan Company.
Foucault, M.
  1963 The Birth of the Clinic: An Archeology of Medical Perception. London: Tavistock Publications.
Fox, R.
  1979 The Autopsy: Its Place in the Attitude-Learning of Second Year Medical Students. In R. Fox, Essays in Medical Sociology: Journeys into the Field. Pp. 51-77. New York: John Wiley and Sons.
Freidson, E.
  1973 Profession of Medicine: A Study of the Sociology of Applied Knowledge. New York: Dodd-Mead.
Good, B.J. and M.J. Good
  1980 The Meaning of Symptoms: A Cultural Hermeneutic Model for Clinical Practice. In L. Eisenberg and A. Kleinman (eds.), The Relevance of Social Science for Medicine. Pp. 165-196. Boston: D. Reidel Publishing Co.
Gorlin, R. and H.D. Zucker
  1983 Physician's Reactions to Patients: A Key to Teaching Humanistic Medicine. New England Journal of Medicine 308:1059-1063.
Gregg, A.
  1957 For Future Doctors. Chicago: University of Chicago Press.
Hahn, R.A. and A.D. Gaines (eds.)
  1985 Physicians of Western Medicine: Anthropological Approaches to Theory and Practice. Boston: D. Reidel Publishing Co.
Halasz, N.A.
  1972 A Clinical Core Course in Gross Anatomy: Design and Experiences. Journal of Medical Education 47:568-572.
Helman, C.
  1985 Communication and Primary Care: The Role of Patient and Practitioner Explanatory Models. Social Science and Medicine 20:923-931.

Hobart, D.G.
    1984   A Dissector of Human Anatomy: Emphasizing the Musculo-skeletal Sys-
            tem. New Hyde Park, N.Y.: Medical Examination Publishing Co.
Holton, G.
    1984   Do Scientists Need a Philosophy? Times Literary Supplement. Nov.
            2:1231-1234.
Illich, I.
    1975   Medical Nemesis: The Expropriation of Health. Toronto: McLelland
            and Stewart.
Jonsen, A.R., M. Siegler and W.J. Winslade
    1982   Clinical Ethics. Toronto: Collier-Macmillan.
Kern, S.
    1975   Anatomy and Destiny: A Cultural History of the Human Body. Indian-
            apolis: The Bobbs-Merrill Co.
Lassek, A.M.
    1958   Human Dissection: Its Drama and Struggle. Springfield, Ill.: Charles C.
            Thomas.
Laurenson, R.D.
    1968   An Introduction to Clinical Anatomy by Dissection of the Human Body.
            Philadelphia: W.B. Saunders Co.
Lief, H.I. and R. Fox
    1963   Training for Detached Concern in Medical Students. In H.I. Lief, V.F.
            Lief, and N.R. Leif (eds.), The Psychological Basis of Medical Practice.
            Pp. 12-21. New York: Harper and Row.
Lifton, R.J. and E. Olson
    1974   Living and Dying. New York: Bantam.
Lloyd, G.
    1970   The Need for Reform as Seen Through Student Eyes. In J.R. Krevans
            and P.G. Conliffe (eds.), Reform of Medical Education: The Effect of
            Student Unrest. Pp. 89-99. Washington, D.C.: National Academy of
            Sciences.
Lock, M.
    1980   East Asian Medicine in Urban Japan: Varieties of Medical Experience.
            Berkeley: University of California Press.
Lock, M. and J.W. Lella
    1986   Reforming Medical Education: Towards a Broadening of Attitudes. In
            T. Hughes (ed.), Proceedings of the Second International Conference
            on Illness Behavior. Pp. 47-58. New York: Plenum Publishing Corp.
Marks, S.C. and S.L. Bertman
    1980   Experiences with Learning about Death and Dying in the Undergraduate
            Anatomy Curriculum. Journal of Medical Education 55:48-52.
Maulitz, R.C.
    1980   Pathology. In R.L. Numbers (ed.), The Education of American Physi-

cians: Historical Essays. Pp. 122-134. Berkeley: University of California Press.

McGuire, F.L.

  1966 Psycho-Social Studies of Medical Students: A Critical Review. Journal of Medical Education 41:424-445.

McKeown, T.

  1979 The Role of Medicine: Dream, Mirage, or Nemesis. Oxford: Basil Blackwell.

Menken, M. and C.G. Sheps

  1984 Undergraduate Education in the Medical Specialties: The Case of Neurology. New England Journal of Medicine 311:1045-1048.

Morinis, A.

  1985 The Ritual Experience: Pain and the Transformation of Consciousness in Ordeals of Initiation. Ethos 13:150-1174.

Mulkay, M.

  1979 Science and the Sociology of Knowledge. London: George Allen and Unwin.

Neame, R.L.B.

  1984 The Preclinical Course of Study: Help or Hindrance? Journal of Medical Education 59:699-707.

Newman, C.

  1957 The Evolution of Medical Education in the Nineteenth Century. London: Oxford University Press.

Olin, H.S.A.

  1972 Proposed Model to Teach Medical Students Care of the Dying Patient. Journal of Medical Education 47:564-567.

O'Malley, C.D. (ed.)

  1970 The History of Medical Education. UCLA Forum in Medical Sciences, No. 12. Berkeley: University of California Press.

Osler, W.

  1903 The Master-Word in Medicine. In C. Roland, 1972, William Osler's The Master-Word in Medicine: A Study in Rhetoric. Pp. 3-34. Springfield, Ill.: Charles C. Thomas.

Pagel, W.

  1945 The Speculative Basis of Modern Pathology: Jahn, Virchow, and the Philosophy of Pathology. Bulletin of the History of Medicine 18:1-43.

Penney, J.C.

  1985 Reactions of Medical Students to Dissection. Journal of Medical Education 60:58-60.

Piersol, G.A. (ed.)

  1906 Human Anatomy Including Structure and Development and Practical Considerations. Philadelphia: J.B. Lippincott Co.

Platt, M.
    1975   The Living Contemplate the Dead: Looking at the Body. Hastings Center Report 5:21-28.
Polhemus, T. (ed.)
    1978   Social Aspects of the Human Body: A Reader of Key Texts. Middlesex, England: Penguin Books.
Puschmann, T.
    1966   A History of Medical Education. (facsimile of 1891 ed.) New York: Hafner.
Reiser, D.E. and A.K. Schroder
    1980   Patient Interviewing: The Human Dimension. Baltimore: Williams and Wilkins.
Renshaw, D.C.
    1979   Death and the Doctor. Chicago Medicine 82:153-160.
Robinson, G.C.
    1957   Adventures in Medical Education: A Personal Narrative of the Great Advance of American Medicine. Cambridge, Mass.: Harvard University Press for the Commonwealth Fund.
Rosenberg, C.P.
    1984   A 10 Year Freshman Support Group Program: Leader Review. Research in Medical Education, Proceedings of the 23rd Annual Conference, Washington, D.C.: American Association of Medical Colleges.
Saunders, J.B. and C.D. O'Malley
    1950   The Illustrations from the Works of Andreas Vesalius of Brussels. Cleveland: The World Publishing Co.
Sauerland, E.K.
    1984   Grant's Dissector. (9th ed.) Baltimore: Williams and Wilkins.
Shneidman, E.S. (ed.)
    1984   Death: Current Perspectives. 3rd Edition, Palo Alto, California: Mayfield Publishing Company.
Schupbach, W.
    1982   The Paradox of Rembrandt's "Anatomy of Dr. Tulp": Medical History. (supplement no. 2) London: Wellcome Institute for the History of Medicine.
Selzer, R.
    1981   Mortal Lessons: Notes on the Art of Surgery. London: Chatto and Windus.
Singer, C.
    1946   A Word on the Philosophic Background of Vesalius. In M.F. Ashley Montagu (ed.), Studies and Essays in the History of Science and Learning. Pp. 75-84. New York: Schuman.
    1956   Galen on Anatomical Procedures. London: Oxford University Press.
    1957   A Short History of Anatomy and Physiology from the Greeks to Harvey.

New York: Dover Publishing.

Smith, R.C.
   1984   Teaching Interviewing Skills to Medical Students: The Issue of Counter-
          transference. Journal of Medical Education 59:582-588.

Starr, P.
   1982   The Social Transformation of American Medicine. New York: Basic
          Books.

Temkin, O.
   1977a Metaphors of Human Biology. *In* O. Temkin, The Double Face of
          Janus and Other Essays in the History of Medicine. Pp. 271-283. Balti-
          more: Johns Hopkins University Press.
   1977b Basic Science Medicine and the Romantic Era. *In* O. Temkin, The
          Double Face of Janus and Other Essays in the History of Medicine. Pp.
          345-372. Baltimore: Johns Hopkins University Press.

Terry, J.S.
   1985   The Humanities and Gross Anatomy: Forgotten Alternatives. Journal of
          Medical Humanities and Bioethics 1:90-98.

Tobin, C.E. and J.J. Jacobs
   1981   Shearer's Manual of Human Dissection (6th ed.). New York: McGraw-
          Hill.

Wiesser, R.J. and F.J. Medio
   1985   The Patient as Teacher. Journal of Medical Education 60:63-65.

World Health Organization.
   1964   Report of the Working Group to Study New Ways and Methods of
          Improving the Teaching of Anatomy. Regional Office for Europe of the
          World Health Organization.

Wright, P. and A. Treacher (eds.)
   1982   The Problem of Medical Knowledge: Examining the Social Construction
          of Medicine. Edinburgh: Edinburgh University Press.

ANDREA SANKAR

PATIENTS, PHYSICIANS AND CONTEXT: MEDICAL CARE IN THE HOME

INTRODUCTION

Most discussions of the physician-patient relationship mention only two parties: the physician and the patient. In describing and analyzing this relationship, careful attention has been paid to such concerns as the linguistic nuances of the dialogue (for example, Mishler 1985; Cicourel 1981; Paget 1982), the roles that are enacted (Parsons 1951); socialization into these roles (Fox 1957; Becker et al. 1961; Merton et al. 1957; Roberts 1977); the perceived social status of the participants (Cartwright 1964); their variant and shared meaning systems (Kleinman, Eisenberg and Good 1978; Kleinman 1980; Good and Good 1981); and both the professional and personal cultures of the actors (Lock 1985; Gaines 1982; Hahn 1985). While approaching the physician-patient relationship only as a dyad has been criticized on a number of grounds, for example, that there are also the family, the profession, the society to consider, an additional character receives insufficient attention -- the *context* of the physician-patient relationship. As we will see in the following account, the context exerts a tremendous influence on what transpires between physicians and patients: it can strongly "dictate" power relationships, what kinds of information are visible or invisible, and how physicians and patients know and experience each other.

The power of context is often invisible because it is usually taken for granted. Most patients are seen in the hospital or clinic (though this is a relatively recent development in medical practice). This influence, however, becomes more visible when the normal context is no longer operating, as when the parties find themselves in a new setting. This in fact was what I observed in a study of the experiences of medical students in a Home Care Program. Not only did the power of the home as a context for medical care become evident, but indirectly the experiences mirrored the characteristics of the "normal" context of doctor/patient relationships -- the clinic or hospital setting. In this study I examine the interaction between medical students and elderly chronically ill patients receiving home health care and how the change in the environment of the medical encounter affects student physicians' perceptions of patients, chronic illness, and control.[1]

M. Lock and D. R. Gordon (eds.), Biomedicine Examined, 155–178.
© 1988 by Kluwer Academic Publishers.

A STUDY OF HOME CARE BY MEDICAL STUDENTS

While studying the attitudes of American medical students toward elderly patients in the context of a Home Care Program clerkship in 1980, I discovered that this clerkship proved to be a powerful and often troubling experience for the students. This led me to question why medical students respond so negatively to the apparently simple task of caring for a patient at home (Eisdorfer 1981).

Accordingly, I chose a medical school where the curriculum requires participation in a Home Care program during the requisite two month clerkship in Family Medicine. The Home Care program at this university has been in existence for 15 years and, while based within the Division of General and Internal Medicine, is part of the Family and Community Medicine clerkship. Because of the teaching requirements of the Department of Internal Medicine, patients chosen for participation in the program are more seriously ill -- that is, they have a higher level of "patient acuity" -- than those selected for a Family Medicine clerkship.[2] A faculty staff, consisting of a medical director, nurse, social worker, and pharmacist, oversee the program.[3]

Teams, consisting of a fourth year medical student and fourth year pharmacy student, were assigned responsibility for the primary care of two or three seriously ill housebound patients. The team visited each patient once a week, taking at least one afternoon of student time. The students visited more frequently if deemed medically necessary and were on twenty-four hour emergency call.

In what follows, I will discuss two significant aspects of context: (1) the information exposed by or excluded from the context and its particular salience in the care of the chronically ill; and (2) the power relationships vested in the context and their effect on the physician-patient relationship. The following case studies of four students as well as descriptive material gathered from others illustrate the significance of context and the different ways students responded to the home environment. It is worth noting from the outset that although the intense experience which many students underwent can in part be attributed to their novice status, the two faculty physicians whom I observed also altered their interactions with the patients in the home. Unlike the students, though, these faculty did not experience anxiety concerning their professional efficacy.

#1 Bob

When Bob was asked what he had expected to get out of the clerkship, he responded, "Myself, as soon as possible." In a more serious vein he went on to state, ". . . my immediate bias is that I won't find it stimulating." During the first three weeks Bob focused purely on medical problems. An elderly male patient suffering from end-stage chronic obstructional pulmonary disease tried repeatedly to get some reassurance or hope for himself and regularly asked Bob if he had a miracle with him that day. In the third week this exchange took place:

Bob :   Give me a call if you need me.
Patient :   I'm callin' you now.
Bob :   There's not a whole lot I can do.

After the visit he commented, "I'm going to radiology, medicine is too depressing." In the fourth week he responded this way to the patient's request:

Bob :   If anything comes up, call me.
Patient :   What good can you do?
Bob :   We can talk.
Patient :   That would be moral support.
Bob :   Sometimes moral support helps a lot

In the car after the visit Bob commented, "Mr. S. needs more P.T. (physical therapy), but what he really needs is time to talk about his problems." During the next three visits over half of the exchanges involved nonmedical topics, while the actual numbers of medical topics remained the same. The proportion of nonmedical topics discussed by Bob and his patients increased for the reminder of the clerkship.

From this visit on Bob decided that he had, ". . . become familiar enough with Mr. S.'s medical problems to relax a bit and concentrate on his nonmedical problems." He even felt it was appropriate to just "shoot the breeze." While Mr. S. occasionally still asked for miracles, his anxiety dropped noticeably. He explored with the young physician new areas for his treatment about which he had read. One topic discussed was a form of oxygen therapy, new to the student, which was later integrated into the therapeutic regimen.

By the sixth week Bob was developing a warm personal relationship with Mr. S. and his other patients, inquiring into their dreams, fears, and memories as well as following their medical problems. After this set of visits he discussed his attitudes toward his elderly patients. In the initial interview he expressed the view that working with elderly patients was, ". . . like pediatrics, just talk to them nicely," and that the elderly, ". . . put themselves in that position by being dependent, like kids." By this point in the clerkship his views had changed significantly:

There is a lot of strength in our patients. They are very proud. A lot of elderly people have great pride but it's broken down by the way society treats them. Illness also breaks down pride. Patients in the hospitals are infantilized. The nice thing about homecare is that you can get therapeutic value without the infantilization.

Bob, like two thirds of the students at the end of the clerkship, was able to turn his attention to the nonmedical problems of the patients so that eventually he could provide integrated medical care for both the psychosocial and physical aspects of his cases. In commenting about the long term care of patients he said:

The long-term prognosis in very sick patients comes up more acutely in homecare. A lot of times you see patients in the hospital and you know they are not very healthy in general and your goal is to get them out of the hospital and once you get them out of the hospital, even if it's to a nursing home, you feel there is some degree of success there and you also don't really have to confront the issue of long-term prognosis head on . . . [in homecare] there is much more of a need to address somehow, in your mind, what the long-term prognosis is.

Despite his initial cynicism, Bob ended his seventh visit having developed strong ties with his patients. This experience was not, however, easily acceptable to him. Bob avoided his final visits with his patients, stating that he did not really need to see them. He added he would try to call them and say good-bye, but since they would be getting a new student next week it probably did not matter. He was one of two students who avoided this last visit.

#2 Diane

Three of the students concentrated on the patient's social and emotional problems in response to their feelings of being overwhelmed by the home care experience. Diane, a twenty-seven year old student, was planning to enter surgery. She was definitely unhappy about the upcoming clerkship. In the initial interview she said:

I don't enjoy working with chronic illness. I like to confront a goal and accomplish it. I've never worked with adults and chronic illness before. I just can't see getting into adult chronic illness. It's not intellectually interesting to me.

After two visits, her attitude had changed dramatically:

My ideas of illness have changed since coming on homecare. I went back to see

Jimmy (a thirty year old muscular dystrophy patient) for the second time today. I went in very cheerily like you do with a patient in the hospital when you know they'll get better. Then I had to stop and get control of myself. I realized nothing had changed. That the room was exactly the same, that Jimmy was exactly the same, lonely and sick, getting sicker. This is a different kind of illness than I'm used to.

By the fifth visit, Diane moved to focusing completely on her patient's nonmedical problems. She commented in the following way about an extremely ill but very articulate diabetic woman who was going blind, losing any sense of touch, had severe digestive problems, and was recovering from a four-month hospital stay because she had broken both legs simply by standing up: "It's not necessary to check her over very closely because she's such a damn good historian."

Diane enthusiastically made plans to get this patient into a swimming program, get a lift for the bath, find a way for her to learn Braille and to consider a new physical therapy program. These were the concerns which the patient had identified as most important and Diane worked with her to realize them. When the patient's blood sugar level hit 744, the medical director had to caution Diane not to get so involved in attending to social problems that the diabetes "sneaked up and hit her from behind." Indeed, Diane did feel that although the patient suffered from brittle diabetes ". . . 99.9 per cent of her problems were people problems." Realizing the significance of the nonbiological in chronic illness, Diane went to an extreme and developed a kind of medical nihilism. She came to feel that medical interventions could do little to help her patient and in fact they might just increase her suffering. Through her involvement in her patients' lives, she developed a deeply satisfying emotional attachment to them. She ended the rotation by deciding to switch her specialty to family practice. In her final interview she stated:

I thought I was going to learn a lot about diabetes. I learned a lot more. This is very corny, but I learned a lot more about how you really have to look at the whole person because in fact diabetes is a great problem in P's life, but it was such a minor part of the last eight weeks. It was dealing with how it had limited her which I wasn't expecting . . . I really enjoyed the emotional aspects of the relationships which formed. It's sad in the other extreme though, because you see people with illnesses that have altered their lives drastically. It makes you real thankful you don't have something like that.

#3 Arthur

Arthur was a twenty-nine year old student who had planned to specialize in oncology. Despite his proposed interest in oncology, he equated the elderly patients assigned to him with the chronically ill about whom he commented:

I don't like dealing with the chronically ill. When I see a patient I like to be able to make a difference, to fix someone, to cure. But to just wait around for an acute episode is a waste of time.

His sense that there was nothing "medically" to do for the chronically ill except when they experienced an acute exacerbation of their illness also underlies much of the frustration experienced by other informants. His alienation went deeper, for he perceived treating the chronically ill as a threat to his skill. After one visit he commented:

When taking care of someone who is chronically ill, you can't help but make a mess sometimes. It's easy to overlook a problem. You get sloppy in your exams, doing them every week, it breeds mistakes. If I had a hundred patients like Mrs. E., I'd be a lousy doctor in a year.

It would seem, following his reasonable concern that his skills and expertise would get "rusty" if not practiced and his statement that, ". . . someone else should take care of the chronically ill, not the doctors," that his patients suffered from simple, uncomplicated, unthreatening complaints. Ironically, among his patients were some of the most acutely ill in the program. One woman recently discharged from the hospital suffered from eighteen different medical problems for which she took sixteen different medications. Yet the considerable expertise and competence required to adequately care for such a patient and to prevent rehospitalization did not qualify as either sufficiently challenging in Arthur's opinion to maintain his skills or worthy of his expertise. Eight of the students voiced similar attitudes during the initial two weeks of the clerkship.

This sense that there was nothing for him to do or that it could be done by "someone else, not a doctor," changed fairly quickly for Arthur as he became overwhelmed by the complexity of his patients' lives. For the first few weeks he resisted acknowledging this complexity as a legitimate concern of medicine. After two visits he commented:

The goal of homecare visits is not to examine chronic illness, but [should be seen] in terms of the patient's needs. It's different than treating chronic illness. There's sort of a hidden agenda there in terms of what you have to do for the patient.

This clear distinction he drew between treating chronic illness and patients' needs was eventually broken down.

Within the initial two week period of the clerkship, Arthur had tried to maintain a doctor-dominated relationship. He would insist that the patients or their children turn off the TV during his visits. A tall man, he towered over his patients and maintained this distance by never sitting near them, but instead bending down to examine them. He tried to ignore the social turmoil which swirled around him in one patient's home. His aloofness caused considerable anxiety for it was interpreted as resignation to the hopelessness of the case.

By the third visit, Arthur's assessment of the key problems had begun to change and he remarked that, ". . . anxiety is the most common cause of what Mrs. W. suffers from." On the fourth visit Arthur walked into Mrs. W.'s home and she greeted him saying, "Give me a shot to kill me, doctor, I don't want to live anymore." Then the fight which she and her daughter, the main caregiver, were having erupted once again with Arthur in the middle. He stood there, his hands raised, and watched helplessly as Mrs. W.'s breathing worsened.

Arthur came to appreciate the intense emotional, social and psychological problems created when a family must care for a sick member at home and maintain the lives of the rest of its members. Arthur acknowledged that these strains exacerbated Mrs. W.'s precarious condition, but his response was one of helplessness: "the most important factor seems to be her emotional, her life situation, which I really don't think we can fix." This sense of being overwhelmed and helpless led him to stop doing even what he felt he could do, which was to monitor the patient's health status and care for her surgical wounds. Crippled by a sense of futility, he allowed a problem to pass which resulted in a medical crisis.

With the help of the Home Care Social Worker and faculty physician, Arthur overcame his sense of being overwhelmed and assisted the family in developing a plan to care for Mrs. W. at home, in getting some relief for the caregiver daughter, and in arranging for Mrs. W. to attend church.

In evaluating his homecare experience, Arthur remarked on the serious limitations in his own education:

The emphasis in Home Care Program is probably where it should be in medicine, but where it's tenuous -- caring for the patient . . . *really caring for the patient* -- (emphasis his) that is what they try to do and it's really something that's been neglected for a couple of years in my medical education. You can't teach it with a lecture course or seminar, you have to teach it with patients.

He then added, ". . . medical schools produce doctors like me." Arthur informed me later that, recognizing his limitations with "the social thing," he decided to change from oncology to a specialty -- such as intensive care, emergency medicine, or anesthesiology -- which entails considerably less involvement with patients or their families.

#4 Louise

Louise was a twenty-six year old student, who at the time of this study was planning to specialize in family practice. Of all the students, she adjusted least well to the home environment. With considerable success, she sought to establish and maintain a doctor-dominated form of interaction within the home. Louise set the tone of her encounters by addressing her elderly patients by their first names, something few other staff members or students did. On her first meeting with an eighty year old man suffering from chronic obstructive pulmonary disease (COPD), she found a chair and pulled it up to the bed without it being offered and began a long and extensive "systems review," despite the fact that the patient was obviously having a great deal of trouble breathing. If other people were present, Louise directed her questions to them rather than the patient, who was completely alert. For example, when a grandson was present Louise asked, "Do he and his wife eat together?" "Ask him," replied the grandson; whereupon Louise dropped the question. When she wished to discuss the patient's status or to consult on medications with the pharmacy student, she lowered her voice to a medical stage whisper, clearly audible to the patient.

Throughout the interviews she maintained her focus on strictly medical problems. When the patient asked for support the response was medical:

Patient :   Why don't you say something about my lungs?
Louise :   They are about the same.
Patient :   Well, I expected that. Why don't you say something supportive?
Louise :   We can't do anything about emphysema. We can about bronchitis. If you take your drugs and drink water to cough up the phlegm, we can help the bronchitis.

Although she would not be drawn into recognizing the patient's request for reassurance, compassion and empathy, later she acknowledged that this exchange had been especially difficult for her: "I had a harder time talking to him this time. I don't know if it's him or me; I'm depressed. He makes me depressed."

But she quickly regained her composure by searching for a medical explanation for the troubling encounter. "There's nothing to suggest heart failure," she went on, except that he's gaining weight. It's hard to judge his symptoms. As far as living conditions, there is nothing you can offer him when you see his emotional status. I'd like to do more for him physically. I'm not sure there is anything else to do for him.

By restricting her gaze to the purely biomedical aspects of the patient's problems, Louise was sometimes led to miss significant information. In the following dialogue Louise ignored the crucial datum concerning the patient's sudden weight gain.

Patient : The more I try to get away from salt, the more I get. My wife gives me peanut butter with salt. I am afraid I'll have to go to the hospital to get salt-free food.
Louise : Are you having trouble with your urine?
Patient : No, but it's uncomfortable. But what worries me is about the salt.

As it turned out, long-standing animosity between the patient and his wife was exacerbated by his homebound condition and she was seeking retribution by giving him a high salt diet, while he was helpless to do anything about it.

Throughout the clerkship Louise was concerned with maintaining control. When another patient jokingly reminded her how shocked she had been by the patient's condition when Louise had first met her, she became exceedingly flustered and denied being upset. "That isn't true, that isn't true," she repeated insistently. The knowledge of the transparency of her emotions to others was threatening to Louise's composure, for, in fact, at the interview after that visit she had been extremely upset. In the evaluation of the Home Care clerkship Louise singled out the threat to control as a major upsetting factor:

Mrs. E.'s wanting to maintain control over her [medical] exams was frustrating to me and threatening. I don't think it was particularly an ego thing. But I think to be effective [as a physician] you have to maintain a certain amount of discipline and control in the situation. I felt sometimes I was lacking control.

Louise managed to contain the relationship with her patients within the limits of biomedicine. She restricted her compassion to their physical problems about which she could possibly do something, thus she commented that she was ". . . exceedingly sorry for my patient's physical condition and so forth." Anything beyond the patient's medical problems she judged to be a spiritual matter and suggested her patients read the Bible. Consistent with the limitations of her interventions, she restricted the information she sought from the home visit to that directly relevant to physical status, yet she was not impressed with its quality once elicited.

This information she said was, ". . . not necessarily the same as someone coming into your office, but the difference was not necessarily a broad one."

Louise's behavior would not be the least remarkable when viewed in the context of the clinic. Yet she was noteworthy among the students because her struggle to maintain control and professional authority led her to block out much of the information and influence of the home context and to emerge with a singularly limited set of information and an experience which she found "somewhat bothersome."

DISCUSSION OF CASE STUDIES

By relinquishing the control associated with physician-patient interactions in the clinic, the medical students gained access to considerable information about the patients. With over half of the students, this increase in information also created the basis for a more intimate personal relationship with the patient than is often possible in traditional medical care settings. Integrating the new information into the data base, and reconciling an expanded understanding of the patient and the patient's problems with the narrowly defined goals associated with the acute care model of medical care was a challenging task for the students, exciting some, overwhelming others.

Bob began the clerkship cynically and ended it by denying his experience. Yet, during the course of the visits and within actual encounters, his responses to patients were some of the most moving. He allowed himself to be drawn into the patients' world far enough to be able to offer emotional support and assistance to them. He did this in a way that was both professional and sincerely empathetic. This was a complicated and stressful development for him; in abandoning the traditional doctor-centered relationship, he could not resolve the resulting conflicts between the demands for intimacy and those for distance, and in the end opted for denial.

Diane finished the clerkship with serious concerns about the proper role of medicine. Her decision to abandon surgery resulted in part from these doubts, but the questions raised about the efficacy of traditional medical interventions were not resolved.

Arthur was uncomfortable and often upset throughout the clerkship. He commented that it was not quite, but almost, as bad as castor oil. The numerous exogenous undismissable factors which confronted him in the homes of the patients forced him to reconsider his definition of medical efficacy. The experience confirmed for him that he was not suited for, and therefore should not seek, direct involvement in patient care.

For Louise the experience was tense and frustrating. She struggled continually to restrict the interactions to those appropriate to the clinic context and maintained a doctor-centered relationship. To maintain control, she utilized rigid structures such as calling elderly patients by their first name and moving about the patients' homes as if they were her own office. When the plight of her patients did penetrate, she separated it out from the illness and attributed it to spiritual malaise.

THE SETTING

From these and other cases several important characteristics of the setting stand out. In the clinic the physician interacts with the patient in a combination of two or three carefully choreographed interactions. At the initiation of the visit, the patient is dressed sitting in the examining room. The physician enters. The patient's problem is briefly discussed and the physician decides if the patient should put on a gown. If so, the physician leaves and the patient, perhaps with the assistance of a nurse, will undress, put the gown on and get up on the examining table. If it is clear at the outset of the visit that a gown will be necessary, the patient is usually dressed appropriately during the initial discussion with the physician. During the exam itself the patient is either seated or lying prone on the table. The physician stands over the patient and proceeds with the exam. That the physician is dressed, standing, and moving about at will while the patient is usually undressed, prone, and relatively motionless creates an immediate social distance between the two. This distance may be heightened by the introduction of various medical technologies such as bright lights, special viewing aids, stirrups, or a stethoscope. The final part of the visit may take place in the examining room or the physician's office. The physician and the patient discuss the findings of the exam and the proposed course of treatment. Except for the occasions when this last interaction takes place in the physician's office, the setting is usually austere, denoting professional authority and technological efficacy.

By contrast, in the home there was usually only one setting for the entire interaction -- the living room, bedroom, or dining room where the physician initially encountered the patient. If the physician wished to examine the patient, he or she had to assist in the disrobing. If the physician or patient desired an examination in privacy, then it was usually the physician who had to negotiate this with the other people present, sometimes successfully, sometimes not.

As a result of the change in the setting the proxemics of the encounter were not fixed as they would be in the clinic. In the clinic the presence of the examining table or the physician's desk or the chairs in the examining room determine the physical space and often social distance between physician and patient. These relationships were not fixed in the home. Some student informants were keenly aware of the problems posed by the spatial relationships of the home context. Maria, a student planning to practice internal medicine, was perplexed by her choice of where to sit in the patient's home. To sit across the dining room table from the patient where she could maintain eye contact meant she then has to stand up and move around to do the examination. Sitting beside the patient was awkward for communication. Sitting at the head of the table was most conducive to the interview but implied a social position with which she felt awkward.

Cindi, a would-be surgeon, was aware of the explicit implications of the proxemics of the encounter when she remarked in her final interview:

The difference in control was most apparent when I had to choose a position in the house, like choose a chair. The whole issue of whether one sat had implications for who would control the conversation. When someone comes into your office, by definition you are behind the desk and the patient sits in the chair.

The physical examination was another item taken for granted in the clinic, but which had to be negotiated in the home. Louise commented about a 99 year old patient: "She was wanting to maintain, to at least some degree, too much control over her own exam. You could only undress her so far and no further."

The assumptions which underlie the smooth routine of the clinic encounters simply did not hold in the home. Patients did not offer to undress themselves. If the student judged that removal of garments was essential to effectively assess the patient, he or she had to request this, had to specify which garments, and usually had to assist in the disrobing. Sometimes patients would resist or counter-offer and suggest they remove only part of their clothes. Patients were able to exert some control over the situation because there was nothing in the contextual environment which assured that they would positively respond to a relative stranger's request to remove their clothes. In fact the assumptions were very much the opposite; namely, that in the patient's own home in the presence of a stranger the medical student would not remove their clothes. The medical necessity for removing clothes had to be compelling and evident to the patient for him or her to readily comply.

Students faced negotiations over the length and the pacing of the visit. In the clinic the physician enters and leaves the patient's presence at his or her choosing. In the home the patient was in most cases waiting for the physician's arrival but the similarity stopped there. Socially appropriate behavior was observed even by those students who successfully maintained doctor-centered interactions. Simply getting situated in the patient's home required more exchange of non-biomedical information than introductory conversations in the clinic. Initiating the actual exam required that the student execute a thoughtful verbal and often physical transition as he or she moved from a position in the room appropriate for an incoming guest to one more conducive to a medical examination. Students who were not skilled in executing this transition could become stuck in the introductory pleasantries and extend the visit way beyond what they felt was appropriate. Most patients acquiesced to the physician's pacing of the examination itself. Leave-taking involved the most complex negotiations. Patients saw the examination as a kind of service performed and they in turn desired to reciprocate. In the home the patient had the opportunity to act on this desire by inviting the physician to have a cup of coffee or a snack, by conversing with him or her about non-medical topics.

TRANSITIONS IN STUDENT BEHAVIOR

Gradually this different setting had an effect on the students' behavior and their relationship to their patients. Students brought to the initial encounters the practices and methods of control and objectification characteristic of clinic and inpatient settings. Many students initially called patients by their first names, regardless of age or circumstances. They appropriated the private space of the patient by electing to sit before being asked, by looking for and using the sink to wash their hands without asking directions or permission and by searching through and sometimes disrupting the arrangement of medications. They further sought to sustain the clinic context by intending to conduct extensive physical exams and by developing what they deemed medically appropriate treatment goals, such as changes in medication, reduction in angina attacks, increased mobility, specialist consults, and improved nutrition.

Within two to three weeks, 15 of the students had changed their behavior in the encounters. Many had changed the mode of address in compliance with patients' stated or implied wishes. Some began asking where to sit and a similar number ceased handling the medications without permission. Thirteen changed some or all of their treatment plans, incorporating those suggested by the patients.

By the fifth or sixth week, over half of the students had developed a more intimate personal relationship with their patients than they were accustomed to. A kind of equality, often marked by reciprocity, emerged. The reciprocity was both material and conceptual. Students accepted, although at first reluctantly, the patients' hospitality, their offers of orange juice, tea, or homemade wine. Students reciprocated by responding to questions concerning their own personal lives. Although the nonmedical content of the encounters increased, this did not usually reflect a reduction in the medical content. Instead, most students could both treat the patient for the medical problem from which he or she suffered and respond to the patient as another individual whose life extended far beyond the narrow confines of the medical encounter.

By the last two or three weeks, two-thirds of the students attempted to intervene in the nonmedical problems of the patients. In numerous ways students sought to address the problems and needs which the patients had identified as key and which they had come to assess as having an impact on the patients' health status. Whether it was something close to medicine such as nutrition, or something more removed like a new kitten to replace a much beloved cat which had been a patient's only companion, or assistance with the Veterans Administration (VA) in securing income for a patient, students tried to be of help. Where they felt the actual intervention was beyond their area of competence, as it often was, they sought the appropriate referrals.

An examination of the age of the students (26-39 years old; mean age 31.5), future specialty (four internal medicine, three each in surgery and pediatrics, two in family medicine, and the remaining in various other specialties), or sex (10 men and eight women) does not help account for the change in the physician/patient relationship. More than half of the medical students at first articulated a sense of loss of control: of these all but one experienced significant changes in their attitudes toward chronic illness and its sufferers. Altogether, more than two thirds of the students experienced a change in the attitudes they had initially expressed regarding care for the chronically ill elderly and finished the clerkship with a more positive attitude towards the care of such patients. For six of these the change was dramatic because they had begun the clerkship by articulating negative attitudes about the prospect of caring for the elderly chronically ill patient. Four of this very negative group were men. Out of those who experienced change and who had initially stated negative attitudes, four (3 men, 1 woman) also considered changing their specialty by the end of the clerkship. These four were among those who were most overwhelmed by the experience. Another informant considered changing specialty, but from a positive motivation of concern for and interest in the elderly.

ANALYSIS

In the hospital . . . you have social problems where you want them . . . there is no social . . . whereas at home, there are all kinds of interactions that you really have no control over (Arthur, future radiologist).

Treating patients at home was illuminating not only for what it said about the patient but also for what it revealed about the clinic. Two particular aspects of the experience were emphasized by the students: (1) the sense of being overwhelmed by the amount of information present in contrast to the restricted context of the clinic and (2) the loss of control over the encounter. For many students these two aspects of the experience gradually changed the relationship they had with their patients; others began to reevaluate the appropriate role of medicine.

Context and Chronic Illness

This clerkship has provided me with more of a gut feeling of what chronic illness is rather than previous ones that were more intellectual. You can know that a patient is chronically ill and take care of them and keep them in the hospital, but you never really get the gut sense of what it's like for this person to live in his home or wherever with a chronic illness, so I think it just changes my perspective from a purely intellectual one to a more gut level one (Bob, future internist).

The medical students interviewed learned that chronic illness could not be reduced to a biomedically-defined pathology or dysfunction. The complexity of the disease process -- the way it affects and is affected by a patient's daily life -- could neither be ignored nor reduced to a simple diagnosis.

The students also learned that it is in the home that chronic illness is lived. They came to understand that in many instances it is very difficult to separate the social, psychological and biological factors that interplay in long-term disease. In light of these findings the student doctors realized that in many cases it is difficult to determine the complexity of the effect of a routine medical intervention or when that intervention may do more harm than good.

This "gut level" appreciation that they developed for the complexity of chronic illness echoes findings in other current research which have indicated that social and psychological factors may have more of an impact on outcome in chronic illness than the underlying disease process. For instance, Yelin et al. (1980) have demonstrated that conditions of the work place rather than biomedical tests and measurements are a better predictor of return to work. Other researchers have demonstrated that chronic pain syndrome, a major cause of disability in the United States,

is encouraged and maintained by marital, family, work and health care
relationships rather than biological dysfunction (Turner and Chapman
1982; Keefe 1982; Sternback 1974).

In the home the interactions between the social, psychological and
biological aspects of a chronic disease were no longer merely academic
questions. Students were confronted in a concrete and sometimes dramatic fashion with the impact of non-biological factors on the illness process.

Just seeing a patient walk in and out of an office room who had trouble walking,
you don't really get the full impact of what that means in their life. All you see is
how they come into the room and sit down and walk out (Cindi, a future surgeon).

The following example illustrates how many students came to feel that
it was not enough to simply monitor the patient's vital status and adjust
their medicines but that in fact, without understanding the social context,
they could make a medical mistake.

It was a real eye-opener for me to realize the thing giving Mrs. C. her breathing
attacks was the problem with her VA checks and not her disease process. If I'd
monkeyed with her meds, as I'd planned, I could have made her condition much
worse (Steve, a future internist).

In the car, following each visit, students were frequently overwhelmed by
the amount of information about the patient which they had found out in
the home. They were also sometimes overwhelmed by the patient's life
situation.

It's the little things; the things you'll never forget, like walking into Mr. S's house
and there's that horrible smell, and realizing there's nothing he can do about it.
He's talked about it before. "I married her and I thought her housekeeping would
get better, but it's gotten worse," he told me. Now I understand his position. He
could tell you that in an office and you'd never understand, but I understand – boy,
do I understand.

The home presented the information the clinic setting actively
excluded. Everything about the context bespoke the patient and the family; smell, color, clothes, design, lay out, furnishings, food, and neighborhood were all concrete details of the patient's identity. The context also
bespoke the disease albeit through a narrower range of detail: smell, furniture arrangement, medical supplies and technology, and clothes. The
sheer amount of this information helped the students see the patients as
people.

This detailed and intimate understanding of the patient to which most
students were unaccustomed was overwhelming for many of them. The

enhanced appreciation of the patient and disease often was accompanied by a realization of the realistic limits of medicine in the care of the chronically ill.

Context and Power

When asked to compare their experience caring for patients in the home with the clinic, fourteen of the students chose to emphasize the "loss of control" they encountered in the home. They described the loss of control as follows:

It's strange how much power you have over patients in the hospital. You can tell them what to do and they let you do anything to them. In the home the patient has control, you must ask them and get permission (David, future internist).

In the home the patients are more in control than I am. In the hospital I could do to them what I wanted. The patient is sort of a captive in the hospital. Here I had to adjust to what they really did (Arthur, future radiologist).

The main difference is the power structure. When you visit them in their home, it's their turf, they are in control. You have to recognize their territory (Sam, future oncologist).

When you are in the home, you are completely alone. You don't feel the same kind of immunity that you do when you handle them [the patients] in the medical center (Lisa, future gynecologist).

Although the frankness with which the students identified control and power as key issues is somewhat startling, it should not be surprising. In a review of the literature on doctor-patient relations, Hauser (1981) concludes that the modal form of relationship demonstrated by the various studies under review was characterized by "doctor-centered" or "doctor-dominated" interaction in the encounter. As we have already observed, this kind of structural control was not present in the home. In fact, noted health manpower expert, Eli Ginzberg (1984), has argued that the American physician's need to maintain control over the patient is in part responsible for retarding the development of home care services in the United States as compared with other Western countries.

Researchers have speculated that this control arises from several possible sources, e.g., socialization (Parson 1951), status (Freidson 1961), protection against the uncertainty physicians face (Fox 1957), a buffer against potentially embarrassing situations like the gynecological exam (Emerson 1973), or reflections of the larger political structure in which the interaction of the patient and physician is embedded (Taussig 1980).

Observations of physicians and student-physicians interacting with patients in the home suggest another source of support for physician control, the context itself.

The students themselves best illuminated this aspect of control when nine of them suggested "turf" as a synonym for the "loss of control" experience. "Turf" is a particularly appropriate term because if connotes power as well as place. "Turf" implies control over a particular domain. It has similar power connotations to the concept of territoriality in studies of animal behavior. In medicine it implies territorial domain, as in "Ward C is Nurse X's turf," or a conceptual domain, as in "this problem is cardiology's turf." It can also connote responsibility such as "let's turf this patient to psychiatry" which means "let's find a psychiatrist to assume responsibility of this patient whom we can't figure out." Using the term "turf" to describe the home implied a clear recognition that this was the patient's domain.

Although the novice status of the students may account for their anxiety regarding functioning as physicians in the home and for their doubts about their professional efficacy (which we will discuss below), it does not account for this acknowledgment of the patient's "turf," for the faculty physicians also recognized the patients' "turf" and experienced a loss of control. A better explanation for this recognition of the patient's "turf" and the consequences for the relationship which followed from it is quite simple. In the home the patient was the host. The rules and expectations of the deep-seated and excessively familiar guest/host relationship came to dominate and structure the interaction. In the patient's home the implicit power of the physician's social and professional prestige is limited by rules governing the guest-host relationship. Those students who initially violated these rules by, for example, moving about the patient's home without asking permission, or rearranging furniture or medicine, changed their behavior as the definition of the situation and their own lack of control became clear. Students, who in previous professions (for example, as a social worker or a reporter) were accustomed to entering people's homes while in a professional capacity, experienced the least sense of loss of control.

The guest/host dynamic can be thought of as the social expression of the territoriality and power expressed in the concept "turf." As such it helps explain the tenacious power of doctor-centered or doctor-dominated relationships in the clinic where, in addition to all the other more subtle operative factors enumerated above, the physician is also the host and the patient, the guest.

IMPLICATIONS OF HOME CARE FOR THE PHYSICIAN-PATIENT RELATIONSHIP

The world's population is rapidly aging. With this demographic transition comes an increase in chronic illness. Although most physicians are unlikely to undertake home care of patients in the near future because of financial constraints, the experiences of these home-visiting physicians and medical students are significant for the insights they provide in the care of the chronically ill.

The increased information which the students obtained in the home had several positive results. For some it made chronic illness more "intellectually interesting," thus dispelling the boredom commonly associated with the care of the chronically ill. For many, it allowed them to develop a more complex understanding of the patient, the family and the struggle they faced in carrying on life burdened by a severely debilitating illness. A greater emotional intimacy developed with the patients and their families than was possible in the clinic. This proved personally satisfying for the students. These relationships with patients were marked by a clear mutual negotiation of goals, and appreciation and respect for the patient's "agenda." Thus mutuality was in part a reflection of the more egalitarian guest-host relationship and of the greater appreciation for the patient as another person which the student gained in the home. In the closing interview, Sam, a 26 year old future oncologist, put it this way:

Home Care is a whole different ball game. You're in their home, on their turf and it's difficult for you to dictate to them . . . the result is if you want to accomplish anything you are forced to develop a more significant type of rapport with the patient. You can't just get away with the surface gloss that you frequently get away with in the in-patient setting or even in a clinic.

Not only did students accommodate patient's goals, many came to question the appropriateness of their own goals. Pamela, a future pediatrician, described this change in her final interview.

I learned to ask myself, "why do we want to do this test?" or "what are our goals?" and that sort of thing. Rarely are goals discussed in a hospital setting, it's clear, it's tacit, and that wasn't the case here. I really had to think about it: this person has had this disease for so long; they have had these treatments, and what is realistic? And what is in their best interest, psychologically as well as medically?

Some students came to modify their treatment plans to incorporate the needs and expectations of the patient. This in turn could lead to improved compliance. After one visit, Lisa, a future gynecologist, commented:

When you deal with patients on their own terms they seem more compliant, more willing to listen, the interaction seems more complete in the sense that you can more freely engage with the social as well as the medical aspects [of the patient].

Although the emergence of a more mutual relationship was surprisingly satisfying to over half of the students, as the case of Diane demonstrates, it could be a problematic course. Four students became unwilling or unable to assert their professional role. There was a tendency for a nihilism about medicine to emerge in which they focused only on the problems the patient identified. In conversations following separate home visits, two students remarked:

I became less inclined to do something intrusive, to say, "would you take off your clothes so I can examine your testicles and do a rectal," even though both my patients needed it. Instead I began focusing on problems they identified as important (Len, a future internist).

When I first met Mr. S. I was most concerned with the COPD. Now I am less concerned with his Alupent inhalers and his Tributalene than I am about getting his dentures fixed and getting Mrs. S. to stop putting salt in his food (Josh, a future surgeon).

"Medical nihilism" refers to the students' assessment that medicine had little effective role to play in aleviating the suffering of the chronically ill and thus it seemed more appropriate to intervene medically. Many patients suffered severe constraints on their lives from the secondary effects of the illness -- loneliness, poverty, depression, lack of physical stimulation, boredom, social conflict. It appeared to the medical nihilists that a more significant impact could be made by intervening in the social and psychological aspects of the patient's life.

What most of the students (even those who did not adopt a nihilistic stance) had difficulty with, and here novice status becomes most apparent, was balancing the medical and non-medical interventions. The faculty physicians were able to just "shoot the breeze" with patients if the situation demanded it, while still remaining "on top" of the patient's medical problems. They could do this because they were convinced of the therapeutic value of apparently non-medical interventions like "shooting the breeze" and were confident of their professional expertise in caring for the patient. In contrast, the students were not comfortable with the balancing. A small number of students failed to balance at all and nihilism about medicine or rigid adherence to the medical model resulted. The majority of students engaged in both medical and non-medical interventions yet many felt that they were not acting as physicians. This particular problem is one which more exposure to appropriate role models and more experience in medical care should ameliorate. The deeper tensions in the physician-patient relationship in chronic illness do not readily lend themselves to solutions.

The value of understanding the "patient as a person", the "whole patient," is clear in medicine and in particular for chronic illness. Such understanding enhances accuracy in assessing patient problems and improves patient participation in treatment regimens. For physicians it enhances their intellectual interest, emotional satisfaction and often the degree of involvement.

Understanding the whole patient, however, seems close to impossible in the traditional hospital/clinic context. As this study indirectly shows, one of the functions of the medical context is to restrict and control for certain types of information and physician-patient relationships, that is, those that best meet the demands of the biomedical model. Treating the "whole person" is essentially impossible in the clinic/hospital context. In fact, physicians are not expected to take on this responsibility in daily practice.

It is likely then that the ideology of "treat the whole person" will remain only an ideal rather than a reality so long as medical training and practice depend exclusively on the clinic/hospital setting. The results obtained from this study indicate that a real commitment to understanding the non-biological aspects of disease and illness can be fostered through the encouragement of medical training and treatment in everyday environments.

NOTES

1. See also Sankar 1986.
2. In the current American medical system these patients would fall into the category of "sub-acute." This new category has emerged as a result of the prospective reimbursement system which offers a financial incentive for the early discharge of patients. The majority of the patients on the Home Care program were, as the staff put it, "stably unstable."
3. The research on which this work is based was conducted over fourteen months between 1980 and 1981. During this period I attended all staff meetings and accompanied four teams of each clerkship on their regular home visits and on emergency visits whenever possible and observed two consecutive medical directors on numerous home visits. Throughout the research I observed roughly the same group of patients. The composition of this group changed when a patient died or was discharged from the program and new patients entered. Teams were chosen by their assignments to these patients. The eighteen medical students were interviewed at length before and after

the eight week clerkship.

The initial research design had been to conduct semi-structured interviews with students before and after the clerkship and to observe a sample of home visits. This design was changed during the pilot study as it soon became apparent that the students I was observing were undergoing an intense and sometimes troubling experience. In changing the design to capture this experience, I retained the interview schedule because the broad questions it contained (e.g. "describe your expectations for this clerkship") would allow me to capture the wide range of individual experience I was observing. I drastically altered the observation strategy and instead of conducting only a sample of home visits, I accompanied the selected teams on all home care visits. I soon discovered that the conversations with the students in the car prior to and following each visit were very valuable sources of information. Although I took notes on interactions in the patients' home, so as to not be obtrusive, in most cases I tape recorded and later transcribed these car conversations.

REFERENCES

Becker, H., et al.
    1961   Boys in White: Student Culture in Medical School. Chicago: University of Chicago Press.

Cartwright, A.
    1964   Human Relations and Hospital Care. London: Routledge & Kegan Paul.

Cicourel, A.
    1981   Language and Medicine. In C.A. Ferguson and S.B. Health, (eds.), Language in the U.S.A. Pp. 407-429. New York: Cambridge University Press.

Eisdorfer, C.
    1981   Care of the Aged: The Barriers of Tradition. Annals of Internal Medicine 99:256-260.

Emerson, J.
    1970   Behavior in Private Places: Sustaining Definitions of Reality in Gynecological Examinations. In H.P. Dreitzel (ed.), Recent Sociology 2. Pp. 74-97. New York: MacMillan Co.

Fox, R.
    1957   Training for Uncertainty. In R. Merton, G. Reader and P.L. Kendall (eds.), The Student Physician. Pp. 207-241. Cambridge, MA: Harvard University Press.

Freidson, E.
    1961   Patient Views of Medical Practice: A Study of Subscribers to a Pre-Paid

Medical Plan in the Bronx. New York: Russell Sage.

1970 Profession of Medicine: A Study of the Sociology of Applied Knowledge. New York: Dodd, Mead, and Co.

Gaines, A.D.

1982 The Twice-Born: "Christian Psychiatry" and Christian Psychiatrists. Culture, Medicine and Psychiatry 6(3):305-524.

Good, B. and M.-J. Delvecchio Good

1981 The Meaning of Symptoms: A Cultural Hermeneutic Model for Clinical Practice. In L. Eisenberg and A. Kleinman (eds.), The Relevance of Social Science for Medicine. Pp. 165-196. Dordrecht, Holland: D. Reidel Publishing Co.

Hahn, R.A.

1985 A World of Internal Medicine: Portrait of an Internist. In R.A. Hahn and A.D. Gaines (eds.), Physicians of Western Medicine. Pp. 51-11. Dordrecht, Holland: D. Reidel Publishing Co.

Hauser, S.T.

1981 Physician-Patient Relationships. In E. Mishler, et al. (eds.), Social Contexts of Health, Illness and Patient Care. Pp. 104-141. Cambridge: Cambridge University Press.

Keefe, F.J.

1982 Behavior Assessment and Treatment of Chronic Pain. Journal of Counseling and Clinical Psychology 50(6):896-911.

Kleinman, A.

1980 Patients and Healers in the Context of Culture. Berkeley: University of California Press.

Kleinman, A., L. Eisenberg and B. Good

1978 Culture, Illness and Care. Annals of Internal Medicine 88:251-258.

Lock, M.

1985 Models and Practice in Medicine: Medicine as Syndrome or Life Transition? In R.A. Hahn and A.D. Gaines (eds.), Physicians of Western Medicine. Pp. 115-139. Dordrecht, Holland: D. Reidel Publishing Co.

Merton, R. et al. (eds.)

1957 The Student Physician. Cambridge, MA: Harvard University Press.

Mishler, E.C.

1985 The Discourse of Medicine: Dialectics of Medical Interviews. Norwood, NJ: Ablex Publishing Co.

Paget, M.

1982 Your Son is Cured Now: You May Take Him Home. Culture, Medicine and Psychiatry 6(3):237 259.

Parsons, T.

1951 The Social System. Glencoe, IL.: Free Press.

Sankar, A.

1986 Out of the Clinic into the Home: Control and Patient-Physician Commu-

nication. Social Science and Medicine 22(9):973-982.
Sternback, R.A.
    1974   Pain, Patients, Traits, and Treatment. New York: Academic Press.
Taussig, M.
    1980   Reification and the Consciousness of the Patient. Social Science and
           Medicine 14B:3-13.
Turner J.A. and L.R. Chapman
    1982   Psychological Interventions for Chronic Pain: A Critical Review (Parts I
           and II). Pain 12:1-21; 23-46.
Yelin, E., et al.
    1980   Toward an Epidemiology of Work Disability. Milbank Memorial Fund
           Quarterly 58(3):386-414.
Young, A.
    1982   The Anthropologies of Illness and Sickness. Annual Review of Anthro-
           pology 11:257-285.

PAUL ATKINSON

DISCOURSE, DESCRIPTIONS AND DIAGNOSES: REPRODUCING NORMAL MEDICINE

INTRODUCTION

This paper deals with some specific aspects of medical students' acquisition of competence in contemporary medical discourse. It is focused on the transmission of clinical methods in the course of "bedside teaching," and is based on data collected in a British medical school in the early 1970s. Many aspects of that work have been reported elsewhere (see Atkinson 1976, 1981a,b,c) and no attempt will be made to account for the totality of socialization in the medical school.[1] The specific theme of this chapter is an examination of some of the strategies whereby students are coached to recognize and to describe the manifestations of "disease." This is enacted in the course of small-group discussions at or near the patient's bedside and in exercises in history-taking and diagnosis.

Commentators on medicine and medical knowledge have repeatedly acknowledged that the phenomena of health and ill-health are "socially constructed." The perspectives of sociology, anthropology, history, philosophy and psychology variously attest to this. The recognition of specific manifestations as evidence of "disease" is a matter of cultural conventions and not of invariant biological phenomena. The assemblage of such evidence into the diagnosis of named disease categories or labels is equally a matter of cultural particularity rather than natural universals (Hahn and Gaines 1985; Bury 1986; Kleinman 1980, 1981; Mishler et al. 1981; Wright and Treacher 1982).

There have been a substantial number of studies of medical education in the United States, the United Kingdom, and elsewhere. While they have in their various ways purported to tell us something of "how medical students become doctors," they all tell us remarkably little about the transmission and assimilation of medical knowledge. There has been so consistent a preoccupation with the latent functions and unintended consequences (the "hidden curriculum") of medical training that the "manifest curriculum" is paid relatively little attention. Paradoxically, therefore, we have studies of medical education with little or no mention of medical knowledge. (For a sustained discussion of this aspect of the sociology of professional socialization, see Atkinson 1983.)

No attempt is made here to account for the entire range of medical

M. Lock and D. R. Gordon (eds.), Biomedicine Examined, 179–204.
© 1988 by Kluwer Academic Publishers.

understanding and decision-making; nor is this paper intended to imply that we can simply explain medical science and practice in general by reference to its acquisition in training. "Medicine" is not a unitary corpus and is not internalized *in toto* in the medical school, only to be applied unaltered in subsequent practice. Nevertheless, there are major features of clinical medicine which are encountered early in a medical student's career and which establish a potent set of norms, expectations and frameworks of understanding.

The early encounters with clinical medicine -- at the patient's bedside, in hospital clinics, in operating theatres and so on -- are fundamental experiences in the medical student's personal and intellectual career. In this context the medical student is presented with vividly dramatic and firsthand representations of clinical method. Here the underlying assumptions and procedures of clinical reasoning are enacted. From the outset, the medical student is incorporated into the discourse of contemporary medicine.

The general features of such bedside encounters in the medical school have a long history. The instruction of medical apprentices as they "walked the wards" is time-honoured. But bedside instruction has an especially privileged and important role to play in the reproduction of modern "biomedicine." By this latter term is meant an approach which conceptualizes the task of medicine in the following terms: it is reductionist in form, seeking explanations of dysfunction in invariant biological structures and processes; it privileges such explanations at the expense of social, cultural and biographical explanations. In its clinical mode, this dominant model of medical reasoning implies: that diseases exist as distinct entities; that those entities are revealed through the inspection of "signs" and "symptoms"; that the individual patient is a more or less passive site of disease manifestation; that diseases are to be understood as categorical departures or deviations from "normality." (For discussions of "biomedicine" as a cultural system, see: Hahn and Kleinman 1983; Mishler et al. 1981, among many others.)

While this is by no means the only approach to clinical medicine that is recognized and practised by contemporary "orthodox" physicians, it continues to hold a dominant position. Certainly in the context of major hospital-based specialties such as internal medicine and surgery it is paradigmatic; it is "normal" medicine in that sense. It is not necessarily taught in terms of an explicit paradigm, however, but is implicit in the taken-for-granted occupational culture. Medical students encounter it through exemplars and assimilate it through the accumulation of firsthand "experience." It is thus incorporated into the "practical reasoning" of the novice and rapidly becomes part of his or her stock of knowledge and assumptions.

The purpose of this paper is not to account for all the many ways in which medical students are inducted into the paradigm of medical orthodoxy. Rather, it addresses a limited, but central, set of concerns -- in particular, the organization of *discourse* in the framing of clinical understanding and experience.

TESTING AND REPRODUCING CLINICAL KNOWLEDGE

The first issue to be considered rests on a type of teaching talk which is by no means unique to clinical instruction, but which is very characteristic of it. This is a type of "Socratic" cross-questioning whereby the teaching physician or surgeon elicits from students -- either individually or as a group -- lists of relevant signs, symptoms, differential diagnoses and so on. This may be done at the bedside of a patient, or elsewhere: after visiting the patient in person the clinician and students may retire to another room, or cluster in a corridor to discuss the findings, the diagnosis and their implications.

The discourse has a distinctive three-part structure, the elements of which are commonly referred to as Elicitation (by the teacher), Response (from the student) and Evaluation (from the teacher) (Sinclair and Coulthard 1975; Stubbs 1983; McHoul 1978; Mehan 1978; Edwards 1980; Atkinson 1981c). Sequences of this type generate chains of talk in which turns alternate regularly between teacher and one or more students.

While teachers may be said to *impart* knowledge in many ways, these "Socratic" sequences are constructed primarily in terms of two complementary functions or effects. In the first place, they are used for the *testing* of students' knowledge. Teachers expect to elicit prior knowledge from the students they question in this fashion. The teacher's "questions" in this and similar contexts have been described as "pseudo-questions" (Stubbs 1983). While they have an interrogative form, they are not requests for information in the sense that most run-of-the-mill questions are.

The functions of the teacher's elicitations and evaluations are somewhat specialized and context-dependent. They are used to discover whether students are in possession of the relevant information and/or can draw appropriate inferences from information that is at hand. If responses are judged inadequate or are not forthcoming, then the teacher may treat it as an occasion for "repair" and instruct students in the correct knowledge. In this catechism of question and answer, individual students may be "put on the spot." Ignorance or lack of confidence will be exposed to the gaze of the teacher and of fellow students. Thus the three-part format of public teaching talk is a very effective vehicle for the

display and legitimation of the teacher's authority and superior knowledge. This teaching talk achieves yet more. It furnishes a particularly
useful way not only for eliciting, but also for *organizing* the knowledge.
Students' responses and the teacher's elicitations collectively order the
elements of knowledge into lists which may further be ordered according
to criteria of relevance, importance or contrast.

In many instances, the teaching exchanges are used to construct brief
typologies of biological and clinical "facts." The following examples are
characteristic of this pedagogical-cum-organizational strategy. The first
demonstrates the collective generation of a *list* of factors. It is taken
from a discussion of the post-operative management of surgical patients.
The surgeon had introduced the notion that losses of fluids and electro-
lytes must be monitored and corrected. He asked the students for sites
of fluid loss:

Student A : Urine.
Dr. :   Yes. (Writes on blackboard). How much?
St.A   A litre and a half.
Dr. :   (Writes on blackboard).
St.B   Skin, sweating.
Dr. :   How much?
St.B :   Up to two litres.
Dr. :   In the tropics, yes.
                (Student laughter)
St.C :   A litre.
Dr. :   (Writes on blackboard: 900 ml.) Any other sites of fluid loss?
St.D :   The site of the operation.
Dr. :   Under NORMAL circumstances.
St.B :   Vomiting.
Dr. :   Under NORMAL circumstances.
St.E :   Stool.
Dr. :   In the stools. How much is in the stools?
St.E :   About a hundred mils.
Dr. :   A hundred to two hundred mils. (Writes on blackboard)

The surgeon then added up the total on the blackboard, concluding that one
needed to give the patient about two and a half litres.

                                            (Surgery notes)

Two features are immediately discernible in this sequence. The surgeon
elicits a list in a stepwise fashion; that is, comprised of volunteered
responses from a series of students. The list is treated by him as a

complete one, when he adds up the total fluid loss. Further, it is constructed on the basis of a contrast between "normal" fluid loss and abnormal circumstances.

The *list* of parts or attributes, and the use of *contrasts* are major elements of the discourse of medical teaching. Together, they are used to construct the semantics of "normal biomedicine." The pervasive use of contrast here recalls Smith's detailed discussion of *contrast structures* in the construction of plausible factual accounts (Smith 1978). Commenting on a lay account ascribing mental illness, Smith documents how the account formulates descriptions of behavior in such a way that the hearer/reader can find in them evidence for anomaly: "the first part of the contrast structure finds the instructions which select the categories of fitting behaviour; the second part shows the behavior which did not fit." In the same way, the contrast structures of normal biomedicine provide "instructions" for the identification of "normal" values and observations, and for finding deviation from them. They thus provide expression for one of the fundamental tenets of contemporary biomedicine: that disease resides in the unequivocal deviation from universal biological normality. Such presuppositions are deeply, pre-consciously embedded in bedside discourse.

The following example shows how the listing generates another "naming of parts." It is taken from later in the discussion of fluid balance in the management of surgical cases. The surgeon had moved on to discuss the fluid absorption and output of the gut. On the blackboard he sketched a diagram of the alimentary tract. He began by asking one of the students:

Dr. :	What is the input of secretions into that?
St.A :	Saliva.
Dr. :	How much?
St.A :	About fifteen hundred mils.
Dr. :	(Writes on blackboard.) Which is the next organ to secrete into the alimentary tract?
St.B :	Stomach.
Dr. :	How much?
St.B :	Two litres.
Dr. :	(Writes on blackboard)
St.C :	Pancreas.
Dr.	How much?
St.C :	About fifteen hundred mils.
Dr. :	(Writes on blackboard) Bile? Anybody seen a patient with a T tube? (pause) No?
St.B :	A litre?

Dr. :   About a litre, yes. And the next?

St.D :   Duodenum.

Dr. :   Yes, that secretes a little bit. But let's take the small bowel as a whole. How much? It may have been written out of the physiology textbooks since I read them. A lot — about three litres. Have you ever heard of succus entericus? I'll call it small bowel secretion.

(Surgery notes)

Again, the surgeon added up a grand total from his annotations on the blackboard. It came to about nine litres. He explained that it is mostly re-absorbed by the time it reaches the ileum:

Dr. :   Have any of you seen a patient with an ileostomy?

There is no reply. The surgeon goes on to explain briefly what an ileostomy is and what it is used for.

Dr. :   We really think there's something wrong if an ileostomy patient is putting out more than about a litre.

Here, then, the "normal" values are supplemented by reference to a "normal" type of patient. The naming of parts thus includes anatomical sites, physiological values and patient types. Their interrelationships are coded implicitly in the organization of doctor-student interaction.

The following examples display some of the ways in which the semantics are built up in the course of these exchanges. The first shows a part-to-whole, or general-to-specific relationship:

Dr. :   Where does bilirubin come from?

St. :   Blood.

Dr. :   What part of blood?

St. :   Haemoglobin.

The next two extracts, from the same surgical teaching episode, show how definitions are generated through the local application of oppositions and contrasts. The surgeon and the students had briefly inspected a patient's post-operative wound. It had appeared to be rather pink and puffy. The students had interpreted this as signs of infection. After this brief inspection of the patient the surgeon embarked on a question-and-answer exchange on infections in general. This was conducted at the patient's bedside.

Dr. :   What are the signs of infection?

St. :   The wound being red and swollen.
Dr. :   Those are the LOCAL signs.
St. :   There's pyrexia.
Dr. :   Pyrexia, tachycardia . . .

(Surgery notes)

Dr. :   What is the source of wound infection?
St. :   From [one's] own bacteria.
Dr. :   Which is?
St. :   Self-infection.
Dr. :   Or may be?
St. :   From the outside?
Dr. :   Which is?
St. :   Exogenous infection.

(Surgery notes)

Here, therefore, perceptions and definitions are built out of local taxonomic arrangements. Infection, for instance, can thus be classified according to generic and specific categories (local/systemic; exogenous/endogenous). Students do not necessarily learn huge all-inclusive taxonomies reminiscent of eighteenth century taxonomic nosologies. Nevertheless, the logic of clinical medicine is implicitly reproduced through partial, shallow taxonomic devices (such as those reproduced above) that are embedded in the discourse of clinical instruction.

The construction of teaching talk episodes may readily facilitate a discursive movement away from a particular patient in order to "fill in" alternatives and possibilities. Again, the underlying logic is often one of taxonomic categorization. In terms of pedagogic strategy this is often accomplished by the introduction of *hypothetical* questions and possibilities. This is one aspect of a yet more general feature of clinical instruction: a shift in the discourse from the individualized patient towards decontextualized medical knowledge, based upon types of patient and categories of condition. There is thus a transition in the teaching talk. A sequence of questions and answers can be initiated by the doctor in terms of, "What if . . .?" "What are the other causes of . . .?" Shifts of this sort may well coincide with a physical shift from the patient's immediate bedside to another room or elsewhere on the wards. The following exemplify shifts and expansion of this sort:

Dr. :   What would you immediately think of if you saw a man of Mr. R's age in hospital? . . .

Dr. :   Yes, if she had been a middle aged man, like Mr. S., what would she have presented with? . . .

Dr. :   While he's doing that [i.e. examining the patient] let's go round and talk
             about the reasons for scrotal swellings . . .

(Surgery notes)

The next extract shows how the teaching clinician expands the frame
of reference. Following a presentation of an elderly female patient, the
physician went on to question the students about a range of possible reasons for a condition such as hers. This was conducted in a separate
teaching room:

Dr. :   What sort of thing, in younger age groups? Any prodisposing factors?
St. :   Hypertension.
Dr. :   Anything else?
St. :   An aneurism.
Dr. :   She in fact had a lumbar puncture - she had no blood there. What else
             would predispose?
St. :   Arteriovenous (inaudible)
Dr. :   Arteriovenous?
St. :   Fistulae.
Dr. :   I don't think I've ever heard of that.

(Medicine notes)

Note how in these examples the teaching doctor hypothetically varies
some of the patients' personal characteristics -- age and gender. The
implicit logic of systematic contrast between types and categories is
invoked and exemplified.

It is worth noting in this context that hypothetical shifts of this sort
can create interpretative problems for the students (and, occasionally, for
their teachers). Ambiguity can arise as to precisely *which* frame of reference is "in play" at any given time -- the individual patient and his/her
problem or the broader frame of generic typologies. (On the problematic
character of "framing" medical encounters, see Evans et al. 1986.)

The individual patient is thus located within a discursive framework --
a sort of semantic space -- in which persons, signs and symptoms, and
the *differentia specifica* of disease categories or labels are arrayed in relationships of similarity and contrast. The medical students thus enter into
the collaborative reproduction of this semantic system. The shared talk
at the bedside or in the teaching room organizes a joint display of clinical
reasoning. It is, indeed, a very powerful means for the reconstruction
and transmission of "normal biomedicine."

These principles are also discernible in the students' initiation into bedside history-taking and diagnosis. Hitherto the discussion has concentrated on doctor-student interaction and the question-and-answer exchanges between them. Clinical instruction importantly includes working more directly with patients. In the course of bedside teaching encounters the junior medical students are required to observe and "notice" *signs* and to elicit *symptoms*. These do not present themselves self-evidently to the student -- nor indeed to the experienced practitioner. They are recognized and organized in terms of the interpretative schemes and frameworks of biomedical knowledge. Signs and symptoms are gathered and classified in terms of disease entities and typologies.

Noticing and Eliciting: Signs and Symptoms

In the course of bedside teaching the physician or surgeon prompts, monitors and reformulates the students' observation of patients and their talk with them. The doctor intervenes to direct the students' exercises in history-taking, and to propose interpretative frames. For one of the most pressing problems for the novice medical student is this: he or she often lacks the appropriate repertoire of interpretative frames in which to locate possible manifestations of disease. Indeed, in the absence of such a repertoire, it may prove very difficult to decide what will "count" as relevant information; the student may find it hard to determine the actual or potential significance of his or her "findings." Likewise, the student may find difficulty in eliciting patient responses which "fit" and illuminate appropriate diagnostic routines. Bedside teaching episodes can thus become complex social encounters as students and patients search for frames of reference, and teachers attempt to steer the interaction in the "correct" directions.

Some of the complexities of bedside teaching discourse can now be explored in rather more detail. The following case captures a number of features: the elicitation and recognition of clinically "noticeable" phenomena by the students; the use of *contrast structures* and the use of disease categories in ordering the inquiry. Throughout, the relative incompetence of the novice students is manifest: in fact, the episode occurred very early in their first clinical year.

The consultant physician begins by introducing the patient and inviting one of the students to take a history.

Dr. :	This is Mr. M. I wonder if you would like to start, Mr. R., by taking a history from him.
St. :	Er er hello.
Pt. :	Hello.

St. :   I wonder if you could tell us in your own words what it is that's brought you into hospital this time.

Pt. :   Well. For over a year I've been bothered with pains in my stomach and I went to my doctor complaining about it and he sent me up to the (Hospital) about six months ago for an X-ray on my stomach. I went up and it showed a small ulcer on my stomach and ever since that I've been bothered with the pain in my stomach and it's got, seemed to get worse ever since. And I -- it -- the only time it really bothers me is before my meals. I get a pain across my stomach and when I do eat it seems to settle it, and it comes and goes at each meal time.

St. :   I see. Anything else?

Pt. :   Er, well not real er.

St. :   I see.

Pt. :   I take er the doctor gave me pills and medicine for this as well, and that seemed to settle it as well. But it's mostly before meals that I get this pain and when I do eat meals it goes away again.

St. :   Huhum. How long ago was it that you first felt this pain?

Pt. :   Well it would be about a year anyway, about a year when I first got the pain.

St. :   Could you describe this pain to us?

Pt. :   Well, it's confined just to my stomach. It's a short stabbing pain, that's all, nothing exciting or anything like that, just a short stabbing pain across my stomach.

St. :   Mhm, mhm. How long does it last?

Pt. :   About two or three minutes, a short stabbing pain.

St. :   Is there anything that makes it go away?

Pt. :   Well, eating. Eating and taking the medicine seems to settle my stomach altogether, and then it will gradually go away again, like supper time the pain starts to come again.

St. :   Do you ever get it at any other time apart from before meals?

Pt. :   No, I've never any bother at all, just before meals.

St. :   And has it been getting worse?

Pt. :   Yes, it has been getting worse.

Dr. :   Okay, fair enough. Now I would like you, in turn, to ask one question each, trying to get further into his history, and I think it is only fair to say that so far you have not elicited all the main symptoms. What other questions are you going to ask? You know, this is not the diffuse interrogation of what we have now got. And speak up.

St. :   How long is it before your meal that you feel the pain, Mr. M.?

Pt. :   Roughly, about thirty minutes before my meal the pains come and they seem to get worse and worse before my meals and after my meal it starts to go away. But before my meals the pain is there.

St. :   And how many meals in a day do you have?

Pt. :	Three main meals a day, breakfast, dinner and tea.
St. :	And do you get it before each of them?
Pt. :	Yes.
St. :	Can you tell us whether or not you have this pain at night?
Pt. :	No it never bothers me at night at all.
St. :	And you have never got up at night with this pain?
Pt. :	Never at all.
Dr. :	Now, can I come back to Mr. M. Is there any other question you would like to ask, Miss J? You've obviously made a provisional diagnosis. You have been told that you have an ulcer. You've got symptoms that you can attribute to an ulcer. Any other questions you would want to ask specifically about your ulcer?
St. :	Can I ask you if there was anything that happened a year ago which might have brought on this, or anything else associated with this pain?
Pt. :	I can't think of anything now. Nothing physical or anything like that. It just seemed to come naturally or something like that, you know. It just came. I had no symptoms at all.
St. :	Is there anything else that you feel, symptoms that you get with this pain?
Pt. :	No it's just the pain I feel. That's all, nothing else.
Dr. :	Is that actually strictly true. You know, is there anything, I think the question really is, is there anything which is happening recently?
Pt. :	Well, apart from the pain I seem to have been drinking, lot of water, milk, things like that. Because of this, I seem to go to the toilet a lot more than I used to do . . . That's over a period of three to four months that this has happened.
St. :	Why have you been drinking a lot?
Pt. :	Well, I seem to get a dryness in my throat and round about my mouth and the drink it seems to help it.
St. :	How long has this been going on for?
Pt. :	The drinking, you mean?
St. :	Having to drink?
Pt. :	About three to four months anyway.
St. :	How much would you say you're having to drink now?
Pt. :	Well I drink quite a lot during the day.
St. :	Can you give us any sort of rough idea of what you're drinking?
Pt. :	Well I'm (pause) sometimes I used to drink two or three bottles of juice a day 'n, 'n, at my work I used to drink a pint of milk every day.
St. :	Hm. Hm.
Pt. :	Eh then at teatime it's two or three glasses of water anyway (pause) before I go to my bed and that seems to check the dryness anyway and then during the night I usually keep a glass of milk beside my bed and I'll wake up and I'll usually feel the dryness coming on again and I'll

	drink that.
St. :	Ye-es.
Pt. :	That seems to help it.
St. :	Huhum.
Dr. :	There is one other symptom er which you initially mentioned. Now I wonder if on this story if there are there specific questions you'd like to ask? Er, here's a chap who's complaining of thirst and polyuria -- are there any other specific questions rather than just generally that you want to ask? (Pause)
St. :	Had you been taking any drugs aspirin or like this before you er got this pain?
Pt. :	No never no aspirins or nothing like that at all.
St. :	Have you ever taken anything for anything before?
Pt. :	No. Well occasional headache 'n' one or two aspirins, to cure that, that's all.
Dr. :	What were you thinking of?
St. :	That aspirin might have induced this ulceration.
Dr. :	Hummm.. I can think of something else Mr. B.
St. :	Hmm.. Mr. M., could you tell us um if you've ha had any changes in your weight?
Pt. :	Yes.
St. :	In the last few years?
Pt. :	Yes (pause) my weight has decreased er lately.
St. :	Can you tell us when you first noticed this?
Pt. :	Well when I first come for tha' chest X-ray about six months ago I started I lost a lot of weight over the past two or three months I've lost a considerable amount of weight, about two stone anyway in weight I lost.
St. :	How heavy are you now?
Pt. :	Eight stone thirteen.
Dr. :	Okay hold it. He's lost two stone (pause). Now what specific questions, hmm, no before that could you summarise for us what we have so far found out?
St. :	Well Mr. M. for the last year has been having abdominal pain which has been diagnosed clinically as peptic ulcer. Ummm more recently he seems to be having polyuria polydipsia, umm with (pause) weight loss at the same time (pause). Thirst.
Dr. :	Okay, humm (thanks a lot?). Well what specific questions are you going to ask what once you know three lines of attack? You've got the weight loss; you've got the pain; you've got the polyuria. What specific questions are you going to ask? Let's take the weight loss.
St. :	Umm. Can you tell us exactly when the weight loss started with respect of your abdominal pain?
Pt. :	Well, my weight loss started over a period of two to three months ago,

	I've lost the weight.
St. :	Yes and your abdominal pain was ummm
Pt. :	About six months.
St. :	Ten months.
Pt. :	Six months to anyway six months to a year I started getting pain.
Dr. :	I want you to ask questions which make me realise you're thinking of a possible cause of his ab -- er of his weight loss.
St. :	Well, I'm I'm I can think of one but I can't
Dr. :	Well
St. :	Formulate the exact question
Dr. :	Well I want you to.
St. :	Ummm (pause). We I I have to have to go as off on a tangent something that would
Dr. :	Go on then.
St. :	Well have you noticed any change in your breath at all?
Dr. :	What are you thinking of?
St. :	Well I'm thinking perhaps is one open to mention
Dr. :	Go ahead.
St. :	Well I'm thinking of diabetes.
Dr. :	Okay now what specific other questions are you going to ask him with regard to the possibility tha' that he would be diabetic?
St. :	Well. I would want to ask him if he's got any past family history of diabetes.
Dr. :	Ask him.
St. :	Has anyone in your family
Pt. :	Yes I have yeh my father has diabetes and my uncle is
St. :	Huh hum.
Pt. :	Diabetic as well.
Dr. :	Okay, next specific question to pursue this diabetic hare (pause). What d'ya see (pause) when you look at his face? (pause) A beautiful handsome chap? (pause -- Heh humm Hmmm) Anything?
St. :	Several skin lesions.
Dr. :	Yeh we-ell okay. What yer going to ask him?
St. :	Have you always been troubled with er small spots?
Pt. :	Yes I have. Yes.
St. :	On your face?
Pt. :	On my back as well.
St. :	Huhummm, have you noticed them getting worse lately?
Pt. :	Not really er. I remember having these over a period. Lately er I used to have small sort of blotches on ma face.
St. :	Ummmmm
Pt. :	And these seem (pause), but I've still got them on my back and I've got one on my face.

St. :   But you haven't noticed any change in the number that you've had?

Pt. :   No, not er, not recently.

Dr. :   Okay. Anything else? There's a, there's a symptom of glycosuria that patients are aware of.

St. :   Hhhuh.

Dr. :   Now.

St. :   Have you noticed any white marks on your shoes at all Mr. M.?

Pt. :   Yes, I have.

St. :   How long has this been going on? (pause)

Pt. :   Well . . . roughly two -- two to three months anyway 'n I get the occasional one on my trouser leg as well that's . . .

St. :   Huhum.

Pt. :   Y'know as well.

Dr. :   Fair enough. You know I would say that we've probably hit the jackpot (student coughing) but let's pretend we haven't. (St: Heh) And let's go on asking him questions with regard to his weight loss. There are three possibilities. I want you to ask a question which makes me realise you're thinking of them. (Long pause)

By this point in the bedside encounter, then, the physician and students seem to be certain that they have arrived at a consensual understanding of some possible diagnoses. Having started with the patient's history of an ulcer they have added diabetes. For the latter disease attribution they have, as the physician put it, "hit the jackpot."

Overall, these early sequences in the teaching episode are remarkable for the pedagogical work of the physician. On several occasions he "interrupts" the flow of questions and answers between students and patient in order to comment on, reformulate and redirect the talk. It is apparent that in the absence of clearly defined interpretative frameworks, the novice students have difficulty in eliciting or observing pertinent manifestations. For instance, it apparently requires the intervention of the teaching physician for the students to take proper account of the small blemishes on the patient's face. Without an appropriate frame of reference they were, presumably, visible but not "seen": unremarkable features on a young man unless given special significance by the co-presence of further manifestations.

I have written at some length elsewhere that teaching physicians can often direct the encounter not simply by virtue of superior general expertise, but because they already have acquaintance with the individual case (Atkinson 1981b). In this teaching episode the physician intervenes to warn the students that they have not yet elicited all the main symptoms, and so to direct more pointed elicitations from them. Likewise, he intervenes to reformulate a question from one of the students in order to prompt the patient to divulge a "missing" symptom – his thirst. (Without this important clue, of course, the students would be very unlikely to "discover" the diagnosis of diabetes.)

It is noteworthy that the physician's formulations and stage-management of the talk are largely driven towards a "classic" biomedical model. His own prior identification of the patient's problems informs an implicit agenda for the entire teaching episode. The students are to discover a series of disease categories, each of which is unequivocally revealed through the observation of the "right" collection of signs and symptoms. Just as we saw from earlier, brief examples, the teaching is based on the identification of "normal" categories and of "normal," predictable deviations from the normal.

The combination of contrasts and disease categories in ordering the talk is evident in the following extract, taken from a little later in that same teaching episode. The students have continued to explore possible reasons for the young man's weight-loss:

Dr. :   Okay, sure now what're going to ask?

St. :   Have you, (hesit) d'you f'na, em, heat tolerance, do you find that er heat bothers you at all?

Dr. :   Nya, again you see you you've got it the wrong way round, haven't you? The first question you should ask is, y'know, have you noticed any change, you know, in your (hesit) attitude to the weather? Er, if you then say does hea-heat bother you the inference is you would like//

St. :   Ye-es.

Dr. :   You know you would//

St. :   Yes//

Dr. :   The answer should be yes pleasing the camera.

Sts.:   (Various) Huh heh yeh. Hhh.

St. :   Well can I, can I ask you then has your attitude towards the weather changed at all?

Pt. :   No. I wouldn't say that at all.

Dr. :   Alright just roll it down the thyrotoxic pathway. What other questions are you going to ask?
(pause)

St. :   Do you have sweaty palms?

Dr. :   Do you notice INCREASED sweating?

St. :   Have you, have you noticed increased sweating?

Pt. :   No, well ma plams o' ma hand occasionally, but that's all.

St. :   How about generally during the day?

Pt. :   No, nothing like that at all. (pause)

Dr. :   Anything else? (pause)

St. :   Can I ask about (?)

Dr. :   Sure, course you can, I mean it could be very important.

St. :   Do you think you've been more excited than normal? Have you sort of
          found any change in your mood over the past year or so?

Pt. :   No (pause) cheery, plenty to laugh, never moody or anything like that
          er . . . (pause)

St. :   Have you had your usual energy, Mr. Marshall, or have you noticed
          yourself being tired?

Dr. :   Well I do find tha I've a tiredness, yes, and at work I seem to during the
          day I seem to have to take things that little bit easier 'nd when I come
          home, from ma work at night I find that I've been goin' to bed early 'nd
          dr-dropping right off to sleep.

At various points in these and similar interactions we can see the physician engaged in various acts of "metacommunication" -- commenting on and directing the talk of others (the students). Several of the doctor's promptings and the student's ensuing questions to the patient are based on a particular version of contrast structure -- a contrast based on the temporal organization of the patient's perceptions and experiences. The elicitations and their implied responses are ordered in terms of the following general format:

Time 1 ----------> Time 2
Change

As the students take up this line of questioning they collectively build up a series of "findings" based on the patient 's reponses to a string of contrast devices:

. . . Has your attitude to the weather changed at all?

. . . Have you noticed increased sweating?

. . . Do you think you've been more excited than usual?

. . . Have you . . . found any change in your mood? Have you had your
usual energy . . . or have you noticed yourself being tired?

Each of these questions provides the patient with a format for reporting that he had "noticed" a number of sensations. The implied contrasts between Time 1 and Time 2, usual versus unusual, thus provide the possibility for the observed sensations being "newsworthy" in some shared frame of reference between the students, the doctor and the patient.

Despite the strategy involved in the students' questions, it so happens that Mr. M. does not report having "noticed" all the possible changes. It is, however, instructive to look a little more closely at one of the "positive" symptomatic reports from the patient. Reporting his increased "tiredness," he locates the symptom within a biographical framework. Thus he formulates his "tiredness" with reference to the routine activities of work and his daily schedule. In his reply the patient is heard offering an "historical" warrant for his "noticing" and his "finding" of a candidate symptom.

In other words, the student's question proffers a method for organizing and presenting an account of the patient's experiences, specifically in terms of an elementary contrast. The patient elaborates on such a device in the display of his symptom:

. . . I do find that I've a tiredness, yes -- and at work I seem t- during the day I seem to have to take things that little bit easier, 'nd when I come home from ma work at night I find that I've been goin' to bed early 'nd dr-dropping right off to sleep.

This response capitalizes on a contrast-structure in a particularly nice way. Having provided a preliminary reply acknowledging that he experiences "tiredness," the patient goes on to elaborate on that symptom in such a way as to confirm its contrast with normal states of affairs. The account is constructed in terms of linked categories of *time* and *activity*, comparing day/night and work/home. The contrast of day and night is used to show first, the remarkable observation of being tired during the day. The account implies that normality consists of having energy by day and being tired only at the end of the working day. The pairing of "day" and "tiredness" is thus discrepant (within the terms of the account itself) and a disruption of normal states of affairs. Tiredness, the patient's story implies, impinges on the proper work of the day. At the same time "home" and "night" are used to reinforce that noticing: going to bed early and going straight to sleep again contrast with the implied normality which links "tired" with "late" rather than "early." Overall, the contrast-structure implied here is based on the following:

Work	Home
Day	Night
Early	Late

"Tired" should thus "normally" be an attribute of the home/night/late collection of terms. The account presented by the patient thus provides "instructions" for reading it simultaneously as picturing "normal" states and a noticeable deviation from normality.

By means of such discursive strategies, then, the medical students, the physician and the patient collaboratively work at the production of "symptoms." These reflect the organized "noticings" indicative of candidate deviations from normality. The clinical teacher coaches the students in the production of such findings. It is through that coaching that students gain access to methodical procedures for generating biomedical findings.

Signs, symptoms and findings are not self-evident. It is apparent from time to time in this teaching episode -- and in others like it -- that the novice students are not competent at hypothesizing diagnostic frames of reference. On several occasions the physician is found establishing frames of relevance for the students' observations and elicitations. For example, the students are physically able to see the little lesions on the patient's face, but they do not "see" their potential significance unless and until prompted by the physician. Both "signs" and "symptoms" have to be organized into descriptive and discursive frames. In the absence of any organizing perspectives we find the students questioning the patient in a manner their teacher at least judges aimless. He redirects and refocuses their inquiries ("All right, just roll it down the thyrotoxic pathway"; "Next specific question to pursue this diabetic hare").

St. :   'nd how long has this been going on?
Pt. :   Well just lately about month, month and a half. I've found that I'm, that I'm finding a tiredness.
Dr. :   Again this is such a nonspecific symptom that it's not going to help us very much. Um, there are obviously other causes of weight loss you know. I think there's one you should consider, you know, quite seriously. Get it in many, you can get weight loss in many chronic wasting conditions or inflammatory conditions.
St. :   Such as//
St. :   Such as//
Both: Tuberculosis.

In the absence of clinically relevant frames of reference then the inexperienced medical student may have difficulty in "seeing" the most "obvious" of physical manifestations, or of appreciating the potential significance of patients' volunteered and elicited symptoms. This is illustrated in the following section of the paper. It is based on a single incident during which a student "failed" to observe a patient to the satisfaction of a teacher. The failure does, however, help to illustrate further some features of diagnostic reasoning in this context.

The Problem of the Grey Hair

The novice medical student faces numerous problems, or potential problems, of interpretations at the patient's bedside. As noted at the outset of this paper he or she -- like any other student -- has to try to interpret and respond to the teacher's questions. Moreover, in order to do so the medical student may have to question, inspect or examine the patient, searching for a possible answer to the teacher's elicitation (cf. Hammersley 1977; French and MacLure 1979).

This is illustrated in the following case, which encapsulates a number of features already alluded to in this paper. First, it illustrates how medical students must search their "world-within-reach" for a suitable reply to a doctor's question. Secondly, it shows how, in so doing, they must attempt to formulate an adequate "clinical" description. Thirdly, it returns to the notion of contrast structures to illuminate a student strategy for searching and reporting. The case helps to throw these things into relief, by virtue of a student's "failure" to bring off the exercise competently.

The brief interaction was observed in the course of a ward round on a surgical unit. The surgeon was taking a small group of students from bed to bed, pausing for just a few minutes with each patient. While the students were not totally inexperienced in the routines of clinical instruction, they were still very much novices. The patients visited on the round were recent admissions: the students had not encountered them before, and had no prior knowledge of their conditions (beyond the reasonable assumption that they had been admitted as surgical cases). In the course of this teaching round, the consultant surgeon stopped the group at the foot of one bed, turned to one of the students and, without preamble, asked him:

Dr. : What do you observe about this patient?

Before looking at the student's answer and the ensuing exchange, I wish to comment briefly on the nature of this elicitation. It shows, perhaps in an extreme form, that students often have to work at their teachers' questions. A teacher's elicitations may be extremely elliptical and open-ended. The teacher will normally have in mind a range of possible answers, or at least a frame of reference, but may not spell them out explicitly. In order to produce an adequate response to the elicitation the student will often have to grasp the teacher's unspoken assumptions. (As a corollary of this, the student who fails to intuit the frame of reference, or who assumes an inappropriate set of relevances, may produce an answer that appears to be very wide of the mark.)

The surgeon's elicitation here was obviously of this sort. While the student's task was focused on the observation of the patient in question, in principle the task was very broadly defined. It is clear that in the normal run of things there are many possible characterizations that can be offered up as "correct" and "factual" descriptions of a given person. Candidate descriptions could be formulated in terms of a number of social categories, for instance. These categories consist of a number of terms, and a correct identification would depend on the selection of the appropriate term for each category. For example, there is the category *sex* (or gender) with the contrasting terms *male* and *female*. Similarly for *age* we have a more complex (and less clearly defined) collection of terms including: *baby, infant, toddler, child, teenager, adult, middle-aged, senior citizen, old, veteran* and so on. Appropriate terms from such collections can therefore be combined to formulate a social "profile" – such as "an elderly middle-class man," "a young white woman" and so on.

The social classifications referred to here are examples of "Member Categorizations." They can be grouped to comprise what are referred to as *Membership Categorization Devices* (MCDs). By MCDs is meant

any collection of membership categories, containing at least one category, which may be applied to some population containing at least a member, so as to provide, by the rules of application, for the pairing of at least a population member and a categorization device member (Sacks 1972).

Members of populations can always be described by reference to some MCD, and more than one MCD will always be available as a possible description. The crucial question for any social actor will be *what* description will be appropriate for practical purposes on a given social occasion. The notion of categorization employed here does not imply that ascribers engage in a quasi-computational process in arriving at descriptions and diagnoses. Rather, it refers to the resources of practical reasoning whereby actors commonsensically produce plausible descriptions of persons and events. The selection and juxtaposition of MCDs may have considerable practical and moral force in context (cf. Jayyusi 1984; Watson 1978).

If we return now to the surgeon's question we can note that the student could, in principle, respond by employing one or more of these categorization devices. Indeed, such designations are often perfectly proper prefaces to case presentations ("Mrs. X, an Italian lady of forty years of age, who is married with two children, complains of . . ."). But these observations would normally have to be tied to the more newsworthy information concerning the present complaint, the past history, the clinical findings and so on. A description such as a "a young white male" would hardly suggest, in and of itself, a display of skilled clinical observation on the part of the student.

The student, confronted by the requirement to "observe" the patient, must therefore search for some possible description which will be found clinically *relevant*. The student must try not simply to "observe" any old thing about the patient, but search for clues as to why the person might be "a patient" at all. In some cases -- though not this one -- there may be a "foot of the bed" diagnosis to be carried out. That is, there may be one observable sign which is regarded as pathognomic.

In this instance the student's reply was: "He's got grey hair."

How can this be understood, in the light of my remarks on MCDs, as a candidate formulation of the man's "patienthood"? The patient did indeed have grey hair -- a very full head of grey hair. Although his precise chronological age was not revealed, he gave every indication of being a youngish man, in his thirties. It appears that we should treat the student's response as a candidate formulation of "abnormality."

Now, member categorizations imply what may be called Category Bounded Activities (CBAs) (Sacks 1972; Watson 1978). They imply a range of doings that commonsensically seem to be tied to particular categories: they naturally "go together." Category Bounded Activities provide resources for reading off what persons may "typically" be expected to do -- and hence for producing accounts of a normal, predictable and recognisable social world. There are also Category Bound Attributes -- stocks of everyday knowledge about typical characteristics. In general, CBAs may be taken to include both activites and attributes.

The student's suggestion can therefore be heard in this way. "He's got grey hair" refers implicitly to the MCD *age*, where having grey hair is an attribute (CBA) of advanced, or at least later middle age. If the person in question appears from all other criteria to be a young man, then "he's got grey hair" can be heard as a candidate account of "abnormality" by reference to a discrepant co-occurrence of Membership Category and Category-Bound Attribute.

Now it may be objected that having grey hair is not usually a very remarkable medical fact. Such an objection misses the point for this analysis. Given that such a response notes an apparent MC/CBA discrepancy, then the student may offer it as a *possible* observation. Nevertheless, the surgeon was clearly unsympathetic to the student's attempt. He replied, "So have I." This was greeted with laughter from the other students. It came off as a somewhat sarcastic come-back at the student, being uttered in a very "dry" manner. It was clear to all concerned that the student's response had been evaluated as falling very wide of the mark.

The surgeon's reply was a particularly apt put-down. It implied that while he had grey hair there was clearly nothing wrong with *him* (and would a very junior student dare suggest otherwise?). Moreover, since he was a doctor, his normality contrasted rather strikingly with the presumptive abnormality of the patient. The student tried to justify himself by spelling out more clearly the grounds of his reasoning: "No -- I meant he was a young man with grey hair." He reaffirmed the rationality of his perceived discrepancy, but this cut no ice with the surgeon, who went on to seek other responses from the rest of the students.

Although the other students did not volunteer any suggestions which were treated as so manifestly incorrect, none was able to produce the surgeon's preferred answer. Indeed, they proved very reluctant to offer any suggestions at all. In the end the surgeon had to remedy the situation by pointing out himself that the patient was manifestly *in pain*. He was, it was revealed, suffering from a torsion of the testicle.

In practice, when required to "observe" a patient in the manner outlined above, it is not uncommon for students to "miss" noticing that the patient appears to be uncomfortable, or breathless, say. Frequently they search for more specific or "dramatic" manifestations. As in this case they can thus fail to observe anything at all, or offer inappropriate suggestions. In the absence of a more clearly framed set of relevances, they search in vain for their teacher's unspoken agenda. Indeed, we know from other studies of teacher-student interaction that "routes to right answers" can prove very problematic for students (Hammersley 1977; French and Maclure 1979). The "problem of the grey hair" throws into relief the extent to which manifestations of ill-health are not self-evidently given to naive observation, and the extent to which a shared framework of perceptions and descriptions is fundamental to the "social construction" of disease.

CONCLUSION

In this paper, I have tried to show some of the ways in which the identification of "disease" and orthodox biomedical knowledge is socially accomplished. In the course of clinical teaching, "normal medicine" is inscribed in the very use of language. The discourse of bedside teaching is organized in such a way as to produce lists of categories and "facts." By means of devices such as contrast structures, descriptions of normality and deviation are constructed. The "social construction" of disease and illness is not the outcome of individualistic or arbitrary labelling. It is grounded in the shared, collaborative work of teachers, students and indeed patients.

Secondly, the "objective" manifestations of disease ("signs") as well as "symptoms" do not present themselves unambiguously to the clinical gaze of the medical student. There is a relationship of mutual reinforcement between individual manifestations and a presumed Gestalt, or pattern (usually corresponding to a named disease category). The underlying logic which is reproduced in clinical instruction is based on the attribution of such disease categories. The clinical instruction typically proceeds *as if* the process were essentially inductive in nature: as if the "facts" of the case were waiting to be discovered, and as if the disease categories existed independently of the discourse of medicine.

The discourse of bedside teaching is thus a very powerful means for the reproduction of orthodox biomedicine. It reproduces a domain of knowledge in which discrete categories of normality and abnormality reside, and where abnormality is equally represented in categorical terms. The complex practical reasoning of clinical understanding is *mis*represented as an unproblematic task. The prompting and stage-management of the clinical teacher, far from exposing the conventional methods for assembling a diagnosis, reinforce the hidden curriculum of orthodox biomedicine by ensuring the "successful" outcome of the students' inquiries.

It might be tempting to relate these observations directly to studies of clinical "decision-making" among physicians and students (e.g., Elstein et al. 1978; Feinstein 1967). That body of research has attempted to characterize general patterns of information-processing in the solution of medical puzzles and problems. Elstein and his colleagues, for example, generate models of hypothesis formation and testing in the course of diagnostic reasoning. Based on "naturalistic" simulations, they account for decision-making in terms of psychological processes. We cannot, however, map the present observations directly onto such psychological, individualistic information-processing models. In the first place, there is no warrant for assuming that these pedagogical displays and student exercises are homologous with diagnostic processes in more general terms. The practical reasoning involved in a student's solution to a teacher's puzzle cannot be assumed to be identical with, nor a direct precursor of, the competent practitioner's methods. Secondly, the emphasis in this paper has been on the interactive performance of shared descriptions; the study of bedside instruction demonstrates how biomedical phenomena are rendered observable, recognizable and describable. No appeal is made to generalised mentalistic models of "information-processing" and "decision-making." Thirdly, Elstein and his colleagues -- and others writing in the same vein -- retain an essentially normative stance. They identify various sources of "bias" and "error" in physicians' methods of reasoning and inference. I imply no normative evaluation or irony: students and physicians clearly use a diverse array of methods to account for medical events or states. This paper has documented just some of the devices used in achieving those accounts in bedside teaching encounters.

ACKNOWLEDGEMENTS

For their comments on earlier drafts of this chapter I am grateful to Sara Delamont and Ruth Davies. Irene Williams and Martin Read provided invaluable help in retrieving a word-processing disaster. I am grateful to the staff and students of the Edinburgh medical school who allowed me to observe their work. The research would not have been possible without the help of Professor H.J. Walton and Professor A.S. Duncan.

NOTES

1. For major studies which deal with other aspects of medical education, see: Becker et al. 1961; Bosk 1979; Broadhead 1983; Bucher and Stelling 1977; Coombs 1978; Gerber 1983; Light 1980; Merton et al. 1957; Miller 1970; Mumford 1970; Shuval 1980.

REFERENCES

Atkinson, Paul A.
    1976 The Clinical Experience: An Ethnography of Medical Education. Unpublished Ph.D. dissertation, University of Edinburgh. Edinburgh.
    1981a The Clinical Experience: The Construction and Reconstruction of Medical Reality. Aldershot: Gower.
    1981b Time and Cool Patients. In P. Atkinson and C. Heath (eds.), Medical Work: Realities and Routines. Aldershot: Gower.
    1981c Inspecting Classroom Talk. In C. Adelman (ed.), Uttering, Muttering: Collecting, Using and Reporting Talk for Educational Research. London: Grant Macintyre.
    1983 The Reproduction of the Professional Community. In R. Dingwall and P. Lewis (eds.), The Sociology of the Professions: Lawyers, Doctors and Others. London: Macmillan.
Becker, H.S., B. Geer, E.C. Hughes, and A.L. Strauss
    1961 Boys in White: Student Culture in Medical School. Chicago: University of Chicago Press.
Bosk, Charles
    1979 Forgive and Remember. University of Chicago Press.

Broadhead, R.S.
    1983   The Private Lives and Professional Identity of Medical Students. New Brunswick: Transaction Books.

Bucher, Rue and Joan G. Stelling
    1977   Becoming Professional. Beverly Hills: Sage.

Bury, Michael R.
    1986   Social Constructionism and the Development of Medical Sociology. Sociology of Health and Illness 8(2):137-169.

Coombs, R.H.
    1978   Mastering Medicine. New York: Collier Macmillan.

Edwards, A.D.
    1980   Patterns of Power and Authority in Classroom Talk. In P. Woods (ed.), Teacher Strategies: Explorations in the Sociology of the School. Pp. 237-53. London: Croom Helm.

Elstein, A.S., L.S. Shulman and S.A. Sparfka
    1978   Medical Problem Solving: An Analysis of Clinical Reasoning. Cambridge MA: Harvard University Press.

Evans, David A., Marian R. Block, Erwin R. Steinberg and Ann M. Penrose
    1986   Frames and Heuristics in Doctor-Patient Discourse. Social Science and Medicine 22(10):1027-34.

Feinstein, Alvan, R.
    1967   Clinical Judgment. Baltimore: Williams and Wilkins.

French, Peter and Margaret MacLure
    1979   Getting the Right Answer and Getting the Answer Right. Research in Education 22:1-23.

Gerber, L.A.
    1983   Married to their Careers. London: Tavistock.

Hahn, Robert A. and Atwood D. Gaines (eds.)
    1985   Physicians of Western Medicine: Anthropological Approaches to Theory and Practice. Dordrecht, Holland: D. Reidel Pub. Co.

Hahn, Robert A. and Arthur A. Kleinman
    1983   Biomedical Practice and Anthropological Theory: Framework and Directions. In Annual Review of Anthropology. Pp. 305-333. Palo Alto: Annual Review Press.

Hammersley, Martyn
    1977   School Learning: The Cultural Resources Required to Answer a Teacher's Question. In P. Woods and M. Hammersley (eds.), School Experience. Pp. 57-86. London: Croom Helm.

Jayyusi, Lena
    1984   Categorization and the Moral Order. London: Routledge and Kegan Paul.

Kleinman, Arthur A.
    1980   Patients and Healers in the Context of Culture. Berkeley: University of

California Press.

1981 The Meaning Context of Illness and Care. *In* Everett Mendelsohn and Yehuda Elkana (eds.), Science and Cultures: Sociology of the Sciences. Volume V. Pp. 161-176. Dordrecht, Holland: D. Reidel Pub. Co.

Light, Donald
1980 Becoming Psychiatrists. Toronto: Norton.

McHoul, A.
1978 The Organisation of Turns at Formal Talk in the Classroom. Language in Society 7 (2):183-213.

Mehan, H.
1978 Structuring School Structure. Harvard Educational Review 48 (1): 32-64.

Merton, R.K., Reader, G.G. and Kendall, P.L.
1957 The Student Physician. Cambridge, Mass.: Harvard Unversity Press.

Miller, S.J.
1970 Prescription for Leadership. Chicago: Aldine.

Mishler, Elliott, Stuart Hauser, Ramsay Liem, Samuel Osherson and Nancy Waxler
1981 Social Contexts of Health, Illness and Patient Care. Cambridge: Cambridge University Press.

Mumford, Emily
1970 Interns: From Students to Physicians. Cambridge, Mass.: Harvard University Press.

Sacks, Harvey
1972 On the Usability of Conversational Data for Doing Sociology. *In* D. Sudnow (ed.), Studies in Social Interaction. Pp. 31-74. New York: Free Press.

Shuval, Judith T.
1980 Entering Medicine: The Dynamics of Transition. Oxford: Pergamon.

Sinclair, John and Coulthard, Macolm
1975 Towards an Analysis of Discourse: The English of Teachers and Pupils. Cambridge: Cambridge University Press.

Smith, Dorothy
1978 K is Mentally Ill. Sociology 12 (1):23-53.

Stubbs, Michael
1983 Language, Schools and Classrooms (2nd ed.). London: Methuen.

Watson, D.R.
1978 Categorization, Authorization and Blame. Sociology 12 (1):105-113.

Wright, Peter W.G. and Andrew Treacher (eds.)
1982 The Problem of Medical Knowledge: Examining the Social Construction of Medicine. Edinburgh: Edinburgh University Press.

PART IV

MEDICINE EVOLVING, MEDICINE ADAPTING

DAVID ARMSTRONG

SPACE AND TIME IN BRITISH GENERAL PRACTICE

For general practice, time and space appear to be objective and external constraints. Time is a constant limit to practice activity governed by the inexorable ticking of the clock; space too is "given" in that the patient population already spreads itself across a geographical area and surgery premises are contained within pre-existing walls. Certainly there can be adaptations to these constraints, in that catchment areas can be redrawn, surgeries rebuilt or extended and practice routines changed, but space and time still provide, in the experience of the GP, the seeming external backcloth to practice activity.

However, despite the GP's personal experience of space and time as external and objective, there are grounds for examining them as social constructions. A sociological tradition going back to Durkheim's observation that time and space were social in origin (1915), would suggest that the spatial and temporal features of GP's perceptions of their work were themselves integrally bound up with the social organisation and activities of that work. Thus a rebuilt health center or new surgery hours need not simply be seen as alternative solutions to external and "real" temporo-spatial constraints, but more as particular ways of reconstructing the very characteristics of space and time. This paper examines such reconstructions through some recent developments in the spatial and temporal organisation of general practice.

RECONFIGURING SURGERY SPACE

The British GP practices from a physical space, which for historical reasons[1], is called a surgery. The surgery can take many shapes and forms, but, other than to those interested in architecture and building design, it is usually reduced to a non-specific three-dimensional space which encloses the much more profoundly interesting doctor-patient interaction. The surgery however is not simply a space which constitutes the physical backdrop to social interaction: it is also a space with internal and external social boundaries which in their turn are intimately linked to the events which occur within them.

While there would seem to be a large variety of different buildings described as surgeries, three types are identifiable which follow a roughly chronological order. The first is the room in the doctor's own house

M. Lock and D. R. Gordon (eds.), Biomedicine Examined, 207–225.

which was the dominant type during the inter-war years and earlier. Secondly, beginning in the post-war years but rapidly expanding from the 1960s, there is the specifically designed or built separate surgery premises. This type has perhaps its archetypal form in the "health centre" though more common is the "group practice premises."[2] Finally, between the inter-war and health centre models is a type which can be described as "transitional" in that it embodied elements of both; this form is most clearly represented in the surgery constructed as a separate annexe of the GP's own home.

These three types of surgery seem to show a historical development with the inter-war form being replaced by the transitional which in turn gives way to the health centre. Nevertheless this change is a relatively slow process and these three models have co-existed during the last few decades, although the inter-war type is now becoming quite rare. It is therefore possible to take three roughly contemporary accounts of these types of surgery to illustrate their different spatial characteristics.

In his James Mackenzie Lecture of 1957 Dr. David Hughes (1958) described the inter-war practice of an 83 year old doctor which he had joined as a young GP in the mid 1930s. The setting, he presumed, had probably hardly changed in the 58 years that the old doctor had been there. "The surgery was a room leading off from his smoke room; the walls were distempered in a dirty dark red; the floor was of bare boards and the room ill-lit by a small gas jet from his own plant. It contained a desk which was rarely used, half a dozen chairs, on one wall a dresser-like collection of shelves . . . There was no examination couch and no washbasin . . . the old patients adored him and gladly waited for hours -- half a day or longer if necessary, sitting on a stone bench around the pump in the yard, if the weather was fine, or on the chairs in the surgery if wet."

The following year, in 1958, Dr. Handfield-Jones (1958) described the surgery he himself had created on appointment to a single-handed practice some 4 years earlier. He had purchased a house which had two rooms at the back with a separate entrance: these two rooms became the consulting room and the waiting room. "The waiting room is furnished with folding wooden chairs and a table for magazines. The walls are painted with white emulsion paint and the floor covered with light coloured linoleum. There are bright red plastic curtains with a floral design at the windows."

In the consulting room could be found the doctor's desk "placed to get the best light from the two windows . . . Behind the desk is the filing cabinet in which the record cards are kept; it can be reached from the chair. The couch lies along the wall. There is a space between the desk and the couch for the patient to undress, and a curtain on runners screens off this area from ceiling to floor . . . the room is decorated in a restful shade of grey emulsion paint, the floor is of wooden blocks. Two big windows look out on to the garden and have gaily patterned curtains" (1958: 205).

Some 4 years later Dr. Richard (1962) described an early yet fairly typical health centre which had opened in 1955. "The general practitioner unit consists of five suites of rooms, each suite comprising waiting, consulting and examination rooms. At the entrance to the corridor leading to the suites is a well-equipped dressings room with a nurse on duty . . . Each general practitioner's surgery is furnished with a desk, three chairs and all the usual diagnostic instruments . . . the examination rooms are equipped with an examination couch, stool, table for instruments, and a washhand basin . . . The clerical and administrative work is conducted from the small office staffed by a secretary and two clerks, all record cards are filled here" (1962: 256). Such a configuration of space was rapidly to become the model for all future health centre and group practice premises.

How was it that general practice could, within a relatively short period, so radically transform its spatial arrangements? General practice could have remained the same; doubtless Dr. Hughes' old partner in his inter-war practice never dreamed that it could be otherwise. Yet it did change. Contemporary accounts justify the shift from the dark corner of the GP's own home, through the "transitional" arrangements, to the modernity of the health centre in terms of progress, increasing efficiency and improvements in the quality of care. But such justifications, understandable as a Whiggish triumph of the new over the old, simply beg the question. What was the nature of this "Renaissance" in general practice?[3] What were the criteria embedded in the new "progressive" regime which enabled improvement to be identified with such confidence?

A NEW INTERMEDIATE SPACE

Perhaps one of the most striking changes in the spatial organisation of the GP's surgery in the above three accounts is the relationship to the domestic domain. Dr. Hughes' inter-war surgery was an integral part of the old doctor's house: patients gained access through the GP's smoke room and without a couch or washbasin there was probably no means for a stranger to recognise that the ill-lit room was in fact a surgery. Dr. Handfield-Jones' surgery marked a transitional phase: it was still a part of the GP's house, yet by having a separate entrance for patients had become more of an appendage than an integral component. Moreover despite the clear "domestic" affinities in terms of decor and homeliness the two rooms which the patients used were specifically designed as medical and so did not appear to have any alternative domestic uses; in addition, by the use of lobbies and doors marked "Private" a strong physical barrier was erected between medical and domestic space. Dr. Handfield-Jones himself was quite clear on the separation he had achieved: "In these circumstances the practice does not intrude unduly into private life" (1958: 211). In Dr. Richards' health centre, on the other hand, home and surgery were completely separated such that there was not even the possibility of patients crossing the boundary between medical and private space.

Seemingly independent of this change in surgery arrangements, and yet exactly paralleling it, the GP also began to move away from the patient's own domestic space. For example, in the 1950s about one-third of all GP-patient contacts were represented by "visits" by the GP to the patient's home; by the mid-1960s it was down to nearer 20% and most recent reports suggest a current figure closer to 10%.[4]

Until the immediate post-war years in Britain, there had been two distinct spaces of clinical practice, namely the domestic and the hospital. Within this medical order general practice had been a domestic activity taking place in either the GP's or the patient's home. With the post-war growth of health centres and separate practice premises, and with the decline in home visiting, a new space of work was established intermediate between home and hospital. When patient and doctor had met in the home, illness had been subjected to an analysis in its natural context; by contrast the hospital was a neutral space which allowed the true features of illness to show themselves uncontaminated by extraneous domestic influences (Foucault 1963: 109). In similar fashion, in the intermediate space of the health centre a new analysis of illness became possible.

The outline of the new analysis of illness is clearly visible in the novel internal spatial differentiation of the surgery which accompanied the movement from domestic to intermediate space. In Dr. Hughes' inter-war surgery there was both minimal functional differentiation between, on the one hand, the doctor's own house and the room in it which doubled as a surgery, and, on the other hand, within the surgery space itself. To be sure, "around the pump in the yard" seemed to have functioned as an informal waiting area but even this was dependent on the weather as when it rained patients would sit on chairs in the same room in which the old GP would do his consulting. In contrast, Dr. Handfield-Jones separated off a specialised area for "waiting" from the consulting room proper although its provisional status was signified by the temporary seating in the form of folding wooden chairs. In Dr. Richard's health centre, however, space was even more differentiated in that every GP had his own waiting room.

Such functional specificity in spatial distributions can be further illustrated in the working arrangements of Drs. Handfield-Jones and Richard. For example, a whole series of activities in Dr. Handfield-Jones' surgery took place within weakly classified spaces in his consulting room.[5] Patient notes were stored in a record cabinet "within easy reach"; patients were examined on a couch which was temporarily separated from the rest of the room by means of a curtain; dressings and equipment for minor treatments were to be found in a cupboard alongside the drugs which the doctor would himself dispense. In Dr. Richards' health centre each of these activities occurred in a completely different room, if not building.

Differentiation was not only found in physical space in the health centre but also in occupational space. Thus whereas Dr. Handfield-Jones carried out all general practice activities himself, in the health centre those same activities were both physically and occupationally separated. The nurse, doctor, receptionist, pharmacist, etc. had separate quarters and separate functions. The health care team, whose growth paralleled the spatial reordering of general practice, had, in effect, emerged to populate and administer the new intermediate space.[6]

The boundaries and their markers which in Dr. Handfield-Jones practice had mainly served to separate medical from domestic, were now deployed to sub-divide and order the new space. Thus, the patient and the illness had become fragmented into a series of modules. "On entering the centre the patient states his name and address, and the name of his doctor. He then proceeds to the appropriate waiting room and his record card is delivered by one of the office staff through a letter box of the doctor's consulting room" (Richard 1962: 258). The patient would then be called into the consulting room to provide a history, sent into the examination room if an examination was judged necessary and directed to either the nurse in the dressing room or the pharmacist in the chemist's shop for treatment, or the administrative office again if another visit was called for.

In summary the spatial realignments of British general practice in the post-war years subjected illness to a new analysis: this focus involved the separation of illness from the domestic and its subsequent fragmentation. Illness itself was thereby reconstructed as a new phenomenon. The old general practice provided a service to individual familiar bodies: they sat chatting in the yard or on the folding wooden seats while they waited. Illness was located within each of these separate "domestic" bodies. The intermediate space of the health centre however was only used to analyse separate bodies to the extent that it compartmentalised them. The space it mapped onto was not primarily the physically discrete patient's body but a new target in the form of "the community."

It is important to be nominalist about "the community." The

community is not a phenomenon which exists independently of the social perceptions that construct it, even though from the vantage point of the present it is possible to speak of the community in the past. The problem is partly one of disentangling the changing meaning of the word community in the context of its contemporary usage; but also, and especially for general practice, it is one of identifying when the word itself began to be used as a descriptive term on a general basis.[7]

Certainly in the 1950s, when Drs. Hughes, Handfield-Jones and Richard were describing their surgery building, the term community was not in general use, GPs referring rather to their "practice" and its physical location.[8] Even when the word community was used it tended to refer to a specific "domestic" space. Thus for example, in their account of the Peckham Experiment, Pearse and Crocker stated that a community is "a specific organ of the body of society and is formed of living and growing cells -- the homes of which it is comprised" (1943:292). Indeed the more common use of the terms "family doctor" and "domiciliary care" perhaps best reflect the dominant domestic orientation of British general practice in the immediate post-war years.[9] This can be seen for example in the 1963 report of the Gillie Committee on "The field of work of the family doctor" which seldom made reference to the community and when it did so seemed to be using the word simply as a synonym for society (Ministry of Health 1963). Yet by the time of the 1971 report on "The organisation of group practice," the expression "medical care in the community" was used with seeming confidence (Ministry of Health 1971).

This transition in the space of illness is also well illustrated in two national surveys of British general practice. In the first, published in 1967, Cartwright devoted a chapter to "Family and domiciliary care" (1967); in the second survey, published in 1981, the term domiciliary care had disappeared (to be replaced by "home visits"): furthermore it was reported that younger doctors were less likely than older to value the notion of "family care" and more likely to want greater emphasis on "community care" (Cartwright and Anderson 1981). Marinker's claim, of 1976, that "family medicine, far from being of the essence of whole-person medicine, is actually inimical to its practice" (1976: 115), can be seen as a stark manifestation of the new perception.[10]

The notion of the community as an origin, space and therapy for illness was not of course exclusively a fabrication of the new general practice[11] but in this case, its particular morphology -- in the form of a space coterminous with the practice population -- certainly was. The GP's list of patients was not an amorphous substrate in which flashes of illness periodically and randomly showed themselves within the private world of domestic space, but was itself the space of illness. "Each population," noted Scott in his 1964 Mackenzie Lecture, "is in effect a population at risk" (1965: 15). Illness no longer struck unexpectedly at an individual body but haunted an entire population.

In a regime of sporadic disorganised illness each patient was, within their domestic space, a separate incident; under a regime which viewed the community as the space of illness then illness became a calculable and summable risk. Surveys of case-load and morbidity in the 1950s marked the beginnings of this new perception[12]; practice disease indices in the 1960s and the spread of age-sex registers in the 1970s signalled its general extension.[13]

TEMPORALISING PRACTICE ACTIVITY

In a "Coronation Issue" of *The Practititoner* published in 1953 a photograph of a modern waiting room was accompanied by the caption: "the whole scheme is bright, clean and radiates efficiency" (Practitioner 1953: 567). The photograph might well be said to show a bright and clean room but how could a particular spatial arrangement of tables and chairs "radiate efficiency?" The problem was that the photograph could only capture in its two-dimensional form the three-dimensional room: what it could not portray was the reconstructed temporal space which pervaded the apparent solidity of the room and its contents.

"Efficiency devices" and "practical efficiency" were the slogans used to justify and explain the new spatial arrangements of general practice (Arnold and Ware 1953).[14] The constant imperative in the post-war years was "to economise medical time and skill" (Ministry of Health 1963). Work study techniques from industry were advocated as "a method of using time and effort more economically" (Jeans 1965: 279). Consulting, examining, investigating, diagnosis and treatment were all possible in the same room, recorded the Ministry of Health's planning booklet on buildings for general practice, "but it was more efficient to have separate spaces" (Ministry of Health 1967).

In a recent paper on the effect of time pressures on GPs' work, Horobin and McIntosh (1983) have argued that time is socially constructed in relation to practice tasks which have to be performed: the effect is that shortage of time has become a structured feature of practice work. The problem of "pressure of time" has therefore become a major characteristic of modern general practice and indeed, as Huntington (1981) has suggested, has produced a particular time orientation amongst GPs which differs from other occupational groups in the health care system. The discourse on "efficiency," which everywhere seemed to mesh with the spatial realignments of post-war general practice, can be seen in the context of this "pressure of time"; but this modern time orientation was itself

underpinned by a more fundamental reconstruction of the temporal
aspects of practice activity.

The shift from the home-based practice to the health centre marked
the strengthening of the boundary between the GP's own domestic/private
life and general practice work. This barrier between private and public
lives was further consolidated by changes in the temporal organisation of
practice activity. It is apparent from Dr. Hughes' account of his inter-
war practice that time had a different meaning: there were no regular
hours or specific times designated as "off duty" for the GP, and his
patients who sat and gladly waited "perhaps for half a day" to see him
seemed equally oblivious to temporal pressures or boundaries. By Dr.
Handfield-Jones' 1950s transitional practice there had been some explicit
acknowledgement and consequent management of time as a manipulable
variable in that morning and evening surgeries together with specialist
clinics such as ante-natal were accorded specific hours. Yet these hours
only represented a limited organisation of temporal periods in that the
remaining practice occupied the undifferentiated time: "the rest of the
time is available for visiting." Compare this pattern with the now more
familiar one at Dr. Richard's health centre in which a rota system operated to cover half-days, holidays and Sundays.

In Horobin and McIntosh's account of time and routine in a group of
Scottish GPs they note that perception of time pressures and constraints
were mainly found in urban practices; the more rural and remote GPs
"spoke about time in quite different ways" (1983: 312). These latter, predominantly single-handed GPs reported no shortage of time although they
did note the absence of proper off-duty. Horobin and McIntosh attribute
this unhurried style to a deliberate choice on the part of the rural GPs to
take on "a set work roles which were time-consuming." But if, at a more
general level, the last few decades has seen the reconstruction of space
and time within British general practice, then the different time perceptions of some remote, rural, single-handed GPs might also represent the
vestiges of an older regime of general practice.[15]

"Off-duty" time periods, as shown in the spread of rotas and deputising services, have become an integral component of general practice and
the expectations of GPs.[16] In this way "GP" and "non-GP" periods are
identified and separated. Equally the development of a formal 3-year
training period in recent years to replace a system of casual entrance has
also served to erect a temporal barrier to separate "non-GP" from "GP"
activities.[17] Thus the "GP identity" has been effectively separated by
these temporal barriers from both non-GP/hospital and non-GP/private
identities. There is a remarkable congruence here with the new spatial
configuration of general practice. Just as the health centre is an attenu-
ated structure lying between, yet separate from, the hospital and the

domestic, so the GP's own identity *qua* general practitioner, has been effectively demarcated by temporal barriers from previous hospital work and his or her own private/domestic life.

The medical space between hospital and home which has emerged in the new general practice has been broadly delineated by temporal markers but that same space has also been subjected to an intensive temporal analysis. This is perhaps most salient in the rapid spread of appointment systems in recent years: in 1964, for example, Cartwright (1967; Cartwright and Anderson 1981) reported that 15% of patients said their doctor had an appointment system, while in the follow-up study of 1977 this proportion had risen five-fold to 75%. Whereas in the inter-war years Dr. Hughes' patients waited on their stone bench "four hours – half a day or longer if necessary," the health centre is meticulously concerned with the placing of patients both in space (by guiding their passage through a highly differentiated series of rooms) and in time through the temporal distributions of the appointment system.[18]

In the natural domain of the home time had passed unnoticed; in the neutral domain of the hospital time had crept in as an important variable but only in the form of the timetable as a system of ordering staff activity and hospital routines (Zerubavel 1979).[19] In general practice there was certainly an element of pacing introduced by the appointment system but the temporal analysis was not so restricted. Time was not a fixed and linear measure but something which was manipulable; time could be used efficiently or inefficiently; time could be conserved or expended. Thus, the temporal space delineated by the appointment system could be further analysed into "activities" by means of "time and motion" studies (Wood 1962), time spent with patients could be titrated against their time needs[20] or measured in relation to the social characteristics of patient or problem (Westcott 1977; Raynes and Cairns 1980). Perhaps the Balint group's *Six minutes for the patient* of 1973 best illustrates the possibilities for extracting more time from an apparently limited consultation period (Balint and Norell 1973).

Undoubtedly the rhetoric of "efficiency" played a large part in this new analysis of time. But to see the changes simply in terms of increasing efficiency would be to miss the fundamental reconstruction of the temporal dimension. Efficiency only became a problem when time became a central concern of post-war general practice. "Time is the most pressing need for any GP," claimed Townsend in 1962, "time to spend with and on his patients and time for reflection, thought and reading" (1962: 501). It was of vital importance, Batten argued in his James Mackenzie Lecture of 1961, to secure time: "time to listen, time to think and to talk" (1961: 5). To that extent post-war general practice was centrally concerned with an economy of time, with a constant balancing of temporal credits and debits. Time could be wasted in establishing a diagnosis or time could be "spent" to allow a diagnosis to emerge; time might be conserved in rapid treatment or it might be allowed to pass as a therapeutic tool in its own right.[21] Time, as Harte (1983) noted, was a working dimension not a linear measure.

ILLNESS AS A TEMPORAL PROBLEM

The traditional pathological lesion was revealed at a single point in time: the diagnosis was an event. True, it had a past history and a prognosis, and the diagnosis may, for some reason, have been delayed but these temporal elements were essentially subsidiary to the immediacy of the lesion. But in the post-war years when general practice became governed by a temporal economy, time became another dimension, independent of the three-dimensional localisation of the lesion, which invested illness.

Illness was not an event but a process whose context and essential nature was contained within a temporal trajectory. Illness became less the momentary revelation of the clinical examination and more the process of becoming ill, or reacting to illness and to treatment, and to becoming well. The ultimate event had been the finality of death: in the new general practice it was successive temporal stages of the process of dying which required negotiation and the expenditure of time.[22] Time became a central attribute of illness; illness was a phenomenon which occupied a temporal space. The emergence therefore of chronic illness as one of the major morbidity problems in post-war general practice marked the crystallisation of both illness and time in a common space.[23]

As the episodic nature of illness was replaced with a temporal characterisation, so the work of general practice was reconceptualised and reorganised. Early studies of morbidity by the Research Committee of the College of General Practitioners noted the difficulty of measuring illness episodes as the end of an episode was impossible to define. The assessment of morbidity in general practice, they concluded, depended on continuing observation (Research Committee 1958). While in the realm of research, illness required continuing observation, in patient management, it demanded continuing care. In 1965 the College, reporting on "Present state and future needs of general practice," pronounced that the GP "provided continuing and long-term care" (Council of the College of General Practitioners 1965).[24]

An ideology of continuity of care however posed two problems. On the one hand it was difficult to explain how such a fundamental component of the GP's task had only recently been "discovered." This problem was tackled by claiming that it was ever-present but that its value had remained implicit: "Continuity in general practice is so important and so all-pervading that paradoxically it tends to be over-looked" and "The significance of continuity only became widely appreciated when general practice was first studied as an independent discipline" (Journal of the Royal College of General Practitioners 1973:749). The second problem was that an ideology of continuity of care had emerged at precisely the point at which care was being fragmented. Group practice and the health care team had ensured that patients and their illnesses were rarely followed through by the same doctor.[25] Pinsent, for example, noted in 1969 that "a stable relationship between patient and doctor is unusual" (1969: 226).[26] The solution to the problem of this fragmentation of illnesses was the medical record.

The opportunity for a record of patient illnesses to be kept had first been introduced for National Health Insurance patients: in 1920 the medical record envelope was introduced and has changed very little to this day. However it would appear that until relatively recently very few records were kept in these envelopes or, if they were, their haphazard nature undermined any claims to continuing observation. Indeed as late as 1978 an editorial in the Journal of the Royal College of General Practitioners confessed that "medical records are the Achilles heel of general practice and reveal the current state of disorganisation bordering in some cases on chaos" (1978: 521).

However, it was not so much that records were bad than that standards by which record keeping were evaluated had been changing. One of the criteria of good practice became a record system adequately filed, every contact recorded, letters stored, summary sheets, regular up-dates, and so on.[27] "Only systematic records," claimed Pinsent, "can replace the series of memories in which a patient's medical history may at present reside, in greater or lesser detail" (1969: 226).

Before records, every patient, every "contact," was a singular event; there may have been a "past history" in the consultation and indeed the doctor might have remembered a significant past occurrence but past and present were different domains of experience. With the record card, however, which marked the temporal relationship of events, time became concatenated. Clinical problems were not simply located in a specific and immediate lesion but in a biography in which the past informed and pervaded the present. The medical record was another device for the manipulation of time.

As a central element of illness began to be seen as a time dimension, medical intervention aimed itself not so much at a specific lesion as at this "temporal space of possibility." This new perception first manifested itself in the notion of early or presymptomatic diagnosis. By 1963 Ashworth, in a paper on presymptomatic diagnosis, could note that "the role

of the GP lies increasingly in the field of preventive medicine is the view
of many leading thinkers" (1963: 71). McWhinney's *The early signs of illness* (1964), Hodgkin's *Towards earlier diagnosis* (1966) and a Symposium
on early diagnosis in 1967 (Council of the College of General Practitioners 1967) all demonstrated this growing emphasis. The role of the GP,
observed the government committee set up to examine the organisation
of general practice, included the detection of the earliest departure from
normal of the individual and families of his population (Standing Medical
Advisory Committee 1971). However, while this approach to prevention
and earlier diagnosis might be labelled as part of "a new approach to disease" (Council of the College of General Practitioners 1965), it could also
be seen, Crombie suggested, as an implicit part of all general practice:
"The cry for more preventive medicine and presymptomatic diagnosis
comes most often from those who have clinical responsibility and who do
not appreciate that practically everything which the general practitioner
does for his patients contains an element of prevention and presymptomatic diagnosis in the widest sense" (1968: 345).

Early diagnosis was a part of a preventive outlook which pushed the
identification of illness or its precursors back in time. But equally, intervention at the earliest stages of illness was justified in terms of protecting
the future. Thus early diagnosis and prevention merged almost impercep-
tibly along the temporal dimension with health education and health promotion. By, as it were, intervening in the past the future could be made
secure because the past and future were directly linked. "We can see
'prevention'," stated the working party on health and prevention in primary care, "as measuring 'care with an eye to the future' or 'anticipatory
care'" (Royal College of General Practitioners 1981).[28]

A TEMPORAL ECONOMY

Some two centuries ago a revolution occurred in medical perception
(Foucault 1963). Illness was seen as coterminous with the body of the
patient and to make that body legible and accessible it was placed in the
"neutral" medical space of the hospital where it was subjected to a three-
dimensional spatial analysis by means of the newly deployed techniques
of physical examination. Of course these changes were justified in contemporary writings in terms of progress, efficiency and enlightenment but
these claims are not the only parallels with the changes in British general
practice documented in this paper.

At the end of the 18th century a physical investigation of the human
body required a new medical space removed from the patient's home. In
similar fashion the new temporal analysis of recent general practice has

demanded a space neither domestic nor hospital. That space has been opened up in the group practice premises or health centre. During the 19th century the study of the body and its illness within the hospital brought about the development of the new skills, instruments and investigations of the physical examination. In the intermediate space of general practice can be found a parallel series of techniques deployed around the patient but on this occasion around a process. The introduction of continuing care and observation, of efficiency in the consultation, of time as a tool, and of a variety of records attest to the inventive vigour of this new perception.

Then there is the relationship between a perception and its objects. In the 19th century and also in the 20th the dominant "medical model," in as much as it identified disease in the three-dimensional space of the human body, has tended to try to reduce all illness to an organic lesion. In the new general practice, as it has shifted from a spatial to a temporal model of illness, a novel set of problems, techniques and possibilities has been identified which increasingly dominate "progressive" general practice thinking. Thus emphasis on a population constantly "at risk," chronic illness, prevention, health promotion, anticipatory care, early diagnosis, and so on, have functioned to celebrate the existence of a temporal trajectory to almost all illness.

With the old dichotomy of home and hospital there was always a domestic space, and a biographical time which escaped clinical surveillance. True, there was the old general practice but that itself was a domestic activity inseparable from the bodies it treated and barely differentiated from lay domestic care. Today there is a new general practice, confident, efficient and humane, concerned about biography, subjectivity and community and yet curiously divorced from the objects of its attention. In effect, just as the domestic body could be made legible in the neutral space of the hospital, so, in the last few decades, it is the domestic biography which has been objectified and scrutinised in the intermediate space and temporal vacuum of the surgery.

ACKNOWLEDGEMENTS

I am grateful to Bill Arney and Gordon Horobin for helpful comments on an earlier draft of this paper.

This paper is based on a contribution to a conference organised by the British Sociological Association and Royal College of General Practitioners held in November, 1983. It was originally published in Social Science and Medicine 20: 659-666, 1985 and is reproduced by kind permission of Pergamon Press Ltd.

NOTES

1. The forerunner of the British GP in the early 19th century was the apothecary surgeon.
2. Health centres and group practice premises tend to differ in who actually owns the building but for the purposes of this paper this difference is unimportant; hereafter the two types of buildings are used synonymously. For the growth of health care centres see Clarke (1972).
3. There seems some agreement among GPs that general practice was transformed for the better after the nadir in the 1950s. See, for example, Hunt (1972).
4. Handfield-Jones (1959) reported a not atypical 1762 visits to 5677 surgery attendances in 1959; Cartwright (1967), in her national survey of 1964, found 22% of contacts were in the home and in the follow-up survey of 1977 (Cartwright and Anderson 1981) this had fallen to 13%.
5. Bernstein (1975) has provided a useful model linking strength of spatial classification with permitted social activity.
6. In the late 1959s there were less than twenty "attached" paramedical staff in the whole country; by 1969 there were several thousand (Watson and Clarke 1972).
7. The word "community" only became a medical subject heading in *Index Medicus* in 1967 when the terms "Community health services" and "Community mental health services" were introduced to replace "Public health."
8. For example, "the practice covers nine villages in an area of two and a half miles radius" (Handfield-Jones 1958: 205). Taylor, in a contemporary classic study defined a key criterion of a GP as looking after "People in a well-defined area" (1954: 546).
9. Taylor (1954), for example, having observed that the Medical Officer of Health also worked within a well-defined area argued that "The GP can claim that his relation to his patients' home is virtually unique" (Taylor 1954: 546).
10. It is perhaps ironic that at the very time British general practice was turning its back on "family medicine," it was becoming the ideological cornerstone of U.S. primary health care. This position is however under challenge (Schwenk and Hughes 1983).
11. For alternative points of its emergence see Armstrong (1983); also Arney and Bergen (1984).
12. The new College of General Practitioners was particularly active in this field through its Research Committee. See, for example, College of General Practitioners (1962).
13. Such data on the practice population is now considered as essential requirement for "good" general practice; an age-sex register, for example, is expected of all practices which take GP trainees.
14. See also: the report of a conference on practiced accommodation, equipment

and management (Journal of the College of General Practitioners 1963); "unless the GP organises his work day he will never get through his work" (Journal of the Royal College of General Practitioners 1969: 67).

15. Further data on Horobin and McIntosh's GP sample also supports a "domestic" orientation of these rural GPs. They categorised their sample into "family doctors" and "primary physicians": all but two of the rural GPs were in the family doctor category (Horobin: personal communication).

16. Witness the recent vehement reaction of GPs to a government proposal to limit the use of deputising services (British Medical Journal 1984: 172).

17. "Curious [that the] College paid so little attention in its first 10 years to the idea of training doctors for general practice and it has only recently begun to think seriously about it" (Journal of the College of General Practitioners 1964: 303). Now, of course, general practice is the only branch of medicine in Britain which has a statutory entry requirement involving a three year vocational training (see Horder and Swift 1979).

18. Space and time are of course inter-related: for example, "the waiting room only needs to be half the size in a practice which works an appointment system than it would need to be in the same practice without an appointment system" (Whitaker 1965: 267).

19. Time in hospital might also of course be organised to structure patient activity (Roth 1963).

20. "Time must be rationed according to the patient's needs" (Hull 1972: 241).

21. The use of the "time scale which is peculiar to general practice" as a diagnostic and therapeutic tool was explicitly made in *The Future General Practitioner* (Royal College of General Practitioners 1972).

22. See, Journal of the Royal College of General Practitioners 1980; also, Arney and Bergen (1984: 101).

23. The Gillie Report (Ministry of Health 1963), for example, had emphasised the importance of chronic illness in the GP's morbidity spectrum. On the emergence of chronic illness as a problem category see Armstrong (1983: 87); Arney and Bergen (1984: 105 and 1983: 1).

24. This identification of general practice with a temporal dimension was also used to separate general practice from hospital medicine: see, for example, Caldwell (1964). It is however, of interest to note that the hospital staff itself was contemporaneously undergoing its own spatial reorganisation as large "Nightingale" wards were replaced by smaller ones and "observability" came to replace "supervision" (Thompson and Goldin 1975).

25. Stott (1984), in recognising this point, differentiates between continuing care and continuity of care: the former represents temporal continuity while the latter means seeing the same doctor (see also Pinsent 1969).

26. See also Aylett (1976).

27. Again, like age-sex registers, well-organised records are indicative of good general practice (Journal of the Royal College of General Practitioners

1984); see also [13] above.

28. *Health and Prevention in Primary Care* (Royal College of General Practitioners 1981) was the first of a series of reports on preventive aspects of general
   practice commissioned by the Council of the Royal College of General Practitioners.

REFERENCES

Armstrong, D.
   1983   Political Anatomy of the Body. Cambridge: Cambridge University
          Press.
Arney, W.R. and B. Bergen
   1983   The Anomaly, the Chronic Patient and the Play of Medical Power.
          Sociology of Health and Illness 5:1-24.
   1984   Medicine and the Management of Living. Chicago: Chicago University
          Press.
Ashworth, H.W.
   1963   An Experiment in Presymptomatic Diagnosis. Journal of the College of
          General Practitioners 6:71-73.
Aylett, M.H.
   1976   Seeing the Same Doctor. Journal of the Royal College of General Prac-
          titioners 26:47-52.
Balint, E. and J.S. Norell, eds.
   1973   Six Minutes for the Patient: Interactions in General Practice Consulta-
          tion. London: Tavistock.
Batten L.W.
   1961   The Medical Adviser. Journal of the College of General Practitioners
          4:5-18.
Bernstein, B.
   1973   Class, Codes and Control, Volume 3. London: Routledge & Kegan
          Paul.
British Medical Journal
   1984   Editorial: Deputising Services: A Serious Blunder. 288:172.
Caldwell, J.R.
   1964   The Management of Inoperable Malignant Disease in General Practice.
          Journal of the College of General Practitioners 8:23-44.
Cartwright, A.
   1967   Patients and Their Doctors. London: Routledge & Kegan Paul.
Cartwright A. and R. Anderson
   1981   General Practice Revisited. London: Tavistock.
Clarke, M.
   1972   Health Centres: An Evaluation of Past and Present Developments.

Update 5:1185-1189.
College of General Practitioners
    1962   Morbidity Statistics from General Practice. London: HMSO.
Council of the College of General Practitioners
    1965   Present State and Future Needs (Reports from General Practice #2). Torguay: The College of General Practitioners.
Crombie, D.L.
    1968   Preventive Medicine and Presymptomatic Diagnosis. Journal of the Royal College of General Practitioners 15:344-351.
Durkheim, Emile
    1915   The Elementary Forms of Religious Life. London: George Allen & Unwin.
Foucault, Michel
    1963   The Birth of the Clinic. London: Tavistock.
Handfield-Jones, R.C.P.
    1958   The Organisation and Administration of a General Practice. Journal of the College of General Practitioners 1:205.
    1959   One Year's Work in a Country Practice. Journal of the College of General Practitioners 2:323-337.
Harte, J.D.
    1973   The Long Approach to General Assessment. Journal of the Royal College of General Practitioners 23:811-814.
Hodgkin, K.
    1966   Towards Earlier Diagnosis. Edinburgh: Livingstone.
Horder, J.P. and G. Swift
    1979   The History of Vocational Training for General Practitioners. Journal of the Royal College of General Practitioners 29:24-32.
Horobin, G. and J. McIntosh
    1983   Time, Risk and Routine in General Practice. Sociology of Health and Illness 5:312-331.
Hughes, D.M.
    1958   25 Years in Country Practice. Journal of the College of General Practitioners 1:5-22.
Hull, F.M.
    1972   Diagnostic Pathways in General Practice. Journal of the Royal College of General Practitioners 22:241-258.
Hunt, J.
    1972   The Renaissance of General Practice. Journal of the Royal College of General Practitioners 22(Suppl. 4):5-20.
Huntington, J.
    1981   Time Orientations in the Collaboration of Social Workers and General Practitioners. Social Science and Medicine 15A:203-210.

Jeans, W.D.
    1967   Work Study in General Practice. Journal of the College of General Practitioners 9:270-279.
Journal of the College of General Practitioners
    1963   Editorial: The Surgery. 6: 1-3.
    1964   Editorial: Training for General Practice. 7:303-304.
    1967   Report of a Symposium on Early Diagnosis. 14: Supplement 2.
Journal of the Royal College of General Practitioners
    1969   Editorial: The Medical Adviser. 17:67-68.
    1973   Editorial: Continuity of Care. 23:749-750.
    1978   Editorial: Medical Records in General Practice. 28:521-522.
    1980   Report of the Working Group on Terminal Care: National Terminal Care Policy. 30:466-471.
    1984   Editorial: Record Requirements. 34:68-69.
Marinker, M.
    1976   The Family in Medicine. Proceedings of the Royal Society of Medicine 69:115-124.
McWhinney, U.R.
    1964   The Early Signs of Illness. London: Pitman.
Ministry of Health
    1963   The Field of Work of the Family Doctor (Gillie Report).
    1967   Buildings for General Medical Practice. London: HMSO.
    1971   The Organisation of Group Practice. London: HMSO.
Pearse I.H. and L.H. Crocker
    1943   The Peckham Experiment. London: George Allen & Unwin.
Pinsent, R.J.F.H.
    1969   Continuing Care in General Practice. Journal of the Royal College of General Practitioners 17:223-226.
Practitioner
    1953   British Medicine: A Pictorial Record. 170:567.
Raynes, N.V. and V. Cairns
    1980   Factors Contributing to the Length of General Practice Consultations. Journal of the Royal College of General Practitioners 30:496-498.
Research Committee
    1958   The Continuing Observation and Recording of Morbidity. Journal of the College of General Practitioners 1:107.
Richard, J.
    1962   The Stranraer Health Centre. Journal of the College of General Practitioners 5:256-264.
Roth, J.A.
    1963   Timetables: Structuring the Passage of Time in Hospital Treatment and other Careers. Indianapolis: Bobbs-Merrill.

Royal College of General Practitioners
  1972   The Future General Practitioners. London.
  1981   Health and Prevention in Primary Care. London.
Schwenk, T.L. and C.C. Hughes
  1983   The Family as Patient in Family Medicine: Rhetoric or Reality? Social Science and Medicine 17:1-16.
Scott, R.
  1965   Medicine in Society. Journal of the College of General Practitioners 9:3-16.
Standing Medical Advisory Committee
  1971   The Organisation of General Practice. London: HMSO.
Stott, N.C.H.
  1984   Primary Health Care. New York: Springer.
Taylor, S.
  1954   Good General Practice. London: Oxford University Press.
Thompson, John D. and G. Goldin
  1975   The Hospital: A Social and Architectural History. New Haven: Yale University Press.
Townsend, E.
  1962   Future Trends in General Practice. Journal of the College of General Practitioners 5:501-547.
Watson, C. and M. Clarke
  1972   Attachment Schemes and Development of the Health Care Team. Update 5:489.
Whitaker, A.J.
  1965   A Study of Purpose Built Group Practice Premises. Journal of the College of General Practitioners 10:265-271.
Westcott, P.
  1977   The Length of Consultations in General Practice. Journal of the College of General Practitioners 27:552-555.
Wood, L.A.C.
  1962   A Time and Motion Study. Journal of the College of General Practitioners 5:379-381.
Zerubavel, E.
  1979   Patterns of Time in Hospital Life. Chicago: Chicago University Press.

ANTHONY WILLIAMS AND MARY BOULTON

THINKING PREVENTION: CONCEPTS AND CONSTRUCTS IN GENERAL PRACTICE

INTRODUCTION

General Practitioners (GPs) often report that it is difficult to "think" preventively in the context of their routine consultations with patients; the immediate complaints of the patient are given priority in a clinical encounter and this is reinforced by a long tradition in which being "responsive" to patients' expressed needs is highly valued. Evidence from recent studies of the content of general practice consultations (Boulton and Williams 1983; Tuckett et al. 1985) shows that prevention does not figure prominently in these transactions, although some have since argued that this overlooks an increasing trend among GPs to raise preventive issues in the consultation (e.g. Stott 1986).[1] These findings are at variance with the current ideology of general medical practice in Britain which places considerable emphasis on the notion of anticipatory care as an integral feature of the service GPs provide to patients, and which portrays general practice as a key locus for such activity in the primary care sector (see Royal College of General Practitioners 1981 and 1983).

From an anthropological perspective, the emergence of a preventive orientation and the conceptual system that underpins it, presents an intriguing area for inquiry since it is intimately connected to aspects of human culture and society that have long been a central concern of anthropologists -- namely, ideas about the causation of illness and the contexts of misfortune from which they arise, and the relationship between the physical and social bodies. However, while much effort has been invested in the examination of non-western etiologies and preventive regimens, relatively little attention has been devoted to these subjects within western cultural settings. This reflects the historical bias in anthropological fields of study, as well as the fact that the re-emergence of a preventive orientation in western health policy is a recent phenomenon. It may also be indicative of the problematic nature of such enquiry, in particular the difficulties of carrying out ethnography "at home" and of gaining access to the members of a professional elite for this kind of research. Nevertheless, as recent analyses show (Hahn and Kleinman 1983; Hahn and Gaines 1985), this field of enquiry offers rich and varied opportunities for the application of anthropological approaches to the analysis of

M. Lock and D. R. Gordon (eds.), Biomedicine Examined, 227–255.

contemporary biomedicine.

For these reasons, the central concern of this chapter is with the concepts of prevention and associated cultural values elicited from a small sample of general practitioners in south east England. This analysis is set in the context of medical views of the general practitioner's responsibility for prevention expressed in medical texts and formal documents, and the interpretations placed upon these views by social scientists. The aim is to promote discussion of the relationship between the "rhetoric" of prevention and the "cultural construction of clinical realities" (Kleinman 1980) among individual practitioners, and the influence of the personal meaning systems of practitioners on these constructs. The theoretical approach that underlies this is essentially a hermeneutic one (Good and Good 1981; Good et al. 1985) which looks at the interplay of "text" and context in relation to the current discourse on prevention, and which draws upon the sociological literature on medical practitioners' use of "typifications" in making sense of clinical work (Horobin and McIntosh 1983; Hughes 1977; Stimson 1976).

MEDICAL VIEWS OF PREVENTION IN GENERAL PRACTICE

Identifying prevention as one of the key responsibilities of the general practitioner is not a particularly recent phenomenon. In 1950 a British Medical Association report envisaged the GP as "specialising" in "continuous and preventive" care and a subsequent report of the association's General Practice Steering Committee strongly argued for these being seen as "unique" and "positive" features of general practice in the future (British Medical Association 1952). Interest in this area gained further momentum in the 1960s with the translation of the notion of continuity of care into a practical concern with early or "anticipatory" diagnosis (McWhinney 1964). Some GPs were already arguing that their main role was essentially a preventive one (Ashworth 1963) and increasingly, general practitioners began to identify key features of their position in the medical services that gave them a particular advantage in mediating preventive medicine to the population. The establishment of practice "lists" of patients and the capitation fee system of renumeration for GPs under the National Health Service (NHS) helped to consolidate these perceptions. By the 1970s a number of authors were pointing to the particular attributes of general practice that made it the "natural setting" for preventive medicine and health education. Among the features most frequently mentioned are that patients consulting the doctor expect advice and view GPs as credible and trustworthy authorities, that their reasons for consulting often render them susceptible to suggestions about health behavior modification and that the GP has access to a large population on a regular basis which, from a preventive viewpoint, can be monitored through time by primary care "teams" (Boulton and Williams 1983; Calnan and Johnson 1983).

Reinforcing this trend has been a growing awareness of epidemiological research into the causes of major acute and chronic disease in which attempts have been made to quantify risk in relation to different groups within a population. Perhaps the most succinct statement of the medical view of the potential for prevention is that by Doll and Peto (1983) which summarises the key evidence for links between mortality and morbidity and "risk factors" (social, environmental and cultural) and indicates the main areas of intervention. The importance of this work in providing general practice with a firmer biomedical sub-stratum on which to build a preventive re-orientation is reflected in one recently expressed view of an "explosion" in knowledge concerning the causes of disease:

Almost every popular lifestyle factor promoted in the mid-1980s was not fully recognised by the scientific community 25 years earlier. The phenomenal pace of change in our understanding of disease pathogenesis and hence the ever increasing list of "risk factors" has been a source of disbelief to many lay people but the trend is showing no signs of slowing down (Stott 1985:3).

Among the influences which led to an emphasis on prevention, one is especially noteworthy: that of the involvement of all British doctors as the subjects of a 20-year research study into the effects of smoking on morbidity (Doll and Peto 1976). The close links this research established between doctors' smoking behaviour and the incidence of lung cancer, cardiovascular disease and bronchitis not only provided further evidence of smoking as a major risk factor, but also may have contributed to the transformation of British doctors into a predominantly non-smoking population upon which a new rhetoric of prevention could be mounted.

A third set of influences that can be detected in the general practice literature are those which arose in response to changes in health policy, both national and international, during the last decade. As Davies (1984) observed, prevention came into an altogether new prominence with the publication, starting in 1976, of a series of official reports from the Department of Health and Social Security on this topic (DHSS 1976, 1977). This concern was later consolidated at an international level by the major switch to prevention and health promotion in WHO thinking (1978). In response to these initiatives, the Royal College of General Practitioners (RCGP) set up its own working party to consider the issues in the context of general practice. Five reports were eventually produced, the first of which looked generally at the role of GPs in prevention (RCGP 1981a) and a further four at specific substantive areas (RCGP 1981b,c,d and 1982). Whilst not constituting "policy" documents in the strictest sense, these reports were an attempt by the College to provide a clear statement of what it believed the field of interest should be in general practice. Three main "arenas of activity" were identified: opportunities arising in consultations with patients; screening and case finding in the practice population; and health education and health promotion undertaken outside of the surgery, but "within the community." Interestingly, most of the subsequent discussion focuses on the consultation, reflecting the individual and family centered approach of general practice and the difficulty many doctor's experience in conceptualising patients as part of a wider social group or network. The authors of the report also acknowledged that the "promotion of health is the part of prevention which is furthest from the doctor's traditional habits and thoughts" and stated that to translate it into practice GPs will need to develop "new ways of "thinking and behaving" (RCGP 1981a:4).

SOCIAL SCIENCE PERSPECTIVES ON PREVENTION IN GENERAL PRACTICE

Social scientists writing on the new preventive ideal have understandably been more concerned to elucidate the social and political origins of this interest. This has resulted in a number of critiques of the move to prevention in health policy in western society generally (for example, Crawford 1977; Labonte and Penfold 1981) and a number of analyses that focus specifically on prevention in British general practice (for example, Davies 1984). The main criticisms that have been aired with respect to the latter are that it is another example of creeping medicalisation, that it is a further attempt by general practice to enhance its professional status in primary health care and that the Royal College of General Practitioners' documents largely conceive of prevention in terms of personal health behaviour which overlooks the influence of social class, gender and ethnic differences.[2] In addition, it has been suggested that the emphasis in the College approach to prevention in the context of the one-to-one consultation may be vitiated by the current trend towards larger group practices and the use of deputising services which it is claimed may increase the impersonality of GPs in the eyes of their patients (Calnan et al. 1986).

A number of analyses have also sought to explain the emergence of the preventive ideal in general practice in terms of the historical structures and processes that underpin the profession's development. One recurrent theme in these writings is the relationship between the emergence of a preventive orientation and the professionalisation of general practice over the past three decades. Jeffreys and Sachs (1983) conclude that interest in prevention was part of a wider response to the ambiguities and uncertainties that general practitioners experienced in the 1950s and 1960s. The crisis of identity they faced, following the creation of the NHS and the consolidation of hospital medicine, prompted a major re-thinking of the GP's role. For a significant proportion of the profession, the future viability of general practice was seen to lie in a "holistic" approach which combined cure and prevention and argued for a change of emphasis from a responsive, demand-orientated service to the prevention of illness and the promotion of health (Jeffreys and Sachs 1983: 242).

The appearance of a new cognitive approach in general practice, of which the preventive focus was a part, is given more explicit and system-atic consideration by Armstrong (1979). What he identifies as "biograph-ical" (and others more loosely refer to as holistic) medicine developed in response to the historical dominance of the hospital-derived paradigms of disease pathology and sickness management (ibid.). This entailed a reconceptualisation of the medical problem, in particular a reassessment of the view that patients frequently consulted GPs for "trivial" com-plaints, and provided the profession with an alternative notion of medical work. A key aspect of this reformulation of the GPs role was a changing perception of time in relation to morbidity which resulted in a shift from an episodic to a continuous view of illness (Armstrong 1979). The grow-ing awareness that GPs were in a privileged position to monitor illness across time and to provide anticipatory care contributed to the rationale for seeing prevention, health education and health promotion as key activities of future general practice.

The central role that Armstrong attributes to the internal cognitive structures of medical knowledge in the genesis of a preventive orientation is not wholly shared by others working in the field. In a recent examina-tion of the approach of the Royal College of General Practioners to pre-vention in relation to children, Davies (1984) challenges the idea that it can be explained purely in terms of seeing professional ideas "as a simple reflection of the interests of professionals," or indeed that professional ideologies and strategies are somehow "entirely concocted from within" (p. 285). Accordingly, the emergence of a "preventive ideal" must also be seen in the context of external influences such as the persistent attempts to constrain the costs of the National Health Service since its formation, particularly in relation to hospital acute services, and a "phil-anthropic" tradition of prevention that developed amongst the middle classes at the end of the 19th century. These latter "lay ideals," she argues, permeated into medical thinking and influenced the current ideol-ogy of prevention in general practice, albeit shorn of the earlier specific focus on improving the health of women and children from the poorer classes (Davies 1984).

Whilst this analysis may underestimate the impact of GPs' contractual arrangements on the way they conceptualise prevention,[3] it points to the influence of lay ideas about prevention on medical thinking and illustrates the degree of overlap and interchange that exists between popular (lay) and professional conceptions of health (Kleinman et al. 1978). However, it begs the question as to why these ideas gained prominence in the first place and as to how they have been assimilated into the current general practice paradigm. For Davies, the explanation can be found in the formation of the NHS which swept away much of the philanthropic ideology with which prevention had become entangled and replaced it with a universalistic system based on equal rights of access to medical care (Davies 1984). But again the question arises as to the nature of the conceptual system that constituted these earlier and more recent ideologies of prevention. In Armstrong's view, the key to this lies in the cognitive transformation of certain aspects of medicine in the early 20th century, in particular changing conceptions of the body, of patienthood, of the nature and locus of disease, and of health:

The old regime (of health) treated the body as an object and attempted to control its relationship to the natural environment by identifying particular aetiological factors and promulgating sanitary laws of health which governed its activities . . . Under the *new hygiene* the natural environment was not of itself dangerous, but merely acted as a reservoir. The danger now arose from people and their points of contact. It was people who carried ill-health from the natural world into the social body and transmitted it within . . . Preventive medicine was therefore no longer restricted to environmental questions and sanitation but became concerned with the minutiae of social life (Armstrong 1983:10-11).

In general practice the particular manifestation of these new cognitive structures arose in the context of a growing awareness of the value of morbidity surveys and records keeping which injected a temporal view of illness susceptibilities in patients and their families into the GPs' work (Armstrong, this volume). All patients could now be seen as being, to differing degrees, "at risk." Whereas previously in the medical view, the presence or absence of symptoms denoted the presence or absence of disease, henceforth disease was increasingly seen as latent in all members of society.

Armstrong's analysis is resonant of a number of themes central to anthropological inquiry, in particular those which conceive of the body as a cultural object mediating between the natural and social orders. As Douglas has observed, the body serves as a symbolic domain through which experience is organised and given meaning in a culture:

The social body constrains the way the physical body is perceived. The physical experience of the body, always modified by the social categories through which it is known, sustains a particular view of society (Douglas 1970:65).

Whilst conceptions of nature vary cross culturally, most societies exhibit interest in the control of the natural and social orders and avoidance of the unpredictable threat each potentially poses to the other. Concepts of bodily control and the formulation of body processes are frequently used to convey ideas about social control and reflect prevailing notions of accountability within society (Douglas 1980). In line with Armstrong's analysis, the focus of accountability in prevention has moved away from the "social hygiene" concepts which sought to control the passage of polluting substances between the body and the natural environment, to the inhalation and ingestion of dangerous substances in the context of social interaction. Implicit in this is a shift in emphasis from the natural to the social causation of disease, although as recent research shows (Pill and Stott 1982), lay concepts of aetiology frequently give salience to the role of the natural order. The gradual shift of the nexus of disease causation and health protection to individuals and personal health behaviour is also one that accords with the values and structure of western capitalist society in which the viability of the individual (as opposed to the group) is stressed and the acceptability of sickness and the sick role is diminished (Douglas 1978). Further evidence of this change in focus is to be found in the semantics of the new preventive rhetoric. As Coreil and her co-authors (1985) have pointed out, the term "lifestyle" (whose origins can be traced back to Weberian sociology and Adlerian psychology) has now become a central concept in western health policy, epidemiology and health education, but has in the process taken on a much narrower set of meanings which minimise the importance of context. Illness has come to be viewed in social terms more as a form of punishment that is often self-inflicted (Crawford 1977).

At this point, however, anthropological interpretations, which generally rest on detailed ethnography, and Foucaultian analyses tend to diverge. Whilst the latter treats these issues largely within the domain of medical thought, in which notions of discourse and power diminish the voice of the individual, anthropological analysis (particularly those in the British and American traditions) have been more concerned to elucidate the "context of use" of these ideas (Shapin 1979:45-46) and their negotiability in respect of social interest and relationships (Bourdieu 1977; Douglas 1978). It also points to the value of viewing notions of health and illness in a wider metaphoric framework which references key aspects of the cultural system. As Comaroff (1982) has observed, illness frequently occasions awareness of deep-seated contradictions in the socio-cultural order. Extending this argument, Crawford (1984) suggests that bodily experience is also structured through the symbolic category of health. It evokes reflections on the quality of physical, emotional, and social existence and provides a means for personal and social evaluation. As he points out, health has become a "generative concept, a value suggestive or attached to other cardinal values" in western societies which is elaborated through several discourses -- professional-medical, political, literary, popular reformist, everyday-personal (Crawford 1984:62-3).

The contribution that an ethnographic approach can make to these issues is illustrated by Crawford's study. Interviews with a sample of adult Americans showed that most individuals expressed two (potentially conflicting) notions of health. On the one hand health was equated with notions of discipline, self control, mastery, and will power which were seen as an essential part of the ability to work and remain productive. On the other hand, health was viewed as a form of release, associated with the ethos of consumption (an equally vital ingredient of American society) which stressed the notion of indulgence and feeling good (Crawford 1984). This points to an inherent contradiction implicit in the new preventive ideal and further illustrates that individual conceptualisations are not simply reflections of an "official" or dominant cultural model. Instead it would appear that these dominant models are part of a wider framework for transforming popular and medical ideas about health into individual systems of meaning. This ethnographic-interpretative approach also had potential for the investigation of individual medical practitioners' conceptions of prevention and their preventive role, since they are also social actors partly enculturated within the popular domain of health and its associated meanings, as well as professionals trained within a more systematised biomedical paradigm.

PRACTITIONER CONCEPTS OF PREVENTION

For a number of reasons, then, internal and external to medicine, general practice has progressively incorporated prevention and health promotion into its domain of clinical responsibility. From the preceding analysis it is clear that this new emphasis has proved controversial and is potentially open to different interpretations. Biographical medicine which gives a central place to prevention is a relatively recent phenomenon and is still evolving. Although it seems to be the dominant ideology of the Royal College of General Practioners, the extent of its acceptance within "field" general practice is probably more limited (Armstrong 1979:6-7; Roland 1986). In spite of this, relatively little attention has been given to the attitudes of individual general practitioners to a greater preventive role or the concepts and models that inform these attitudes.

This may reflect a predominant concern with the patient's view and, in the case of anthropology, may also be a consequence of the tendency to view disease and clinicians' models as unproblematic. Kleinman, for example, contrasts medical explanatory models, based on "a single causal train of scientific logic" with lay models, characterised by vagueness, multiplicity of meanings, frequent changes and lack of sharp boundaries between ideas and experience (1980:105, 107), although he subsequently appears to have revised this interpretation (see Hahn and Kleinman 1983).[4] More recently, however, there has been an attempt among medical anthropologists to introduce a hermeneutic approach (Good and Good 1981) into the study of physicians' clinical realities and explanatory models. One aspect which is emphasised in this approach is the interpretation by the clinicians of a given theoretical or cultural model in terms of their own life histories and personal meaning systems. Thus, clinical models although they appear to be the same, may be linked to "quite different contexts of personal significance" leading to differences in the interpretations placed upon them (Good et al. 1985:195). Examples of how this approach can be developed for different clinical phenomena and settings are to be found in Lock's (1985) study of medical thinking and individual physicians' models of the menopause and Helman's (1985) elucidation of the plurality of medical models, and the interplay between lay and medical interpretations in a single case of a patient with symptoms of angina.

Applying this approach to the study of practitioners' concepts of prevention, however, is not easy since the preventive field is, by definition, wide and not restricted to a particular illness syndrome or affliction. Further distinctions are required which can potentially differentiate between individual models "for reality" and "of reality" (Geertz 1973). One approach to investigating these different levels of social reality is that formulated by Caws, which distinguished between *representational* models which correspond "to the way the individual thinks things are" and *operational* models which correspond "to the way he practically responds or acts" (Caws 1974:3). Whilst this does not entirely resolve the problem of the direction of causality between operational "models" and actions performed (Holy and Stuchlik 1983), it does provide a framework for exploring notions that are abstractions of clinical practice and which reflect wider socio-cultural discourses on prevention. In the context of the rest of this discussion, these are referred to as general practitioners' representational constructs of prevention. This distinguishes this level of thinking from the more situational constructs that can be elicited through direct observation and discussion of individual consultations with patients. As will be seen, there are overlaps between this level of thinking and specific instances of practice; nevertheless it is on the representational knowledge that this analysis tries to focus.[5]

Sample and Interviews

The material presented here is drawn from interviews with 34 experienced general practitioners in the Oxford and South East Thames Regions in England. These doctors are senior members of the groups responsible for postgraduate and continuing education in the two regions and exercise an influence on future trends and developments in general practice. The interviews were based on a standardised non-scheduled questionnaire approach which explored the views of the physicians on the potential for prevention in general practice and the differing constraints they perceived in putting this into effect. Informants were asked about what concepts and meanings they considered the term prevention to cover, and about the kinds of opportunities that arose to practice prevention in their clinical work; how they saw their role in relation to other health professionals, including physicians, what they regarded as the principle aims and objectives of preventive activity, and the kinds of issues and approaches they saw as important. The interviews were conducted in a manner that allowed interactive exchanges and discussion to occur in order that exploration of the underlying conceptual orientations in the context of current general practice could emerge.[6]

Whilst the majority of the informants were familiar with the recent debates about prevention within their own profession, they interpreted their role in quite different ways. By abstracting from their detailed comments it proved possible to identify four groups of practitioners in terms of their different approaches and general orientations to prevention. These groups differed on the basis of moral considerations, as well as their perceptions of medical knowledge and of patients. In the following analysis, the four groups are examined in a sequence which reflects a range of older and more recent traditions in general practice.

(a) Traditional Clinical Orientation

The first category of doctors identified by this analysis -- representing 32% of the sample -- tended to view their role in prevention in *ad hoc* terms, and as a fairly low priority in relation to other clinical activities. Prevention and health education played some part in their work, but were mainly defined as problem-related interventions on smoking and weight.

We have an opportunistic role, one within the consultation, but that will vary according to what the patient has, like with a hypertensive: "lose weight" is helpful. I'm an anti-smoking fanatic myself so that's a bit of health promotion that gets thrown in. Then I suppose immunisation I would define under health promotion, but that's within the practice as a whole rather than an individual consultation (Dr. Thomas).

Explanations were conceptualised solely in terms of improving compliance in order to exercise firmer control over the management of disease and patients:

The problem is compliance. The consultation offers opportunities at quite a simple level, e.g., making sure they understand what you want them to do. We use jargon sometimes and they may not understand what we are trying to say, so the first thing is to make sure my wishes, or what we have decided, is properly understood (Dr. Parsons).

Some doctors said they carried out limited screening exercises, but few were willing to get personally involved in more systematically organised forms of health promotion, particularly when the issues were apparently unrelated to the problem a patient presents. In part, this was attributable to the way they viewed prevention, but in some instances it also reflected a lack of awareness of recent initiatives. For example, two doctors said that they had read none of the Royal College of General Practitioners' documents on prevention nor any of the articles in medical journals which preceded or followed them (see Stott and Davies 1979; Smail 1982). In other cases the lack of a preventive orientation was a more direct consequence of the perceived difficulties in undertaking this role. The pressures of time, uncertainties about the medical evidence to support prevention and the disruptive effect on the doctor-patient relationship were presented as immutable obstacles:

I couldn't do it without considerable alteration in my workload. I've got 2700 patients. The BMA [British Medical Association] are talking about a list of 1700. I consult for 2 hours twice a day. I start at 8:15 and finish at 7:00. It's time-consuming, health education. It's a two-way process. It's not like a sore throat -- "How long?" "Three days." "Open your mouth. We'll give you penicillin." That can be over in 4 or 5 minutes (Dr. Johnston).

I remember having all the middle-aged ladies with diverticular disease on low residue diets throughout the 60s, up to the mid-70s when Burkitt came along with his fibre and everyone with diverticular disease went on to high fibre. You feel a bit of an ass. We're going to find some terrible things that fibre does to you some time (Dr. Johnston).

You've got to give some dignity to what they've come about. Once you've got into paying attention to what they've come about, it is in fact quite difficult to switch to these matters. If they show signs of going fairly rapidly, do I let them go with relief or do I say, "Well, while you're here, I want to reinforce what I said about smoking." Some people say, "Oh, don't start on about smoking again. I can't stand it." I laugh. The relationship is more important than the advice (Dr. Burgess).

The doctors in this category also admitted that they personally found health education dull and boring. For some it conjured up images of endlessly repeating the same tedious formulae to uninterested patients, a prospect which discouraged their making any real attempt at health education:

To the average GP, health education is rather boring. Anything that is routine and not part of the drama of medicine can rapidly become boring. Anything you don't see the results of can become boring (Dr. Lloyd).

Others simply "forgot" health promotion in most consultations, relying instead on the patient's presenting problem to remind them to ask about aspects of his or her lifestyle:

I think it would be appropriate but I just don't remember to do it. I think the reason for not remembering is that I've got other things on my mind as well . . . I need something that *reminds* me. So the patient's problem *reminds* me (Dr. Keating).

In view of these comments, one might suspect that the obstacles described above are more post-hoc rationalizations than stumbling blocks encountered in trying to adopt a new approach. In keeping with this interpretation, the aims for prevention, health promotion, and health education these doctors advanced were less precise than those reasons advanced by doctors advocating other approaches, and indicated a lack of interest in the area rather than a well-worked-out position:

. . . making people feel happy, content and confident . . . the amount of physical illness one can prevent is basically limited (Dr. Hughes).

. . . to help each patient's life to be pleasant, rewarding and worthwhile (Dr. Parsons).

A more active role in health education and prevention was left to others in the practice, particularly the health visitors. It was generally assumed that this was a central feature of their role, though many GPs had little idea of what exactly they did:

Health visitors are far more concerned with health promotion than we are (Dr. Dixon).

I've always felt the health visitor was the king-pin in health education (Dr.

Crawford).

With health visitors as the mainstay in health promotion and prevention, the doctors in the first category felt able to limit their efforts largely to the key advice of "give up smoking" and "lose weight" when their patients presented problems which suggested that it was advisable.

(b) Lifestyle and Screening Orientation

About a quarter (27%) of the doctors interviewed saw prevention mainly as a form of intervention on lifestyle and risk factors to which they brought specific technical and managerial skills. They had a clear idea of its aims, as promoting behaviour change in individual patients and encouraging greater self-care:

Not having people with preventable strokes, congenital rubella effects, heart disease, etc. and that in routine clinical care people should accept responsibility for their own health (Dr. Gray).

In addition, their views of prevention reflected both the way they had evolved such activity as a form of practice policy, including the computerisation of records and comprehensive "prompt" lists, and the importance of personal motivation:

Well, health education. We are very strong on smoking, weight, advising about rules of CHD [coronary heart disease] advice, breast examination and cervical cytology. We have a computerised record system to assist with this and screening and the four of us [partners] are very committed to prevention. We have a very positive attitude to smoking particularly, and all of us spend considerable time with people who smoke and can produce a list of patients who have stopped. There's a good organisational structure that gives us a wide prompt list in consultations (Dr. Gray).

Revealing that their own habits and those of their peers exert an influence, Dr. Gray also commented:

Our attitude to alcohol is improving. It's not uncommon now to find doctors drinking non-alcoholic drinks at some stage in the evening. Exercise is more difficult. The evidence is not so definite. One doesn't know about doctors' exercise habits as one does their smoking, drinking and eating unless you regularly play some form of game or sport (Dr. Gray).

Personal commitment to prevention thus played an important part in the way they viewed health education and in their willingness to become involved with it *alongside* other members of the primary care team. In most practices, the doctors worked closely with both health visitors and practice nurses, involving them in developing practice policy, performing screening procedures and managing cases. A "team" approach was seen as essential for practical as well as professional reasons. Well-organized screening clinics run by nurses and health visitors under the doctor's direction avoided problems of limited time in consultations and of diverting attention away from the patient's problem:

If you raise these sorts of issues, you've got to recognise that a lot of time will be involved. I think we're going to have to consider the involvement of ancillary staff. We've just employed a practice nurse whose doing a number of prevention strategies. She's very involved with obese patients now in a general way. She's beginning to take blood pressures routinely on people who haven't had it. It does depend on having people with enthusiasm (Dr. Lloyd).

Further potential barriers that other practitioners mentioned were simply not seen as major obstacles by the doctors in this group:

Maybe I'm over-zealous [but] I think doctors need to take a moral stand, they are one section of the community that ought to . . . Our job is not only to treat, but also to prevent disease and if our patients are persisting in things that are harmful to health, then I think we are neglectful if we don't try and advise them to correct these things (Dr. Hughes).

We are ideally placed to do prevention and health education because we've built up a rapport with our patients over the years. They trust us and will listen to us. I do it well because I've got severe emphysema from smoking too much years ago. I say, "look at me, I wouldn't be here if I had kept on smoking" (Dr. Richards).

In summary, these doctors saw a clear and unproblematic role for the GP in terms of lifestyle and risk factor intervention. Explanations designed to persuade or motivate a patient, and the role of the patient's health beliefs in this, were seen as important, although only for the preventive issues the doctor had defined on his own agenda. In relation to other aspects of the presenting problem, they were viewed as separate or something the GPs felt uneasy about, either because of uncertainty about the benefit of explanations or because of their impact on the doctor's use of time.

(c) Counselling Orientation

The third category of doctors saw prevention and health education principally in terms of being responsive to the patient's presenting and ongoing problems, and being sensitive to the context in which they arose. In discussing the aims of health promotion, they emphasised increasing patients' coping abilities and promoting a greater sense of control through informed choice:

I hope to help the patient achieve greater control over their lives in the sense that if they know what the possibilities are and they've got information, they can choose to act on it or not. Because I do believe they ultimately have the right to reject everything (Dr. Nicholas).

For them - again approximately a quarter (27%) of the sample -- prevention and health promotion mainly represented explanations that dealt with these issues and helped patients to come to terms with a particular problem:

Health education and prevention relate far more to what we do all the time rather than trying to put over a specific "message" to someone at a specific time . . . it's about a patient coming in with a cold and the doctor doesn't even examine their chest, but takes time to give them a better understanding of what health and illness are and what needs intervention and what doesn't (Dr. Hepburn).

The GPs in this group were not unaware of other aspects of prevention, but in their own consultation work they chose to emphasise the responsive mode which related to the immediate concerns of the patient, rather than imposing a view of longer term risks. This choice was often based on a considered rejection of opportunistic advice on lifestyle and of risk factor screening, arising from a number of reservations about its effects. Dr. Sinclair, for example, wished to avoid projecting an authoritative view of how people should live their lives:

I would be very wary of giving "advice" -- of thinking that I knew what the "correct" advise is -- because I'm not sure what the correct advice is (Dr. Sinclair).

Some doctors were also worried that health education and health promotion conveyed hidden messages which were not altogether desirable. By seeking to increase awareness of health and healthy living, some believed they could also create anxiety, dependence, and a restricted life:

I have a bit of a bee in my bonnet about prevention and health education in that I think that on the whole what you can do is fairly limited and has side effects which you must take into account. For example, I endorse the view that you should check blood pressure and, where necessary, pick people up and treat them, but you pay a price for that. You focus people's attention on themselves as *potential invalids*. If you find raised blood pressure you turn a fit man into a sick man. If you have to treat him, you make him a permanent appendage to your practice . . . The more interference, the more we do to people, the more attached they become to the idea of themselves as wearing out machines needing continuous engineering (Dr. Hayes).

Others believed that health education and prevention were essentially intrusive and moralistic and therefore at odds with the GP's role as morally neutral and technically specific.

There's a tremendous drive to paternalism once one takes up this position. It's justifiable paternalism to try to inflict on a person their own standards. What is very dubious is inflicting on someone one's own standards and we're pretty much in that court with health education. The middle classes succeed by postponing satisfaction. They rein back a bit now for benefits in the future. They're the non-HP types [that is, they do not use charge accounts]. In health education we're really trying to inflict that approach on working-class people (Dr. Ainsworth).

It was also the doctors in this group who conveyed the strongest awareness of social and cultural factors -- ethnic and social class differences, the "cultural" distance between the doctor and many of his patients, multiple deprivation effects in the community and so on -- and a view that pathology exists for many in the very conditions of social existence:

I'm conscious of the Marxist view which is that people smoke because of what society does to them. There is a point here -- you understand after a while why people smoke and drink, without necessarily approving of it (Dr. Hayes).

Health education interventions at the individual level were therefore often seen as both inappropriate and ineffective:

At this stage, all the public is aware that smoking is harmful. A lot of people are smoking from way back and either can't do something about it or don't want to. I'm worried about sending them out with guilt feelings about their smoking which they are still going to continue. I think it's reasonable to discuss it, but by and large it's up to them. I can't bring myself to brow beat my patients . . . What worries me is that it has a *punitive* aspect, a need to control and play God (Dr. Gibson).

The limited screening and preventive activities that were carried out in the practices of these doctors were delegated to others in the primary health care team, who were seen as having different forms of contact and interactions with their patients which made them appropriate to carry out such activities.

(d) Anticipatory Care Orientation

The last category of doctors (15% of the sample) were those who endeavoured to take a broad view of their role in prevention. They saw prevention and health education as relevant to all aspects of presenting and continuing problems, and also as a new service they could offer in identifying risks and discussing the effects of individual lifestyles. They related this to recent developments in approaches to teaching trainee GPs (for example, consultation analysis and communication skills training) and the way they thought general practice should develop. For them, the main aim was one of improving patients' health knowledge and understanding, either as an outcome in its own right or as an essential precursor to other sorts of change:

Helping patients to learn about health and increasing people's health understanding is the most important (Dr. Robertson).

It is also in this group that one encountered a greater familiarity with the Royal College of General Practioners' approaches to prevention:

Every patient one senses is an opportunity. There are "at risk" groups -- antenatal, adolescent smokers, middle aged men and women. And activities -- rubella, hypertension, smears, immunisation. *Then* in every consultation I see opportunities to alter or increase people's health understanding -- about themselves, diseases, problems they've got, tablets they are taking, life they are leading -- not just physical, but mental and social (Dr. Robertson).

They were, however, in a minority, representing less than a sixth of all the doctors interviewed. Perhaps significantly, they identified fewer constraints and barriers interfering with their functioning in this way and many of those mentioned by others were perceived to be surmountable by good practice organisation and by new approaches to the consultation and skills in communication:

I think its fundamental that patients understand what is going on. It helps them to make decisions and may influence their behavior. It shows you care enough to give them information, even if the fact that they are lonely, unemployed or live in a tower block means they may not have the capacity to change their smoking or alcohol consumption (Dr. Selwyn).

You need skills to intervene outside of the presenting complaint and I mean acquired communication skills. It's sometimes quite difficult. For example, when a woman comes in with a cold, to then talk to her about a cervical smear needs a change of gear. You have to discuss it in a way that doesn't make it appear to be a moral crusade (Dr. Young).

Of particular interest here is their view of the status of patients' ideas and the way the doctor responds to them:

I have always felt that explanations are important and finding out what the patient thinks is wrong with them is essential, and then going on and explaining that to the limits of my knowledge whether it is something preventive or routine. I don't think there are areas of medicine which people shouldn't know about (Dr. Elliot).

The importance of developments at an organizational level, however, were also underlined, particularly with respect to the employment of practice nurses with a specific preventive brief and the shared preventive care of patients between doctors, nurses and health visitors:

The practice nurse is very important, she does all the family planning and cervical smears. She's also doing a few blood pressures and is getting involved in a lot of explanation and discussion of patients' ideas about hypertension, about the contraceptive pill and periods. The health visitor has responsibility largely for the children and the elderly, but is also involved in prevention projects with high risk groups (Dr. Robertson).

The practice nurse is wonderful. She does screening and information gathering with all new patients in relation to smoking, weight, blood pressure, diabetes, family history and "social assets," such as family support and income -- or the lack of it (Dr. Selwyn).

As the doctors in the "lifestyle and screening" category also argued, these organisational advances facilitated a proactive approach to health promotion involving systematic screening and comprehensive advice on lifestyle issues, which the doctors felt was potentially effective in changing their patients' behaviour. Health education was once again conceived as a collaborative endeavour, involving various members of the primary health care team. Some of the GPs in this group, however, also pointed to (at least potential) links between the primary health care team and other sectors of society interested in promoting health, including other groups in the district health authority, patient organisations and self-help groups, social services, schools and the government.

We have close links with the health education department. We are meeting with a dietician to talk about some dietary advice we give in the practice. We are preparing health education packages for different groups of people, with the help of a psychologist (Dr. Robertson).

I see one of my roles as a facilitator of contact with different groups, particularly of the voluntary and self-help kind. This is a major area for improvement in general practice (Dr. Young).

For a small minority of doctors, then, prevention was conceived of as a complex enterprise involving many groups and individuals working in different ways and at different levels. The concept of anticipatory care, and the incorporation of some of the more subtle points in the College documents about educating patients into this, generated an approach which integrated the longitudinal perspective of middle and long term disease risks with being responsive to the patient's immediate complaints.

DISCUSSION

The inconclusiveness of research into the causes of major diseases in Britain, particularly in relation to the role of risk factors and appropriate forms of medical intervention, leaves the field of prevention open to widely differing interpretations. The above accounts illustrate this diversity and provide insight into the relationship between the rhetoric of prevention in general practice, as expressed in medical texts, and its conceptualisation among individual medical practitioners. They also demonstrate how the "construction of clinical reality" is shaped in ways that have implications for patient care. The focus of the analysis has been on physicians' representational constructs of the field of prevention which are abstractions of their clinical experience, although they often refer to patients and patient encounters. It is possible that these differences would disappear (or be replaced by new distinctions) if the issues had been investigated in terms of specific patient cases with each doctor. However, a companion study (Health and Prevention Project 1987) of the "operational" models of prevention and health education among GP trainers in the two regions suggests that these conceptualisations are reproduced more widely in the local sub-cultures of general practice, albeit with a greater tendency for a polarisation between those who advocate a lifestyle approach and those who remain sceptical and argue for restricted involvement. Caution should be exercised, however, in extending these findings to general practice as a whole, since those GPs with responsibility for postgraduate and continuing education are often regarded as a new elite which is unrepresentative of grassroots thinking in the profession.

With these qualifications in mind, the different orientations that emerged in the interviews illustrate that general practitioners' concepts of prevention and their views of their role in carrying it out are far from a simple replication of the current medical model. The Royal College of General Practioners documents describe what the field of interest *should be* for GPs (see discussion earlier), but from the proceeding accounts it is apparent that the majority of practitioners recognise only the first arena of activity and many do not approach it in a systematic way. Organised screening was accepted as important by only half the doctors and involvement in health promotion work in the community was beyond the experience and imagination of most of the them. The focus on the consultation and on individuals and their families that is apparent in the Royal

College of General Practitioners documents is even clearer in the inter-views with the GPs and reveals more deep seated ideas about the causes of disease and of accountability in the cultural and social domains. They also suggest that the influences which shape individual practitioners' views of prevention are, in many respects, a reflection of the diversity of views expressed by different interest groups in society more widely.

Reviewing the different orientations in turn, it is clear that those doc-tors who took a traditionalist view of their role in prevention were more conservative in their approach and were wary of innovations. Accord-ingly, they questioned whether general practice was the "natural" setting for prevention and questioned the evidence of "progress" in medical knowledge about risk factors in the aetiology of disease. This was allied to their scepticism about the relevance of prevention and health to clini-cal work, which they conceived of as treating illness. However, they did not criticise the move to prevention in terms of the arguments by social scientists and historians of medicine. Implicit in their conceptualisations is a view of disease as an autonomous biological process which is only marginally influenced by social and cultural factors. Their view of dis-ease aetiology was then largely conceived of in terms of physiopathology, with limited recognition of more complex lines of causation that ramify back into the social order. Modern diseases deemed to be preventable by many of their peers were relegated by them to the natural domain, wherein medical science had little power to predict or control the occur-rence of illness events in individual members of society.

Conversely, those who advocated a lifestyle approach had taken on board the rhetoric of prevention as set out in the Department of Health and Social Security and Royal College of General Practicners documents. They emphasised the characteristics of general practice that made it the most appropriate setting for anticipatory care and justified this with argu-ments that closely resemble those outlined in the general practice litera-ture. Their emphasis on rational organisation and teamwork illustrates the growing influence of a consumerist perspective on their thinking. Several of them viewed general practice as a kind of small business enter-prise which could now offer prevention as a new technical service, with the added incentives that it would largely be carried out by ancillary staff and could be self-financing within current contractual arrangements. For them, the main causes of major disease, such as coronary heart disease and cancers, were firmly located in the realm of the social order. Unhealthy behaviour was deemed to be cultural in origin, but in a way that conceives of culture as "personal baggage" inherited through family practices or acquired through peer group interactions. Thus, encouraging greater personal responsibility for health was a key element of their thinking and reflected a view that much contemporary disease is largely self induced.

Whilst a greater emphasis on patients' understanding of health issues distinguished the anticipatory care group from the lifestylists, in certain respects their approach was a more sophisticated version of the latter. In addition to taking on board the main tenets of the Royal College of General Practioner documents, they had also incorporated some of the more subtle points about educating patients and encouraging "informed choice." They were sensitive to the dangers of victim-blaming in a preventive orientation and were aware of the complex range of factors influencing health behaviour. Nevertheless, the focus of their preventive interventions rested with individual patients, implying that they viewed the principal causes of major diseases as being within the control of the individual. They had few reservations about the evidence for linking health behaviour with major disease, and expressed confidence in the growing ability of medical science to predict and influence the course of disease.

Those who advocated a counselling approach were more aware of and sympathetic to the criticisms levelled against the new preventive ideal by social scientists and others. They were more sceptical of medicine's ability to influence the course of disease in any significant way and focussed their preventive responsibility on helping patients to acquire a greater sense and means of control over their social circumstances. Attempting to modify patients' *health behaviour*, therefore, took a secondary place, especially in view of their scepticism about the effectiveness of such interventions. The aetiology of major disease was for them a complex interaction between the biological organism and the cultural background and socio-economic circumstances of their patients. Unlike other informants who tended to separate illness off from other kinds of misfortune, the doctors in this category saw the former as part of the latter, as a process greatly influenced by life events and social and economic conditions.

This leads briefly to the question of the various formative influences on the doctors' preventive constructs. As Katz's (1985) observational study of American surgeons has shown, non-medical criteria play an important part in decisions about clinical intervention. Lock (1985) similarly identifies a large number of variables (for example, physician personality, age, sex, training, practice population) which influence the production of clinicians' working models of the menopause. The present study did not systematically investigate all of these but we point to three main influences that emerged in the interviews with the GPs and from other data collected on them and their practices. Firstly, many of the GPs were trained and socialised in medicine in the immediate post war "era" and, as an older cohort, had been subjected to images of prevention characteristic of the "old system" of disease causation and personal hygiene (Armstrong 1983:10) which they associated with terms such as health education and health promotion.

Among the younger doctors, most of whom had undergone postgraduate training in general practice and had been subjected to more of the epidemiological evidence on risk factors and disease, these associations were not apparent. Secondly, the doctors who were most sensitive to the cultural context of health and the social and environmental constraints upon behaviour, practiced either in inner city areas amongst predominantly working class populations where ethnic differences, employment problems and multiple deprivation effects are most evident, or in areas whose practice populations included a large number of elderly patients. Conversely, those advocating the lifestyle approach practices in areas containing a high proportion of middle class patients who, in their contacts with doctors, may well convey a greater acceptance of the currently popular social values of "fitness," "healthy lifestyle" and personal control over health (Pill and Stott 1982).

Whilst the social characteristics of patients in the practice populations and the era of professional socialisation were identifiable influences on practitioners' thinking, others can be discerned particularly at the level of regional differences which are attributable to the local networks and ideologies that had developed among the doctors. In the Oxford region, the GP informants were in a course organisers' group that met regularly to discuss issues and to formulate policy. At the time of the study, the energies of this group were devoted to generating a consensus view of policy and strategies for GP training with an emphasis on the preventive role of the family doctor (RCGP 1985). It was also apparent that the Oxford doctors had closer links with the College and identified with its approaches.

Their counterparts in South East Thames Region did not see their role as one of policy making and emphasised the autonomy of individual course organisers in respect of the content of their courses. The GP advisers saw their role in terms of "bringing issues to the attention of the course organisers" and creating a forum for discussion which stimulates people's imagination as opposed to "laying down a prescriptive programme." This non-directive approach was justified in terms of the large number of course organisers and the geographical size and social diversity of the region which generated different local problems and needs. It also proved to be a reflection of a more implicit philosophy among the more senior doctors, several of whom attached great value to a counselling approach in general practice that developed from the psychotherapeutic approaches to consulting (Balint 1957; Balint and Norell 1973) which dominated ideas about professional development in the 1960s and early 1970s. This contrasted with the dominant approach among doctors in the

Oxford region who emphasised a "task" orientated strategy to consulting, in conjunction with acquired communication skills (Pendleton et al. 1984).

CONCLUSION

Just as lay people's perceptions of prevention do not wholly conform to dominant societal models and often show active dissent from them (Crawford 1984), so do practitioners draw upon a range of knowledge, personal and professional values, and clinical experience to interpret their preventive responsibility. The presence of several dissenting voices from the currently dominant risk factor and surveillance approach provides encouraging evidence for those fearful of an uncritical adoption by general practitioners of a highly organic, reductionist view of prevention.

The doctors' perceptions of constraints on practicing prevention, for example, the differing rationalisations about the degree to which time is "externally" imposed, are indicative of the interest and priority individual doctors attach to their preventive function. As Horobin and McIntosh (1983) observe, time is an abiding preoccupation of general practitioners, but research shows it has little relationship to patient list size, consultation rates, or home visits, but varies instead with ideas about work routines, residual (non-serious) diagnoses, and differing perceptions of the risk of a missed or wrong diagnosis. Similarly, the extent to which the doctors perceived constraints on prevention in terms of the characteristics of patients, rests on stereotypes that are more broadly used by doctors to control the consultation, rather than on the elicited views of patients in relation to health and lifestyle (Tuckett et al. 1985). The barriers mentioned in relation to the uncertainty of medical knowledge may also be contextually derived, since uncertainty is a generalised problem for medical practice (Calnan 1984) but appears to be given particular salience in relation to preventive medicine. However, concern about the medicalising effects of prevention -- in particular, its equation by some of the practitioners with a new moral order with attendant risks of iatrogenesis and unwarranted social control -- raises more profound questions about the role of medical practitioners in society. One detects a trend among many of the doctors who were interviewed to take a broader view of prevention and health promotion, than of activities in relation to lifestyle and medical risk factors. Many now see a role in educating patients in general about health and illness, with a view to encouraging them to take a more active and informed part in their own health care. This conception of health education and health promotion, with a greater emphasis on explanation, counselling, and shared understanding between doctor and patient, may prove a more productive direction for the development of prevention in general practice, and one ultimately more acceptable to both general practitioners and patients.

NOTES

1. In this paper the term prevention is used generically to cover the interrelated concepts and activities of preventing disease, promoting health and health education. This is intended purely as a literary convention, rather than being of analytic significance.

2. For a more detailed review of the various criticisms levelled at the preventive approach in general practice, see Calnan, Boulton and Williams (1986).

3. It may also be the case that a preventive orientation which encourages greater personal responsibility for health is attractive to GPs for more pragmatic reasons than those advanced by Davies (1984). The patient capitation system of renumeration for general practitioners has made a "solely demand-responsive approach absurd" and a preventive approach attractive to some as a mechanism for "controlling or reducing patient demand" (Dr. Hayes, informant in this study).

4. An exception to this is to be found in the field of psychiatry. For example, Lazare (1973) argues that clinical psychiatry commonly makes use of four conceptual models -- medical, psychological, behavioural, and social -- in evaluating and treating patients. The particular emphases and biases of the clinician can lead to radically different outcomes for the patient. This view is also evident in the work of Eisenberg (1977), but until recently does not seem to have been generalised to other areas of clinical activity. In a recent review, Atkinson (1987) suggests that the use of these models is more typical of anthropology than sociology, since sociology has for some time treated medical knowledge and practice as problematic.

5. The "operational models" of prevention were explored in the consultations of small sample of GPs under the aegis of the same study. For ethical and practical reasons these could not be the same doctors as those who were interviewed about their representational models. The use of the term "model" may also be misleading in the sense that it can convey the impression of ideas that are formally organised and systematised. Whilst this was evident in the way some of the doctors in the study articulated their views on prevention, others gave a more *ad hoc* set of responses which are less formalised and closer to personal constructs.

6. All the interviews were tape recorded and later transcribed. When individual doctors are quoted, their names have been changed to ensure anonymity.

REFERENCES

Armstrong, D.
   1979   The Emancipation of Biographical Medicine. Social Science and Medicine 131:1-8.
   1983   Political Anatomy of the Body. Cambridge: Cambridge University Press.
Ashworth, H.W.
   1963   An Experiment in Presymptomatic Diagnosis. Journal of the Royal College of General Practitioners, 6, 71.
Atkinson, P.
   1987   Review of R.A. Hahn and A.D. Gaines (eds.) , Physicians of Western Medicine. Social Science and Medicine 24 (19):787.
Balint, M.
   1957   The Doctor, The Patient and His Illness. London: Pitman.
Balint E. and Norell, J.
   1973   Six Minutes For The Patient. London: Tavistock.
Boulton, M. and A. Williams
   1983   Health Education in The General Practice Consultation: Doctor's Advice on Diet, Alcohol and Smoking. Health Education Journal 42 (2):57-63.
Bourdieu, P.
   1977   Outline of a Theory of Practice. Cambridge Studies in Social Anthropology, 19. Cambridge: Cambridge University Press.
British Medical Association
   1950   General Practice and the Training of the General Practitioner. London.
British Medical Association
   1952   A College of General Practitioners, Report of the General Practice Steering Committee. British Medical Journal 2:1321.
Calnan, M. and B. Johnson
   1983   Influencing Health Behaviour: How Significant is the General Practitioner? Health Education Journal 42 (2):34-45.
Calnan, M.
   1984   Clinical Uncertainty: Is It a Problem in the Doctor-Patient Relationship? Sociology of Health and Illness 6:1:75-85.
Calnan, M., M. Boulton and A. Williams
   1986   Health Education and General Practitioners: A Critical Appraisal. In A. Watt and S. Rodmell (eds.), The Politics of Health Education. London: Routledge and Kegan Paul.
Caws, P.
   1974   Operational, Representational and Explanatory Models. American Anthropologist 76:1-10.

Comaroff, J.
    1982   Medicine: Symbol and Ideology. *In* P. Wright and A. Treacher (eds.),
            The Problem of Medical Knowledge. Edinburgh: Edinburgh University
            Press.
Coreil, J., J. Levin and E. Jaco
    1985   Lifestyle -- An Emergent Concept in the Socio-Medical Sciences. Cul-
            ture, Medicine and Psychiatry 9:423-437.
Crawford, R.
    1977   You are Dangerous to Your Health: The Ideology and Politics of Victim
            Blaming. International Journal of Health Services 7:4:663-680.
    1984   A Cultural Account of "Health": Control Release and the Social Body.
            In J.B. McKinlay (eds.), Issues in the Political Economy of Health.
            London: Tavistock.
Davies, C.
    1984   General Practitioners and the Pull of Prevention. Sociology of Health
            and Illness 6:3:267-289.
Department of Health and Social Security (DHSS)
    1976   Prevention and Health: Everybody's Business. London: Her Majesties
            Stationary Office.
    1977   Prevention and Health: Reducing the Risk: Safer Pregnancy and Child-
            birth. London: Her Majesties Stationary Office.
Doll, R. and R. Peto
    1976   Mortality in Relation to Smoking: 20 Years Observations on Male Brit-
            ish Doctors. British Medical Journal 2:1525-1536.
    1983   Prospects for Prevention. British Medical Journal 286:445-452.
Douglas, M.
    1970   Natural Symbols. Harmondsworth, England: Pelican Books.
    1978   Cultural Bias. Occasional Paper No. 35: Royal Anthropological Insti-
            tute of Great Britain and Ireland.
    1980   Evans-Pritchard. Glasgow: Fontana Paperbacks.
Eisenberg, L.
    1977   Disease and Illness: Distinctions Between Professional and Popular
            Ideas of Sickness. Culture, Medicine and Psychiatry 1:9-23.
Geertz, C.
    1973   The Interpretation of Cultures. New York: Basic Books Inc.
Good, B.J. and M.D. Good
    1981   The Meaning of Symptoms: A Cultural Hermenuetic Model for Clinical
            Practice. *In* L. Eisenberg and A. Kleinman (eds.), The Relevance of
            Social Science for Medicine. Dordrecht, Holland: D. Reidel Publishing
            Co.
Good, B.J., H. Herrera, M.D. Good, J. Cooper
    1985   Reflexivity, Countertransference and Clinical Ethnography. *In* R.A.
            Hahn and A.D. Gaines (eds.), Physicians of Western Medicine.

Dordrecht, Holland: D. Reidel Publishing Co.

Hahn, R. and A. Kleinman
  1983   Biomedical Practice and Anthropological Theory. Annual Review of
          Anthropology, 12: 305-333.

Hahn, R.A. and A.D. Gaines (eds.)
  1985   Physicians of Western Medicine. Dordrecht, Holland: D. Reidel Pub-
          lishing Co.

Health and Prevention Project
  1987   Final Report. London: Health Education Council.

Helman, C.
  1985   Disease and Pseudo-Disease: A Case History of Pseudo-Angina. *In*
          R.A. Hahn and A.D. Gaines (eds.), Physicians of Western Medicine.
          Dordrecht, Holland: D. Reidel Publishing Co.

Horobin, G. and J. McIntosh
  1983   Time, Risk and Routine in General Practice. Sociology of Health and
          Illness 5:3:312-331.

Hughes, D.
  1977   Everyday and Medical Knowledge in Categorizing Patients. *In* Ding-
          wall, R., Heath, C., Reid, M., Stacey, M. (eds.), Health Care and
          Health Knowledge. London: Crom Helm.

Jeffreys, M. and Sachs, H.
  1983   Rethinking General Practice: Dilemmas in Primary Medical Care. Lon-
          don: Tavistock.

Katz, P.
  1985   How Surgeons Make Decisions. *In* R.A. Hahn and A.D. Gaines (eds.),
          Physicians of Western Medicine. Dordrecht, Holland: D. Reidel Pub-
          lishing Co.

Kleinman, A., Eisenberg, L., Good, B.
  1978   Culture, Illness and Care. Annals of Internal Medicine 88:251-258.

Kleinman, A.
  1980   Patients and Healers in the Context of Culture. Berkeley: University of
          California Press.

Labonte, R. and S. Penfold
  1981   Canadian Perspectives in Health Promotion: A Critique. Health Educa-
          tion. April: 4-8.

Lazare, A.
  1973   Hidden Conceptual Models in Clinical Psychiatry. New England Jour-
          nal of Medicine 288:345-351.

Lock, M.
  1985   Models and Practice in Medicine: Menopause as a Syndrome or Life
          Transition? *In* R.A. Hahn and A.D. Gaines (eds.), Physicians of West-
          ern Medicine. Dordrecht, Holland: D. Reidel Publishing Co.

McWhinney, I.
    1964 The Early Signs of Illness. London: Pitman.
Pendelton, D., T. Schofield, P. Tate, P. Havelock
    1984 The Consultation: An Approach to Learning and Teaching. Oxford:
          Oxford University Press.
Pill, R. and N. Stott
    1982 Concepts of Illness Causation and Responsibility: Some Preliminary
          Data From a Sample of Working Class Mothers. Social Science and
          Medicine 16:43-52.
Roland, M.
    1986 The Family Doctor: Myth or Reality. Family Practice 3 (3):199-204.
Royal College of General Practioners
    1981a Health and Prevention in Primary Care, 18. London: RCGP Report
          from General Practice.
    1981b Prevention of Arterial Disease in General Practice, 19. London: RCGP
          Report from General Practice.
    1981c Prevention of Psychiatric Disorders in General Practice, Report from
          General Practice, 20. London: RCGP Report from General Practice.
    1981d Family Planning – an Exercise in Preventive Medicine, Report from
          General Practice, 20. London: RCGP Report from General Practice.
    1982 Healthier Children – Thinking Prevention, Report from General Prac-
          tice, 22. London: RCGP Report from General Practice.
    1983 Promoting Preventive Medicine, Occasional Paper, 22. London: RCGP
          Report from General Practice.
Shapin, S.
    1979 Homo Phrenologicus: Anthropological Perspectives on an Historical
          Problem. In B. Barnes and S. Shapin (eds.), Natural Order. London:
          Sage Publications.
Smail, S.
    1982 Opportunities for Prevention: The Consultation. British Medical Jour-
          nal 281:231-5.
Stimson, G.
    1976 General Practitioners, Trouble and Types of Patients. In M. Stacey
          (ed.), The Sociology of the NHS. University of Keele: Sociological
          Review Monograph 22.
Stott, N.C. and Davies, R.H.
    1979 The Exceptional Potential in Each Primary Care Consultation. Journal
          of the Royal College of General Practitioners 29:201-205.
Stott, N.C.
    1983 Primary Health Care. Berlin: Springer Verlag.
    1986 The Potential for Health Education in Primary Health Care. In M.
          Boulton and A. Williams (eds.), Health Education in General Practice.
          London: Health Education Council.

Tuckett, D., Boulton, M., Olson, C. and Williams, C.
  1985   Meetings Between Experts: An Approach to Sharing Ideas in Medical
         Consultations. London: Tavistock.
World Health Organization (WHO)
  1978   Declaration of Alma Ata. Report on the International Conference on
         Primary Health Care, Alma Ata, USSR. Geneva.

DEBORAH R. GORDON

CLINICAL SCIENCE AND CLINICAL EXPERTISE: CHANGING BOUNDARIES BETWEEN ART AND SCIENCE IN MEDICINE

INTRODUCTION

While science may be considered a symbol of legitimacy and source of power for the medical profession, physicians' clinical expertise may be regarded as their personal power and private magic. For years this expertise has been left relatively unchallenged: science moved more and more into the "lab" (laboratory) to develop basic science theory, leaving patient care decisions in the clinic to physician judgement (Starr 1982; Feinstein 1983a).

The terms of this coexistence however are changing. The increasing development of "clinical science" (scientific guidelines for patient care) presents physicians with a serious test of their exclusive personal authority (Armstrong 1977). And a growing number of physicians want to increase this challenge by developing a "basic science of patient care" (Feinstein 1983a; Sackett, Haynes, and Tugwell 1985). Science, they argue, has contributed much to our understanding of disease causation and diagnosis, but has added little to patient care decisions, long considered the art of medicine.

The challenge does not stop here. Not only is there a call for further development of clinical scientific knowledge to guide physicians' clinical decisions, there is also pressure to make the decision and reasoning process of physicians more explicit, formal and quantitative -- more "rational." Once protected by a status of "inexplicable art," clinical judgement is increasingly seen as ordinary, teachable reasoning, something that even computers can or will be able to do. More and more physicians are asked to explain their decisions.

Thus, much as the patient became visible for medicine (Foucault 1975) and continues to be so (Armstrong 1983; Arney and Bergen 1984), physicians are being asked to make themselves and their practice more visible. This phenomenon both reflects and supports the increasing surveillance and monitoring -- both internal and external to medicine -- of physician practice (Starr 1982). Science, while the ally of the profession as a whole, presents a threat to the individual clinician's autonomy.

In this paper I will briefly discuss two dimensions of what I am identifying as a movement to make medical practice more scientific: (1)

257

M. Lock and D. R. Gordon (eds.), Biomedicine Examined, 257–295.
© 1988 by Kluwer Academic Publishers.

efforts to develop more and better "clinical science" for medical practice – a heterogeneous field I will refer to as "clinical epidemiology"[1]; and (2) efforts to make clinical judgement more rational, explicit, quantitative, and formal – also a heterogeneous field that I will refer to as "medical decision making."[2] These efforts embody particular interpretations about the nature of clinical expertise and of human rationality in general. I will counter some of these common assumptions with a different paradigm of knowledge, "practical knowledge," and of clinical expertise -- the "Dreyfus Model of Skill Acquisition" (Dreyfus and Dreyfus 1986). Contrary to a growing medical opinion that "intuition" should be replaced by explicit, rational calculation, the Dreyfuses argue that practical expertise is characterized by a replacement of analytic reasoning with intuition. In fact, intuition, based on vast concrete experience, is the hallmark of expertise.

The symbols "art" and "science" have been large protective shields in medicine, hiding much that is not very scientific nor very artful. I will argue for the need to understand better the strengths and limits of both clinical science and clinical expertise, and more specifically of analysis and intuition. Both have their place. To "disqualify" intuitive knowledge would be to threaten an irreplaceable human resource. Further, I will critically examine the proposed expansion of scientific rationality into medical practice. The dominance and spread of this type of knowledge into more and more areas of life and of medicine reflect reasons beyond its potential efficacy: it supports our identification of our humanity with will, consciousness, and the mind. This predisposes us to appreciate scientific knowledge as well as to find in it our way of coping with uncertainty. As Fox (1980) suggests, the current concern with medical uncertainty may also serve to express our social and cultural uncertainty about meta-medical existential issues.[3]

TWO TYPES OF MEDICAL KNOWLEDGE: "CLINICAL SCIENCE" AND "CLINICIAL EXPERIENCE"

Studies of medical practice, particularly of medical education, regularly describe two sources of medical knowledge and medical legitimation: "clinical science" (or simply "science" or "the textbook") and "clinical expertise" (or "clinical judgement" or "clinical experience," (Becker, et al. 1961; Bosk 1979; Carlton 1978). While the latter is also referred to by other terms[4] which represent different emphases, I will use the above three terms interchangeably for all of them.

Physicians differ in their allegiance to clinical science or to clinical expertise. Certainly research clinicians are more knowledgeable of and biased towards science and legitimize an action more by citing "clinical evidence" rather than "clinical experience" than do regular clinicians (K. Taylor, this volume).

In clinical practice, as many medical students learn, senior physicians may overrule the "textbook" on the authority of their "clinical experience" (Becker, et al. 1961; Bosk 1979; Carlton 1978). Students who justify an action by citing the textbooks are sometimes surprised when the attending physician (clinical faculty) says, "Well, Harrison (a famous textbook of internal medicine) may say that, but in my experience . . ." (Carlton 1978). In fact, experiences are counted and cited (Sudnow 1967). Some veteran clinicians, like internist Barry Siegler (Hahn 1985), are highly sceptical about scientific findings: "The stuff you read in the literature," he says, "90% is bullshit, maybe 95%, maybe even 98% . . . you can quote me" (p. 82). Siegler, like others, however, argues "from the literature" to resolve an ambiguous situation (Hahn 1985:82; Gifford 1986; Becker, et al. 1961).

Art and Science, the Clinic and the Lab

The two types of clinical knowledge are linked to two dominant metaphors in medicine -- "art" and "science." Their relationship is often depicted as physicians learning "basic science" principles (theory, universals), which they then "apply" to the care of individual patients, which is where the "art" comes in. Science has been most intensely used for diagnosis and etiology (explication), while patient care "management" decisions have received significantly less attention. In fact, over the past years a distance has grown between researchers or producers of medical knowledge and its users (Starr 1982; Feinstein 1983a; de Santis 1980).

Clinical judgement has traditionally been regarded as in the "art" camp, considered relatively unanalyzable and inscrutable. It is also highly valued. One study found clinical judgement rated as the foremost attribute desired by physicians (reported in Elstein 1976:696). Science on the other hand is associated with medicine's rise to social and cultural dominance (Starr 1982). The claim that medicine is scientific serves almost as a "covering law," symbolizing a universal, "objective" truth and legitimizing the authority of the medical profession.

Clinical science and clinical expertise, however, have important differences and significant social implications. As a form of knowledge, clinical science is characteristically explicit, universal, abstract, and public (Armstrong 1977). Making the criteria for practice public allows for more democratic medicine by potentially exposing practitioners to the judgement of peers, junior physicians, external agencies, and patients (ibid).

Clinical expertise, on the other hand, consists of practical knowledge of concrete particulars (Gorovitz and MacIntyre 1976); it is the personal knowledge of a physician, passed on mostly by apprenticeship (Polanyi 1958), oral culture, and the case method. Often implicit, ineffable, and tacit, clinical knowledge is less open to public scrutiny and outside surveillance; it cannot be reduced to rules and supports a hierarchy based on expertise (Armstrong 1977; Bosk 1979). Physician clinical judgement is the basis of "physician autonomy." For years it has remained virtually beyond vision and judgement (Freidson 1970; Pellegrino 1979).

Current Medical Practice: Not Very "Scientific," Not Very "Artful"

Metaphors and symbols notwithstanding, the literature on medicine documents that both medical science and practice are often neither very scientific nor very artful -- even by their own standards. For example, in medical science and theory, important lacunae remain. Many new diagnostic tests and treatments lack sufficient evaluation and guidelines for their use. Supposedly "scientific" studies are criticized for being too narrow, poorly done, or methodologically unsound because they attend to too few variables, or too short a time period, or are very costly, or are outdated by the time they are completed (Feinstein 1983a,b,d; Kennedy 1983).

Much as medical science is less than always scientific, physician practice has been found wanting in both art and science. Evidence cited for this assertion includes a notable lack of concensus among physicians on the recognition of physical signs, such as heart sounds (Feinstein 1967; Berwick, Fineberg, and Weinstein 1981); on the interpretation of symptoms, signs, and diagnostic test results (Zir, et al. 1976); and on the treatments of preference for the same problem. Physicians have been described as erring in their "intuitive logic" (Borak and Veilleux 1982); or as not using all the available data for making a decision (Elstein, et al. 1978); or as drawing conclusions not supported by the data (Berwick, Fineberg and Weinstein 1981).

Further, much evidence documents that physicians do not practice "state of the art" "scientific" medicine even when it is available and well documented. Diffusion of new treatment findings is often very uneven (Carlton 1978), such that proven improved treatments are not adopted and may meet with resistance (Israël 1978; Scriven 1979; Gabbay 1982; Light 1980). Unproven practices, on the other hand, may be adopted readily, sometimes dependent more upon colleague's anecdote than on science or clinical trials (Banta 1984:76). In fact, it has been found that the results of controlled clinical trials fail to influence practice (Banta

1984; Chalmers 1974; Bunker, et al. 1978), and that the published literature seems to have little influence on practice (Banta 1984). Similarly, practices that have been "scientifically" shown to be useless or to be bettered by others are kept alive by many physicians (see Sox 1986; Gabbay 1982).

TOWARD "A SCIENCE OF THE ART OF MEDICINE"

Among the potential explanations for many of the problems of contemporary medicine, the physicians in the movement I am describing generally attribute them not to medicine being too scientific, but to it not being scientific enough. A leader in the field, Alvan Feinstein (and as a leader he is not necessarily representative, but is influential), defines the situation this way:

> . . . technologic controversies and humanistic distress both have the same cause -- the limitations of basic biomedical science . . . The solution that will be proposed for this defect, however, is not to attack basic science, but to expand it (Feinstein 1983a:394).

Why This Movement? The Social and Cultural Climate

Before proceeding, however, we must first consider why there is increasing interest in and demand for making medical practice more scientific through the particular approaches to be discussed below. The reasons are complex, and space allows for only brief mention of outstanding points.

The rising concern for the weak scientific basis of medical practice outlined above is clearly one important triggering factor. It is, as Fox (1980) notes, part of a more pervasive questioning and confronting of the limits of human rationality evident since the 1970s. "Uncertainty" and "risk" are central concepts in the theoretical frameworks of medical decision making and clinical epidemiology, an indication perhaps, of a paradigm change in medicine from a mechanistic, causal model to a probabalist one (Bursztajn, et al. 1981; Elstein 1982; Fessel 1983; Sox 1986). From this perspective, symptoms, for example, are no longer seen to be *causally* related to underlying diseases, but *probably* related. There is always an element of uncertainty. In fact, one important contributing force in medical decision making came from those applying theory and methods of "decision analysis" -- "choice under uncertainty" (Raiffa 1968) -- to medicine (Lusted 1968; Albert 1978).

The increased attention to "uncertainty" also reflects changes in medical certainty: the burst of new knowledge and technology have made diagnosis and treatment less and less automatic processes, as they have created new ambiguous states, such as "benign disease" (Gifford 1986; Diamond and Forrester 1983) and new ethical dilemmas.

This movement to make medical practice more scientific may also be seen as a result and response to increasing demands for medical account-ability. The years of extreme medical autonomy are ending (Starr 1982). Where once medical practice could be described as "no one is keeping score" on medical outcomes (Scriven 1979), score is increasingly being kept by patients, lawyers, physicians, and external agencies. "Cost con-tainment" perhaps leads the list of concerns, and as medical technology leads the list of costly services, criteria for its appropriate use are sought.[5] Explicit criteria allow for more evaluation. The demand by patients for greater participation in their care, "informed consent," and medical malpractice threats all pressure physicians to explain their ration-ale for diagnostic and treatment decisions. Not unrelated, medical prac-tice in the United States is becoming more standardized.[6]

Another force for this movement comes from physicians concerned with teaching new physicians the skills of medical decision making. Many argue that while the stakes of medical decisions have increased, physicians receive virtually no training in how to make them (Elstein 1976; Israël 1982).

A final contributing factor is the rising impact of the digital computer in medicine and society and the concomitant interest in developing medi-cal "expert systems" (artificial intelligence) to use in medical practice (cf. Gottinger 1984). Such systems depend on physicians making their rea-soning process explicit and formalizable.

One last note: I am singling out two aspects of this movement -- "clin-ical epidemiology" and "medical decision making" -- even though some consider medical decision making to be a part of clinical epidemiology (Sackett, Haynes, and Tugwell 1985; Fletcher and Fletcher 1983).

CLINICAL EPIDEMIOLOGY: DEVELOPING "BASIC CLINICAL SCIENCE"

Clinical epidemiology is a growing discipline in the United States and England, primarily in academic centers.[7] Special programs, departments, and courses for medical students, post-graduate and continuing education physicians have been established (see Fletcher and Fletcher 1983).[8] The number of books and articles on the topic is growing rapidly.[9]

The general goals of clinical epidemiologists[10] may be summarized as follows: (1) to expand the role of science in medicine from that of prima-rily explanation (disease etiology) to include diagnosis, prognosis, and prediction; (2) to apply epidemiological methods traditionally used on general populations in the community to sick populations in the clinic or hospital setting; and (3) to expand the use of quantitative methods from primarily only etiology to include all phases of illness (UCSF brochure).

More specifically clinical epidemiologists aim: (1) to improve clinical research by improving research methods, training physicians for careers in clinical research, critically evaluating medical studies and practice, and consulting on other research projects; and (2) to help physicians practice more scientifically by helping them to understand and evaluate critically the growing medical literature and by establishing guidelines for more appropriate use of technology. Some programs emphasize these goals to different degrees.[11] Let me briefly elaborate on them.

1. Improve Clinical Research and Develop a "Basic Science of Patient Care"

Reflecting their premise that clinical medicine is not reducible to basic biomedical sciences (science developed in the laboratory), some clinical epidemiologists aim to develop a "basic science of patient care" that is founded on the study of live, sick people in "natural" settings. This is necessary, they argue, because much of current medical science is inappropriate, in part because "the lab" of basic science is not "the clinic," and studies of animal or diseased human parts in the laboratory are not the same thing as studying live human beings in natural settings, like the hospital (Feinstein 1967, 1983a; Sackett, et al. 1985; Engelhardt and Erde 1980). The clinic, they argue, *is* amenable to scientific study that considers not only death and morbidity outcomes of illness, but also "softer outcomes," such as disability, discomfort, activity level, satisfaction or "subjective" data (Feinstein 1983a,b,d; Fletcher and Fletcher 1983). This soft data can in fact be measured and usually quantified.

This would require, it is often argued, changing the current interpretation of what is "scientific." The epitome of science, the laboratory experiment, cannot be the only model, as many medical treatments cannot be tried through experiment. Even the randomized clinical trial, considered the "gold" of clinical methodology, is limited (Feinstein 1985). For example, Toth and Horwitz (1983) attribute the inconclusiveness of numerous clinical trials on the preferred treatment of asymptomatic mild hypertension to the insufficiently-varied and overly-sensitive study samples. They recommend more "realistic" studies:

To ensure scientific validity, randomized trials require restrictive eligibility criteria . . . In adhering to this statistical paradigm, therapeutic trials avoid bias by avoiding the complexities of clinical reality . . . studies are needed that describe the benefits and risks of therapy in actual clinical practice, complete with all its complexity (Toth and Horwitz 1983:487).

Clinical trials are particularly poor in situations of multiple illnesses (which applies to many patients).

"Scientific rigor" through good careful observations in the clinic is what many think is required (Feinstein and Horwitz 1982). The hallmarks of science -- reliability, validity, predictability -- are what make something scientific, not the context of research. Much clinical epidemiological work is directed at identifying and eliminating bias (Harvey, et al. 1984; Sackett 1979) and at improving measurement by applying quantitative methods -- particularly probability measurements -- to all the phases of the illness career of the patient (Feinstein 1983; Fletcher, et al. 1982; UCSF brochure).

2. Make Medical Practice More Scientific

The limited training most physicians receive in statistical methods and criteria for judging medical research is a second major concern of those working in this field. Physicians are taught to practice medicine (some would say only scientific medicine, Cassell 1976a) but not how to evaluate or to produce medical science (Feinstein 1985). As Fletcher and his colleagues point out, none of the over 800 books recommended by the American College of Physicians for Internists is devoted to methods of clinical observation (1982:16). In order to change this situation, more and more is being published on such topics as research methods (for example, Research Development Committee 1983); how to critically read medical journal articles (Department of Clinical Epidemiology and Biostatistics 1981;Gehlbach 1982; Sackett, Haynes, and Tugwell 1985); and how to use and interpret diagnostic tests in the medical literature (Sox 1986; Griner, et al. 1981).

Finally, many feel that the process of clinical reasoning or judgement can and should be made more scientific by improving observations, clarifying the variables observed, eliminating bias, increasing reliability, using probability measurements, and quantifying the judgement process. This topic I will discuss in the following section.

MEDICAL DECISION MAKING: MAKING CLINICAL JUDGEMENT MORE "RATIONAL"

The second branch of the movement to make medical practice more scientific focuses on the judgement or decision making process of physicians themselves. While once a mystique surrounded clinical judgement, more and more attention has been paid in the past 20 years to understanding, studying, teaching, modeling and improving it by people from both inside and outside of medicine.[12]

Elstein describes two dominant approaches to medical judgement -- "clinical" and "statistical" (1976). The clinical approach "means any of the artful, informal, qualitative, or not explicitly quantitative strategies generally employed by clinicians for their task" (1976:696). Statistical he defines as "any method that relies on formal quantitative techniques or formulas to reach diagnostic decisions" (ibid). A third approach, sometimes included under "statistical," is found among those using an information processing model (Tanner 1983).

While the "clinical" approach dominates in practice, and while in fact the statistical one has not been well received by practicing physicians, a medical opinion is growing that the traditional informal approach should be made more formal and "rational." Despite their heterogeneity, most working in this field share this and other assumptions: that medical practice and clinical judgement can and should be improved by replacing or supplementing intuition with more rational, explicit, formal, and, for many, quantitative analyses; that medical decision making is "choice under uncertainty," and therefore involves probabalist reckoning; and that clinical judgement should be formally taught to physicians. In fact, many maintain that to date, clinical judgement has essentially been left untaught (Elstein 1976):

Although decision making is the pre-eminent function of the physician, medical education has paid remarkably little attention to the nature of the decision-making process . . . nowhere is the student or graduate physician exposed to a systematic exposition of procedures for good decision-making. Rather, it is generally believed that with experience the physician will somehow acquire the precious attribute of "clinical judgment" (Schwartz, et al. 1973:459).

Before considering these assumptions further, I will briefly describe the decision analysis and information processesing models.

Decision Analysis

After the introduction of decision theory in medicine approximately seventeen years ago, it is increasingly being taught to physicians (see Elstein, et al. 1981) and gaining acceptance in practice (Kassirer et al. 1987). The Society for Medical Decision Making began in 1980; the journal *Medical Decision Making* in 1981. Major medical textbooks now incorporate chapters on the topic in their revised editions and the Association of American Colleges recently recommended adding it to the undergraduate curriculum (Kassirer, et al. 1987:275); it is also being tested on licensing examinations (Fox 1980).

Medical decision making is recommended for improving physicians' judgement and containing costs. It is applied to topics such as screening for and prevention of illness, the interpretation of test results, and treatment decisions of individuals and groups. Through mathematical and statistical models and decision trees, decision analysis is used to describe and prescribe decisions (Albert 1978). While the specifics of decision analysis are beyond the scope of this paper (see Albert 1978; Weinstein and Fineberg, et al. 1980; Pauker and Kassirer 1980; Cebul and Beck 1985; Zarin and Pauker 1984; Tanner 1983), I will briefly outline some of the important points. First, decision analysis proceeds by breaking down a decision situation into parts (such as "cue acquisition" and "cue interpretation") and into sub-decisions, computing the latter in a linear progression. Decisions are analyzed in terms of concepts such as: "risks," "benefits," "chance," "action," "uncertainty," "utilities" (values), "probabilities" and "subjective probabilities" (the probability the decision maker feels about the situation). Through systematic analysis or decision trees, often with quantification, the choice having the most probability of meeting the goals of the actors is identified and then compared with physicians' choice (Weinstein and Fineberg, et al. 1980; Albert 1978:364). Not infrequently, this "physician decision" is based on a written case example (e.g. Elstein, et al. 1986). For example, a given treatment is considered to have a probability of bringing certain risks and benefits to a given patient. The probabilities of these are computed and combined with other factors such as the physician's own feelings about risk, which are assigned mathematical values. This process is followed through the various subsets of diagnoses and treatments.

Despite the unquestionable spread of medical decision making, it is not readily used by physicians in their practice (Kassirer, et al. 1987:286) and opposition to it persists (Schwartz 1979; Kassirer, et al. 1987). Its precision, appropriateness, feasibility, and validity are questioned; in fact, the latter remains untested (Kassirer, et al. 1987).

While decision analysis represents a particular approach to decision making, other methods have developed which use probability as a central concept and which describe decision making in the idiom of conscious or unconscious "gambling" (Bursztajn, et al. 1981), since it is assumed that chance and uncertainty are inevitable.

Information Processing Model

The information processing approach places more emphasis on describing and explaining actual reasoning processes rather than prescribing a decision. Usually with simulated cases, practitioners are asked to simultaneously or retrospectively describe their reasoning (Tanner 1983; Larkin, et al. 1980). An information processing model of the mind often provides the framework of those in this field (Tanner 1983). Decisions, like computers, are considered to consist of an input, a process, and an output (Wulff 1976). Novice and expert performers are compared in their approach to clinical problems, often with the aim of modeling experts' knowledge for a computer (Larkin, et al. 1980). Many consider knowledge to be organized in "rules," such as "rules of thumb," e.g., "if the patient has a rash, ask about allergies," or in "heuristics" -- "the rules by which we use the information stored in frames (a set of associated facts or ideas about a central concept) to guide our thinking" (Evans, et al. 1986:1027).

Common Assumptions About Clinical Judgement and Expertise

Despite their heterogeneity, many working in this field share some assumptions about what clinical judgement is, how to improve it, and what the relationship is between formal modeling and actual practice. In reviewing a sample of works in this field,[13] we may summarize the common understanding of clinical judgement as: *cognitive, mental, or intellectual reasoning, that involves combining data through inference, logic, probability statistics, or decision rules, based on a series of conscious steps.*

There is also much agreement about how to improve clinical judgement: *that it should not be "left" to intuition; that it can and should be made more scientific and "rational" by making the implicit explicit, the informal formal; and that numbers are a key for moving from ambiguous to certain diagnositic situations.*

How is the relationship between formal modeling and clinical expertise considered? Most assume that what is implicit can essentially be made explicit through formal modeling and that there is no transformation in the process.

Some, like Elstein, however, note that a real clinical expert may supersede the ability of formal models to represent the expertise: "Experienced, competent practitioners of an art may well know more than formal theories encompass" (Elstein 1976:699-700). As I will argue below, formal models are *inherently* limited in their ability to represent fluid, contingent practice.

Finally, what are the main assumptions about the process and the development of clinical judgement? While it was initially believed and taught that physicians behave inductively like scientists in their clinical work -- that they collect data, analyze it, and only then come to a diagnosis (Schneiderman 1979; Wulff 1976) – more recent research has challenged this version. Many maintain that experienced clinicians formulate a hypothesis/es (from 4-6) very early on (within the first 10 minutes, Elstein, et al. 1978) which they then systematically eliminate. This "hypothetico-deductive" approach, as it is called (Wulff 1976; Cambell 1976), may be more an assumption than a well documented process, however, as there is much indication that experts behave otherwise (Groen and Patel 1986). Finally, the expert is not seen so much as possessing any mystical, non-learnable "art" but is considered to be better and quicker at "encoding," or interpreting, the cues. In other words, expertise is seen to develop on a continuum.

PRACTICAL KNOWLEDGE AND CLINICAL EXPERTISE

General Assumptions About How We Know

The attempt to make clinical judgement more explicit, formal, precise, and thus more "rational" by using some type of analytic medical decision making reflects a particular approach to knowledge in general. Briefly, this approach considers knowledge to be representation, and reason to be calculation. By placing conscious, explicit, formal knowledge as an ideal, this "rational" approach accepts the idea that there is a detached, universal, "truth" about reality (Dreyfus 1979; Polanyi 1958). This ideal of rationality can be traced first to the Greeks, later to the Enlightenment, and more recently to the impact of the digital computer (Dreyfus and Dreyfus 1986; Turkle 1984; Taylor 1985a). It is found not only in medicine but in much of western society. In fact, with the spread of the digital computer these views of knowledge and of the human mind that they embody are spreading (Turkle 1984). I will subsequently refer to this dominant approach to knowledge as "knowing that" or "theoretical knowledge" (Polanyi 1958; Dreyfus 1979; Benner and Benner 1979).

Theoretical knowledge contrasts with another kind of knowledge -- "knowing how" or "practical knowledge." A practical knowledge paradigm emphasizes knowledge as embodied "know-how" that derives predominantly from extensive encounters with real, concrete situations and their outcomes. I use "experience" here in a particular way, following

Benner and Wrubel (1982) and Gadamer (1970): the adjustment of expectations in relation to real situational encounters. Thus it is not synonymous with longevity though it is of course related to time.

By reference to these paradigms, we can then examine some of the assumptions about clinical expertise outlined above, as indeed many of them warrant questioning. In the following discussion I reiterate five (in italics), then challenge them with an alternative view of knowledge that draws heavily on the works of Dreyfus and Dreyfus (1986); Dreyfus (1979); Benner (1984a); Heidegger (1962); Kuhn (1979); Polanyi (1958); Merleau-Ponty (1962); and my own research (Gordon 1984a,b, 1986; Gordon and Benner 1985).

1. Clinical judgement is primarily intellectual, cognitive, analytic: it is "knowledge" as opposed to "skill"

As discussed above, clinical judgement is nearly always referred to as some type of "cognitive"/mental/intellectual/analytical reasoning. A more accurate view, however, is that much of a clinician's expertise is "practical knowledge," the sort of "know-how" or skill one uses in getting around in the world (like driving a car, using a language). Skills, in fact, do not *necessitate* thinking or calculating, as implied above. Many, such as swimming, bicycling, or practicing science, are learned and practiced without knowledge of the rules or theories that explain them. Instead they move from practice to practice without ever entering discourse (Bourdieu 1977; Polanyi 1958; Dreyfus and Dreyfus 1986). Abstract thinking may be used mostly at the early stages of skill learning (see next section) or where usual responses do not work (such as in a breakdown), or to reflect on one's "natural" response. Although science requires that the skilled performance be described according to rules, these rules need in no way be involved in producing the performance (Dreyfus 1979:253).

Similarly, much of clinical and practical knowledge is "embodied" knowledge -- knowledge sensed through and with the body. This includes senses of sight, sound, touch, smell, as well as emotions and more general senses, such as "feeling that a situation makes sense," having a "gut feeling" or a sense of salience (Benner 1984a; Gordon and Benner 1985). Polanyi and Prosch describe this training of medical students:

They are training their eyes, their ears, and their sense of touch to recognize *the things* to which their textbooks and theories refer. But they are not doing so by studying further textbooks. They are acquiring the skills for testing by their own bodily senses the objects of which their textbooks speak . . . (Polanyi and Prosch 1975:31).

Finally, rather than calculation, clinical expertise/judgement is more "understanding," a kind of "getting it," as one "gets" a joke (Wartovsky 1986). It is experiencing things as "making sense" or not, without having to explicitly spell out that sense. This constrasts with a common view that considers meaning to be consciously, mentally "assigned" rather than "grasped"; we *normally* experience our world as meaningful without having to consciously calculate its meaning (Heidegger 1962).

> 2. *Knowledge, such as clinical problem solving, proceeds from part to whole; it can be analyzed into elements and then reunited.*

In fact, much evidence indicates that rather than proceed from the part to the whole, clinicians, like performers in other areas, *first* perceive a whole, a gestalt (like a diagnosis), and *then* perceive parts or elements based on that gestalt (like "signs" of disease). This involves a kind of "situational understanding," that is a global, implicit grasp of a situation as a whole. Things appear salient not according to some objective *a priori* criteria but depending upon the particular situation and concerns of the actor.

[B]eing concerned in a certain way or having a certain purpose is not something separate from our awareness of our situation: it just is being aware of this situation in a certain light . . . (Taylor in Dreyfus 1979:261).

[W]hat counts as an object or is significant about an object already is a function of, or embodies, that concern (Dreyfus 1979:261).

> 3. *Knowledge is organized in rules, formulas, or some form of abstract representation. Accordingly, the best way to describe and generalize knowledge is through rules or formulas that combine the essential elements of a situation.*

As I discuss below, abstract rules and formulas, while usually essential to beginner performance, are generally replaced by concrete experiences which seem to be remembered through pattern recognition and seem to be organized as whole, real situations like gestalts, paradigms, exemplars, images, or prototypes rather than as intellectual categories, decontextualized elements or rules. Kuhn discusses this idea in terms of paradigms and exemplars:

Paradigms may be prior to, more binding, and more complete than any set of rules for research that could be unequivocally abstracted from them (Kuhn 1970:46).

When I speak of knowledge embedded in shared exemplars, I am not referring to a model of knowing that is less systematic or less analyzable than knowledge embedded in rules, laws, or criteria of identification. Instead I have in mind a manner of knowing which is misconstrued if reconstructed in terms of rules that are first abstracted from exemplars and thereafter function in their stead (Kuhn 1970:46).

Context is very important because in practice, meaning is inseparable from context and timing: "It is all a question of style, which in this case means timing . . . for the same act . . . can have completely different meanings at different times . . ." (Bourdieu 1977:6).

Behaviour then can be orderly without recourse to rules (Dreyfus 1979) or conscious deliberation. Like jazz music, for example, it can be fluid, "organized improvisation" (Sudnow 1978).

4. *Objective knowledge is the ideal knowledge, best gained in a detached, neutral, universal stance in which the neutral knower receives stimuli from an external reality.*

Knowledge is also gained in an involved, committed stance. The knower perceives through "foreknowledge" (Heidegger 1962), never neutrally, and has an active role in that knowledge. Polanyi discusses this in terms of "personal knowledge," maintaining that the sharp lines between subjective and objective forms of knowing are inaccurate:

I regard knowing as an active comprehension of the things known, an action that requires skill . . . Clues and tools are things used . . . and not observed in themselves. They are made to function as extensions of our bodily equipment and this involves a certain change in our own being . . . into every act of knowing there enters a passionate contribution of the person knowing what is being known, which is a vital component of his knowledge (1958:vii).

Sensitivity to signs, for example, is a function of involvement. One must be open to being solicited by a situation (Benner 1984b; Benner and Wrubel 1982; Wrubel 1985). The person who anticipates is most likely to perceive. Indeterminacy is more quickly resolved by expectations and involvement. In real situations, people are invested in the outcome of the decision.

In this sense, it is important to remember that most medical decision making analysis and information processing studies take place either in hypothetical situations or paper-and-pencil presentations. Such contexts eliminate the effect of being in a real situation. Instead, the subject "is reduced to reasoning out how he believes he might feel about a hypothetical future situation and thereby loses all contact with the feelings that would be elicited by real situations" (Dreyfus and Dreyfus 1986:183).

> 5. *Explicit knowledge is better than and equal to implicit knowledge.*
> *Knowledge is truer and better when it is put into words or num-*
> *bers. Inexplicable knowledge is likely to be only belief (Dreyfus*
> *and Dreyfus 1986).*

In this traditional paradigm of theoretical knowledge, rational understanding is linked to articulation (Taylor 1985b:137). Yet there are several things wrong with this assumption. First, intelligent behavior is often
not translatable into representation but is often ineffable, as Polanyi
writes:

Although the expert diagnosticians . . . can indicate their clues and formulate their
maxims, they know many more things than they can tell, knowing them only in
practice as instrumental particulars, and not explicitly as objects. The knowledge
of such particulars is therefore ineffable . . . This applies equally to connoisseur-
ship as the art of knowing and to skills as the art of doing. Therefore, both can be
taught only by aid of practical example and never solely by percept (1958:88).

In fact at higher levels of skill, analytic knowledge can be surpassed
by intuitive response. Following the Dreyfuses (1986) and Benner
(1984a), I use intuition here to refer to situational understanding that
occurs effortlessly due to perceiving similarities with prior experiences
but without necessarily knowing exactly why. It is an automatic, implicit,
non-rationalized response, that is neither mystical nor guessing, but
rather the type of "know how" based on extensive experience that people
use all the time as they go about their everyday activities (Dreyfus and
Dreyfus 1986:29). As I discuss below, while "calculative rationality" is
used in the earlier stages of skill acquisition, it can be replaced by an
intuitive response. One can then stand back and look at the facts from
another perspective (what the Dreyfuses call "deliberative rationality"),
but this differs from arriving at one's response through calculation. In
fact, as anyone who suddenly becomes self conscious of his/her typing
knows, consciousness often interferes with smooth performance.

The preference that many in the medical decision making movement
have for precise, explicit, objective knowledge that minimizes human
judgement also deserves questioning. As argues Merleau-Ponty (1962),
indeterminacy is essential in order to perceive a foreground against a
background. It is by "all things being equal" or "characteristically and
for the most part" that allow one situation to be flexibly perceived as similar to another, without having to spell out explicitly, "similar in what
way" (Kuhn 1970; Dreyfus 1979). Interpretation is unavoidable:

But no formulas can foretell the actual readings on our instruments and there is no
rule – and can be no rule – on which we can rely for deciding whether the

discrepancies between theory and observation should be shrugged aside as observation errors or be recognized, on the contrary, as actual deviations from the theory. The assessment in each case is a personal judgment (Polanyi and Proesch 1975:30).

The Nature and Development of Clinical Expertise: The Dreyfus Model of Skill Acquisition

The Dreyfus Model of Skill Acquisition (Dreyfus and Dreyfus 1986) is based on studies of airplane pilots, chess players, language learners, and car drivers. Benner (1982, 1984a) found it described the development and nature of nursing expertise (see also Gordon 1984a,b, 1986). The Dreyfuses describe 5 stages in the acquisition of skill: Novice, Advanced Beginner, Competence, Proficiency, and Expertise. The transitions are marked by a series of shifts: (1) from reliance on abstract principles to the use of past experience in the interpretation of problems; (2) from a piecemeal to a holistic grasp of situations; (3) from analytic reasoning to an intuitive response; and (4) from a detached to an involved stance. Below I illustrate the model, which refers essentially to unstructured problem situations, with examples taken mostly from autobiographical accounts of western clinicians.

1. Novice: The novice, lacking any practical experience with situations, is taught to recognize objective attributes ("context-free") of the environment and to act on them, usually with the help of straightforward, "context-free" rules. Numbers are particularly good friends to novices if they have specific rules for how to interpret them. One novice medical student, for example, wanted to look at a patient's laboratory report in order to determine whether his patient had a heart attack or not (he was then schooled by his preceptor to look instead at the patient). Questionnaires, such as disease history, or lists are essential maps which novices tend to depend on. They lack, however, a sense of the developing game (such as the normal course of a disease), and thus, in effect, are without strategy and goals. In many ways they are like the foreigner Bourdieu describes, who must rely on a map to find the way:

It is significant that "culture" is sometimes described as a map; it is the analogy which occurs to an outsider who has to find his way around in a foreign landscape and who compensates for his lack of practical mastery, the prerogative of the native, by the use of a model of all possible routes . . . (Bourdieu 1977:2).

The contrast between the structured "maps" presented in textbooks, where what one needs to notice is already selected, and the unstructured situations of real practice, is captured in the following account of a new intern ready to do his first appendectomy. After having reviewed in the textbook and in his mind the steps he would take, Nolen found himself before the real thing:

Suddenly my entire attitude changed. A split second earlier I had been supremely confident . . . now with the knife finally in my hand, I stared down at Mr. P's abdomen and for the life of me could not decide where to make the incision. The "landmarks" had disappeared. There was too much belly (1970:51).

Feeling, touching, seeing, all have to be learnt by the novice. To continue with the above example: Nolen, finally having made an incision in his first appendectomy, was told to reach in and pull out part of the bowel:

I stuck my right hand into the abdomen. I felt around. But what was I feeling? I had no idea . . . Everything felt the same to me. How did one tell them apart without seeing them? (1970:48)

2. *Advanced Beginner*: With practice and encounters with real situations, the actor learns to recognize meaningful recurring situational elements ("aspects"): the look of a dying man (Doctor X 1965:94), or as Polanyi describes, the meaningful shadows on an X-ray:

Think of a medical student attending a course in the X-ray diagnosis of pulmonary diseases. He watches in a darkened room shadowy traces on a fluorescent screen placed against a patient's chest, and hears the radiologist commenting to his assistants in technical language, on the significant features of these shadows. At first the student is completely puzzled. For he can see in the X-ray picture of a chest only the shadows of the heart and the ribs, with a few spidery blotches between them. The experts seem to be romancing about figments of their imagination; he can see nothing that they are talking about. Then as he goes on listening for a few weeks, looking carefully at every new picture of different cases, a tentative understanding will dawn on him; he will gradually forget about the ribs and begin to see the lungs. And eventually, if he perseveres intelligently, a rich panorama of significant details will be revealed to him: of physiological variations and pathological changes, of scars, of chronic infections and signs of acute disease. He has entered a new world. He still sees only a fraction of what the experts can see, but the pictures are definitely making sense now . . . (Polanyi 1958:101).

New physicians at this phase come to recognize a large number of meaningful patterns and have some general guidelines to follow (instead of the more specific rules). Significantly, however, they lack a sense of what is salient -- the ability to differentiate the essential from the non-essential. This inability to "zero in" on the important possibilities is frequently compensated for in medical training by an emphasis on thoroughness, the hope that by covering everything, the important will be attended to. This helps explain "intern's commandments" like, "Thou shalt leave no stone unturned" (Doctor X: 1965:24).

3. Competence: The competent performer is able to recognize whole situations and proceeds through a conscious, deliberate plan. For example, whether a patient "looks ill" or not (Wulff 1976:24) involves a recognition of a whole situation. In fact, one of the main things that a number of physicians said they learned in their first year of residency was to recognize whether a patient was "really sick" or not (Koenig, personal communication).

Having a sense of the whole picture, such as the trajectory of an illness or a sense of an operation, allows clinicians to have a plan or perspective and in turn, to sense salience quickly. This explains experienced clinicians' oft-noted efficiency: they generally conduct shorter interviews, adapt their questioning to the patients' problem, and ask questions that are most likely to solve the problem. While novice practitioners tend to follow a prescribed routine, geared for the universal patient, experienced clinicians tailor their work to the particular individual and rarely do a routine physical examination, not because they are taking "short cuts," but because they are thorough *only* in the relevant region (Cambell 1976:16-17).

4. Proficiency: With repeated encounters with similar situations and their possible outcomes, the performer becomes able to sense, *without deliberation*, the best plan for a particular situation. Understanding moves from being elemental to being holistic and intuitive (see earlier definition). This ability is illustrated in the following example in which a nurse describes the transformation in her practice after having worked on an Intensive Care Unit for 4 years.

I used to have to sit down and list things. I would sit and write the patient's hemoglobin, blood studies, and I'd try to look at the picture and get all the pieces together and come up with what's wrong. And then I had two experiences the same week. One was with a little baby, maybe 6 or 7 months old. He had a catheter that was up his inferior vena . . . And there was something screwy about this kid. And I finally decided that what he had was peritonitis -- which flashed in my mind. I practically saw the word in neon lights: PERITONITIS. So I called the physician. And the kid's catheter had slipped out of the vein and it was putting all this medication and fluids into his belly. And he did have peritonitis (Gordon 1986:954).

While *recognition* of situations at this stage is intuitive, *decision-making* still may be guided by maxims.

5. Expertise:

This is what they used to describe in medical school all the time as "sixth sense". . . Never knew what the hell that was. And it finally dawned on me after being in practice that there is nothing like a "sixth sense"; it's just your experience, what you've seen . . . ("Barry Siegler" in Hahn 1985:59).

The expert can now not only intuitively recognize a situation but also intuitively sense what to do, since he/she has such a large reserve of experienced situations to draw upon. "Nothing less than vast experience with concrete, real world situations can produce expertise," the Dreyfuses maintain (S. Dreyfus 1982:146). In fact, rather than using any abstract rules, generalizing principles, or inference, the expert is very likely recognizing thousands of individual cases (Dreyfus and Dreyfus 1986).

While the experts may intuitively sense the meaning of a situation and what to do, they often reflect on these intuitive responses. This "deliberative rationality," however, differs from the "calculative rationality" of the earlier stages in that it entails shifting perspectives rather than calculation, which provides the expert a check against snap judgements or tunnel vision.

Not infrequently, experts are inarticulate as to why they do what they do. Things often "look right" or "wrong," "make sense" or "feel" complete, or do not. This inability to articulate suggests that they are likely not perceiving "elements" nor calculating reasons.

What separates the masters and real experts from others the most, however, seems to be their special ability to "zero in" on the most significant features of a given situation.

Loeb was a master diagnostician . . . he could walk on a ward and recognize, by some kind of instinct, each of the patients in whom something deeply serious was going on (Thomas 1983:72).

The speed of "zeroing in" is related to the meaning that particular appearances or sounds have, such that they are perceived as signs. Selzer, for example, writes:

[I] was informed by a man's kneecaps that he was going to die. Flashing blue lights, they teletyped that he was running out of oxygen and blood. As soon as I got their cyanotic message, I summoned his family for a last vigil (1982:14).

Selzer's teacher himself was an expert on "reading" the human body:

Long before it became visible to anyone else, he could detect the first sign of granulation at the base of a wound, the first blue line of new epithelium at the periphery that would tell him that a wound would heal or the barest hint of necrosis that presaged failure. This gave him the appearance of a prophet. "This skin graft will take," he would say, and you must believe beyond all cyanosis, exudation, and inflammation that it would" (1982:21-22).

The use of the body and its senses and feelings are intimately involved in clinicians' expertise. Having a sense of a meaningful picture or of one that doesn't make sense is an important capacity that practitioners develop (Hahn 1985). This understanding is usually intuitive, sometimes described as a nagging feeling that something is not right, that something "feels" funny. For example, the protagonist of *Extreme Remedies* describes his hesitation to declare a patient dead, even when the clinical signs and the political context justified it, and as he said:

The corpse in front of me was obvious enough, but something obscure still bothered me about the case . . . The man was dead, and his kidneys would save a life. This logic seemed impeccable, but I still lacked an inner intuitive certainty about the case. Something wasn't right (Hejinian 1974:130). (In fact, the patient turned out to be drugged, not in an irreversible coma.)

DISCUSSION

In light of the above discussion, I now want to reconsider the movement to make medical practice more scientific, by making clinical judgement more "rational" and formal and by further developing clinical science. Then I will briefly discuss the dominance of scientific rationality, its implications and some of the reasons for it, specifically as it supports our view of humanity and provides a means for managing uncertainty -- both medical and existential.

Medical Decision Making and Clinical Expertise: Strengths and Weaknesses

In evaluating the medical decision making movement, especially in light of an understanding of clinical expertise as practical knowledge and the Dreyfus Model of Skill Acquisition, we need to distinguish several potential roles of medical decision making: as an ideal, as a model of human thinking, as a potential replacement of clinical expertise (artificial intelligence), or as a support and complement to physician expertise.

Perhaps the most serious implication of the medical decision making approach is the ideal it posits: explicit, analytic, precise and quantified knowledge. This ideal disqualifies intuition and masks embodied knowledge, both unable to meet the standards of what is considered rational. Further, by objectifying physician "subjective" assessments, it presents impersonal rather than involved rationality. The patient, the situation and the physician are all abstracted.

This ideal contrasts with the essence of clinical expertise which is more intuition than analysis or calculation, more an implicit grasp (understanding) of the meaning of a whole situation than of parts or variables; more "getting it" (as in a joke) than calculating it; more pattern recognition of particular cases than inference from principles or rules; more "embodied" knowledge than abstract "mental" reasoning; and more involved rather than detached rationality.

Thus, making expert clinicians decompose their understanding of a situation into variables and cues -- often in an artificial setting -- articulate their "reasoning process," and give rationales for their actions, very likely forces the clinicians to a lower level of skill than he or she actually uses. Expert understanding is thus sacrificed and replaced with formalization. Formal modeling of expert knowledge represents only an abstraction of expert understanding, which resembles more inexperienced beginners' approach than real expertise. Medical "expert systems," then, are not really expert.[14]

Further, while medical decision making may be useful for teaching, it is noteworthy that many in this movement argue that there has been little training in clinical judgement and that formal education is the answer. As discussed above, expertise may consist more of specific know-how about the concrete real world than of a general skill. This leads one to believe that traditional medical education, while not providing *formal* training in decision making, does provide an important basis for clinical judgement by inundating physicians with patient experiences.

This however points to an important limitation of clinical expertise -- it is specific knowledge, intimately tied to practical experience. It thus is not a given trait of a profession, nor of an individual physician, nor even of an expert in another area. When experience is lacking, it is unlikely there is clinical expertise to rely on: "The most critical factor in establishing the boundary between use and abuse [of practical expertise] is familiarity with and understanding of the problem situation" (S. Dreyfus 1982:134). Real clinical expertise based on sound, concrete, situational understanding must be distinguished from arbitrary subjectivity, guessing, mystical intuition, instinct, routine, or habit.

Clinical expertise can also be cost effective. One of the characteristics of experts is the speed with which they move into the correct problem region. Anecdotal stories of physicians suggest that physicians who lack an intuitive sense of patients tend to be among those who order more tests, as they search somewhat blindly for what is wrong. An expert quickly senses the most likely possibilities and thus may need fewer tests.

Finally, medical decision making is sometimes defined as a technique or a procedure. The decisions made, however, inevitably involve value judgements. Quantifying these judgements cannot eliminate the moral dimension of medical practice, although the use of quantification may be encouraged for this very reason.

Will More Guidelines Make Medical Practice More Scientific?

One of the assumptions of clinical epidemiology is that with more clinical science, physicians will practice more scientifically and use technology more prudently. But, as noted above, physicians already are not practicing "state of the art" "scientific" medicine. One must question, then, the assumption that *producing* more science will make medical practice more scientific, and its corollary, that the lack of scientificness is due primarily to ignorance or to the lack of guidelines. This seems to presuppose a "rational man" among physicians (Young 1981) and assumes that scientific knowledge is the major factor determining medical practice.

To assume this, however, ignores the many non-scientific factors that shape physician practice and their use of technology. For obviously, factors other than scientifically-proven efficacy account for much medical practice. This is clearly evident in observing how practice is patterned by such factors as subspecialty, geographical area, and the number of surgeons in a given area (Hahn and Kleinman 1983). Physicians' practice may be based on informal knowledge, such as the experience of a gynecologist's wife (Lock 1985), on the power of routinization (see Koenig, Barley, K. Taylor, this volume; Toth and Horwitz 1983), or on fads or fashion (Millman 1978; Scriven 1979). It may also fulfil magical or religious needs (Comaroff 1984; Gordon 1987a,b). Perhaps further study needs to be directed at understanding the non-scientific mechanisms that contribute to physicians' and society's overuse of technology.

Secondly, scientific guidelines still cannot dictate scientific care. Science is geared toward the typical and the general. It not only requires a judgement, but does not have the flexible adaptation of human "situated understanding" needed to respond to changing situations. Guidelines present "the law," not the "spirit of the law," which requires an understanding of context. Knowing which reference group to take as relevant is at the heart of clinical skill, and it is a judgement that no decision analysis, computer, or scientific guideline can make (Dreyfus and Dreyfus 1986:200). Guidelines are only as good as the users' judgements (Gordon 1984a).

This problem is highlighted with the increase in group statistics that clinical epidemiology proposes. The translation of these statistics, such as "at risk," to the case of a particular individual patient remains a major stumbling block for physicians (Gifford 1986). A slippage, and with it the possibility for error, cannot be eliminated (Gorovitz and MacIntyre 1976). For these reasons, some like Gorovitz and MacIntyre (ibid.) and Cassell (1986) propose that medical research devote more time to developing a "science of particulars."

Finally, can quantification and traditional science significantly improve patient care? While clinical epidemiologists seek to include subjective and "softer" data through measurement or quantification, the inherent limitations of this model of science make it insufficient for providing essential data on patient care, therapy and recovery. It favors study of events rather than process (Kaufert, this volume), and thus fails to address important dimensions of illness. Furthermore, many studies assume what is therapeutic (albeit allowing room for the "placebo effect" to explain the rest). Yet studies have shown that "clinically useless laboratory tests are found to be therapeutically beneficial to patients" (Sox 1986). Longitudinal, "naturalistic," ethnographic studies of recovery and non-recovery of patient illness courses as they develop over time is a need still unmet by the program to build "a basic science of patient care." Undoubtedly more knowledge is necessary, but it is very unlikely that the traditional or even revised (Feinstein 1983b, for example) scientific model can provide all of it. The decomposing of situations, of patients, of patient-physician decisions into parts and more parts not only poorly replicates how we usually live our lives, it undoubtedly contributes to abstract, disembodied, atomistic medicine. Research models that do not decompose or objectify, but present illness, disease, and medical care from the inside are needed (see Kleinman, Eisenberg and Good 1978; Kleinman 1980; Good and Good 1980; Benner 1984a). Furthermore, much as studies of illness in natural settings are needed, research of physician judgement in real rather than in artificial settings could contribute importantly to our understanding of clinical expertise (see Benner 1984a for a model). In fact, Brehmer et al. (1986) found that the "errors" that have been identified in traditional studies of medical decision making are because the studies usually take place in artificial, experimental environments and that these "errors" have close to no effect in real, practice situations. This highlights the normative character of much of medical decision making which uses a norm based on an abstract ideal model of rationality rather than that needed in the face of a particular situation. As Dreyfus (1979) notes, behavior need be only as orderly as the situation requires.

To return, then, to an appraisal of this movement: while medical decision making presents distinct limitations as an ideal, model, or substitute for clinical expertise or intuition, it may be useful in situations that are either structured -- in which all the answers are known (rather than unstructured ones where a "grasp" of the situation is first necessary) -- or novel, such that expertise is lacking, or good for routinization. Formal modeling may also increase awareness and systematic consideration of the factors involved in judgement and encourage accountability. Once a problem has been found and defined, rational calculation may help arrive at the best treatment. It serves as a prompter and a check against tunnel vision (Dreyfus and Dreyfus 1986). As a learning tool or an aid and support it likely has a place and can help manage the burst in knowledge and technology.

In an effort to end the abuse of "clinical expertise," and in patients' demand for physicians' rationales for medical decisions, we must not lose sight that a real expert may not be able to explain all he knows (Dreyfus and Dreyfus 1986; Benner 1984a). While the language of "it felt right," "in my years of experience," has obviously been abused, intuitive clinical knowledge is too valuable to be discarded. Both analysis and intuitive knowledge have their place. Perhaps we may better understand their relationship in terms of stages of skill. Clinical science may be useful and even essential for the beginning stages, both for the profession as a whole which lacks experience with new therapies and illnesses, and for the individual practitioner, whose experience cannot always serve as a guide (Gordon 1984a,b, 1986). It provides a map for those who do not know the way. But the map may be superseded with experience to arrive at more particular, contextual knowledge that works more as intuitive embodied intelligence than calculative reasoning. Let us hope that an increase in analytic discourse about clinical judgement does not constrict practice to the level of advanced beginner, thereby stunting or masking real expertise.

The Dominance of Scientific Rationality in Western Culture

Gladwin (1964) has contrasted the European and Trukese methods of navigating an open sea. As these two approaches are relevant to this discussion, I quote at length Berreman's paraphrase:

. . . [T]he European navigator begins with a plan . . . which he has charted according to certain universal principles, and he carries out his voyage by relating his every move to that plan. His effort throughout the voyage is directed to remaining "on course." If unexpected events occur, he must first alter the plan, then respond accordingly . . . The Trukese navigator begins with an objective rather than a plan. He sets off toward the objective and responds to conditions as they arise in an ad hoc fashion. He utilizes information provided by the wind, the waves, the tides and current, the fauna, the stars, the clouds, the sound of the water on the side of the boat, and he steers accordingly . . . If asked, he can point to his objective at any moment, but he cannot describe his course. The European may not know where his objective is relative to himself at a given moment, but he knows his course and can quickly compute his location on the course . . . The European can verbalize his navigational techniques whereas the Trukese cannot. The European's system is based on a few general principles applied to any given case. The Trukese's system is on a great many cues, interpreted as they arise . . . it takes apprenticeship to learn them (Berreman 1966:347).

Western society obviously favors the European navigators, and in fact even find it difficult to consider the Trukese way a type of "intelligence" or rationality (Galdwin 1964). Yet we practice like the Trukese in most aspects of our lives -- in driving a car for instance. One would likely find much in common between expert clinicians and the Trukese, only that physicians appear to use a combination of both approaches. While science has been the official knowledge of medicine, art and clinical expertise were long considered legitimate. It is the current tendency toward considering intuitive knowledge illegitimate that is of concern. Foucault writes about "subjugated knowledges": "a whole set of knowledges that have been disqualified as inadequate to their task or insufficiently elaborated: naive knowledges, located low down on the hierarchy, beneath the required level of cognition or scientificity" (1980:82). While medical knowledge is not exactly low on the hierarchy, corporate and external control is increasing in medicine. One means and response to this is to increase scientific knowledge in medicine, produce more universal and replicable knowledge. In the process, intuition is becoming disqualified. As Wartovsky notes, "Our age of scientific objectivity has made us suspicious of unjustified intuition as a dangerous form of arbitrary subjectivity" (1986:88).

The reasons for and implications of the dominance of scientific rationality are many, including the possibility it allows for visibility and control over medical practice. This probably accounts in part for the resistance of many physicians to this movement. The externalization of medical knowledge decreases both medical authority and autonomy. Informal knowledge, Young notes, (or "embedded knowledge" as he calls it) "is a form of professional investment (absorbing money and professional time) and is integral to his [the physician's] reputation and capacity for continued productive work" (1981:324). This point is echoed by physicians' response to medical decision making:

We are asked to lay out to the public view both the structure of our problem-solving and the values, both probabilities and utilities, that we have used. Even a junior student can challenge the most senior professor (quoted in Schwartz 1979:559).

While it seems clear that better guidelines for the use of medical technology are needed, and that much has been hidden under the name of clinical expertise, intuition may also be devalued because it appears mystical. Its mystique relates to the preoccupation in western culture with conscious, explicit, precise knowledge, yet intuition defies these qualities. Intuition thus often appears mystical because it does not respond well to the scientific demands of conscious control and because we tend to associate intelligence with rational calculation.

Although scientific rationality is assumed to be unbiased, it too is a particular approach to reality, albeit a particularly powerful one, that is as committed to a particular set of values as any other approach. The demand for precision and predictability, the hallmarks of science, are not neutral because certain specific measurements are selected for while other types of information are rendered unimportant or irrelevant. As the Dreyfuses assert: "Computers are more precise and predictable than humans, but precision and predictability are not what human intelligence is about" (1986) when facing complex, fluid situations in which values are emerging and evolving.

Medicine and Meta-Medical Issues

Medicine is surely not alone in this process of rationalization. How we understand and value knowledge and the process of knowing reflect much larger issues -- our sense of ourselves, our place in the world, in the universe, our security with our culture.

As long as humans are identified with their consciousness, with control, and with their minds, we can expect to see a very high premium placed on explicit, rational, precise, and replicable knowledge. As long as truth is seen to reside not in the particular human condition but in the universal, we can expect to see the hegemony of scientific knowledge. This forces us to consider how much are we turning to quantitative and formal techniques to resolve moral questions of value. Are moral questions being framed in technical terms (Cassell 1976b; Fox 1980) and decided upon through deductive analyses? Is it as Kass suggests, that "right and wrong" are translated into "benefit and risk," "good and bad" into "rights and promises," and "purity and sin" into "gains and costs" (Kass 1979, quoted in Fox 1980)? It appears, as Fox maintains, that the language of risk and uncertainty in medicine speak to larger concerns:

Our current preoccupation with medical uncertainty, error, risk, and harm is a symbolic language through which we are communicating some of our deepest questions about the cognitive, moral, and the metaphysical foundations of our cultural tradition and outlook (Fox 1980:45).

Mystery and uncertainty are here to stay. So too is our need for certainty. Current attempts to increase certainty by eliminating human judgement will not only fail but threaten an invaluable human resource. Effective treatment of patients requires understanding, and entails moral judgements that no technical application can eliminate. Clinical science and clinical expertise both have their essential place in medicine. There is potential danger in attempting to replace one by the other, in banning intuitive knowledge from the realm of the "rational" and in placing explicit, quantitative, calculating technique over implicit, intuitive human understanding as the ideal for clinical medical knowledge.

ACKNOWLEDGEMENTS

I would like to thank Margaret Lock, Sharon Kaufman, Patricia Benner, Edith Jenkins, Allaman Allamani and Victoria Kahn for their helpful comments and suggestions on earlier drafts of this article.

NOTES

1. Also referred to as "clinical investigation," "clinical research," "clinical bio-statistics," "clinimetrics" (Feinstein 1983a), and "clinical reasoning" (Fletcher and Fletcher 1983).
2. Also referred to as "clinical or medical problem solving," "clinical judgement"; "decision analysis."
3. This essay is based on a review of the literature on clinical epidemiology and medical decision making and three brief interviews with clinical epidemiologists in the United States. The proposed alternative model is based on my ethnographic study of nursing expertise over several years (Gordon 1980, 1984a,b, 1986; Gordon and Benner 1985), much of it based on Benner's (1984a) use of the Dreyfus Model of Skill Acquisition as an interpretive framework for nursing expertise (see part three of this paper). In addition to a few informal interviews with physicians regarding clinical expertise, I have also observed physicians frequently in practice in other contexts. For the rest, as I indicate, I have relied on autobiographical and ethnographic accounts of physicians.
4. "Clinical sense" (Armstrong 1977), "clinical judgment" (Feinstein 1967); Engelhardt and Spicker 1979; "clinical reasoning;" "diagnostic reasoning" (Feinstein 1973-74); "medical or clinical problem solving" (American Board of Internal Medicine 1979; Norman, et al. 1985; Elstein, et al. 1978; Connelley and Johnson 1980); and "clinical acumen" (Becker, et al. 1961; Bosk 1979; Freidson 1970; Hahn 1985).

5. Several journal series, for example, are funded for the interests of cost containment. Sox (1986) is commissioned by the Blue Cross-Blue Shield Medical Necessity Project in which commonly used diagnostic tests are reevaluated critically. Griner, et al. (1981) is sponsored by the Ad Hoc Committee on Cost Containment of the American College of Physicians.

6. Such as DRGs (Diagnostic Related Groups) which standardize treatments and costs for particular diseases.

7. The term "clinical epidemiology" was first used by Paul in 1938 and later in 1958, with reference to the field of public health. Feinstein used it in 1968 to refer to the use of epidemiological methods in the study of diseased populations. Sackett used it also in a publication in the following year (1969).

8. Programs exist at a minimum at McMaster and McGill in Canada and at University of North Carolina, Yale University, University of California, San Francisco, and Johns Hopkins and Buffalo in the United States (Fletcher, Fletcher, and Wagner 1982:6). The Robert Wood Johnson Foundation has provided substantial support in the United States for this discipline. No survey, however, was made for this study.

9. For example, Fletcher, Fletcher, and Wagner 1982; Sackett, Haynes, and Tugwell 1985; Feinstein 1985; Weiss 1986; Cox 1986; Griner, et al. 1981; Feinstein 1983a,b,c,d).

10. In order to ease the discussion I will refer to people working in this general and somewhat amorphous field as clinical epidemiologists, even if many do not call themselves that.

11. At one extreme is pioneer Alvan Feinstein, an internist and epidemiologist at Yale, who seeks to train new researchers, to establish new approaches to medical care, including medical taxonomies, and to improve clinical research (1985, 1983 series). Other programs stress an adjunct role with researchers and physicians, while others focus on the goal of providing novice physicians a background for making patient care decisions, training them in how to apply the basic science principles they have learned to practice (Fletcher, et al. 1982).

12. In fact, much of the stimulus for its study in the last 20 years has come either from external pressures -- from students in the psychology of reasoning, decision theory, or artificial intelligence, who try to produce expert knowledge -- or from concern for teaching medical students judgement skills they will later need.

13. See for examples: Borak and Veilleux 1982; Larkin, et al. 1980; Elstein 1982, 1976; Schwartz, et al. 1973; Feinstein 1973; Wulff 1976; Jelliff 1973; Kassirer, et al. 1987; Weinstein and Feinberg, et al. 1980; Diamond and Forrester 1983; Sackett, et al. 1985.

14. While it is beyond the scope of this paper to systematically and critically review the progress made in artificial intelligence in medicine, having studied expert nurse clinicians in practice and having studied the Dreyfuses' works

closely I agree with their discussion of the limits of formal modeling in capturing practical expertise (see Dreyfus 1979; Dreyfus and Dreyfus 1986; also Wartovsky 1986).

REFERENCES

Albert, Daniel A.
    1978   Decision Theory in Medicine: A Review and Critique. Milbank Memorial Fund Quarterly 56:3:362-401.

American Board of Internal Medicine
    1979   Clinical Competence in Internal Medicine. Annals of Internal Medicine 90:402-411.

Armstrong, David
    1977   Clinical Sense and Clinical Science. Social Science and Medicine 11:599-601.
    1983   The Political Anatomy of the Body. Cambridge: Cambridge University Press.

Arney, William Ray and Bernard J. Bergen
    1984   Medicine and the Management of Living. Chicago: University of Chicago Press.

Banta, H. David
    1984   Embracing or Rejecting Innovations: Clinical Diffusion of Health Care Technology. In S. Reiser and M. Anbar (eds.), The Machine at the Bedside. Pp. 65-92. Cambridge: Cambridge University Press.

Becker, Howard, et al.
    1961   Boys in White: Student Culture in Medical School. Chicago: University of Chicago Press.

Benner, Patricia
    1984a   From Novice to Expert: Excellence and Power in Clinical Nursing Practice. Menlo Park, CA: Addison Wesley.
    1984b   Stress and Satisfaction on the Job: Work Meanings and Coping of Mid-Career Men. New York: Praeger Scientific Press.

Benner, Patricia and Richard Benner
    1979   The New Nurse's Work Entry: A Troubled Sponsorship. New York: Tiresias Press.

Benner, Patricia and Wrubel, Judith
    1982   Clinical Knowledge Development: The Value of Perceptual Awareness. Nurse Educator 7:11-17.

Berreman, Gerald D.
    1966   Anemic and Emetic Analyses in Social Anthropology. American Anthropologist 68(2):346-354.

Berwick D.M., H.V. Fineberg and M.C. Weinstein
  1981 When Doctors Meet Numbers. American Journal of Medicine
        71:991-8.
Borak, Jonathan and Suzanne Veilleux
  1982 Errors in Intuitive Logic Among Physicians. Social Science and Medi-
        cine 16:1939-1947.
Bordage, G. and R. Zacks
  1984 The Structure of Medical Knowledge in the Memories of Medical Stu-
        dents and General Practitioners. Categories and Prototypes. Medical
        Education 18:406-416.
Bosk, Charles L.
  1979 Forgive and Remember. Chicago: University of Chicago Press.
Bourdieu, Pierre
  1977 An Outline of a Theory of Practice. Cambridge: Cambridge University
        Press.
Brehmer, Berndt, et al. (eds.)
  1986 New Directions in Research on Decision Making. Amsterdam: N. Hol-
        land.
Bunker, J.P., et al.
  1978 Surgical Innovation and its Evaluation. Science 200:937-41.
Bursztajn Harold J., et al.
  1981 Medical Choices, Medical Chances. New York: Delacorte.
Cambell, E.J.M.
  1976 Basic Science, Science and Medical Education. Lancet i:134.
Carlton, Wendy
  1978 "In Our Professional Opinion . . ." Notre Dame: Notre Dame Press.
Cassell, Eric J.
  1976a The Healer's Art. New York: Penguin.
  1976b Preliminary Explorations of Thinking in Medicine. Ethics in Science
        and Medicine 2:1-12.
  1986 Towards a Science of Particulars. Hastings Center Report (October):
        12-15.
Cebul, Randall D. and Beck, Laurence H. (eds.)
  1985 Teaching Clinical Decision Making. New York: Praeger.
Chalmers, T.C.
  1974 The Impact of Controlled Trials on the Practice of Medicine. Mt. Sinai
        Journal of Medicine 41:753-9.
Comaroff, Jean
  1984 Medicine, Time and the Perception of Death. Listening: Journal of
        Religion and Culture 19:155-169.
Connelley, D.P. and P.E. Johnson
  1980 The Medical Problem Solving Process. Human Pathology 11:412-419.

Department of Clinical Epidemiology and Biostatistics. McMaster University Health Sciences Centre
　1981　How to Read Clinical Journals. I. Why to Read Them and How to Start Reading Them Critically. Canadian Medical Association Journal 124:555-8.
de Santis, Grace
　1980　Medical Work: Accomodating a Body of Knowledge to Practice. Sociology of Health and Illness 2:2.
Diamond, George A. and James S. Forrester
　1983　Metadiagnosis: An Epistemological Model of Clinical Judgment. The American Journal of Medicine 75:129-136.
Doctor X
　1965　Intern. New York: Fawcett.
Dreyfus, Hubert L.
　1979　What Computers Can't Do. The Limits of Artificial Intelligence. Revised Edition. New York: Harper and Row.
Dreyfus, Hubert L. and Stuart E. Dreyfus
　1986　Mind Over Machine. The Power of Human Intuition and Expertise in the Era of the Computer. New York: The Free Press.
Dreyfus, Stuart
　1982　Formal Models vs. Human Situational Understanding: Inherent Limitations on the Modeling of Business Expertise. Office: Technology and People 1:133-165.
Elstein, Arthur S.
　1976　Clinical Judgment, Psychological Research and Medical Practice. Science 194:696-700.
　1982　Comments. Social Science and Medicine 16:1945-6.
Elstein, Arthur, et al.
　1978　Medical Problem Solving: An Analysis of Clinical Reasoning. Cambridge: Harvard University.
　1981　Instruction in Medical Decision Making. Medical Decision Making 1:70-73.
　1986　Comparison of Physicians' Decisions Regarding Estrogen Replacement Therapy for Menopausal Women and Decisions Derived From a Decision Analysis Model. The American Journal of Medicine 80:246-258.
Engelhardt, Jr., H. Tristram and Stuart Spicker (eds.)
　1979　Clinical Judgment: A Clinical Appraisal. Boston: D. Reidel Publishing Co.
Engelhardt, Jr., H. Tristram and Edmund Erde
　1980　A Philosophy of Medicine. In P. Durbin (ed.), A Guide to the Culture of Science, Technology and Medicine. Pp. 364-461. New York: The Free Press.

Evans, David A., et al.
  1986   Frames and Heuristics in Doctor-Patient Discourse. Social Science and
         Medicine 22(10):1027-34.
Feinstein, Alvan R.
  1967   Clinical Judgment. Baltimore: Williams and Williams.
  1968   Clinical Epidemiology. I, II, III. Annals of Internal Medicine 69:4.
  1973- An Analysis of Diagnostic Reasoning. I, II, III. Yale Journal of Biology
  1974   and Medicine 46:212.
  1983a An Additional Basic Science for Clinical Medicine: I. The Constrain-
         ing Fundamental Paradigms. Annals of Internal Medicine 99:393-397.
  1983b An Additional Basic Science for Clinical Medicine: II. The Limitations
         of Randomized Trials. Annals of Internal Medicine 99:544-550.
  1983c An Additional Basic Science for Clinical Medicine: III. The Challenges
         of Comparison and Measurement. Annals of Internal Medicine
         99:705-712.
  1983d An Additional Basic Science for Clinical Medicine: IV. The Develop-
         ment of Clinimetrics. Annals of Internal Medicine 99:843-848.
  1985   Clinical Epidemiology: The Architecture of Clinical Research. Phila-
         delphia: W.B. Saunders.
Feinstein, Alvan R. and Ralph I. Horwitz
  1982   Double Standards, Scientific Methods, and Epidemiologic Research.
         New England Journal of Medicine 307:1611-7.
Fessel, W. Jeffrey
  1983   The Nature of Illness and Diagnosis. The American Journal of Medi-
         cine 75:555-560.
Fletcher, Robert H. and Suzanne W. Fletcher
  1983   Clinical Epidemiology: A New Discipline for an Old Art. Annals of
         Internal Medicine 99:401-403.
Fletcher Robert H. and Suzanne W. Fletcher and Edward H. Wagner
  1982   Clinical Epidemiology: The Essentials. Baltimore. Williams and Wil-
         kins.
Freidson, Eliot
  1970   Profession of Medicine. New York: Dodd and Mead.
Foucault, Michel
  1975   The Birth of the Clinic. New York: Random House.
  1980   Power/Knowledge. Selected Interviews and Other Writings, 1972-1977.
         Colin Gordon (ed.). New York: Pantheon.
Fox, Renée
  1980   The Evolution of Medical Uncertainty. Milbank Memorial Fund Quar-
         terly 58(1):1-49.
Gabbay, John
  1982   Asthma Attacked? Tactics for the Reconstruction of a Disease
         Concept. In P. Wright and A. Treacher (eds.), The Problem of Medical

Knowledge. Pp. 23-48. Edinburgh: Edinburgh University Press.

Gadamer, Hans Georg
1970 Truth and Method. London: Sheer & Ward.

Gale, Janet and Marsden, Philip
1983 Medical Diagnosis: From Student to Clinician. Oxford: Oxford University Press.

Gehlbach, S.H.
1982 Interpreting the Medical Literature. A Clinician's Guide. Lexington, Mass.: D.C. Heath and Co.

Gifford, Sandra
1986 The Meaning of Lumps: A Case Study of the Ambiguities of Risk. In C. Janes, R. Stall and S. Gifford (eds.), Anthropology and Epidemiology. Pp. 213-246. Dordrecht, Holland: D. Reidel Publishing Co.

Gladwin, Thomas
1964 Culture and Logical Process. In W. Goodenough (ed.), Explorations in Cultural Anthropology. Pp. 167-177. New York: McGraw Hill.

Good, Byron and Mary-Jo Delvecchio Good
1980 The Meaning of Symptoms: A Cultural Hermeneutic Model for Clinical Practice. In L. Eisenberg and A. Kleinman (eds.), The Relevance of Social Science for Medicine. Pp. 165-196. Boston: D. Reidel Publishing Co.

Gordon, Deborah R.
1980 A Portrait of Two Expert Nurses in Practice. Paper presented at the AMICAE Workshop. San Francisco: January.
1984a The Use and Abuse of Formal Models in Nursing Practice. In P. Benner, From Novice to Expert. Pp. 225-243. Menlo-Park: Addison Wesley Publishing.
1984b Expertise, Formalism and Change in American Nursing Practice: A Case Study. Ph.D. Dissertation. Medical Anthropology Program, University of California, San Francisco/Berkeley.
1986 Models of Expertise in American Nursing Practice. Social Science and Medicine 22(9):953-962.
1987a Magico-Religious Dimensions of Western Medicine: The Case of the Artificial Heart. Paper presented at the First National Conference of the Cultural Anthropology of Complex Societies, Rome, Italy, May 27-30, in press in Italian.
1987b Magical Aspects of Biomedicine (Aspetti Magici in Biomedicina.) Psichiatria e Psicoterapia Analitica VI:I:93-98.

Gordon, Deborah R. and Patricia Benner
1985 Intuition in Expert Nursing Practice. Paper Presented at the American Anthropology Annual Meetings, Washington, D.C., December.

Gorovitz, Samuel and Alasdair MacIntyre
1976 Toward a Theory of Medical Fallibility. Journal of Medicine and Philos-

ophy 1:51-71.

Gottinger, H.W.
1984 Computers in Medical Care: A Review. Methods in Information Medicine 23:63-74.

Griner, Paul F., et al.
1981 Selection and Interpretation of Diagnostic Tests and Procedures. Annals of Internal Medicine 94(4):553-582.

Groen, G.J. and Vimla L. Patel
1986 Medical Problem Solving: Some Questionable Assumptions. Medical Education 19:95-100.

Hahn, Robert A.
1985 A World of Internal Medicine: Portrait of an Internist. *In* R. Hahn and A. Gaines (eds.), Physicians of Western Medicine. Pp. 51-111. Dordrecht, Holland: D. Reidel Publishing Co.

Hahn, Robert A. and Arthur Kleinman
1983 Biomedical Practice and Anthropological Theory: Frameworks and Directions. Annual Review of Anthropology 12:305-33.

Harvey M., Ralph Horwitz and Alvan R. Feinstein
1984 Diagnostic Bias and Toxic Shock Syndrome. The American Journal of Medicine 76:351-360.

Heidegger, Martin
1962 Being and Time. New York: Harper & Row.

Hejinian, John
1974 Extreme Remedies. New York: Bantam.

Helman, Cecil G.
1985 Disease and Pseudo-Disease: A Case History of Pseudo-Angina. *In* R. Hahn and A. Gaines (eds.), Physicians of Western Medicine. Pp. 293-331. Dordrecht, Holland: D. Reidel Publishing Co.

Israël, Lucien
1978 Conquering Cancer. New York: Vintage.
1982 Decision-Making: The Modern Doctor's Dilemma. New York: Random House.

Isselbacher, Kurt J., et al. (eds.)
1977 Harrison's Principles of Internal Medicine. Eighth Edition. New York: McGraw Hill Book Co.

Jelliffe, Roger W.
1973 Quantitative Aspects of Clinical Judgment. The American Journal of Medicine 55(4):431-433.

Kass, Leon R.
1979 "Making Babies" Revisited. The Public Interest (Winter): 32-60.

Kassirer, J.P. and G.A. Gorry
1978 Clinical Problem Solving: A Behavioral Analysis. Annals of Internal Medicine 89:245-255.

Kassirer, J.P., et al.
   1987 Decision Analysis: A Progress Report. Annals of Internal Medicine
          106:275-291.
Kennedy, Ian
   1983 The Unmasking of Medicine. London: Granada.
Kleinman, Arthur
   1980 Patients and Healers in the Context of Culture. Berkeley: University of
          California Press.
Kleinman, Arthur, Leon Eisenberg and Byron Good
   1978 Culture, Illness and Healing. The Annals of Internal Medicine
          88:251-258.
Kuhn, Thomas S.
   1970 The Structure of Scientific Revolutions. Chicago: University of Chi-
          cago Press.
Larkin, Jill, John McDermott, Dorthea P. Simon, Herbert A. Simon
   1980 Expert and Novice Performance in Solving Physics Problems. Science
          208:1335-1342.
Light, Donald
   1980 Becoming Psychiatrists. New York: Norton.
Lock, Margaret
   1985 Models and Practice in Medicine. Menopause as Syndrome or Life
          Transition? In R. Hahn and A. Gaines (eds.), Physicians of Western
          Medicine. Pp. 115-139. Dordrecht, Holland: D. Reidel Publishing Co.
Lusted, Lee B.
   1968 Introduction to Medical Decision Making. Springfield Ill: Charles C.
          Thomas.
Merleau-Ponty, Maurice
   1962 The Phenomenology of Perception. London: Routledge and Kegan
          Paul.
Millman, Marcia
   1978 The Unkindest Cut. New York: William Morrow and Co.
Murphy, Edward A.
   1976 The Logic of Medicine. Baltimore: Johns Hopkins University Press.
Nolen, William
   1970 The Making of a Surgeon. New York: Random House.
Norman, G.R., et al.
   1985 Knowledge and Clinical Problem-Solving. Medical Education
          19:344-356.
Pauker, Stephen G. and Jerome P. Kassirer
   1980 The Threshold Approach to Clinical Decision Making. New England
          Journal of Medicine 302:1109-1117.
Paul, J.R.
   1938 Clinical Epidemiology. Journal of Clinical Investigation 17:539.

1958   Clinical Epidemiology. Chicago: University of Chicago Press.
Pellegrino, Edmund D.
1979   Medicine, Science, Art: An Old Controversy Revisited. Man and Medicine 4:43-52.
Polanyi, Michael
1958   Personal Knowledge. London: Routledge & Kegan Paul.
Polanyi, Michael and Harry Prosch
1975   Meaning. Chicago: University of Chicago Press.
Raiffa, H.
1968   Decision Analysis: Introductory Lectures on Choices Under Uncertainty. Reading, Mass: Addison Wesley.
Research Development Committee. Society for Research and Education in Primary Care Internal Medicine
1983   Clinical Research Methods: An Annotated Bibliography. Annals of Internal Medicine 99:419-424.
Sackett, David L.
1969   Clinical Epidemiology. American Journal of Epidemiology 89:125-128.
1979   Bias in Analytic Research. Journal of Chronic Disease 32:51-63.
Sackett, David L. (ed.)
1984   Clinical Research Methods. Boston: Little, Brown and Co.
Sackett, David L., Brian Haynes and P. Tugwell (eds.)
1985   Clinical Epidemiology: A Basic Science for Clinical Medicine. Boston: Little, Brown and Company.
Schneiderman, Lawrence J.
1979   Three Approaches to Problem Solving in Medicine and Some Problems They Create. The Journal of Family Practice 9(3):59-61.
Schwartz, William B.
1979   Decision Analysis: A Look at the Chief Complaints. New England Journal of Medicine 301(10):556-559.
Schwartz, W.B., et al.
1973   Decision Analysis and Clinical Judgment. The American Journal of Medicine 55:459-472.
Scriven, Michael
1979   Clinical Judgment. In T. Engelhardt, Jr. and S. Spicker (eds.), Clinical Judgment. Pp. 3-16. Dordrecht, Holland: D. Reidel Publishing Co.
Selzer, Richard
1982   Letters to a Young Doctor. New York: Touchstone.
Sox, Harold C.
1986   Probability Theory in The Use of Diagnostic Tests. Annals of Internal Medicine, 104:606-66.
Starr, Paul
1982   The Social Transformation of American Medicine. New York: Basic Books.

Sudnow, David
   1967   Passing On: The Social Organization of Dying. Englewood Cliffs, NJ. Prentice-Hall.
   1978   Ways of the Hand. The Organization of Improvised Conduct. Cambridge: Harvard University Press.
Tanner, Christine
   1983   Research on Clinical Judgment. In W. Holzemer (ed.), Review of Research in Nursing Education. Pp. 1-32. New Jersey: Slack Publishers.
Taylor, Charles
   1985a   Human Agency and Language. Philosophical Papers 1. Cambridge: Cambridge University Press.
   1985b   Philosophy and the Human Sciences. Philosophical Papers 2. Cambridge: Cambridge University Press.
Thomas, Lewis
   1983   The Youngest Science: Notes of a Medicine-Watcher. New York: Viking Press.
Toth, Patrick J. and Ralph I. Horwitz
   1983   Conflicting Clinical Trials and the Uncertainty of Treating Mild Hypertension. The American Journal of Medicine 75:482-488.
Turkle, Sherry
   1984   The Second Self: Computers and the Human Spirit. New York: Simon and Schuster.
UCSF (University of California, San Francisco)
   n.d.   Clinical Epidemiology Program Brochure.
Wartovsky, Marx W.
   1986   Clinical Judgment, Expert Programs and Cognitive Style: A Counter Essay in the Logic of Diagnosis. The Journal of Medicine and Philosophy 11(1):81-92.
Weinstein, Milton and Harvey V. Fineberg, et al.
   1980   Clinical Decision Analysis. Philadelphia: WB Saunders.
Weiss, Noel S.
   1986   Clinical Epidemiology: The Study of the Outcome of Illness. Oxford: Oxford University Press.
Wrubel, Judith
   1985   Personal Meanings and Coping Processes. Ph.D. Dissertation, Human Development and Aging. University of California, San Francisco.
Wulff, Henrik R.
   1976   Rational Diagnosis and Treatment. Oxford: Blackwell Scientific Publications.
Young, Allan
   1981   When Rational Men Fall Sick: An Inquiry Into Some Assumptions Made By Medical Anthropologists. Culture, Medicine and Psychiatry

5:317-335.
Zarin, D.A. and Stephen G. Pauker
  1984   Decision Analysis as a Basis for Medical Decision-Making. The Tree of
          Hippocrates. The Journal of Medical Philosophy 9:181-213.
Zir, L.M., et al.
  1976   Interobserver Variability in Coronary Angiography. Circulation 33:627.

PART V

MEDICAL CONSTRUCTION OF LIFE CYCLE PROCESSES

PETER W.G. WRIGHT

BABYHOOD: THE SOCIAL CONSTRUCTION OF INFANT CARE AS A MEDICAL PROBLEM IN ENGLAND IN THE YEARS AROUND 1900

SOCIAL CONSTRUCTIONISM

What does it mean to regard infancy -- or for that matter any other concept employed in medicine -- as a social construction?[1] It is important to
be clear because the social constructionist approach is easily misunderstood. The reason for this, I believe, stems from the inherent difficulty
we all experience in distancing ourselves from, and analysing, familiar,
taken-for-granted cultural concepts. The problem is not simply that we
have to call into question categories which we are constantly using to
organise our everyday experience -- hard though that is -- it is also that
the only tools of analysis which we can use for the task are themselves
also cultural categories -- whether those of natural language or of self-
consciously created theoretical systems. These difficulties become especially daunting when the categories concerned -- as in the case of medicine -- relate to the natural world and form part of a specialised,
technical discourse: a discourse which in modern western culture is
endowed, of course, with a distinctive and privileged status: the scientific. It is not surprising that the social constructionist analysis of medicine may sometimes appear willfully counter-intuitive.

Perhaps the simplest ways of classifying the core elements of a social
constructionist approach is to begin by stating two things that it is not.
The first is this: to regard medical concepts as social constructions is not,
of course, to cast doubt on the material existence of our bodies, of diseases, or of divisions of the life-cycle such as infancy. The proponents of
this approach, like others, want to see illnesses dealt with in the most
effective ways possible. To consider a category as social-constructed is
not to render it illusory, or a figment of the imagination: it is, if anything,
to ground it more firmly by rooting it in the lived experience of members
of a shared culture.

Secondly, social constructionism does not simply amount to an assertion that the social world is infinitely variable and ever-changing and,
since social meanings are context-dependent, that they too are infinitely
diverse. To an extent this may be true -- though trivially so. No two
social circumstances are ever identical in every respect, nor are the social
meanings in use within them. Nonetheless, for many purposes -- including those of social and cultural analysis -- the meaning of particular

M. Lock and D. R. Gordon (eds.), Biomedicine Examined, 299–329.
© 1988 by Kluwer Academic Publishers.

cultural categories may be regarded as constant over time, or from one place to another. Constancy cannot, however, be laid down in advance. One of the abiding tasks of the human sciences, as Wittgenstein reminded us, is to make judgements concerning similarity or difference between various human meanings. In everyday life too, constancy of meaning is endlessly being probed, confirmed, reproduced, negotiated, denied, abandoned -- or whatever. Generalised philosophical doubt and purposive activity such as social constructionist analysis are not the same thing, even if they sometimes involve making the same points. A philosopher who finds it impossible to step into the same river twice may be making an important ontological observation: the pathfinder in the same condition is simply incompetent.

Social constructionism starts from a different basis: from the recognition that all knowledge -- medicine and science not excepted -- is the product of human social activity and is used by human beings to bring into existence their own lives and experience. "Man", as Clifford Geertz (quoting Max Weber) has written, "is an animal suspended in webs of significance he himself has spun" (Geertz 1973:5). Scientific and medical knowledge are important strands within these webs, notwithstanding the features which distinguish them from other components of culture. The fact that science -- and to a lesser extent medicine -- can be used with extraordinary power to transform the natural world, the claim that science has uniquely privileged epistemic access to that world, the esoteric and technical nature of modern scientific and medical knowledge, which demand lengthy specialised training of the practitioner and generate a discourse remote from many other elements of culture: none of these features negate the cultural role of science, or medicine. The fundamental claim of the social constructionist is that science and medicine play a central part -- along with many other facets of culture -- in generating and establishing human experience, even the experience of the majority of members of western societies who are not themselves directly engaged in the technical practice of science or medicine. Thus, they merit the technique of "thick description", the interpretive approach to culture advocated by Geertz. Science and medicine, the social constructionist would argue, resemble religion, art, or popular culture in being constituted and actualized by social forces.

That is not to say that medicine is fashioned socially in the same way as any other particular aspect of culture. But, no more would one assume that the forces at work within the development of law were identical with those shaping the history of music. The specificity, or relative independence, of various cultural discourses is not something to be determined in advance, but to be assessed by investigation. The principle of social constructionism is not that medicine is simply another form of culture just like all the others (whatever that might mean). It is, instead, the precept that medical and scientific knowledge are distinctive social products which are constituted and actualized in social practice. Thus, they must be examined with the tools of social and cultural analysis.

INFANCY TRANSFORMED

This paper is intended as an instance of such an analysis. It aims to demonstrate that over a period of some twenty-five years from 1890 a fundamental transformation occurred in England in the ways in which infancy and infant health were regarded by such public figures as doctors, administrators, journalists and politicians.[2] My thesis is not simply that views and opinions changed, but that a new set of concepts came into being that were distinct from any which had existed before. I contend that by the First World War a new vocabulary of concepts and associated metaphors had become available with which to structure and understand the phenomena of early childhood; that a new conceptual terrain had been created which made possible novel ways of looking at the world, that made visible what had been invisible earlier; and that made the attitudes and conceptions concerning infancy of the previous generation seem jejune and irrelevant – if not inexplicable. In addition, I shall argue that this transformation had significant implications for ideology and modes of social control, for it was an important instance of the extension of a scientized mode of social control into everyday life. It represented the transmutation of a sphere of human experience in which – to use Habermas' terminology (Habermas 1969:93) – "symbolic interaction" was dominant, into one under the sway of "purposive-rational action."

It will be suggested that this change cannot be explained in terms of any single cause, but is related in a complex way to a range of interacting factors which include, among others: the rise of a new field of professional practice – infant welfare; the growing pre-occupation with imperialism in British politics; shifts in the definition of medical knowledge; discoveries in bacteriology; a re-definition of the social position of working-class mothers and the working-class family; and, at the most general level, the acceptance of new implicit, common-sense metaphors for describing and making sense of social events.

The new conception of infancy comprised four elements none of which was completely novel but which served, in combination, to bring about a radical shift in perceptions of the social and cultural significance of the baby in English life that effectively created babyhood as a medical object.

Firstly, the death of infants came to be seen as a problem about which
something ought to be done – rather than an inescapable fact of nature
or, even, as a positively beneficial mechanism for culling the unfit, or dis-
couraging immorality. Expressions of concern about the high incidence
of infant death can be found, of course, in the mid-nineteenth century or
earlier; but it is only in the last years of the century, and the first years of
this, when such deaths come to be generally acknowledged as constituting
a major social and political problem.

Secondly, a consensus came into being that the problem of infant care
was essentially medical in nature and that solutions might be found to it
by the application of medical science under the tutelage of members of
the medical profession. That is to say, child rearing ceased to be simply
part of a moral discourse in which actions were judged against accepted
values and various forms of sanction applied against those who did not
conform; instead, it came to be constituted within a technical, medico-
scientific discourse that sought instrumental control over nature. In the
first, control was based upon value consensus; in the second, on the exi-
gencies of the workings of the natural world.

This development, as we shall see, was particularly uneven and
erratic, both because the social meanings of medical practice and medical
science were themselves being re-negotiated over these years, and
because there was a persistent tendency for arguments cast in a medico-
scientific form to slip back into a more moralistic, retributional shape.
Writers for instance who pointed out that a great number of infant deaths
resulted from bacterial infection or malnutrition frequently failed to pursue the logic of their views which would have been to propose technical
expedients with which to lessen these dangers. They preferred, instead,
to use them as the occasion to deplore the supposed fecklessness of
mothers or their selfishness in working after marriage. Such moralization, of course, never completely disappeared; nonetheless, by the end of
the period, one can detect a distinct change in emphasis. Policies
directed towards reducing infant mortality came to be seen as practical
issues deriving from a medical understanding of infancy. There was no
longer a desire to make a change in morality a pre-requisite for cutting
the number of infant deaths. Rather, when attempting to reduce infantile
mortality, the new policies took as given the attitudes of working-class
families and sought to change them – together with certain aspects of the
baby's environment. A good example of this is breast feeding: although
propaganda continued to be directed towards persuading mothers to
breast-feed their children – as it had been for many years – there was
also a growing recognition that many would not, which led to more attention being given to ways of making it easier for bottle feeding to be carried out safely. This leads us to the third characteristic of the new

babyhood: the form in which the proposals for lessening infant death were framed. Although, as has been suggested, the foundations of the infant welfare policy of that period were medical in the sense that they drew on medical knowledge, were underwritten by the prestige of medicine, and advocated by, at least some, groups of doctors, they depended little on doctors for their execution, and were scarcely concerned at all with therapy. Instead, the problem of infant death was addressed through a series of social channels: by the provision of a national system of trained health visitors and domiciliary midwives; by milk treatment and bottling; by classes and printed information for mothers; by improvements in sanitation and, to a certain extent, towards the end of the period, by infant welfare and ante-natal clinics.

The final element of the new approach which both linked it to many other aspects of early twentieth-century English culture, and gave the whole movement a scientific legitimacy, was the incorporation into it of germ pathogen theory. This provided not simply a scientific and non-moralistic explanation of infant death but also situated it within a new cosmological framework. It was able to transform the meaning of the problem for all those involved, and gave a special legitimacy to medical expertise as the appropriate means for its resolution. No longer did the death of babies have to be seen, for example, as the legacy of Original Sin, or as the working-out of some ineluctable struggle for existence -- or even as the price of social inadequacy. Now, such deaths could be understood as the consequence of alien invasion: an invasion that was in no sense natural, still less related to the normal development of babies. Like all invasions, it could be guarded against, and the attackers could themselves be attacked. In this view, the mothers of babies, like politicians who failed to defend the nation, must take responsibility. Hardly surprising that ". . . the mother who claimed to know all about childbearing and childrearing because she had 'born 12 and buried 8'" became the bête noire "of those who sought to improve maternal and child welfare," in the early twentieth century (Lewis 1980:13). In the imagery of invasion such mothers had been incompetent to the point of treason.

The metaphorical power of the germ pathogen theory, especially when understood in a way that lays overriding emphasis on specific aetiology, has not received much study from those working on the history of medicine even though Canguilhem drew attention to its cultural power nearly half a century ago (Canguilhem 1979 (1943)). In writing on public health and hygiene in the twentieth century, however, it is a persistent undertone; sometimes even, as we shall see, it becomes the explicit theme itself.

The point is not that a piece of medical knowledge was simply taken up by the infant welfare movement and applied; it is rather that infant welfare was one of the fields of social practice out of which the germ pathogen theory was constituted as medical knowledge.

A NEW GENERATIVE METAPHOR -- THE 'PROBLEM OF INFANTILE MORTALITY'

The argument of this paper is that these four elements in combination, particularly the perception of infant mortality as something unnatural and in need of remedy, brought about a redefinition of the state of infancy, and made possible new kinds of thought and action in relation to it. Donald Schön has coined the term "generative metaphor" to describe what happens when a familiar category is reconceptualised in a metaphorically novel way. One of these examples is drawn from the difficulties confronted by product-development researchers trying to design paint brushes made with synthetic bristles.

The difficulty was how to understand why their prototypes, unlike traditional natural bristle brushes, delivered paint in a "discontinuous 'gloppy' way." They were able to resolve the issue, Schön suggests, when one of them succeeded in seeing the familiar paintbrush in a new way: as a kind of pump. This, he argues, led them to an appreciation of the physical factors that affected how brushes worked and ultimately made possible the design of satisfactory synthetic bristly brushes (Schön 1979:257-260).

The new view of infancy served also as a "generative metaphor" in that it enabled babies, and the causes of their deaths, to appear quite differently. It was a gestalt shift, one might say, that involved drawing together what had previously been seen as separate -- infant death and bacterial infection for example -- and of separating what had hitherto been associated -- infant death and natural wastage or climatic variation.

INFANT MORTALITY AS A SOCIAL PROBLEM

To begin with, it is necessary to distinguish changes in general attitudes to children and their worth from changes in attitude that resulted in a high proportion of children dying in infancy coming to be perceived as a social problem. Obviously, these can be connected: if parents were ever callous and indifferent to their children -- as some authors have argued that they were at times in early modern Europe[3] -- that could explain, in part at least, a lack of concern about the wide-spread occurrence of infant death. But even this putative link is open to doubt. There is no necessary reason why the two attitudes should be causally connected. It is equally possible that unfeeling adults could deplore the prevalence of early death on the calculative grounds that it robbed them of the potential economic contributions of their children – the fruits of the resources that had been invested in child bearing and rearing. Conversely, it is also known that parents who were deeply attached to their children, and were griefstricken if they died, could nonetheless accept such calamities as the necessary outcome of the workings of an inscrutable Providence.[4]

It was not, I contend, until the years around 1900 that infantile mortality came to be perceived as a potentially resolvable national problem. That is to say, that it was only then that the literate and vocal came to regard it as unacceptable that some 15% of all children born in England should die in their first year of life (over three times as many in certain districts) and that this was a loss about which something could, and should, be done.

The first signs of public concern about the great prevalence of death among babies begin to become visible in the eighteen-fifties and sixties. Harriet Martineau, for instance, contributing to a medical journal in 1859 (Martineau 1859), wrote of an unnecessarily high rate of mortality (which she put at 40% among children under 5) and advocated the replacement of bottle-feeding by breast-feeding to remedy the problem. Although such recommendations had been made repeatedly for at least two centuries, and certainly owed much to a rhetoric of "naturalness," there was also considerable circumstantial and statistical evidence available by that time to show that the cows milk used in bottle feeding might often be a source of infection or be nutritionally inadequate. It was also obvious that the forms of feeding apparatus on sale (e.g. the long-tube bottle) were very unhygienic. Sir John Simon, Medical Officer to the Board of Health, also became concerned with the issue in the early 1880's and encouraged research into the causes of high rates of infant mortality (Lambert 1963:165-7). The theme was taken up too by Dr. Pye Chavasse in the 1869 edition of his *Counsel to a Mother*, one of a series of such booklets by him which became bestsellers throughout the nineteenth century and beyond (Chavasse 1869:11).

Nevertheless, such concern was by no means general in these years and was far from having achieved the status of a political issue. For one thing, during many of these years the infantile mortality rate literally did not exist, because it was not calculated. Thus, for example, the Registrar General's *Annual Reports*, which commenced in 1839, only began in 1857 to give the number of deaths among children under the age of one year, and did not present this as a rate until 1877 (Armstrong forthcoming:2). Again, although death rates for people in other age groups began to decline steadily in England and Wales after 1870, it was only in the eighties that this became marked, and, apparently, irreversible for all groups

except infants. Only then, was it possible to recognise the latter as anomalous, and potentially susceptible to reduction in the same way as the
death rates for young adults or the middle-aged.

Other barriers too stood in the way of regarding the deaths of babies
as directly comparable to those of adults. There was a persistent tendency to treat infant death as, in some sense, normal and to categorise
its causes in ways that assimilated it to the natural order, making it
appear difficult – if not impossible – to prevent. David Armstrong has
shown how this was reflected in the way that the Registrar General classified causes of death. When the latter attempted, in 1855, to reduce the
proportion of deaths classified under the rather unsatisfactory heading of
"uncertain seat," he did so by creating a new sub-classification "diseases
of growth, nutrition and decay," which grouped together deaths from
"congenital malformation," "prematurity" and "debility," "atrophy" and
"wasting," and, simply, "old age" (Armstrong forthcoming:15).

But how was this apparently unproblematic naturalness called into
question and eventually eroded? How was it that, much later, in the
nineteen-fifties, the rejection had become so great, so irresistible, that
even the remaining unexplained (and probably heterogeneous) cases of
infant death were grouped together under the heading of "sudden infant
death?"

What, in other words, were the forces that brought about the social
construction of a view of infancy, which separated it sharply from the
phenomena of disease, decline and attrition; which presented it not as
one of death's natural habitats but as a terrain in which death was an
obscene intrusion?

Although Armstrong hints at several answers to this question and to
parallels elsewhere (in education, in welfare and so on), it is not easy to
find an explicit explanation. The implication of his work, like that of
much influenced by Michel Foucault, seems to be that medical discourse
evolved as it did as one of a series of related, unspecified and, it seems,
immanent forces. Foucaultian analysis has certainly played an important
part in stimulating a new kind of socio-historical study of medicine; nonetheless, if not taken farther, it tends to become self-validating and to
close off the examination of what seem, *prima facie*, to be interesting
issues. In the case of infant death the Foucaultian approach gives us no
appreciation of process; no awareness of the contingencies and loose
ends that influenced the particular way in which babyhood as a category
took form in Britain. To be aware that there might have been other outcomes is not just idle conjectural history. It would be interesting, for
example, to investigate whether what Armstrong takes as essential
aspects of the "medical gaze" are simply contingent factors incorporated
into it from other cultural and social spheres such as the ideological and

political.

As a matter of fact, the historical data would appear to provide strong justification for raising precisely these questions. Firstly, there is considerable evidence of conflicting attitudes towards the causes of infant death within the medical profession itself in these years (a point that I shall return to later). Secondly, there is much to encourage the view that the rise in the public visibility of the social problem of infant death was linked both to general changes in social imagery and to the particular political circumstances of Edwardian England. It can even be argued that the major factors leading to the perception of infant deaths (or most of them) as unnatural -- and, thus, preventable -- came from outside medicine, and that medical thought simply reacted to these changes in its social and cultural environment. At all events, it is difficult to deny that a complex series of forces were involved, including many that were external to the technical practice of medicine.

THE LATE-VICTORIAN RECONCEPTUALIZATION OF SOCIAL PROBLEMS

To begin with, there are striking parallels between the changes that were occurring in public attitudes towards infant mortality and the shifts in social policy in the last decades of the nineteenth century that have been described by writers such as Gilbert, Semmel, Searle and Stedman Jones. Gilbert, for instance, has identified a "new philanthropy" (Gilbert 1966) which tended to re-conceptualize the ills of industrial capitalism, not simply as the consequences of individual moral failings but, instead, as the more-or-less inevitable costs of progress. From such a standpoint the provision of permanent welfare services would come to seem inevitable. Similarly, stress on the Benthamite principle of "less eligibility"[5] as enshrined by Chadwick in the 1834 Poor Law, and which served to keep the price of labour power at its lowest possible level, came to be replaced by other principles which tended, in contrast, to lay stress on improving the quality of the labour force. A key element in such changes was a movement of emphasis from attempts to bring about the individual moralization of the poor to policies designed to cope with structural, social problems -- often with the use of the technical knowledge of doctors or other professionals. Such developments, it has been argued, were closely related to the rise of imperialist politics and the re-evaluation of the position of the working class which was often associated with it (Semmel 1960).

A remarkably similar development can be distinguished in attitudes to infant welfare. Here, there is a change in emphasis away from lamentations about the immorality, drunkenness or cruelty of the poor, and its

consequences for their children, to a new mode of discourse in which the
high rate of infant death is presented as the outcome of good intentions
and loving care nullified by ignorance of hygiene or the ill effects of
deleterious customs.[6] There is a movement, that is, towards presenting
the problems as ones of technical ignorance rather than of evil intention.
As one English speaker in the 1902 International Congress for the Welfare and Protection of Children rather condescendingly put it:

> . . . our English mothers of the working classes . . . are, with all their ignorance,
> stupidity and superstition, affectionate to their children, jealous to promote their
> comfort and enjoyment, tenderly solicitous for them in sickness, anxious to make
> them honest, polite and considerate for the weak and suffering. It is to our Eng-
> lish mothers, I suggest that we mainly owe the admirable conduct of our soldiers in
> South Africa (International Congress 1902:18).

The Boer War, indeed, is a topic to which most English discussions of
child survival and health in the first decade of this century tend, sooner
or later, to return. It serves as a prism through which are focused a
variety of related concerns touching the quality of the labour and fighting
forces, the integration of the working class, the position of women,
eugenics and the nature of social policy. Although one may distinguish a
variety of reactions to the war, the lessons that are drawn from it most
frequently concern the alleged evidence it supplied of the general ill
health of the British working class. Much play is made of the threat
posed to the British Empire and British industry by the superior level of
physical health which was, supposedly, to be found in other nations. The
topic of national efficiency became a political rallying cry in certain circles. The Inter-departmental Committee on Physical Deterioration of
1904 is one of the better-known reactions to this public anxiety; but there
were many others including official studies of school training in domestic
duties as carried out in other countries, an investigation of the desirable
minimum school attendance age for children, and the famous studies of
infantile mortality published by Newsholme as supplements to the Local
Government Board Annual Reports (Local Government Board
1910:1913).

These investigations may make one wonder whether there may have
been controversies within Edwardian England over the role of working-
class mothers that do not become immediately obvious from the secondary material. Were there, for example, debates over whether mothers
of young children should be discouraged from working outside the home?

Carole Dyhouse has drawn our attention to the existence of a few statisticians who were unwilling to accept the generally assumed view that
there was a close causal link between the severity of the infantile

mortality rate and the proportion of mothers of young children working. What we do not know is whether there were, perhaps, employers who were dependent on the work of young mothers who also expressed scepticism about the connection.

A further factor stimulating anxiety over the frequency of infant death in the period was the realization that the birth rate was falling at the same time as the infantile mortality rate showed no significant decline from the levels of the eighteen-seventies. The additional fact that the fall was taking place almost entirely among the higher social groups still further stimulated concern that something be done to improve the "quality" of the next generation. As one writer put it: ". . . if quality goes down as well as quantity, the outlook is not an enviable one." (Allison 1902:14). The same author inserted as a frontispiece to his booklet entitled *Health in Infancy*, a photograph of three black men standing reverently behind a white boy of about four years old seated on a rocking-horse. Beneath was a verse: "They had never seen many white men/They had never seen one afraid. THE EMPIRE RESTS ON THE INFANT" (ibid:frontispiece). The connection is seldom made quite as explicitly; nonetheless, this booklet illustrates the point made both in Anna Davin's research and in the retrospective observations of leaders of the Infant Welfare Movement such as Newsholme and McCleary[7]: namely, (that),

For many doctors and medical officers in the 1900s the saving of infant life seems to have become "a matter of imperial importance" (Newsholme 1905). (Davin 1978:14).

Faced with such testimony, there seems no doubt that in England the needs of imperialism were certainly the occasion, and probably a principal cause of infant death coming to be perceived as a major, remediable social problem in these years (Wright 1978).

A few passages from Anna Davin's paper make this point clearly:

The connection between the "health of the nation" and the "wealth of the nation" is nowadays comparatively easy to accept, since it has become a basic political tenet in contemporary Britain, so much taken for granted that it is seldom even articulated. But the timing of its emergence is significant, as is its particular focus. The recognition that the population was power, and that quality -- the standard of the physique of that population -- was also important, are clearly part of that background . . .

Healthier babies were required not only for the maintenance of the empire but also for production under the changing conditions made necessary by imperialist competition. The old system of capitalist production (which itself had nourished imperial expansion), with its mobile superabundant workforce of people who were underpaid, underfed, untrained and infinitely replaceable, was passing. In its place, with the introduction of capital-intensive methods, was needed a stable workforce of people trained to do particular jobs and reasonably likely to stay in them, neither moving on, nor losing too much time through ill-health (Davin 1978:49).

The ideological approach to the question of infantile mortality and domestic life can be seen therefore to have a close connection not only with the economic and political problems posed by falling birth rates, but also with new developments with industrial capitalism, in Britain. The barrage of propaganda on the importance of child health, with its bias about motherhood, did provoke official action, enquires, modest legislation, and various provisions by local authorities. It also helped to confirm or create attitudes about the relation between child and family and state, and most of all about the role of women; the influence of such changes was probably more far-reaching than any measures at the time. Where the solutions offered for improving national health were more concrete than the simple exaltation of motherhood, they were generally ones which tended to confirm the family in its bourgeois form and to consolidate the mother's role as child-rearer and home-keeper, as also did improvements in male wages -- the family wage -- and perhaps eventually family allowances (Davin 1978:56).

Thus it seems that public concern over infantile mortality is intimately related to the political and ideological pre-occupations of the last decades of the nineteenth century. The passages cited above indicated that the new conception of infancy cannot be fully explained simply by drawing attention to pragmatic considerations. Factors of this kind may well have sharpened public awareness; but they did not construct the conceptual framework within which such problems took shape. This, I argue, was the product of a conjunction of social factors.

THE MEDICALIZATION OF INFANT MORTALITY

But how was it that the problem came to be cast in a technical medical mould? Up to that time, the doctors seem to have had little to say professionally about infant rearing. When the subject was dealt with by medical writers of textbooks or popular handbooks published before the eighteen-seventies it was done in a manner closely resembling the homilies on the topic by clergymen and other moralists. The advice given was typically vague and non-specific, open to exceptions or, even, internally contradictory. It was expressed in the same terms as lay discourse on the subject and was organized predominantly around notions of naturalness and balance. When breast-feeding of babies by the mother was recommended -- as it usually was -- the justification was generally couched in terms of a norm of naturalness; rather than on the grounds that the practice was effective in preserving the health, or life, of the baby.

Even when medical authors appear at first sight to be providing instrumental advice that concerns the "consequences" of various forms of infant care, closer inspection often suggests that their recommendations were not founded on a rigorous evaluation of evidence but were simply re-enactments of traditional cosmological views or norms of behaviour concerning mothers and babies.

Thus, for instance, C.H.F. Routh (a physician to a women's hospital and FRCS) cautioned, in 1860, against suckling a baby when the mother had recently experienced strong emotion. He did so with the support of an anecdote about a woman who had inadvertently poisoned her child by suckling it just after she had helped to defend her husband against a violent attack (Routh 1860:257). Again, in 1869, P. Chavasse, author of a succession of best-selling medical guides to mothers, and also FRCS, counsels against the employment of a costive wet-nurse. ". . . Like follows like," he warned, harking back to long-current classical notions of harmony and similitude (Chavasse 1869:14). What is clear from such examples is that medical practitioners in the mid-nineteenth century were not claiming to possess a unique expertise in the care of infants. In many fields of life they were coming to be regarded as having a privileged, technical understanding. But on questions concerning infancy, their authority appears to have been no greater than that of the others who made public pronouncements on the question. Support for this assessment can also be drawn from the evidence of what non-medical writers had to say about medical expertise in child rearing. One anonymous mother, for example, wrote that she had,

. . . known physicians who made infants and children and especial study, and who were very clever as regards the ailments of children, quite at a loss as regards the feeding of a young infant, and obliged to rely more or less on the knowledge of a nurse (A Mother 1884:35).

In similar vein, the author of a booklet on child care addressed to young schoolmistresses cites as an authority, when recommending the careful washing of feeding bottles, ". . . the best old nurse I ever knew" (Lonsdale 1885:12).

By the turn of the century, however, the tone and justification for advice on child rearing was beginning to change noticeably. It was becoming precise, authoritative, specific and self-consciously based on scientific knowledge -- in particular on their germ pathogen theory. Some authors, indeed, were anxious to emphasize this shift in order to distance themselves from what had gone before. The medical author of a booklet on infant feeding in 1869, for example, recognised that the topic had frequently been neglected by the medical profession. It was "better not to write at all", he insisted, "than to collect a bewildering mass of facts from others and not know their relative value" (Physician 1869:5).

It seems likely that the writer had in mind books such as those by Chavasse, already referred to. A little further on, he explicitly articulated the new, medical claim to appropriate infancy as a field of exclusive expertise. The "proper feeding of infants," he wrote, "is based on scientific principles to understand which a medical training is absolutely necessary" (loc cit).

But the process by which infancy and the care of infants came to be defined as part of the technical, medical domain was complex. It is not plausible to see it as a simple example of medicalization (Illich 1975; Freidson 1975; Wright 1979); the boundaries of medicine, as they existed in the late nineteenth-century, were not simply extended in order to annex infancy to the realm of medical expertise. On the contrary, the forms of contemporary medical knowledge and practice provided many barriers to medicalization of infancy. I shall try to show that it only became a medical field with the creation of new forms of professional practice and the reconceptualization of the medical categories that had previously been used for the understanding of infant death.

Child Rearing in Victorian Medical Practice

For example, the major sphere of activity for doctors at that time, general practice, seems to have played little significant part in infant rearing -- and scarcely does even today. (In many countries infant welfare is still largely the responsibility of specialised personnel and clinics, not of general practitioners). The evidence strongly suggests that child rearing developed, was legitimated, institutionalized, and transmitted within a new arena of practice. An arena which was constructed from a series of activities that came into being, or were re-constituted, between the eighteen-nineties and the First World War: health visiting, domiciliary midwifery, schools for mothers, infant welfare centres, child care instruction to schoolgirls, and so on. Until these became established there was no practice, or practitioners, taking as a central professional concern the health and survival of infants. This can be seen by considering the major groups within the medical profession.

There is considerable evidence to suggest that general practitioners were not even likely to be brought into direct contact with the mass of the sickness that was to be found among young babies, particularly those of the poor (Smith 1979: 113). When they were consulted, it seems, it was frequently so late that the death of the infant was imminent and unavoidable. What is more, as doctors themselves were ready to admit -- at least to professional audiences -- the treatment of babies likely to succumb to the main causes of infant death was frustrating and disappointing. Why this should have been becomes obvious when one recalls that the main categories of infant death in towns during the Edwardian years were prematurity, wasting, diarrhoea, and convulsions. In many years these constituted 60% of all deaths among children less than a year old and 80% of those occurring in the first three months of life. Effective therapy was available for none of these rather vague and nebulous conditions. Moreover, all the evidence seemed to point to their being caused by environmental factors over which doctors in general practice could have very little direct control. Even when the next two commonest categories of death (bronchitis/pneumonia and infectious diseases) are included the position was not very different. Environmental factors still played a major role and there was little a doctor could do to improve the chances of survival of infants suffering from such diseases.

While the consultants attached to women's or children's hospitals were likely to see a far higher proportion of infants in their practice, and were often to be found among the authors of handbooks on child rearing, their daily experience does not appear to have been one which brought them into regular contact with a full and typical range of babies. They were unlikely to have been able to acquire the familiarity with variations in child rearing that would have enabled them to discern connections between these different practices and variations in the infant mortality rate. Nor is it probable that babies suffering from the ill-defined conditions which were categorised as the major causes of infant death would have formed the majority of their patients. It seems more probable that these were largely composed of those suffering from abnormalities of development, obvious chronic conditions or distinguishable acute diseases.

The only medical practitioners who seem to have regarded infant death as a major element of their work were the Medical Officers of Health (MOsH) and, indeed, it is from these that were drawn most of the early researchers on infant death and the leading advocates of the infant welfare movement. Nonetheless, the institutional position of the MOsH, especially their relationship to the medical profession as a whole, was ambiguous. Traditionally, they had been associated with the wave of public health improvements that had made possible the steep decline in deaths from infectious diseases among the adult population. Their typical mode of action had been to employ public, administrative, means based on epidemiological knowledge to combat conditions whose aetiology might at best be uncertainly, or incorrectly, understood.

MOsH played an important part in stimulating studies of infant mortality with the result that a considerable body of epidemiological research on the question had been carried out by 1900. From this, it became clear that a high rate of infant death was associated with urban areas, poverty,

poor housing, bottle-feeding of babies, unpaved streets, hot, dry sum-
mers, and – probably – the absence of mains drainage. Such results
were interesting (though, perhaps unsurprising), yet difficult to translate
into public health measures. Certainly MOsH could urge street paving
and better sanitation; but they had been doing this for decades without
improvement to the infantile mortality rate.

The difficulty was that many of what seemed to be the major causative
factors in the environment were not susceptible to local public health
measures but could only be remedied by major political and social
changes. This realization was further complicated by the lack of understanding of the precise causative mechanisms that linked the various factors. This, in turn, made it hard to judge which measures would be
effective, or to establish priorities among them.

The Impact of the Germ Pathogen Theory

What was to change this state of affairs, however, was the incorporation
of the germ pathogen theory into the infant welfare debate. Once this
had been achieved one particular set of causative links -- those concerned
with the transmission of infection -- came to dominate discussion and to
serve as the legitimating principle for most of the initiatives proposed.

The influence of this theory developed only slowly among MOsH and
was not always welcomed. Jean Raymond has suggested, that during
most of the eighteen-seventies "germ theory served as a sort of intellectual top-dressing to reinforce doctors' claims to public power and prestige" (Raymond 1985:4). Bacteriological knowledge, she suggests, did not
really become part of the regular practice of public health in Britain until
around 1895. Even then, it was typically employed to investigate unusual
outbreaks of rare diseases such as salmonella or anthrax and to provide
an ancillary technique of diagnosis for general practitioners. It was not
until about 1900 that bacteriological techniques were regularly being
employed to clarify the aetiology of summer diarrhoea and other major
categories of infant death.

There were many reasons for this. One was certainly the residual pre-
dilection of many MOsH towards general environmental explanations: a
tendency reinforced by their daily experience and practice. Thus, for
instance, the MOH for Southport could argue in 1894 that:

The singular persistence with which filth allies itself with diphtheria . . . makes it
difficult to believe that diphtheria has no more than an accidental connection with
dirt. That is not the impression produced by practical experience of sanitary
administration, whatever may be the results of laboratory experiments and artificial
breeding of bacteria (quoted in Raymond 1985:14).

Thus it appears that an awareness of the main features of the germ theory and the ability to understand its potential applicability to the problem of infant death were two different things. As so often in intellectual history, abstract knowledge only became intellectually accessible to the extent that it was incorporated into a practice where it could be utilised. Germ pathogeny, by virtue of its specific association with discrete infectious diseases, was hard to assimilate to a public health routine that centered on environmental action, often of a very generalised kind.

CONCEPTUAL BARRIERS TO THE UNDERSTANDING OF INFANT MORTALITY

Another impediment to linking bacteriological knowledge to infant mortality lay in the manner in which the major causes of this mortality were categorised. "Wasting" was a vague term that could be interpreted to imply some natural process; "convulsions" was simply descriptive and symptomatic; while "summer diarrhoea" was elusive in two different ways. Firstly, the adjective summer tended to draw attention to seasonal and climatic factors and this made it easy to assimilate the condition into traditional models of cosmic disease causation. Even when new bacteriological findings were accepted, the old climatic and environmental views of diarrhoea tended to be reinterpreted in a more restricted way rather than simply abandoned. Thus, for example, Newsholme, who had regarded high temperature as the primary cause of summer diarrhoea in 1885 (Newsholme 1935:140 et seq) later accepted germ causation but could still express himself in 1911 in a way that contained many echoes of his earlier position. "The fundamental condition favouring epidemic diarrhoea," he wrote, "is an unclean soil, the particular poison from which infects the air, and is swallowed, most commonly with food, especially milk" (Newsholme 1935:357). "Soil", "poison" and "air" had each been important components of the earlier discourse.

Secondly, the term "diarrhoea" also created problems since it was treated as a disease entity in classifications of causes of death, although in medical theory it was regarded as only a symptom of disease, not an entity in itself. Doctors, it was said, were unwilling to enter the term on death certificates because they believed that the general public regarded it as a trivial condition: to cite it as a cause of death could thus be read as an admission of professional impotence, if not incompetence (Waldo 1900:1344). To escape such embarrassment a Committee was established by the Royal College of Physicians in 1899 which later proposed that summer diarrhoea be reclassified either as epidemic enteritis or as zymotic enteritis (ibid). Nothing appears to have come of this proposal in the years that immediately followed the Committee's report.

The RCP's recommendation is most revealing for it indicates that when doctors tried to understand infant death in terms of germ causation they did so in terms of one particular model that was already fairly familiar -- that of "enteric fever" (typhoid). This made it hard to understand conditions such as summer diarrhoea which were nebulous and related to pathogenic agents which were numerous and hard to isolate (Waldo 1900).

Furthermore, assumptions about the mode of transmission of diarrhoea based upon an understanding of typhoid might well be misleading. In his Milroy lecture to the Royal College of Physicians in the Spring of 1900, F.J. Waldo, a leading London MOH, argued that familiarity with typhoid had probably encouraged doctors to lay undue emphasis on water and milk as mechanisms for transmitting infection and to neglect "infection spread by elements of the environment affected by stools -- e.g. bedding" (Waldo 1900:1428).

THE ASCENDANCY OF THE GERM PATHOGEN THEORY IN INFANT WELFARE

Despite all such barriers, however, emphasis on the importance of microbial agents in infant death grew rapidly in the Edwardian years until such a point that debates about infant welfare tended to be dominated by the subject of summer diarrhoea and the role of germ pathogens in its causation. So great was this emphasis that, as Dyhouse and Wohl have pointed out, (Dyhouse 1981:91) other causes of death -- even "wasting diseases" -- the biggest category of death throughout the eighteen-nineties -- remained largely neglected. Summer diarrhoea became the paradigmatic example.

Explaining the sudden rise in the germ pathogen model of summer diarrhoea raises a number of important theoretical questions. A social constructionist analysis will not regard as acceptable the once-common view that only scientific "error" requires social explanation since scientific "truth" explains itself by factors internal to the development of science. Sociological explanations must be symmetrical (Bloor 1976:142) and address both those aspects of science regarded as established at a particular time as well as those considered discredited.

For this reason alone, it is not sufficient to try to explain the ascendancy of bacteriological models of infant diarrhoea either by arguing that they were correct science, or demonstrably more effective that earlier ways of conceptualizing the issue. Both claims would in fact be difficult to advance as, even today, there is no consensus among historians on which factors were the most important in bringing about a striking reduction in the infantile mortality rate (from a figure of around 110-130 per thousand live births in the Edwardian years, down to a level of around 70 per thousand at the end of the twenties). F.B. Smith, for example, has written:

I know of no advance in public sanitation which could explain the sudden change after 1902, nor of any major innovation in medicine which affected infant lives in these years (Smith 1979:113).

Smith's conclusion, like those of other writers, is that the improvements were probably brought about by a combination of factors some, at least, of which -- including improvements in general nutrition -- had no direct connection with the infant welfare movement or medical knowledge of the causes of death among babies.

Another possible explanation for the centrality of summer diarrhoea in the Edwardian infant welfare movement is that the condition was becoming more prevalent. Certainly there is some evidence for this. The Registrar General's figures show that the five-year moving average for death from diarrhoea among infants under one year of age peaked at 34 per thousand in 1900, a figure roughly double that of the early nineties. Although the figure then fell, it levelled out at around 23 per thousand in the mid-Edwardian years before falling again by the beginning of the First World War, to just above that of 1890 (Lewis 1980:62). Also, variations in the rate of death from diarrhoea constituted a major element of the differences between areas of relatively high and relatively low infant mortality.

It would be rash, however, to accept these figures at their face value and to assume that the fluctuations in them occurred quite independently of the infant welfare movement. I have already referred to contemporary evidence which suggests that doctors were sometimes unwilling to register diarrhoea as a cause of death. It does not seem implausible that such reluctance lessened as the condition became the focus of national debate. Similarly, one can also conceive of social factors that might have encouraged urban practitioners to employ the term more frequently than their rural colleagues.

Again, the distinction between "diarrhoea" as a cause of death and "atrophy and debility" (which were classified under wasting diseases) was often simply a question of convention or the chance consequence of when a doctor first happened to see the baby concerned. This too, was noted by contemporary observers (Smith 1979:87).

When such contemporary evidence is placed in the context of modern sociological work on official statistics and medical classification it seems unjustifiable to regard variations in the reported rate of death from diarrhoea as a significant cause for the rise to prominence of this topic.

THE NEW INFANT WELFARE PRACTICE

The thesis of this paper is that the major factor explaining the predominance of the germ pathogen theory in the understanding of "summer diarrhoea" was social use. Germ pathogeny provided a rationale for social intervention and a metaphor of social danger that was not otherwise available.

During the twenty years from the mid-eighteen-nineties infant welfare established itself as an institutionalized field of practice. This was the period, for example, in which health visiting was placed on a professional footing. That activity had already been growing with increasing speed since the mid-nineteenth century, and by 1905, over fifty towns in Britain had health visitors – usually middle-class lady volunteers, supported by paid assistants of humbler social background.

In 1905-6, however, these services were institutionalized as a result of two factors: firstly, the requirement that all new health visitors were to have completed a course of medically-based training; secondly, the Notification of Births Act which alerted the authorities to each birth which took place and enabled health visitors to begin early and regular visits to the mothers concerned.[8] Similarly, the Midwives Act of 1902 made medically-based training mandatory for new entrants to this field. Infant welfare clinics were also springing up quite rapidly in the first decade of the twentieth century, most under the aegis of voluntary bodies. By 1916, 160 branches of voluntary organisations and 35 local authorities were running infant welfare clinics (Lewis 1980:34). In addition to such formally organized activity, a wide range of other services were also coming into being: lectures and pamphlets directed towards working-class mothers, depots for the supply of sterilised milk, baby competitions, financial rewards to mothers whose children survived the first year of life, instruction on hygiene and child care to school children, and many others.

This new movement created new audiences and the need for new kinds of explanatory and teaching material. It was necessary to equip the trained health visitors and midwives with skills that would enable them to convince working-class mothers of the need to adopt the precepts of the new child-rearing even when this meant abandoning habits that were deeply rooted in the family and local culture. It was necessary, too, to equip the lecturers and publicists with material that would help them to convey their ideas persuasively.

A distinctive feature of this movement was that it depended scarcely at all on already established institutions: some use, it is true, was made of schools and hospitals in the diffusion of the new infant care, but the key mode of transmission was essentially novel. It was based upon the direct penetration into working-class homes of female exponents of the medically-validated infant welfare. Those who acted as bearers of this knowledge -- midwives and health visitors -- though usually drawn from higher social classes than their clients, were nonetheless far closer to the working-class mothers than doctors, clergymen or, even, sanitary inspectors, were likely to be. What is more, unlike the over-whelming majority of such professionals, *they were women*. Thus a new kind of dialogue was made possible which provided a direct channel of communication to individual working-class mothers, a channel relatively free of mediation by husbands, neighbours or the local community.

Jacques Donzelot has written that related developments in France forced part of the emergence of the new sphere of "the social": that they constituted new ways of classifying and understanding problems that were associated with new techniques for shaping and controlling the family from within, particularly through the mother (Donzelot 1979:xix-xxvii).

Interestingly, however, the developments in England fit Donzelot's thesis more closely than the French experience on which he largely based his argument. In France, no health visiting service was set up and instruction in infant care was largely carried on through clinics and classes for mothers. What is more, this period was one in which the *écoles maternelles* (schools taking children from 2 to 5 years) survived and were even expanded (Boltanski 1969).

In England, there was a far more explicit emphasis on confining both the young child and the mother to the home. Carol Dyhouse has already shown how the reduction of the proportion of married women in employment became a central theme in the infant welfare movement notwithstanding considerable evidence that this might have no effect -- or even a detrimental effect -- on infant mortality (Dyhouse 1981). What is less well known is that the policy was also adopted of reducing the proportion of children under five in schools. The *Report of the Consultative Committee on the School Attendance of Children* in 1908, for instance, laid great stress on ". . . the natural relationship between mother and child" and went on to emphasize that this, along with the other influences of a "good home" were ". . . a moral and educational power which it is of high national importance to preserve and strengthen" (Report 1908:16). In fact, as the *Report* shows, school attendance of under-fives was already falling: it had gone down from a peak of 43.5% of the 3-5 age group in 1899 to 33.76% in 1906 (loc cit).

The superior effectiveness of home visiting over welfare centres in reaching all mothers was explicitly recognised by policy makers. The Local Government Board in 1916 stated that it regarded

the provision of adequate home visiting as the most important element in any scheme of maternity and child welfare (Local Government Board 1916:4).

Further on in the same circular, it is noted that, in contrast, only about a quarter of mothers could be expected to attend infant welfare centres (ibid:6).

From the standpoint of the present day it is easy to overlook the social and political significance of such developments for we may be inclined to take for granted the existence of a multiplicity of channels of communication between the state and individual members of the family. In the period in question the situation was different: working-class mothers were relatively untouched by large-scale networks involved with the transmission of knowledge and values. For many, the only exposure would have come through the school, with perhaps some contact through advertising. The popular press had scarcely been born and was probably most influential among men. Nor, of course, did women share the experiences of the army and navy that some working-class males were exposed to. Although women may have been more inclined to religious practice than men, religious attendance by working-class women in towns was not high, especially among the unskilled.

But the practice of infant welfare was not only distinctive in the channels of communication that it established; the context and authority of the message that it transmitted was also novel. The advice on the care of infants was cast in a quite new mould, different from that of doctors thirty years before. It was, to borrow Habermas' term, presented as purposive-rational action (Habermas 1969). That is to say, its precepts were presented not as social norms which mothers should follow because they derived from desirable values -- what Habermas refers to as communicative action -- but as instrumental recommendations deriving from technical rules. Breast-feeding, for example, was not urged because it was thought natural, divinely-ordained, or even just a proper expression of the mother's role; it was advocated as the most successful way of avoiding the death of a child. The consequence of ignoring such advice was presented as material failure -- the working out of scientific laws leading to the illness or death of the child -- not as social nonconformity which might be punished, as it had been a few decades before.

In all of this, the notion of germ pathogeny played an essential part: it was an exemplar of the scientificity of medicine and spread a cloak of authority over the work of the new infant welfare professionals, giving legitimacy to their injunctions even when there was no logical connection with them. Thus, for instance, Janet Lane-Claypon, Assistant Medical Inspector at the Local Government Board from 1912-16, could attack the "excessive kissing of infants" (Local Government Board 1914:2) and others fulminated against demand feeding and "dummies" (Porter 1900), masturbation (Coolidge 1905:203) and the dangers of the baby sleeping on its right side -- "to avoid pressure of the full stomach on the heart"

(Maynard 1906:20). Similarly, early pot-training and weaning before one year of age were also presented with the same unquestionable, scientized authority.

The stress in published child rearing advice during these years was by no means limited to explanations of germs and sterilization, or to the principles of nutrition in order to improve milk preparation or feeding, it extended into innumerable aspects of the lives of babies and their mothers and continually raised the need for principles such as regularity, order and abstinence. Given that this published material was often directed towards the training of those active in infant welfare -- or even, sometimes the mothers themselves -- it seems probable that the verbal advice given in the home was not greatly different. The flavour of this discourse is conveyed by a penny pamphlet (which was likely to have been directed towards the mothers themselves) written by E.L. Maynard who had been a Sanitary Inspector in Sheffield:

Regularity is the first part of the baby's education . . . [the mother] will find it easier to train it in *clean habits,* for if the child is fed at odd times the bowels will fail to act regularly (emphasis in the original, Maynard 1906:19).

THE CULTURAL POWER OF THE GERM PATHOGEN MODEL

If the germ pathogen model of disease provided a major source of legitimacy for the new infant welfare movement the question still remains as to what were the sources of its power. Unfortunately, little research has been done on this question and many aspects of the impact of germ theory still remain to be investigated. Nonetheless, it is possible to suggest two general types of explanation.

The first is the germ pathogeny represented to the lay public an object lesson in the benefits of the application of scientific knowledge to medicine. Certainly, there is some evidence for this: Pasteur, Semmelweiss, Lister and Koch are celebrated again and again, and the identification of the microbial agents associated with such diseases as anthrax, puerperal fever, gangrene and tuberculosis were events that caught the popular imagination. In this sense, medicine represented one aspect of the rise in prestige of scientific knowledge which seems to have been taking place in these years. In popular writings this tendency sometimes assumed hagiographic form. As English textbook in hygiene of 1916 directed towards older schoolchildren wrote, for example:

Truly after the Great Healer of Mankind, no man had done more to banish disease and death than Louis Pasteur. It cost him much -- at one time almost his own life

. . . he showed mankind its most merciless foe, its most powerful enemy, and all young Commanders who join the crusade that Louis Pasteur began will wish to know the story of the Great Scout (Pasteur), (Hood 1916:179-180).

Part of the power of the germ pathogen theory, then, was indirect: it symbolized the claims of infant welfare advice to scientificity and can therefore only be explained to the extent that the cultural ascent of science as a whole can be explained.

But the other side of the appeal of germ pathogeny was metaphorical: its power to act as an image with which to place human affairs within a convincing framework of meaning.[9] To demonstrate this is not, of course, easy since it involves judgements about the relative plausibility of various frames of reference. Nonetheless, there is much to suggest that Edwardian England was a culture that placed considerable weight on categories related to purity and boundary maintenance and was sensitive to many of the related distinctions to which Mary Douglas' work has made us alert (Douglas 1966:1970). Such emphasis is especially evident in the discourse of politics and social policy. The small change of Edwardian journalism and politics was made up of words and phrases such as "social purity," "mothers of the race," "the imperial race," "British stock," "degeneration" and "purification of the race" (examples quoted in Davin 1978:13-19).

Two characteristic themes can be detected running through such imagery. The first, is the central importance placed on the boundary between inside and out: between the "imperial race" and aliens; the nation and its enemies. The second, is the fear of internal pollution: the danger of degeration, internal decay or dissent. The terminology of germ pathogeny fell into easy harmony with such imagery, strengthening it and vicariously endowing it with some of the added weight of science. Germ pathogeny came to serve as a trope for society and social affairs -- a model for making sense of disease.

These points are well illustrated by a few excerpts from the hygiene reader already quoted, E. Hood's, *Fighting Dirt: the World's Greatest Warfare* (1916):

While the boy and girl yearn to handle a sword and go to the wars, deadly enemies surround them, lurk in their clothes, cling to their flesh, penetrate into their mouths, and only wait a favourable moment to attack their bodies in force, as soldiers have attempted the capture of some great fortress, such as Gibraltar (Hood 1916:16-17).

Your body is a great fortress, and contains yards of trenches in which are ranged countless invisible soldiers to defend it . . . (ibid:18) . . . while the comparatively small submarine can destroy huge vessels, so the tiny germ of disease can lay low the strongest of men (ibid:28).

Thousands of babies die quite needlessly every year from summer diarrhoea. Where the infant draws its food straight from the mother's breast there is no danger of the milk becoming contaminated by flies or filth, but with cow's milk there are scores of opportunities in carelessly kept dairies and homes for the enemy to enter the milk and so be carried into the delicate stomach of the baby. There the usual plan of campaign is adopted — poisons are produced by every member of the invisible invaders, and unless the doctor is at once summoned to check their ravages, the baby soon surrenders its weak little body into the hands of the foe (ibid:154-155).

CONCLUSION

The purpose of this paper has been to consider medicalized child rearing as a social construction. In this construction, it is suggested, many different social forces were involved that happened, contingently, to come together into one particular set of relationships.

Four groups of processes have been distinguished and examined in varying degrees of detail. The first group consists of those which raised child mortality and health to become topics of great political importance and visibility in Edwardian England. These are considered in considerable depth in Davin (1978), Dyhouse (1981) and Lewis (1980) and are only briefly touched on in this paper. The second, comprises the changes that were taking place in professional, medical ideas at this time. Some writers — including, for instance, David Armstrong (forthcoming) — have examined these in terms of the internal evolution of medical knowledge. I, however, have tended to give greater weight to the impact of the organization and content of medical practice on the changing nature of medical attitudes to infancy. In addition, I have also tried to show how differences in the day-to-day experience of doctors may have encouraged different judgements on the relative plausibility of various ways of conceptualizing the causes of infant death.

The principle emphasis of my approach, however, has been on the two remaining groups of processes: the creation of a new sphere of professional activity directed towards infant welfare, in which a new kind of knowledge could come into being and be practised, and the association of infant welfare with the germ pathogen model of disease which led to the establishment of a vocabulary of vivid imagery through which infancy could be apprehended and understood.

Inevitably, this paper does no more than scratch the surface of these last two groups of processes. Both merit sustained research, particularly the question of the cultural and ideological power of medical ideas.

How we may ask, returning to Schön's notion of "generative metaphor," mentioned earlier, does it happen that new social circumstances may suddenly provide a space in which new forms of medical knowledge may rapidly constitute themselves? Or, again, how should it be that an element of seemingly esoteric medical knowledge may suddenly serve as the prism through which groups within a society succeed in making sense of what is occurring around them?

These, I believe, are some of the central problems of social and historical study of medicine; but they are essentially cognitive, not instrumental. They must not be confused with issues of whether medicine "works" in some unproblematic, pragmatic sense. Take, for instance, what John Harley Warner writes about the weakness of the history of medicine in his recent survey of research on the history of American medicine:

> ... there is virtually nothing that analyses the meaning of science from the patient's point of view ... it is plainly necessary to consider why Americans bought what many plainly believed that laboratory science gave them to sell. Even if the ideal of experimental science won the hearts and minds of physicians, the questions of to what extent and why it did the same for the lay man and woman remains unclear especially given the doubtful ability of scientific medicine in the late nineteenth century to deliver the goods in terms of a demonstrably elevated power to cure (Warner 1985:46).

Having rightly drawn attention to the neglect of the important question of why patients support a particular form of medical practice, he then writes as if it is obvious that this neglect is made more serious because medicine in the period in question was "demonstrably" unable to "cure."

But surely those involved in the cultural and historical study of meanings can never allow themselves to make such assumptions. A social constructionist, for example, must always examine how, and why, a particular sphere of knowledge is successful — or not — in asserting its claims to cultural credibility. The question should never be begged simply by assuming that the "rightness" or "effectiveness" of knowledge is its own justification (Bloor 1976).

In trying to explain why medicalised child rearing was successful in establishing its claim to credibility I have argued that it is necessary to give close attention to the part played by medical practice and medical ideas in the constitution and reproduction of ideology.[10]

NOTES

1. Several writers in recent years have studied medicine from a social constructionist standpoint. These include Armstrong (1983), Figlio (1978) (1979), Freidson (1975), Sedgwick (1982) and the contributors to the Wallis (1979) and Wright and Treacher (1982) volumes. For a critical analysis of social

constructionism and a consideration of different approaches within it see Bury (1986).

2. This is not, of course, to imply that the same transformation took place in the way that the mass of mothers saw infancy. From present evidence it is hard to make an informed judgement on this question. Anna Davin (1978), however, has initiated consideration of some of the issues involved. See also Stanton (1979).

3. See writers such as Shorter (1975), Stone (1977) and De Mause (1976).

4. For example, the seventeenth-century diarists John Evelyn (1955 (1620-1706):iv, 463-4) and Ralph Josselin (1976 (1644-83):165-7) were both fathers who seem to have felt a deep — but resigned — sense of loss when their children died prematurely.

5. "Less eligibility" was the principle that those receiving poor relief ought to be worse off than those maintained by their own labour.

6. Although Dyhouse (1981:78) contends that the tendency to lay blame on mothers for infantile mortality became greatest after 1900, I do not believe there is real disagreement between us. My view is that the supposed "faults" of mothers were increasingly perceived as ones of ignorance or lack of skill, not as *moral* failings. Indeed, Dyhouse in the passage cited refers to mothers being blamed for being "ignorant and incompetent in matters of infant care" (loc cit). This is important, for a discourse framed in terms of ignorance is likely to be linked to different remedies from one framed in terms of morality.

7. Dr., later Sir, Arthur Newsholme had been Chief Medical Officer to the Local Government Board from 1908 to 1918. Dr. G.F. McCleary had been Chief Medical Officer to the National Insurance Commission, Chairman of the National Council for Maternity and Child Welfare, and Chairman of the National Association of Maternity and Child Welfare Centres and for the Prevention of Infantile Mortality.

8. There was also explicit recognition that health visiting was directed only towards the working-class mothers and then, specifically at the poorest. Lane-Claypon, for instance, reported that, "better class artisans (were) not visited." (Local Government Board 1914:6).

9. Susan Sontag (1978) has explored the metaphorical uses of illness in literature.

10. This theme has often been explored by Karl Figlio. In a recent paper he writes:

People in Britain go to see a doctor on average about four times per year; for the society as a whole, that amounts to well over 200 million encounters per year in confirmation of natural law. That makes up a massive participation in scientific naturalism; nowhere else does everyone take part to that extent in confirming the natural order of things through medical science. So medicine looms large in the everyday reproduction of ideology as an aggregate of routines through which we learn to accept the organization of nature and society as it is (Figlio 1985:129).

REFERENCES

Allison, T.M.
  1902 Health in Infancy. Newcastle-upon-Tyne: T. and G. Allan.
Armstrong, D.
  1983 Political Anatomy of the Body: Medical Knowledge in Britain in the Twentieth Century. Cambridge: Cambridge University Press.
  The Invention of Infantile Mortality. Sociology of Health and Illness. (forthcoming)
Bloor, D.
  1976 Knowledge and Social Imagery. London: Routledge and Kegan Paul.
Boltanski, L.
  1969 Prime education et Morale de Classe. Paris: Mouton.
Bury, M.
  1986 Social Constructionism and the Development of Medical Sociology. Sociology of Health and Illness 8:137-69.
Canguilhem, G.
  1979 *The Normal and the Pathological*. Dordrecht: D. Reidel Publishing Company. (translation of Le Normal et le pathologique. Paris: Presses Universitaires de France, 1966, which included Canguilhem's doctoral thesis -- from which quotation is made -- first published in 1943).
Chavasse, P.
  1869 Counsel to a Mother (First Edition). London: Churchill.
Coolidge, E.L.
  1905 The Mother's Manual. London: Hutchinson.
Davin, A.
  1978 Imperialism and Motherhood. History Workshop 5: 9-65.
De Mause, L.
  1976 History of Childhood. London: Souvenir Press.
Donzelot, J.
  1979 The Policing of Families. London: Hutchinson. (translation by R. Hurley of La Police de familles. Paris: Editions de Minuit 1977).
Douglas, M.
  1966 Purity and Danger: An Analysis of Concepts of Pollution and Taboo. London: Routledge and Kegan Paul.
  1970 Natural Symbols: Explorations in Cosmology. London: Barrie and Rockliff.
Dyhouse, C.
  1981 Working-Class Mothers and Infantile Mortality in England, 1895-1914.

In C. Webster (ed.), Biology, Medicine and Society 1840-1940. Pp. 73-98. Cambridge: Cambridge University Press.

Evelyn, J.
1955 The Diary of John Evelyn (1620-1706). Edited by E.S. de Beer. Oxford: Clarendon Press.

Figlio, K.
1978 Chlorosis and Chronic Disease in Nineteenth-Century Britain: The Social Construction of Somatic Disease in a Capitalist Society. International Journal of the Health Services 8:589-617.
1979 Sinister Medicine? A Critique of Left Approaches to Medicine. Radical Science Journal 9:14-68.
1985 Medical Diagnosis, Class Dynamics, Social Stability. *In* L. Levidow and B. Young (eds.), Science, Technology and the Labour Process. Marxist Studies Volume 2. Pp. 129-165. London: Free Association Press.

Freidson, E.
1975 Profession of Medicine: A Study of the Sociology of Applied Knowledge. New York: Don Mead.

Geertz, C.
1973 The Interpretation of Cultures. New York: Basic Books.

Gilbert, B.B.
1966 The Evolution of National Insurance in Great Britain: The Origins of the Welfare State. London: Michael Joseph.

Habermas, J.
1969 Technology and Society as "Ideology". *In* idem., Toward a Rational Society. Pp. 81-122. London: Heinemann (translated by J.J. Shapiro from Technik und Wissenschaft als "Ideologie." Frankfurt: Suhrkampf 1968).

Hood, E.
1916 Fighting Dirt: The World's Greatest Warfare. London: Harrap.

Illich, I.
1975 Medical Nemesis: The Expropriation of Health. London: Calder and Boyars.

International Congress for the Welfare and Protection of Children
1902 Report of the Proceedings of the Third International Congress. London: P.S. King and Son.

Lambert, R.
1963 Sir John Simon 1816-1904. London: McGibbon & Kee.

Lewis, J.
1980 The Politics of Motherhood: Child and Maternal Welfare in England, 1900-1930. London: Croom Helm.

Local Government Board
1910 39th Annual Report: Supplement on Infant and Child Mortality (by A. Newsholme). London: HMSO. Cd. 6263.

1913   42nd Annual Report: Supplement on Infant and Child Mortality (by A. Newsholme). London: HMSO. Cd. 6909.

1914   Maternity and Child Welfare: A Memorandum on Health Visiting (by Janet Lane-Claypon). London: HMSO.

1916   Maternity and Child Welfare: Circular of 23 September 1916. London: HMSO.

Lonsdale, M.
1885   The Care and Nursing of Children. London: Hatchards.

McFarlane, A.
1970   The Family Life of Ralph Josselin: A Seventeenth-Century Clergyman (1644-83). London: Cambridge University Press.

Martineau, H.
1859   Breast Feeding. Medical Times and Gazette. 10 September 1859.

Maynard, E.L.
1906   Baby: Useful Hints for Mothers. Bristol: Wright.

A Mother
1884   A Few Suggestions to Mothers. London: Churchill.

Newsholme, A.
1905   Infantile Mortality: A Statistical Study. Practitioner. October 1905: 494.

1935   Fifty Years in Public Health. London: Allen and Unwin.

Physician
1896   Infant Diet and Sterilized Milk. London: Sampson & Low.

Porter, C.
1900   Suggestions as to the Feeding and Care of Infants (For use in the senior classes of girls' schools). Stockport: Thompson.

Raymond, J.
1985   Science in the Service of Medicine: Germ Theory, Bacteriology and English Public Health, 1860-1914. Unpublished paper presented to the Conference on Science in Modern Medicine organised jointly by the British Society for the History of Science and the Society for the Social History of Medicine, Manchester, 19-22 April, 1985.

Routh, C.H.F.
1860   Infant Feeding. London: Churchill.

Report
1908   Report of the Consultative Committee on School Attendance of Children below the Age of Five. London: HMSO Cd. 4259.

Schön, D.A.
1979   Generative Metaphor: A Perspective on Problem-Setting in Social Policy. In A. Ortony (ed.), Metaphor and Thought. Pp. 254-283. Cambridge: Cambridge University Press.

Searle, G.R.
1971   The Quest for National Efficiency. Oxford: Oxford University Press.

Sedgwick, P.
   1982   Psychopolitics. London: Pluto Press.
Semmel, B.
   1960   Imperialism and Social Reform. London: Allen & Unwin.
Shorter, E.
   1975   The Making of the Modern Family. London: Collins.
Smith, F.B.
   1979   The People's Health 1830-1910. London: Croom Helm.
Sontag, S.
   1978   Illness as Metaphor. New York: Farrer, Strauss and Giroux.
Stanton, J.
   1979   Responses to the Problem of Infantile Mortality with Special Reference
          to the Infant Welfare Movement in Oxford, 1902-1918. Unpublished
          MA dissertation, Birkbeck College, University of London.
Stedman Jones, G.
   1971   Outcast London: A Study in the Relationship Between Classes in Victo-
          rian Society. Oxford: Oxford University Press.
Stone, L.
   1977   Family, Sex and Marriage in England 1500-1800. London: Weidenfeld
          & Nicholson.
Waldo, F.J.
   1900   Summer Diarrhoea (the Milroy Lectures to the Royal College of Physi-
          cians presented on March 8/13/15 1900). The Lancet 12 May 1900,
          1344-50; 19 May, 1426-30 and 26 May, 1494-8.
Wallis, R. (ed.)
   1979   On the Margins of Science: The Social Construction of Rejected
          Knowledge. Sociological Review Monograph 27. Keele, Staffs.
Warner, J.H.
   1985   Science in the Historiography of American Medicine. Osiris 1:37-58.
Wright, P.W.G.
   1978   Child Care, Science and Imperialism. Bulletin of the Society for the
          Social History of Medicine 23:15-17.
   1979   Some Recent Developments in the Sociology of Knowledge and their
          Relevance to the Sociology of Medicine. Ethics in Science and Medi-
          cine 6:93-104.
Wright, P.W.G. and Treacher, A. (eds.)
   1982   The Problem of Medical Knowledge: Examining the Social Construction
          of Medicine. Edinburgh: Edinburgh University Press.

PATRICIA KAUFERT

MENOPAUSE AS PROCESS OR EVENT: THE CREATION OF
DEFINITIONS IN BIOMEDICINE

INTRODUCTION

An historical analysis is the more usual approach to understanding the
production of medical knowledge, but in this essay I will examine the
anatomy of a construct which is current. The focus is menopause.
While not denying the existence of an underlying biological reality in
which women age, lose their fertility and no longer menstruate, menopause is a social construct and not a separate, independent, biological
entity. More accurately, there is a multiplicity of constructs parading
under the same label: the feminist versions of menopause have little in
common with the medical (Kaufert 1982). It is with the medical construction of menopause that the following discussion is concerned.

The social constructionists have described medical knowledge as
appearing "under scrutiny to be composed of limited interpretations of
the complex phenomena of illness" (Gabbay 1982). Gabbay contrasts his
view with that of the medical profession which, undisturbed by such Foucaultian insights into the nature of its knowledge base, sees medical
knowledge as an assembly of "proven, timeless objective facts" (Gabbay
1982). This medical world view depends on belief in a reality in which all
is orderly, predictable and stable. A world in which disequilibrium is
materially generated (whether by viruses, bacteria, parasites or some
other cause), can be empirically observed and externally corrected (Berliner 1982). Above all, it is a world which is knowable, but only by those
who honour the rules of scientific method.

The materials used in the construction of this world come largely,
although not exclusively, from medical research. Indeed, seen from
within the medical community, the function of research is the provision
of an array of facts which, when assembled together, constitute biomedical knowledge. The methods used in this research are reflections of philosophic assumptions concerning the nature of the scientific enterprise
which are deeply embedded within the culture of the medical community.
Locke's dictum, for example, that the "two primary functions of language
(are) designation and classification" (Good and Good 1981:181) is
accepted as fundamental to the production of medical knowledge. It follows that acts of definition and the delineation of boundaries separating
the "diseased" from the "non-diseased" state are integral to the

331

M. Lock and D. R. Gordon (eds.), Biomedicine Examined, 331–349.
© 1988 by Kluwer Academic Publishers.

construction of any medical phenomena, be it Miner's nystagmus (Figlio 1982) or asthma (Gabbay 1982) or menopause.

For epidemiologists, "designation" and "classification" are both philosophical principles and rules governing the actual process of research. As the most methodologically compulsive members in the medical research community, epidemiologists will debate endlessly over the selection of a population, the choice of a definition, the appropriateness of a statistical test. This preoccupation is a function of their training as epidemiologists. It arises also from their recognition that these minutiae of the research process have a critical role in structuring the end product: the research results.

Compliance with the rules governing research is seen within the medical community as a scientifically good, but a socially and politically neutral act. From a constructionist perspective, however, no element in the production of medical knowledge is neutral. Yet, while they have exposed the production of knowledge as an ideological exercise, the social constructionists have been more interested in the philosophical and social base of research and in its product – the disease construct – than in those details of the research process which fascinate the epidemiologists.

Epidemiologists give more importance than the constructionists to the link between the methods of research and the results partly because methodological expertise is the stock-in-trade of their discipline. (By comparison, the main consumers of medical knowledge, the practising clinicians, are too often methodologically naive.) But the link is also missed because once produced, medical facts take on a form of independent life. Methodological details, such as the characteristics of the subject population or how a particular variable was measured, are forgotten and ignored. Facts are allowed to float free from their original base in the research process. To understand how a construct came into being, particularly the role of research design in shaping its structure, one must go back and re-attach these "facts" to this initial base. As an illustration of the connection between the methods and the product of research, this essay will use the medical construction of menopause. It will explore the relationship between the methods used in menopause research, the definition of menopause and the social context of being a woman in midlife.

THE CONSTRUCTION OF MENOPAUSE

The "designation" and "classification" of menopause is singularly instructive of the way in which methodological concerns can determine the construction of a medical entity out of the complexities of physiological process. Consequent on its use of a disease/non-disease model of biological reality, medicine prefers boundaries which are discrete, objective, valid and reliable. The biological changes involved in women's aging tend to be ambiguous, continuous and idiosyncratic. They do not easily lend themselves to use as boundary markers, for unlike pregnancy and childbirth, there are no well marked entrance or exit points which can be used to establish the boundaries between being or not-being menopausal. Undaunted, and in an effort to comply with the rules of categorization, medical researchers have looked for indicators of change from one menopausal status to another which would be stable and universal. By general agreement within this research community, the choice has fallen on a woman's last menses, but this decision has brought with it a series of conceptual and methodological problems.

The adoption of the last menses as the definition of menopause makes it an event in time, but my own research suggests that women see menopause as a process over time (Kaufert 1982). I found that Canadian women in Manitoba did not wait until the menses had stopped, but called themselves menopausal if there had been a change in their accustomed pattern of menstruation. This was a self-anchoring definition, being based on each woman's perception of what was, or was not, normal or regular for her. If experiencing hot flushes (the classic sign of menopause in North American culture), some women declared themselves menopausal regardless of whether they were still menstruating or whether their last menses was one or several years in the past. These women were seeing menopause as a process which lasted as long as its symptoms. Neither the physiological symptom -- hot flushes -- nor the interpretation of this experience as a sign of menopause are universals. When these Canadian women were compared with women in Japan, we found that the Japanese women defined themselves as menopausal using a quite different set of physical markers (Lock, Kaufert and Gilbert, in preparation). In both Canada and Japan, however, women relied on their own perception that their bodies were changing, holding to a concept of menopause not as single event, but as a complex and long drawn-out process of physiological change.

Medical accounts of the aging of a woman's body are also phrased as a series of gradual transitions, whether from one hormone level to another, or from one level of disease risk to another. The hormonal changes of aging, for example, are described as a gradual rather than a sudden shift from one level of hormone production to another. According to Rinehart and Schiff (1985:406) "Postmenopausal ovaries remain responsible for approximately 50% of plasma testosterone and 30% of plasma androstenedione" and "may even contribute to plasma estrogen." Not only do the ovaries not stop functioning as women stop menstruating, but they are not the only source of hormone production. Grodin et al. (1973) note that:

The primary estrogen in postmenopausal women is estrone and that it is derived from the peripheral aromatization of androstenedione from the adrenal.

Vermeulen (1976) reported correlations between estrone and estradiol levels and fat mass, suggesting that the pace and degree of hormonal change may vary among women depending on body characteristics. (Other factors, ranging from diet to genetic inheritance, may be involved, although the data are scanty.)

Looking at other aspects of the aging of the female body, the two symptoms positively linked with these shifts in hormone levels -- hot flushes and night sweats -- may start while a woman is still menstruating and continue for some years after she has stopped. The decline in fertility and loss of ovarian function, seen as closing off the end of the reproductive period in a woman's life, can be traced back to the fetal stage (Rinehart and Schiff 1985). Again these changes are gradual, accelerating as women age, but with the pace of this acceleration varying from one woman to another. The two chronic diseases attributed to the loss in estrogen production -- coronary heart disease and osteoporosis -- are long term rather than immediate effects of not menstruating. In sum, the definition of menopause as a single traumatic event is oddly at variance with the way in which these physiological processes are conceived.

Defining menopause by a woman's last menses is not only conceptually problematic, it is methodologically awkward. Women may stop menstruating permanently, or for prolonged periods, for a variety of reasons other than pregnancy or menopause. Women who are excessively fat or excessively thin (the anorexic or bulimic, but also the victim of famine) may stop. Women who exercise excessively or are physically overworked or under excessive stress may stop (women in concentration camps were amenorrheic). The cause may be iatrogenic, the result of surgery or radio- or chemotherapy or the use of depo-provera as a contraceptive. From a methodological standpoint a definition based simply on a failure in menstruation is unsatisfactory, because it takes no account of the reasons, other than the natural process of menopause, which can produce this effect.

Finally, the definition conceals a series of culture-bound assumptions about women and menstruation. Developed from the menstrual career patterns of the North American and European woman, it depends on a notion of an adult woman as someone who menstruates regularly until she reaches menopause. There may be one or two breaks for pregnancy, rarely more, and preferably before she reaches thirty-five. But in some societies menstruation is only an occasional interlude between pregnancy and lactation and a woman may go from a last pregnancy into menopause without any intervening period of menstruation (Beyenne 1986). A model based on the well-nourished women of the white middle class is clearly inappropriate to societies where women are frequently malnourished or constantly overburdened by pregnancy or physical or psychological stress. Under these conditions menstruation becomes sparse and irregular well before the expected age of menopause.

In sum, the definition of menopause as a woman's last menses does not fit with women's account of their own experience or with medicine's own explanation of the biological processes involved. It is methodologically problematic and culturally naive. The question, therefore, is why has an inappropriate definition become the accepted definition. The answer I suggest lies partly in the constraints which the methods used in clinical and epidemiological research exert on the end product. As I will show, issues such as the choice of a definition, the selection of a research population, or the formulation hypotheses are determined less by the realities of a woman's menopausal experience than by rituals of the research process. To simplify the discussion, I have selected a single account from the literature on menopause. My concern is not with the details of this text nor with criticizing its content (although some criticisms emerge coincidentally within the course of the discussion) but with examining the relationship between the facts used in this construction of menopause and certain methodological aspects of the research which produced them.

A MEDICAL CONSTRUCTION OF MENOPAUSE

The account chosen is by Rinehart and Schiff (1985). Both men are leading researchers in the field and their review of the literature on menopause is succinct, but relatively comprehensive. It is because this paper represents the best and most current medical thinking in relation to menopause that it was chosen. The task addressed by Rinehart and Schiff (1985) is to set out the state of medical knowledge in relation to menopause for an audience of practising gynecologists and general practitioners. In doing so, they have drawn widely, if selectively, on the available literature.

In their bibliography, Rinehart and Schiff (1985) list 122 items. As preparation for this essay, each item was read and classified according to the type of research. A note was also made of the definition of menopause used in a study, the size of the study population and the sociodemographic characteristics of this population. After excluding two studies using rats, a single reference to U.S. Census data, two citations to material on the contraceptive pill and another to coronary heart disease among men and eight other reviews of the literature, the remaining 108 items can be roughly divided into epidemiological or clinical research or the clinical case study. These categories represent three approaches to the production of medical knowledge. As Rinehart and Schiff are themselves clinical researchers, they draw more material from that literature than might another researcher -- an epidemiologist for example -- but they also select from outside their own field and the key findings from epidemiological research are included. Each category provides Rinehart and Schiff with a different type of research material and is used by them for different purposes.

Occasional background information on particular studies is provided by Rinehart and Schiff (1985), but as in most reviews of medical literature, their quotations usually give little indication of the type, design or quality of the original project. But the three categories into which these papers can be divided (the clinical case study, clinical and epidemiological research) are methodologically very different and have different objectives.

In the case study category, I included all those papers describing the characteristics and clinical management of a specific patient(s). Six of the papers cited in this section report on a single patient; the other nine report on a handful of cases. Rinehart and Schiff (1985) use the case study as a source of descriptive material on the various pathologies involved in premature menopause, creating an impression that it is abnormal and extremely rare not to menstruate below the age of forty.

By contrast to the "story telling" aspects of the case study, the medical research community sees the work of the clinical researchers as the closest to true science. The latter are particularly convinced of this claim when embarked on a full-fledged, randomized, cross(ed)-over, double-blind(ed) clinical trial. All the papers based on the analysis of hormone levels, whether under "natural" or "experimental" conditions (that is, when another substance -- usually an estrogen -- is introduced into the system and its impact recorded), are assigned to the clinical research category. Most of these studies were laboratory based and most treated menopause as a biomedical event.

Approximately a third of the citations in Rinehart and Schiff's bibliography were assigned to the epidemiological research category. Rather curiously, the clinical trial is also the form of research design most admired by epidemiologists, although least often practised by them. It is unsuitable for most epidemiological research purposes, yet reluctant to see it as the exclusive preserve of the clinical researcher, epidemiologists include the design within their own methodological armamentarium (cf. Lilienfeld 1976; Alderson 1976). The epidemiological research category included all the retrospective case-control studies, the prospective cohort studies, the cross-sectional surveys, plus the large scale trials and the single cost-evaluation study. Rinehart and Schiff use epidemiological data to describe the impact of menopause on the rates, frequencies and statistical probability of chronic disease, particularly osteoporosis, coronary heart disease, breast and endometrial cancer. Except for a reference to a paper by McKinlay and Jefferys (1974), Rinehart and Schiff make relatively little use of epidemiological data describing symptom patterns or the impact of menopause on health care seeking behaviour.

The assignment of papers to one or other of these three categories was occasionally problematic. Grey areas existed between the descriptive art of the case study and the less rigorous of the clinical research papers, or when deciding whether a particular clinical trial should be counted as epidemiology or as clinical research. Occasional use was made of the size of a study population as a quick (if irreverent) rule of thumb when allocating some references to clinical rather than epidemiological research. A comparison based on the size of the populations in each study suggested that clinical researchers believed that truth may be found in a handful of cases. Over half the papers classed as clinical research had less than 50 cases and many had less than thirty. By contrast, epidemiologists preferred a cast of thousands or at least several hundred; for example, Rosenberg et al. (1981) used a data base provided by 121,964 nurses, while 2,873 residents in Farmingham supplied the information for Gordon et al. (1978).

The range of sociodemographic variables considered in any of these studies was very narrow. A few papers (9%) include weight in their list of recorded variables; some added height to weight (3%); others include parity (2%) or smoking behavior (4%). Seven papers record the racial characteristics of their study population, but race is used as a variable in only two of these. Only one paper among the 108 (McKinlay and Jefferys 1974) maps out the sociodemographic characteristics of its population, providing information on age, education, marital status, employment, SES level, domestic responsibilities and weight.

Taken altogether, these 108 papers are the source of the "facts" which Rinehart and Schiff (1985) used as building blocks to construct menopause. Before linking these facts back to their research base, I want to examine briefly the definition of menopause which Rinehart and Schiff offered to their clinical audience.

RINEHART AND SCHIFF'S DEFINITION OF MENOPAUSE

All accounts of menopause, whether they are medical or feminist, psychological or anthropological, have to deal with the same questions. What are the boundaries between menopause and not-menopause and where should they be set? What are the boundaries between natural and not-natural menopause? Is menopause an event or a process?

Rinehart and Schiff (1985) incorporate into their definition both hormone levels and an absence of menstruation, advising clinicians that "the diagnosis of menopause requires an amenorrheic state with elevated gonadotropins" (p. 406). The exact level at which hormones count as elevated is not specified, neither is the exact time a state of amenorrhea must last before a woman becomes post menopausal. Despite a certain vagueness in diagnostic criteria, menopause is presented as a dichotomy; a woman either is, or is not, menopausal: there is no transitional status.

Rinehart and Schiff (1985) make a distinction, however, between "natural" and "not-natural" menopause. Describing three forms of menopause, they write:

Menopause can have a variety of causes which can be broadly classified under the heading of premature ovarian failure, artificial menopause and natural menopause (p. 399).

The loss of ovarian function is common to all three; the distinctions between them are based on either the reasons for this loss or on its timing. Clinicians are provided with the diagnostic criteria for assigning the individual patient to one category or another.

The clinician is told to distinguish between natural and premature menopause by the timing of the event. Rinehart and Schiff (1985:399) define premature menopause as "secondary amenorrhea with elevated gonadotropin levels occurring before the age of 35 and 40." The causes of premature menopause are various, but are presented as inevitably pathological. Rinehart and Schiff (1985) write:

Theories abound about the causes of premature ovarian failure; genetic factors, the destruction of follicles by exogenous agents (e.g., a virus) and autoimmune disorders are all candidates (p. 400).

Certain assumptions about normality and aging are implicit within this distinction between premature menopause and natural menopause. Below the age of forty, the normal woman is the menstruating woman and, therefore, not menstruating is abnormal. After forty, it is equally normal for a woman to menstruate or not menstruate; after fifty, normality lies in not menstruating. Intermittent menstruation may also be abnormal. Clinicians are warned that a resumption of menses after a lapse of six to twelve months is a sign of pathology and should be investigated (Rinehart and Schiff 1985:408).

The other category, artificial menopause, is described as "a condition brought about by either the surgical removal or irradiation of the ovaries" (p. 401). Rinehart and Schiff advise against this surgery in young women, while concluding that "the procedure would seem to be justified" (p. 401) for women in the period immediately preceding or past menopause. Unlike premature menopause, however, no distinction is made by age; a woman is artificially menopausal regardless of whether she is twenty, thirty or forty. The essential criteria is that both ovaries must be lost; if any part of any ovary is retained, a woman is not artificially menopausal. The position of women who lose their uterus, but retain their ovaries is ambiguous. They are not artificially menopausal as they have their ovaries. If these ovaries are producing estrogen, they qualify as premenopausal by their hormone levels, but menopausal because of amenorrhea.

In each definition, menopause is a dichotomy; a woman is either pre- or post-menopausal, regardless of whether her menopause is natural, artificial, or premature. Rinehart and Schiff (1985) do provide clinicians with a third and intervening category, the "perimenopause," which they define as "the time immediately preceding and immediately following a woman's last menstrual period" (p. 399). They also use the term "the climateric," defining it as "the phase in the aging process during which physiologic changes take a woman from the reproductive to a non-reproductive stage of life" (p. 399). A passing reference is made to irregularity of menses as the boundary between the pre- and the perimenopausal woman, but Rinehart and Schiff do not tell a clinician how to separate their peri- from the post-menopausal patients. (By convention, the post-menopause starts twelve months after a last menstruation.) Neither do they set boundaries for the climateric, which is left as a relatively nebulous concept.

Rinehart and Schiff (1985) also deal with the complex processes involved in the aging of a woman's body, clearly describing for their clinical audience the gradual processes of hormonal and other forms of change. (This account was used earlier in this paper.) Yet, they then ask clinicians to use a dichotomous (or at best a trichotomous) structure with the implication that each category represents a relatively homogeneous group. But these categories of menopausal status are methodological artifacts only. A premenopausal group will include women with different hormonal patterns, different levels of fertility, at different points in their progression towards the end of menstruation. The criteria for post-menopause means only that hormone levels have passed a certain point, a definition leaving room for much variation, including differences in the number of years since women last menstruated, their degree of bone loss or the point reached on the risk curve for coronary heart disease. In sum, the dichotomous form conceals a considerable degree of heterogeneity.

These definitions were not the creation of Rinehart and Schiff (1985). They simply took from the menopause literature definitions of menopausal status which had been developed and refined for use in clinical and epidemiological research. It is in this research, therefore, that one must look for an explanation of their form.

THE RULE OF RESEARCH AND THE DEFINITION OF MENOPAUSE

Rinehart and Schiff tell clinicians that a woman is menopausal if she is in "an amenorreic state" or has "elevated gonadotrophins." The use of two criteria, either hormone levels or the cessation of menses, is a product of differences in research design and objectives between epidemiologists and clinicians which lead them to prefer different definitions of menopausal status.

Clinical researchers are interested in the biochemical processes internal to a woman's body, in measuring and assessing the consequences of these changes, and in what happens when they introduce some form of hormone replacement therapy into the system. Clinical research is laboratory-based research and determining hormone levels for the relatively few patients entering clinical studies is not a problem. There is no concern with cause in this definition, but only with whether a specified hormone level is or is not present. Some researchers describe how many women in a study population are artificially menopausal, but do not treat them as a separate category. In at least sixteen of the papers describing some form of hormone analysis, no distinction is made between naturally and artificially menopausal women; others leave unclear whether their subjects include women made artificially menopausal.

Yet, while combining both naturally and artificially menopausal women without any apparent qualm, clinical researchers are reluctant to include perimenopausal women in their study population. The first reason is a problem of measurement and categorization. The perimenopause is characterized by unpredictable and marked fluctuations in hormone levels. Taking a single measurement could lead to miscategorization; but a series of measures done over a series of days would complicate and add to the expense of a project. The second reason is also methodological. Most clinical studies require a population in which hormone levels are relatively stable both between women and in the same woman over time. For example, a clinical trial monitoring the impact of estrogen therapy on hormone levels cannot deal with women whose own hormones are in a natural state of flux.

By comparison to clinical researchers, epidemiologists have a different set of research objectives. Their raison d'etre is to study the distribution of disease. They became interested in menopause when it (or its management through estrogen replacement therapy) was seen as a potential risk factor for an assortment of chronic diseases, including coronary heart disease (Jick et al. 1978), breast cancer (Hulka 1982) and osteoporosis (Hutchinson et al. 1979; Weiss et al. 1980). Only the occasional epidemiologist -- or quasi-epidemiologist -- is interested in menopause as an experience in and of itself, but it is in their work rather than in clinical research that Rinehart and Schiff (1985) find basic data on such issues as age at menopause or changes in menstrual regularity as women age. For example, their source on changes in menstrual pattern over the life cycle is not a clinical study, but a report from the data bank into which Treloar (1981) had collected "35,000 person years of menstrual history," having followed a generation of women from their first years in college through to their menopause.

To establish disease risk in relation to menopausal status, epidemiologists need a large population and one which can be separated into menopausal and not-menopausal women. Their definition of menopause, therefore, must be appropriate for use in the large scale survey or the retrospective audit of medical records. Hormone assays are too impractical, or too expensive, or both, for most epidemiological research. Yet, to simply ask women whether they are menopausal is unacceptable, being seen as insufficiently objective.

In the papers cited by Rinehart and Schiff (1985), most epidemiologists use the definition of menopause as prolonged amenorrhea, categorizing anyone who has not menstruated for twelve months as menopausal. This definition cannot deal with women who do not menstruate for reasons unrelated to natural menopause, such as hysterectomy. A clinical researcher can use a hormone assay to determine whether a woman's ovaries still function, but this option is not usually open to epidemiologists. They must either accept that their post-menopausal category includes an unknown percentage of women with functioning ovaries or, they must in some way determine which women have ovaries and either exclude or treat them as a separate category. (As the disease risk of women with bilateral oophorectomies is different to that of women with natural menopause, current disputes between epidemiologists over the status of menopause as a risk factor for coronary heart disease turn on whether or not this distinction is made.)

Using menstruation as the definition of status does offer epidemiologists a way of dealing with women in the perimenopausal phase. Indeed, they are responsible for the term, having developed the category as a way of separating women whose hormone levels they presume are relatively stable (those in the pre- and post-menopause) from those experiencing major shifts and changes. Creating a perimenopausal category served epidemiologists who wanted to examine the impact of hormone fluctuations on symptoms and behaviour. But the perimenopause is largely irrelevant to those of their colleagues concerned solely with menopause as a risk factor for chronic disease: the latter are content with the simple division between the pre- and the post-menopausal.

Despite the various drawbacks to these definitions of menopausal status, epidemiologists and clinical researchers would presumably argue in their favour. Use of the accepted formularies is seen by most researchers as a matter of practical necessity. The peer review system -- whether of grant proposals or papers for publication -- enforces compliance with current terminology. In menopause research, those opting for other than the standard definition risk having their work rejected. The advantages go beyond the urge of researchers to communicate their findings or to have their project funded. From a medical science perspective, if knowledge and understanding of what happens to women as they age is to be cumulative, researchers must agree on and use the same system of designation and classification. The end of menses and the subsequent shift in hormone levels are the only indicators of menopausal status which are both universal and relatively objective.

While accepting the validity of these arguments for those who make them, it is also true that the choice of definitions for menopause has implications which extend beyond the assignment of women to one category or another. One side effect is that researchers become reluctant to work with a population which is difficult to categorize. Few of the studies cited by Rinehart and Schiff include perimenopausal women. The exceptions are a group of fifteen papers on premature menopause: Treloar's painstaking accumulation of data over a life-time of research, a handful of clinical research studies which include pre-menopausal women in their analyses of hormone levels (e.g. Reyes et al. 1977; Korenman 1982), and the cross-sectional survey of menopausal symptoms and experience by McKinlay and Jefferys (1974). Admittedly, from the research perspective, the methodological reasons for excluding perimenopausal women from a study population are good. The result, however, is that clinical researchers produce few "facts" on the perimenopause and the construction of menopause is skewed by the omission of this phase in women's experience.

The use in clinical studies of populations which combine artificially menopausal with naturally menopausal women helps skew the construct in another direction. Artificially menopausal women are often much younger and have passed through some form of physiological crisis (otherwise they would not have lost their ovaries) not experienced by naturally menopausal women. Most comparisons of the two groups suggest that not only their risk for chronic disease, but their symptoms and their use of medical care are very different; they are alike only in the narrow sense of their hormone levels having passed the same point (Kaufert et al. 1986). The practice of including artificially menopausal women in a study population is widespread throughout these 108 papers, but is not evident when reading Rinehart and Schiff. Most of their statements refer simply to "menopausal women," which is accurate by the terms of their definition, but also misleading. Presented as an account of natural menopause, Rinehart and Schiff have incorporated a quite different experience of hormonal change. The recognition of how many of these data originated with artificially rather than naturally menopausal women came only when going through the 108 papers. These women had been invisible when first reading Rinehart and Schiff, which raised a series of questions about the relationship between the research base and the final construct.

THE RESEARCH PROCESS AND THE CONSTRUCTION OF MENOPAUSE

Part of the fascination in going from account back to the original research lay in seeing how the work of constructing menopause had been carried forward, such as the principles used in the selection or discarding of materials. Like most clinicians, Rinehart and Schiff are prejudiced in favour of clinical knowledge. For example, reasoning as clinicians, they expect a relationship between heart disease and menopause and ascribe a protective role to estrogen. Epidemiological data showing that menopause is not associated with a dramatic increase in mortality from heart disease is discussed; however, respect for epidemiological research is much weaker than the power of clinical logic. When advising clinicians to give estrogen therapy as a protective measure, these epidemiological data are ignored.

This preference for clinical research is a major influence on the choice of materials. Epidemiological data is used more as descriptive or corroborative detail, whereas interpretation and explanation is based on clinical research. The definition of which symptoms are menopausal, for example, comes from a paper in which Utian (1972) reports on symptom experience among women before and after an oophorectomy. Yet, Rinehart and Schiff also had available the general population survey on menopausal symptom patterns by McKinlay and Jefferys (1974) for they quote its figures on the incidence of hot flushes. In choosing Utian's data, they opt for the clinical study and the clinical researcher with the result that the symptoms reported by women just come to major surgery are used as the model of all menopausal symptom experience, whether natural or artificial.

The medical research community often criticizes the quality of its own products, but usually by reference to the mistakes of method or interpretation made by individual researchers (cf. Emerson and Colditz 1983; Sheps and Schecter 1984; Lipton and Hershaft 1985). Yet, the choice of the study by Utian rather than the one by McKinlay and Jefferys was not a mistake in this methodological sense, but a reflection of differences in the value placed on clinical rather than epidemiological knowledge.

These papers are not without methodological flaws and another insight into the construction process came from seeing how flaws become masked in the final construct. A key factor in this masking process is that all facts are presented as if equal in status, regardless of their quality or source. Yet, any medical researcher knows at some level of research consciousness that all facts are not equal. Textbooks for fledgling epidemiologists spend much time on a design hierarchy in which the clinical experiment is at the top, the descriptive study is at the bottom; the clinical case study does not even qualify as scientific research (Armstrong 1977). Yet, this hierarchy is ignored by Rinehart and Schiff as they put together an account of premature menopause out of a handful of case studies. Admittedly, they had no choice as there is little research on premature menopause. My point is that there is no indication that this piece of their construct is based on material which most medical researchers would judge of poorer quality than the rest. Juxtaposed with the results from a double blind trial or carefully controlled hormone analysis, the descriptive details from an individual's case history are transformed as if by osmosis into medical facts and their origins forgotten.

Finally, going from the account back to the original papers, one becomes aware that, although Rinehart and Schiff write as if their statements applied to all menopausal women at all times, their research base is a restricted one. It is narrow both in the sense of what questions had been asked by researchers and in terms of the number and characteristics of the women who had supplied the answers. In the final portion of this essay, I will look at the implications of the sampling procedures for the construction of menopause.

SAMPLING PROCEDURES AND THE CONSTRUCTION OF MENOPAUSE

Questions of sample size and representativeness of study populations are usually confined to the margins of debates over methodology. Yet, in looking at how this particular piece of medical knowledge was put together, I became conscious of the characteristics of the population supplying the facts. Reviewing the papers cited by Rinehart and Schiff (1985), it was apparent that clinical researchers and epidemiologists are similar in their choice of where best to corral a study population. Most used patients, usually alive but sometimes dead (a handful of clinical studies used autopsy material). A favorite source of research subjects was patients attending special clinics for menopausal women. Pathways to the specialist clinics select for women with particularly severe menopausal problems, many as a result of a bilateral oophorectomy. All patient samples are suspect, but these latter women are a markedly unrepresentative group. Using data from this source has the unavoidable effect of skewing the construct towards the extremes of menopausal experience.

While epidemiologists are generally more particular than clinical researchers in their concern with the niceties of sampling, they also customarily depend on patients. Use of their favorite case-control retrospective design, for example, does not necessarily restrict epidemiologists to the hospital wards, but finding their controls there, or after a search through the medical records of their institution, is a common and accepted practice. This is the approach taken by Hoover et al. (1976), when looking at the relationship between breast cancer and estrogen therapy and by Rosenberg et al. (1981), when investigating the link between estrogen therapy and myocardial infarction. Something approaching a general population sample appears in only six of these 108 papers. There are reasons which are good in terms of standard methodologies for not using patient samples when investigating a natural, relatively unmedicalized occurrence, such as menopause. Yet, leaving aside these methodological considerations, these sampling procedures provide some insight into the vision of menopausal women held by these researchers.

The willingness of researchers to generalize from a handful of subjects to the universe of women reflects a certain way of seeing women. More specifically, in clinical research the female body is seen as a series of biochemical processes which vary relatively little from one woman to the next; therefore, to study one body at menopause provides insight into all other bodies. The researcher is quite comfortable with the assumption that one can generalize from the few women in a particular study to the universe of menopausal women. The material on vaginal atrophy, for example, comes from three clinical studies and is based on fourteen (Semmens and Wagner 1982), ten (Schiff et al. 1980) and nine (Yen et al. 1975) patients. Epidemiological research is somewhat different, being based on the expectation that there is variation between individuals, but not between groups. Therefore, as long as one follows the rules of sampling procedure, generalizations can be made from findings in one population of menopausal women to another.

Both clinical researchers and epidemiologists also seem to believe that the variables in which they have no interest, or on which they have no data, make relatively little difference to outcome. They ignore as irrelevant the range of social, cultural, racial and economic characteristics which may distinguish one woman in their study population from another or which distinguish the women they have studied from other menopausal women. For example, the menstrual patterns reported by Treloar (1981) are presented as if universal, yet the women in Treloar's data bank came from the well-nourished, low parity, European descended, Mid-Western, white middle class. The pattern of their menstruation is no more universal than these sociodemographic characteristics are universal. Discussing osteoporosis, Rinehart and Schiff (1985) refer briefly to the lower rates of the disease among Black American women, but do not speculate whether other of the menopausal characteristics they describe might be particular only to the Caucasian woman making up most study populations.

Both epidemiologists and clinical researchers are relatively narrow in the range of their curiosity about the human condition. As judged by the work cited by Rinehart and Schiff (1985), the questions researchers ask women reflect their own perception of menopause as a series of changes in the body chemistry which may -- or may not -- influence the distribution of disease. As a general rule, study populations are socially invisible. Aside form their sex, their age and their menopausal status, the women taking part are as anonymous as laboratory rats. Women are not only invisible, they are usually dumb. Their hormones are measured by the clinical researcher and their patterns of morbidity and mortality recorded by the epidemiologists, but there is rarely any reference to what women say or to what they experience. There is little sense of menopause and its symptoms being a process experienced by women and about which they are the valid reporters.

CONCLUSION

Rinehart and Schiff and most of the researchers cited in their bibliography probably see themselves as scientists. Equally probably, most would agree with Strong and McPhersons' (1982:644) statement that

there is a strong sense in which the findings of natural science are held to be neutral or amoral; whether we like it or not that is the way nature is.

Yet, as the previous discussion has shown, the way in which menopause is defined has very little to do with the way "nature is." This is menopause not as women experience it in all its potential range and variety, but an abstract concept limited in its ability to explain.

ACKNOWLEDGEMENTS

The research was supported by the National Health Research and Development Program through a National Health Research Scholar Award (No. 6607-1213-48) to the author.

REFERENCES

Alderson, Michael
  1976   An Introduction to Epidemiology. London: The MacMillan Press Ltd.
Armstrong, David
  1977   Clinical Sense and Clinical Science. Social Science and Medicine 11:599-601.
Berliner, Howard
  1982   Medical Modes of Production. *In* P. Wright and A. Treacher (eds.), The Problem of Medical Knowledge: Examining the Social Construction of Medicine. Pp. 162-173. Edinburgh: Edinburgh University Press.
Beyenne, Yewoubdar
  1986   Cultural Significance and Physiological Manifestations of Menopause: A Biocultural Analysis. Culture, Medicine and Psychiatry 10:47-71.
Emerson, John D. and Graham A. Colditz
  1983   Use of Statistical Analysis in the New England Journal of Medicine. New England Journal of Medicine 309:709-12.
Figlio, Karl
  1982   How Does Illness Mediate Social Relations? Workmen's Compensation and Medico-Legal Practices, 1890-1940. *In* P. Wright and A. Treacher (eds.), The Problem of Medical Knowledge: Examining the Social Construction of Medicine. Pp. 174-224. Edinburgh: Edinburgh University Press.
Gabbay, John
  1982   Asthma Attacked? Tactics for the Reconstruction of a Disease Concept. *In* P. Wright and A. Treacher (eds.), The Problem of Medical Knowledge: Examining the Social Construction of Medicine. Pp. 23-48. Edinburgh: Edinburgh University Press.
Good, Byron J. and Mary-Jo DelVecchio Good
  1981   The Semantics of Medical Discourse. *In* Everett Mendelsohn and

Yehuda Elkana (eds.), Sciences and Cultures, Sociology of the Sciences
    Yearbook, V. Pp. 177-212. Dordrecht, Holland: D. Reidel.
Gordon, T., W.B. Kannel, M.C. Hjortland and P.M. McNamara
    1978 Menopause and Coronary Heart Disease: The Framingham Study.
          Annals of Internal Medicine 89:157-161.
Grodin, J.M., P.K. Siiteri and P.C. MacDonald
    1973 Source of Estrogen Production in Postmenopausal Women. Journal of
          Clinical Endocrinology and Metabolism 36:207.
Hoover, K., L.A. Gary, P. Cole and B. MacMahon
    1976 Menopausal Estrogens in Breast Cancer. New England Journal of Med-
          icine 295:401-405.
Hulka, B.S., L.E. Chambless, D.C. Deubner and W.E. Wilkinson
    1982 Breast cancer and Estrogen Replacement Therapy. American Journal
          of Obstetrics and Gynecology 143:638-645.
Hutchinson, T.A., S.M. Polansky and A.R. Feinstein
    1979 Post-Menopausal Oestrogens Protect Against Fracture of Hip and Distal
          Radius. Lancet 2:705-708.
Jick, M., B. Dinan and K.J. Rothman
    1978 Noncontraceptive Estrogens and Nonfatal Myocardial Infarctions. Jour-
          nal of the American Medical Association 239:1407-1408.
Kaufert, Patricia
    1982 Myth and the Menopause. Sociology of Health and Illness 4:141-166.
Kaufert, Patricia, Margaret Lock, Sonja McKinlay, Yewoubdar Beyenne, Jean
Coope, Donna Davis, Mona Eliasson, Maryvonne Gognlons-Nicolet, Madeleine
Goodman and Arne Holte
    1986 Menopause Research: The Korpilampi Workshop. Social Science and
          Medicine 22:1285-1289.
Korenman, S.G.
    1982 Menopausal Endocrinology and Management. Archives of Internal
          Medicine 142:1631-1635.
Lilienfeld, Abraham M.
    1976 Foundations of Epidemiology. Oxford: Oxford University Press.
Lipton, Jack P. and Alan M. Hershaft
    1985 On the Widespread Acceptance of Dubious Medical Findings. Journal
          of Health and Social Behaviour 26:336-351.
Lock, Margaret, Patricia A. Kaufert and Penny Gilbert
          Cultural Construction of the Menopausal Syndrome: The Japanese Case
          (In preparation).
McKinlay, S.M. and M. Jefferys
    1974 The Menopausal Syndrome. British Journal of Preventive and Social
          Medicine 28:108.
Reyes, F.I., J.J. Winter and C. Faiman
    1977 Pituitary-Ovarian Relationships Preceding the Menopause. I. A Cross-

Sectional Study of Serum Follicle-Stimulating Hormone, Luteinizing Hormone, Prolactin, Estradiol, and Progesterone Levels. American Journal of Obstetrics and Gynecology 129:557-561.

Rinehart, John and Issac Schiff
1985 Menopause. In D.H. Nichols and J.R. Evrard (eds.), Ambulatory Gynecology. Pp. 399-429. Philadelphia: Harper and Row.

Rosenberg, L., C.H. Hennekens, B. Rosner, et al.
1981 Early Menopause and the Risk of Myocardial Infarction. American Journal of Obstetrics and Gynecology 139:47-53.

Schiff, I., D. Tulchinsky, D. Cramer and K.J. Ryan
1980 Oral Medroxyprogesterone Treatment of Postmenopausal Symptoms. Journal of the American Medical Association 244:1443-1448.

Semmens, J.P. and G. Wagner
1982 Estrogen Deprivation and Vaginal Function in Postmenopausal Women. Journal of the American Medical Association 248:445-451.

Sheps, Samuel B. and Martin T. Schechter
1984 The Assessment of Diagnostic Tests: A Survey of Current Medical Research. Journal of the American Medical Association 252:2418-22.

Strong, P.M. and K. McPherson
1982 Natural Science and Medicine; Social Science and Medicine: Some Methodological Controversies. Social Science and Medicine 16:643-657.

Treloar, A.E.
1981 Menstrual Cyclicity and the Premenopause. Maturitas 3:249-256.

Utian, W.H.
1972 The True Clinical Features of Postmenopause and Oophorectomy and Their Response to Oestrogen Therapy. South African Medical Journal 46:732-736.

Vermeulen, A.
1976 The Hormonal Activity of the Postmenopausal Ovary. Journal of Clinical Endocrinology and Metabolism 42:247.

Weiss, N.S., C.L. Ure, J.H. Ballard, et al.
1980 Decreased Risk of Fracture of the Hip and Lower Forearm with Postmenopausal Use of Estrogen. New England Journal of Medicine 303:1195-1198.

Yen, S.S.C., P.L. Martin, N.M. Burnier et al.
1975 Circulating Estradiol, Estrone, and Gonadotropin Levels Following the Administration of Orally Active 17 B-estradiol in Postmenopausal Women. Journal of Clinical Endocrinology and Metabolism 40:518-523.

JESSICA H. MULLER AND BARBARA A. KOENIG

ON THE BOUNDARY OF LIFE AND DEATH: THE DEFINITION OF DYING BY MEDICAL RESIDENTS

> New technology, enabling the lungs to pump, heart to beat, intestinal tract to digest, kidneys to cleanse, and immunological system to fight long after the body has given out, has transformed dying into a daily dilemma: a dilemma of agonizing decisions and impossible choices (Kleiman 1985).

INTRODUCTION[1]

The care given to terminally ill or dying patients by American physicians has been discussed extensively in recent years in the media, popular press, and scholarly journals alike. Physicians in particular have been criticized for their "aggressive" approach to terminally ill patients: for continuing treatment which has a minimal chance of success; for fruitlessly prolonging a life that is only biologically functional with a plethora of technological equipment and machines; and for failing to allow patients to die peacefully, "with dignity." Although a number of reasons have been advanced to account for these approaches to dying persons -- including the physicians' fear of failure and defeat, fear of death, and fear of malpractice suits -- one facet of this issue which remains largely unexplored concerns the effect of *how* and *when* physicians determine that patients are dying. If patients are not defined as dying until they are very close to physiological death, they are not treated as dying persons and consequently may receive, in the attempt to "save" them, the very intensive technological interventions that conjure up frightening images of horrible, machine-dependent deaths.

In this chapter we will explore how and when physicians-in-training define certain terminally ill patients as "dying." We begin this exploration by examining the processes by which they select information about patients and interpret it to reach conclusions about these patients' futures. Two types of information are particularly significant here: patients' capacity for interaction and the therapeutic possibilities open to physicians. In the discussion which follows we will show how physicians-in-training, through their interpretative activities, construct a clinical reality in which patients are seen as still having "a chance" and therefore

351

M. Lock and D. R. Gordon (eds.), Biomedicine Examined, 351–374.
© *1988 by Kluwer Academic Publishers.*

deserving of extensive medical interventions. While physicians are engaged in the process of offering treatment, the question of patients' potential deaths is forestalled.

This analysis is based on our assumption that in a clinical situation certain data about a patient are identified as relevant, interpreted within a context of meaning, and used to construct a clinical reality that becomes the object of therapeutic endeavor (Kleinman 1980; Good and Good 1981). Central to this discussion is the notion that defining someone as dying is a social process. This is not to deny the physiological realities of medical conditions, but to emphasize that pathophysiological states are given meaning and significance through a process of interpretation. In this process medical practitioners build a clinical reality on which they base their assessment of a patient's status, treatment and future. When a patient is seen as dying, therefore, is not simply a biological "given" but is shaped by physicians' interpretations of certain cues and information. These interpretations provide evidence that the patient is "nearing death" and "there is nothing more to do," or, alternatively, that the patient still has "a chance."

Physicians' treatment of dying patients in hospitals has been examined from different perspectives; among them, the ways in which hospital staff approach dying patients at different points along their trajectories to death (Glaser and Strauss 1965a, 1965b, 1968), the socially organized activities surrounding dead or dying persons (Sudnow 1967), the factors that influence active treatment of the critically ill (Crane 1975), physicians' evasion and denial of death (Kübler-Ross 1969, 1975), and the ethical and legal issues involved in treatment decisions (President's Commission 1983). This discussion will take a different perspective in its focus on how the very definition of "dying" itself becomes problematic in the care of some terminally ill patients.

It is important to note that this chapter is not about those persons who experience sudden death as a result of accident or trauma. Nor is it concerned with those patients who are hospitalized at the end-stage of their disease and about whom there is general consensus among the staff that they are on an irreversible course toward imminent death and will not benefit from further diagnostic and therapeutic procedures. Rather, it is about those individuals with severely debilitating chronic diseases who are highly likely to die but who have not yet been defined by their doctors as dying. It is these patients, as we shall see, who are surrounded with definitional ambiguity. For the purposes of this chapter we are focusing only on the perspectives of physicians, even though the wishes of patients and their families can have a significant impact in some situations upon the care the dying receive in hospitals.

Data for our analysis are drawn from a larger research project on the ways in which residents in internal medicine at a university teaching hospital perceive, learn, and manage the responsibilities of caring for dying patients.[2] The setting is an American medical center of considerable repute where patients with routine as well as serious, exotic, or baffling diseases come for diagnosis and treatment. The day-to-day management of medical patients is carried out by residents (collectively known as housestaff) who are in their first year (the internship), second year or third year of postgraduate training in internal medicine. Although house staff in teaching hospitals are under the supervision of more senior attending physicians, they have the primary responsibility for making many kinds of decisions concerning patient care, including the life-and-death decisions we discuss here.[3]

To establish the socio-cultural context for our discussion, we will in the following section discuss some of the transformations which are taking place in the meaning of death in American society and in the management of dying within biomedical practice in the United States. Next, we present the cases of two individuals and trace their hospital course from admission to death. Through a discussion of these cases we analyze how the physicians' interpretation of information concerning the patients' capacity for interaction and the therapeutic possibilities available to them led to extensive medical interventions and, in the process, served to delay the identification of these patients as dying. We conclude with an examination of the cultural relevancies of these issues within the context of biomedical practice and their implications for the care of dying patients.

DYING IN A BIOMEDICAL CONTEXT

In most general terms, "dying" represents the passage or transition a person makes between the status of life and the status of death. Just as other transitions from one stage to another in the life cycle are often marked by special ceremony, so have all cultures elaborated socially prescribed ways for understanding and handling the *rite de passage* symbolized by dying and death. Different cultures define and treat the role of the dying in different ways, as they do the role of the sick, the injured, or the dead.

In twentieth century America, societal definitions of this transition from life to death are being profoundly altered by demographic, social, technological, and cultural changes which have raised questions about the meaning of death and its management. Dramatic shifts in morbidity and mortality patterns, for example, have occurred with the control of infectious disease, the advent of antibiotics, and certain technological developments. Consequently, most deaths now occur not among the young but the old, and the majority of people will die from chronic rather than acute diseases (Hingson et al. 1981). Furthermore, institutions, particularly hospitals, have supplanted homes as the expected and accepted scene of the normal death, and medical personnel rather than members of a person's immediate social group have taken over the care of the dying (Aries 1981). Today, it is typically the hospital and the hospital staff which provide the context for death.

Physicians in particular have become the gatekeepers of dying and death in contemporary America. The significant pharmacological and technological advances which have taken place in the United States since World War II have given physicians greater control over both the process of dying and the timing of death. Death, which was at one time looked upon as a familiar and timely "surrender of the self to destiny" (Aries 1974), is no longer accepted as a natural and inevitable event. It is seen as something that can be controlled, postponed, and potentially reversed, its timing elective, planned, and managed. Certain terminally ill patients can be "kept alive" through the use of life-sustaining equipment, procedures, or drugs almost indefinitely if treatment is not withdrawn, while others can be revived from the world of the dead as a result of recent advances in resuscitative techniques.

These developments have evoked medical, legal, social, and ethical debates in both professional and popular circles that would have been inconceivable fifty years ago. When should therapeutic measures be used to prolong physiological life beyond the point where a person's social life appears to be over? What level of treatment should be offered to a dying patient? How long should efforts to sustain the life of a dying patient be continued? How should decisions about the use of life-sustaining technology be made and who should make them? The cases and discussion which follow illustrate how the responses to such questions depend in part on the process by which medical practitioners assess patients' futures and come to recognize them as dying.

THE CASES

Tommy Sheldon[4]

Tommy Sheldon, a 35-year-old male, was admitted to the hospital with nausea, vomiting, jaundice, and ascites (accumulation of fluid in the abdominal cavity). He had been seriously ill with hepatitis for several months, for which he had already been hospitalized twice. Upon admission his team of physicians, which included an attending physician, a second year resident, two interns, and a medical student, provisionally diagnosed him with hepatitis B, although they were initially unsure of the type because of unusual blood antigen tests. Over the next few days the patient came to be identified as having viral hepatitis which had progressed to fulminant hepatic necrosis (acute liver failure). He was considered at that point to be seriously, but not terminally ill.

Mr. Sheldon's sixteen day hospital course was marked by increasing hepatic encephalopathy (disturbances of consciousness leading to changing mental status and tremors), worsening kidney failure, shrinkage of his liver, increasing ascites, and gastrointestinal bleeding. In spite of his deteriorating condition Mr. Sheldon did not suffer any permanent mental impairment. Until the time when he could no longer communicate verbally, he remained alert most of the time, joked extensively with the members of his medical team, and participated to the extent possible in decisions concerning his treatment. The housestaff appreciated his witty comments, respected his desire to become involved in his care, and grew to like him as a person as well as a patient. Throughout his time in the hospital, they invested a great deal of time and energy in his care.

Early in the course of Mr. Sheldon's treatment the housestaff were hopeful that if they could keep him alive his liver would spontaneously regenerate and he could return to normal functioning as most people do who have had hepatitis. On the third day of hospitalization, however, he developed what the doctors feared was hepatorenal syndrome, a virtually fatal form of kidney failure that can develop in patients with severe liver disease. At this point, the team resident told the team members that before they gave Tommy Sheldon the diagnosis of hepatorenal, which he called a "diagnosis that portends mortality," they needed to investigate other, potentially reversible sources of his worsening condition.

Failing to find anything that was reversible, the doctors concluded that Mr. Sheldon had in fact developed hepatorenal syndrome. They began to re-evaluate his prognosis: now seen as someone who was terminally ill, he was discussed as "an incipient disaster . . . a man who has a 99% chance of dying from liver failure . . ." In their estimation, his only chance of survival would be the spontaneous regeneration of his own liver. At this point, the gastrointestinal physicians consulting on the case casually suggested the possibility of a liver transplant at a nearby medical center. When they first raised the idea of a liver transplant, which was still an experimental procedure for adults, it was greeted by Tommy Sheldon's doctors with amusement, astonishment, and skepticism. In the first team discussion of the idea, the attending physician spoke out against pursuing the possibility of a transplant. He cautioned the team about the risk of infection, stating that in his opinion the low survival rate of patients with a transplant (5-10%) was not worth the possibility of infecting 10-20 people on the surgical team with hepatitis.

On the basis of these considerations, the medical team initially rejected the suggestion of the transplant. As the days passed, however, and Mr. Sheldon's condition rapidly deteriorated, his doctors suspected they would not be able to maintain him until his liver could regenerate. The transplant idea began to look more attractive to the team resident. He argued that even though they did not have strong data to support the decision, Tommy Sheldon would probably have a better chance of surviving with a transplant than if they did nothing. In spite of the potential risks, the transplant surgeons at the nearby medical center also expressed interest in doing the transplant operation, and were willing to instigate the search for a donor. Mr. Sheldon was supportive of the decision as well, urging his doctors to "go for it." After considerable discussion, the team of doctors decided to give him a transplant if a donor could be found. The attending physician explained their decision in this way:

The more it seemed like he was going to die of his disease -- and since you can only die once, and he definitely did not want to die and was willing to do aggressive procedures -- I thought we would explore the idea [of a liver transplant] . . . We knew what data there was and it wasn't real promising, but as his prognosis got worse and worse, this experimental procedure began to look better and better. Given the only chance he had was either transplant or spontaneous recovery . . . I finally came around to the view . . . that we should support him the best our technology can, given he's a young guy with potentially totally curable disease.

Once they decided to pursue the idea of the transplant, the physicians undertook extensive medical interventions which they hoped would keep the patient alive until the transplant could take place. These interventions included a transfer to the intensive care unit, a central line to monitor the functioning of his heart and lungs, fluids to sustain his blood pressure, a respirator (a mechanical breathing machine) when his breathing failed, and dialysis to support his kidneys when they no longer functioned.

Although Tommy Sheldon's team of doctors finally embraced the idea of a liver transplant, the case was the subject of considerable debate among hospital staff on both medical and moral grounds. Some physicians and nurses were concerned about the risk of infection from his hepatitis. Others questioned the medical team's decision to continue with aggressive medical interventions when the transplant was, in the words of a physician in the intensive care unit, "99% theoretical."

By the sixteenth day of hospitalization Tommy Sheldon had severe liver failure, poor kidney function for which he was receiving peritoneal dialysis, signs of infection, bleeding, hypotension, and worsening encephalopathy. Through constant monitoring of his status and intricate juggling of medications and machines, the medical team worked feverishly to "save" Mr. Sheldon even though they acknowledged the situation was nearly hopeless. The day before he died, the intern reported sadly to the rest of team, "We're daily knocking off another system . . . the lungs, kidneys, and liver have gone now . . ." Finally, Mr. Sheldon began to experience massive upper gastrointestinal bleeding. The physicians administered blood transfusions, but when their usual medical maneuvers to stop bleeding (including inserting a large tube into the stomach) did not work, they acknowledged they had "nothing more to offer him" -- he was dying in spite of their efforts. Since he was not responding to maximal therapy, the decision was made not to transfuse him further and not to resuscitate him in case his heart stopped. Three hours later he died from massive bleeding. The next day the team received a phone call from the liver transplant team. They had found a donor.

Sally Serpa

Sally Serpa was a 53-year-old woman who had experienced multiple hospitalizations for exacerbations of chronic obstructive pulmonary disease (COPD). When Mrs. Serpa was admitted to the hospital with respiratory distress for what turned out to be her final, four-week long hospitalization, the responsibility for her care was assumed by a team of doctors which consisted of an attending physician, a third year resident, two interns, and a medical student. No one on the team had known Mrs. Serpa previously, but when questions arose during her hospitalization about her pre-hospital health status they sought the advice of her clinic doctor who had been actively involved in her care for several years.

In response to Mrs. Serpa's respiratory distress, her doctors intubated her and placed her on a respirator to sustain her breathing. Like the other times when she had been intubated, she was extubated (taken off the respirator) within several days. Instead of improving as she had in the past, however, her condition deteriorated rapidly. She was placed back in the intensive care unit and reintubated, where she remained totally dependent on the respirator. At this point she had to be heavily sedated and was unable to communicate with her physicians or her family.

Complications arose shortly after Mrs. Serpa was placed back on the respirator. She first developed sinus tachycardia (abnormally rapid heart rate), the cause of which perplexed her doctors. They speculated that it may have developed in response to a Staphylococcus sepsis (a bacterial infection) or congestive heart failure (inability of the heart muscle to pump efficiently) or a pulmonary embolus (a clot which lodges in the lungs). Since any stress such as heart failure or pulmonary embolus can result in rapid deterioration, the housestaff decided to undertake several diagnostic procedures, including a perfusion lung scan and the insertion of a Swan-Ganz catheter (a pulmonary artery or PA line), to determine if Mrs. Serpa's current condition was the result of a new development or worsening lung disease.

In describing the team's decision to do these diagnostic procedures, the third year resident explained that they needed the information these tests would provide to obtain a "definitive diagnosis" of the new clinical problems that had developed in Mrs. Serpa. It would tell them if she had a reversible, and therefore treatable, condition in addition to her primary disease, COPD, and what treatment interventions they could then make. The potential reversibility of a condition was an important consideration to the resident:

The reason for [putting in the PA line] was because one thing we hadn't considered that could have been making her appear clinically the way she was was that she had congestive heart failure. And the only way we could figure that out was to do a PA line . . . possibly the only reason we were unable to make any progress with her was that we misdiagnosed an entire condition − that she really didn't have respiratory failure, what she had was heart failure. That's why she was having so much trouble oxygenating. And heart failure is something that's easily reversible with drugs . . . It would have been a tragedy not to know that, because it's such a treatable thing.

When the test results indicated a low probability of heart failure or a pulmonary embolus, Mrs. Serpa's doctors were baffled and continued to look for treatable complications.

The intern on the team who was responsible for Mrs. Serpa's care objected strenuously to this search for treatable conditions. In her opinion, regardless of the new complications, Mrs. Serpa's underlying disease had progressed to such a point there was no hope of recovery and she would never leave the hospital. During one of the daily team conferences she raised the issue of Mrs. Serpa's prognosis: "Prognosis is the question. Is there 'no prognosis' or is there a small chance of her leaving the hospital? If there is 'no prognosis' she has to die from something. Does it matter if it's treatable or not?"

The attending physician's response to this question revealed his uncertainty about Mrs. Serpa's future: "The bottom line is, we don't know her prognosis . . . medically I don't give her 'no prognosis.'" He added that while "statistically" he agreed with the intern that Mrs. Serpa probably would not leave the hospital, he still considered important the information provided by her clinic physician that she was an "active functioning person one week before she came into the hospital." Consequently, she still did not "fit" the portrait of someone with "no prognosis" to him.

In the last days of her life, Mrs. Serpa's condition deteriorated rapidly, which in the view of the doctors required additional diagnostic and therapeutic interventions. She experienced upper gastrointestinal bleeding, her blood pressure dropped, and she developed massive cellulitis (inflammation of the soft or connective tissues due to infection). She also began to appear swollen and puffy from an additional complication: the development of subcutaneous emphysema (presence of air in the subcutaneous tissue resulting from the seepage of air from the respirator). Furthermore, Mrs. Serpa developed new pleural effusions (accumulation of fluid in the lining of the lung.) When withdrawal of some of the fluid indicated the presence of a staph infection, she was diagnosed as having an empyema (infection of the lining of the lung). The housestaff decided to treat this infection with the standard therapy for empyema which includes surgical drainage through the use of a chest tube plus high dose antibiotics.

Although Mrs. Serpa's team of doctors determined a chest tube was necessary since "theoretically any person with empyema should get it," the surgeons, who would actually do the procedure, were less enthusiastic since it was a risky procedure in someone who had no respiratory reserve. Claiming they did not want to be the patient's *modus exitus*, they initially balked at doing the procedure. Furthermore, the nurses taking care of Mrs. Serpa protested the procedure, arguing that it was painful and unnecessary in someone who was about to die anyway. In face of this controversy regarding the insertion of a chest tube to treat an infection in someone who was this gravely ill, the medical team reviewed their decision with the Chief of the Medical Service who concurred that a chest tube was appropriate.

Three days later, on the morning of the day Mrs. Serpa died, the surgeons reluctantly inserted the chest tube. By this point, however, her doctors had begun to realize that they were not going to be able to control the seepage of air from the respirator into Mrs. Serpa's subcutaneous tissue. In spite of their efforts to control the adverse effects of the respirator, she was blowing up like an enormous balloon. The implications of this were not lost on the attending physician:

Mrs. Serpa's prognosis is certainly very grim. This shows how with a critically ill person you can get complications. Our therapeutic interventions are clearly giving her complications. We are circling . . . I can't think of much else we can treat once we're through with this round.

When it finally became apparent that Mrs. Serpa was not responding to either standard treatment or to extraordinary treatment, the team resident admitted there was "nothing more they could do." At this point, Mrs. Serpa was sedated heavily and the respirator was turned off. She died immediately.

THE CASES EXAMINED

Although these vignettes differ in various ways, both represent situations in which non-curative, "life-prolonging" procedures and equipment were employed to keep patients alive beyond the point at which others present felt that the death of these patients was inevitable. While it could be said that these cases represent extreme forms of intervention common in university teaching hospitals but uncommon in the routine treatment of the terminally ill, they nevertheless highlight some of the characteristics of biomedical thought and practice which shape the care of all patients.

In the cases discussed here certain cues, which in other situations would signal impending death, were not interpreted as such because of the emphasis given to other information emerging within the context of treatment. As a result of the ways in which physicians interpreted clues about their patients' capacities for interaction and constructed a vision of available therapeutic possibilities, these patients came to be viewed as individuals who had "a chance" at life and therefore "deserved" extensive biomedical interventions. As long as the medical focus was one of intervention, these patients were not defined as "dying" until the point at which the physicians determined there was nothing more they could do. The social processes involved in constructing these patients' futures are described below.

Interpretation of Interactive Capacity

The capacity of seriously ill patients for social interaction is one of the factors that physicians consider in making judgments about their patients, treatment, and future. An important source of this information for physicians comes from their *actual* interaction with patients. Communication between patient and practitioner yields important diagnostic and prognostic information which lies outside the realm of conventional biomedical understandings of pathophysiology, information from measurement instruments, or statistical data about disease outcomes. In particular, interaction with patients can give physicians cues about patients' responsiveness to their social world or their ability to engage meaningfully with other human beings. On the basis of interaction with patients, in conjunction with information from other sources, physicians make inferences about how well or poorly patients are doing, reach conclusions about what type of treatment to offer, and make estimates about their probable prognoses. In those situations where patients are mentally impaired and unable to communicate with their caretakers, physicians, on the basis of their knowledge of disease processes and outcomes, as well as their familiarity with individual patients, evaluate patients' *potential* for resuming social roles or interacting in a meaningful way with others in their environment (Crane 1975).

In addition to the information it can yield, interaction between physician and patient can also lead to an emotional investment in the patient which influences how that patient is perceived and treated. The vignettes of Tommy Sheldon and Sally Serpa illustrate the relevance of these points in situations where patients are likely to die.

Of central importance in the case of Mr. Sheldon was that he was often mentally alert during his hospitalization, in contrast to most patients who develop hepatorenal syndrome. As a result, until the point when his breathing had to be assisted artificially, he interacted extensively with his doctors. Not only did he discuss his course of treatment with them, but he also engaged them in elaborated and stylized joking behavior about daily developments in his physical condition. Similar in age and background, the housestaff considered Mr. Sheldon a peer — someone whom they liked and with whom they could interact easily and enjoyably. The doctors attached considerable significance to his interactive behavior. The attending physician remarked:

> If he'd been comatose, I would not have dialyzed him . . . I guess it's very hard not to dialyze someone who is talking to you or making jokes. The day we decided about his dialysis was the day he came up with this great line about, "What's a cough between friends?" It's hard to say, "We're going to kill you, or let you die," at that point. But prognosis-wise, statistically, we would have been correct for not dialyzing him.

These comments suggest that the physicians did not rely merely on Tommy Sheldon's physiological status or on their knowledge of his statistical chances for survival in determining their approach to him. They also made inferences about his status and treatment on the basis of his communicative behavior. They selected certain "data" emerging from interaction — his talking and joking behavior — and invested them with meaning: his "jokes" became symbolic of his ongoing interactive ability and intact cognitive function.

Because he was seen as someone who was still a functioning, mentally alert individual for whom a medical intervention might possibly work, it was difficult for the housestaff to conclude that he had "no prognosis" and treatment could therefore be terminated. His doctors felt an obligation — born not of their knowledge of statistical probabilities of disease but of their response to this patient as a human being — to offer him the transplant and other intermediate treatment procedures like dialysis, without which he would die imminently. This perception of him as a functioning person made his death more difficult to accept, as his intern pointed out when he commented how hard it had been for the residents "to watch a young person who was basically 'with it' die."

This suggests that the residents' interaction with Mr. Sheldon involved an interpretation of his interactive ability which went beyond biological understandings of disease and its outcome: the "meaning" of being human. He was viewed as someone who, in spite of his serious illness, was still "truly and fully human" (Fox and Willis 1983), with the capacity to relate to others, communicate with others, reason and feel. In such situations, where individuals are known not only as patients but also as interacting human beings, a patient's capacity for interaction can influence notions of what constitutes "evidence" for continuing treatment. Moreover, it can influence when a patient is identified as dying. If patients are mentally alert and able to engage with others in their environment, physicians are more reluctant to interpret other signs as indicating a steady progression toward death. As an intern commented, "If they can laugh, they're not dying."

The case of Sally Serpa presents further evidence of the importance of the physicians' perceptions of a patient's capacity for interaction. Because she was heavily sedated for much of the time that they knew her, her doctors were unable to make assessments of her future based on their own interaction with her. They did, however, rely heavily on the knowledge of her clinic physician who had been treating Mrs. Serpa for several years. He had seen her recover from several bad episodes before and was impressed by her ability to return to what was for her a normal state after these exacerbations of her illness. This doctor had become very committed to Mrs. Serpa and involved in her care: "I had a lot of empathy for her . . . I wanted her to be well and happy. I really wanted her to do well."

The personal experience and hopes of Mrs. Serpa's clinic physician helped shape the clinical picture that emerged of her. This picture was also based on the doctors' knowledge of, and experience with, other patients with COPD. In this context, Sally Serpa was a statistical anomaly:

I mean, all chronic lungers eventually die. But she, from what it looked like she was doing before she came in the hospital, she didn't statistically seem ready to die, so I had a very hard time applying my first criteria, my statistical criteria, to her. Because every time I did it I came up with she shouldn't die . . . she was a functioning person, according to [her doctor] just right before she came in. She was cooking and cleaning and wandering around the house. So she didn't have the gradual decline: she had an abrupt decline. That's fine -- I mean, that definitely happens. But that's why I wanted to keep looking for things with her . . . It made me uncomfortable because I couldn't say, "Oh, yeah, I've seen a million people like her, and they all do this."

In the estimation of the attending physician, Mrs. Serpa did not "statistically seem ready to die" because she did not exhibit the characteristics of typical COPD patients. Her recent interactive history, her atypical disease course, and her potentially reversible conditions did not constitute sufficient evidence for her doctors to determine that she had "no prognosis" -- that she was, in fact, dying.

To reach this conclusion, Mrs. Serpa's doctors used the evidence of her past functional ability to construct a picture of her as someone who could once again return to her pre-hospital functioning status even though they knew they could not cure her underlying disease. They were, in other words, evaluating her potential for interaction on the basis of what she had been a short while before – in the words of the attending, "an active, functioning person." Because of what she once was, they created a vision of what she could be again. Given this constructed reality, they instituted extensive diagnostic and therapeutic interventions which they hoped would arrest the course of her disease.

In the treatment of both Tommy Sheldon and Sally Serpa, the physicians' interpretation of their patients' capacity for interaction, in conjunction with their perceptions of treatment options, led to an emotional and intellectual investment in these patients which influenced what therapeutic interventions were made and when they were identified as dying. This was particularly evident in the case of Tommy Sheldon. In caring for him, his intern and resident not only invested extensive time and energy in his management but also developed a strong attachment to him. They were willing, even eager, as the attending noted, to fight for him: "Tommy was probably kept alive an extra week by the will power of this housestaff!"

Several factors bear on this expressed commitment to save Tommy Sheldon. As we have seen, the housestaff considered him a peer and a friend. The death of a peer served to these young physicians as an all too close reminder of what could happen to someone like themselves. The residents were also responding to the cultural sense of tragedy which is experienced in contemporary American society when someone is dying who is young, in the prime of life, or endowed with culturally valued attributes – when "the life being saved still holds much unrealized promise" (Parsons, Fox, and Lidz 1972:397). In addition, Mr. Sheldon's initial peculiar antigen findings, his atypical course, and the later possibility of a liver transplant – a new and innovative procedure – made him an "interesting" patient who was diagnostically, therapeutically, and pedagogically engaging for the residents. His situation in many ways represented the "ideal" patient for these physicians-in-training. In taking care of him they could command all that biomedicine has to offer – the intricacies of diagnosis, therapeutic regimens which require sophisticated management, and the promise of a technologically exciting cure.

Thus, because of the housestaff's attachment to Tommy Sheldon, which grew out of their daily interactions with him, and because of their perceptions that his situation was biomedically interesting and challenging, they continued their intensive efforts to keep him alive, while at the same time aware there was little chance of success.

In the case of Mrs. Serpa, the housestaff were not engaged with her on a personal level, but their decisions concerning her treatment and her potential were strongly influenced by the experience and impressions of her clinic doctor. Because he had been committed to her care for some time, and had seen her recover before, he had become heavily invested in her recovery. Until death was imminent, he appeared unable to give up hope that she might get better this time. Moreover, Mrs. Serpa, as we will see in the following section, represented a diagnostic puzzle for the housestaff. The search for what was wrong with her became a challenge in its own right, and provided the impetus for further interventions.

The investment of the housestaff in both Mr. Sheldon and Mrs. Serpa led them to struggle hard to "save" these patients. They worked day and night, discussed their treatment endlessly among themselves, consulted with other physicians, and put an enormous amount of energy into trying many different diagnostic and therapeutic procedures. Through this intensive medical care they expressed their "caring" for these patients, which made it more difficult to acknowledge that "this time" the patients really were dying, and there was nothing more they could do. The degree to which such involvement with a patient can influence perceptions of that person as dying is suggested in the following remark made by an intern, "Someone that you're involved with . . . you're sort of fighting with them. If you're fighting, you don't think they're dying."

Interpretation of Therapeutic Possibility

The cases under consideration here, in addition to demonstrating how interactive data are interpreted by physicians, also illustrate the process by which physicians construct therapeutic possibilities in caring for the seriously ill. In keeping with the biomedical focus on medical intervention, therapeutic solutions ranging from the routine to non-routine are defined, interpreted, and pursued in situations of great uncertainty or near futility. In this process, clinicians' reliance on the hope provided by perceived therapeutic options influences how and when terminally ill patients are defined as dying.

Tommy Sheldon's story provides a compelling example of how a non-routine treatment goal for a seriously ill patient -- in this case an experimental liver transplant procedure -- can be suggested, rejected, reconsidered and finally pursued actively. Liver transplants are generally carried out in patients who are in the end-stage of their disease, who have failed to respond to more fully established, conventional medical and surgical therapies, or who can no longer be expected to benefit from them. In the case of Tommy Sheldon, his physicians were initially opposed to the idea of a transplant but came to consider it as a possibility when they felt his options for survival were narrowed to only one: the transplant. During the process of coming to this conclusion, the transplant procedure was collectively transformed by the housestaff team from a far-fetched idea into a concrete treatment goal.

This shift in orientation was reflected in the different ways the survival statistics of patients who had received transplants were discussed. As we have seen, the attending argued when the idea of the transplant was first suggested that the 5-10% survival rate was too low to warrant pursuing the procedure. A week later, after Mr. Sheldon's condition had worsened, the same numbers were used to indicate that he still had a chance, albeit a very small one, in the form of the transplant. Of interest here is how the "facts" -- in this case the statistical data concerning survival rates of transplant patients -- had been reinterpreted as "evidence" for the "only" thing to do. Where these statistics had been used initially to support the physician's first position against the transplant, they were employed later, when the situation seemed extremely grave, to legitimate the possibility of the transplant.

As Tommy Sheldon went rapidly downhill, his doctors felt they were faced with the choice of his almost certain demise versus the unpredictable outcome of an experimental procedure. In deciding to pursue the transplant option, the physicians weighed the *certainty* of the one outcome (death) if they did nothing against the *uncertainty* of the outcome of the transplant (the possibility of survival.) Although they readily acknowledged the chances of surviving the transplant were small, the physicians, in making this decision, stressed the possibility of life. In this instance, they opted for uncertainty of outcome because the transplant, even though new, experimental, and more than likely to result in death, nevertheless represented a potential therapeutic solution to an otherwise insurmountable problem -- the death of a patient. In this process, the physicians collectively redefined the situation to emphasize that, in spite of all the odds, their technical prowess might be able to save Tommy Sheldon. In the face of overwhelming evidence to the contrary, a therapeutic procedure with only the most tenuous possibility of success was interpreted to represent the "only hope."

This selection of the most positive outcome out of a range of possible outcomes for the focus of their work reflects the tendency of physicians to respond with an "optimistic bias" in situations of great uncertainty (Comaroff and Maguire 1981). By giving weight to evidence that a medical intervention could benefit this patient, the doctors interpreted their patient's status within a framework of possible recovery rather than probable death. Once Mr. Sheldon was considered potentially salvageable, his physicians set on a path of active and aggressive treatment "to save him," in the words of the resident, "for transplant." They took it as their obligation to keep him alive and in as good condition as possible until the time when a donor could be found, in spite of others' criticisms of their actions, and in spite of their own belief that the transplant most probably would not work anyway. This perceived obligation influenced their daily care of Mr. Sheldon in many ways, including placing him in the intensive care unit, using intrusive diagnostic procedures, putting him on dialysis to support his kidneys, and employing a mechanical breathing machine.

The goal of saving Tommy Sheldon for transplant served two important functions. In their daily activities it allowed the housestaff to focus their attention on efforts to aggressively intervene in the downward spiral of his disease, rather than on the probability that their interventions would fail and he would die anyway. Moreover, it gave them a therapeutic goal to strive towards and a focus to their work which supplanted the need to confront the very real possibility of death. The transplant thus became imbued with symbolic meaning. Because it offered the "gift of life," death was seen as an event to which the doctors were not yet willing to accede.

While the housestaff were exercising their technical power to keep Tommy Sheldon alive, they suspended their definition of him as "dying." It was only when they were forced to acknowledge that his body was giving out despite their heroic efforts that he came to be identified as near death. As the attending explained, "He was bleeding, he was hypotensive, he was on maximal therapy. We had nothing more to offer him . . . we had tried all our tricks." When all their "tricks" had failed, Tommy Sheldon was seen as dying. At this point, they stopped giving him blood transfusions, thus finally allowing him to die.

Sally Serpa exemplifies how the search for reversible conditions can lead to the pursuit of aggressive diagnostic and therapeutic interventions in a seriously ill patient. As we have seen, she presented both a diagnostic and prognostic puzzle to the physicians caring for her. Although they knew her underlying disease was chronic obstructive pulmonary disease, her clinical course resisted typification. They did not know if her recent deterioration was due to her underlying disease or to other, still undiagnosed pathophysiological processes which, if they could be reversed, might lead to an improvement in her condition. The only thing they did know was that she was gravely ill and her condition was not changing. In the words of the team resident:

She was just difficult because she made no progress. A lot of times she didn't get any worse, she just didn't get any better. And I got the feeling there was something going on with her that I really didn't figure out yet, something about her we were missing . . . and while looking at the overall gestalt of her situation, she had the worst response of anyone I'd ever seen.

Faced with the uncertainty of Mrs. Serpa's condition, the housestaff, in keeping with the pattern of detective work and puzzle-solving that characterizes biomedical practice, devoted time and effort to the "game of diagnosis" (Atkinson 1984). They elected not to withdraw the life support offered by the respirator and let her die, but to concentrate on finding out and treating what was wrong with her.

Because of the possibility that she might have treatable conditions, a speculation that was reinforced by her atypical disease course, the house staff defined Mrs. Serpa as a patient who still was salvageable. In the words of the resident:

Statistically there's a low chance for Sally Serpa, but there's still a chance. I may not feel that after we finish this little thing [the current round of interventions], but everything warrants a chance.

After establishing the goal of treating her reversible conditions on the grounds that she still had "a chance," Mrs. Serpa's doctors not only undertook diagnostic procedures to figure out what was going on, but therapeutic procedures to correct the situation once they obtained evidence of the underlying abnormalities. When they discovered Mrs. Serpa had an empyema, for example, they decided to insert a chest tube to drain the infection because surgical drainage, along with antibiotics, comprise the standard response to such an infection. In spite of objections from other physicians and nurses, they pursued this course of action because it represented the medically correct response. In justifying the team's decision the team resident referred to his "arbitrary rule" about infections:

Well, everybody has to define their own arbitrary rules for situations when you don't know what's right and what's wrong. One of my arbitrary rules is that if there's an infection, treat it. Because in my mind, infections are treatable diseases. End-stage pulmonary disease isn't, infections are. And if she had an infection, then the only treatment for it was to drain it . . . I sort of look at people who are in the process of dying, to look for reversible things and treat those things as long as they're not . . . horribly invasive things to do . . . a chest tube is not a particularly horrible thing . . . I just look for things that are readily reversible and try to treat those things. That's just the way I approach it.

Medical training, by its very nature, inculcates a set of routinized responses to specific situations – or "recipes for action" (Schutz 1967) – which practitioners draw on to carry out their daily work. These routines help make certain medical actions almost automatic; given certain conditions, predetermined sets of routines and procedures are followed. Reversing conditions that are reversible is one example of a prescribed recipe for action that is considered critical in clinical practice. The following quotation from an intern indicates the strength of these routinized responses:

You're always worried that you're going to miss something, that you're not going to treat something that you could treat, that's reversible even in the short [run]. You know somebody has an infection, you treat infections; you're sure they are going to die but you still have to do it, you still have to treat the infection, you still *want* to treat the infection.

While this adherence to a prescribed course of action is unquestioned and beneficial in the care of most acutely ill persons, the case of Mrs. Serpa suggests how it affects the care of terminally ill patients by reinforcing the practice of seeking solutions in the form of medical interventions even in situations of near futility. As the disease process begins to affect more and more of the body's vital functions, the physicians focus their energies on attempting to treat a gradually diminishing number of possibly treatable conditions. This can result in a restricted vision that denies the very real implications of the rapidly failing body in favor of a progressively narrow focus on what is treatable.

CONCLUSION

This discussion has illustrated how physicians' interpretation of information about patients' capacity for interaction and available therapeutic possibilities shapes their construction of patients as potentially salvageable. We have seen how cues about a patient's past or current interactive ability are marshalled as "evidence" for continuing treatment. We have also seen how the hope provided by perceived treatment options affects physicians' assessment of a patient's status and the treatment decisions they make. Further, this paper suggests that the orientation towards "hope" influences physicians' actions regarding the terminally ill in two ways: it encourages a treatment focus on whatever medical interventions are deemed necessary to forestall death and, in the process, it allows physicians to delay the identification of patients as dying until death is unavoidable. In the remainder of this paper we will examine some of the cultural sources and consequences of this focus on intervention as well as some of its implications for the care of dying patients.

Dying and the Culture of Biomedicine

This discussion of the dying of Tommy Sheldon and Sally Serpa illustrates how biomedicine's commitment to *curative* therapy and *medical* interventions shapes the care of the terminally ill. The work of biomedicine is based on the expectation that its practitioners will be able to stop or alter the process and outcome of disease through the diagnosis of the physical disorder and its treatment. This orientation towards intervention is deeply rooted in the cultural scheme underlying medical theory and practice. Parsons, for example, argues that the basic cultural orientation of modern society towards mastery over the environment, what he terms "instrumental activism," is embodied in the pragmatically activistic and melioristic institutional forms of modern medicine: "We [tend] to accept illness not as a given aspect of the human condition but rather as something which [is] within very wide limits, intrinsically manageable and controllable" (1964:337). This view of controllable illness rests on the assumption that vigorous efforts to intervene in the course of disease by the application and advancement of biomedical knowledge and technology will result in the alleviation of suffering and the betterment of the world in general.

When, as Comaroff (1984) suggests, biomedical knowledge came to focus on disease as the disruption of organic functioning which would be mediated through intervention, life and death became polarized: life came to be synonymous with organic structure and function and the life-span of the physical body, while death came to mean the "end-state," the destruction of life. A major goal of biomedicine emerged as the conquest of death, its inevitability difficult to accept in a society and medical system geared to active control and domination over nature. The physician's mandate became one of protecting life and fending off adventitious, unnecessary, or premature death (Parsons, Fox, and Lidz 1972).

With the many advances in technology and basic scientific knowledge taking place in the realm of biomedicine today, physicians now have the opportunity to intervene at more diverse points and with a greater choice of diagnostic or therapeutic procedures. Many conditions which were once irreversible are now curable or at least capable of being arrested. Consequently, the time at which physicians acknowledge that "there is nothing more we can do" comes at a much later point now for some patients in their trajectory to death than it did for people with the same condition forty years ago.

As more "things to do" continually become available as a result of research findings and clinical innovations, physicians are frequently faced with a choice between continuing to treat a patient who might die or "ending" the life of a patient who just might recover. In situations such as those described here – where doctors are invested in patients, and where they believe that patients are capable of interacting with others or there are potentially beneficial treatment options – physicians tend to respond in culturally patterned ways by opting to continue treating the patient who might die. The stories of Mr. Sheldon and Mrs. Serpa illustrate how the duty of caring for a patient who is seriously ill and very likely dying then becomes defined in terms of the pursuit of medical interventions rather than the cessation of such interventions.

Physicians, when faced with a desperately ill person, will tend to use whatever exists in their pharmacological or therapeutic armory if they construct a picture of the patient as having "a chance." This tendency to pursue a therapy which offers a ray of hope, however slim, is particularly evident at university medical centers where the largest experimental therapies and drug protocols are available to patients. In the background is often some further therapeutic maneuver which can be tried in the attempt to forestall death. Moreover, if some therapy is available, even if the likelihood of its success is low or uncertain, it becomes extremely difficult for them not to use it in life-and-death situations. This was evident in the case of Mr. Sheldon, whose physicians became committed to the idea of a liver transplant even though they were at the same time pessimistic about its outcome.

The consequences of this focus on medical interventions in the approach of physicians to the terminally ill are several. In the first place, the procedures that might give a patient "a chance" often become the preoccupation of the caretakers despite their awareness that the odds are unfavorable. Other considerations – that the procedures may very well be futile, that they may only prolong the patient's life for a short period of time, or that the patient may experience considerable suffering as a result of the procedures – become secondary in face of the physicians' overriding concern with fending off premature death.

In this context, as we have seen in the cases considered here, doctors base their activities on the "fiction of probable recovery" (Glaser and Strauss 1965a) by focusing on the most optimistic outcome, by reinterpreting statistical probabilities, and, as the patient's condition deteriorates, by narrowing their area of concern to the diminishing number of problems which are still treatable. Treatment "plans" gather their own momentum and courses of actions are embarked upon which are difficult to alter or stop. Treatment may ultimately be given more as a ritual commitment to the value of fighting disease and death than out of a rational expectation that it will help the patient (Parsons, Fox and Lidz 1972). Medical interventions, offered as "gifts of life," take on symbolic meaning for physicians that extends beyond the individual patient to represent medicine's battle with death. It is the "fight" itself that becomes important as physicians struggle to find out what is wrong with a patient and to stop or alter the course of disease.

Moreover, when physicians are in pursuit of medical interventions, their evaluation of a patient's proximity to death becomes closely linked to their assessment of what is available in their medical and technological armamentarium that could potentially benefit the patient. To understand how perceptions of treatment options influence the identification of patients as dying, we must explore the meaning for the housestaff of the relationship between physician action and the status of "dying." Patients were generally considered by residents to be dying when the residents determined that there was nothing they could do to reverse the course of the disease and the patient would not recover no matter what they did. In medical parlance, patients came to be seen as dying when "all therapeutic options were exhausted," "there is nothing more we can do," or "we have nothing more to offer." As the following statement by an intern suggests, perceived futility is an important ingredient in these views:

The bottom line is whether there's clinically anything more you can do to help a patient. The fact that somebody has a terminal disease doesn't really even enter my mind until the point at which I realize the whole endeavor of really trying to do anything . . . is futile.

Dying is defined, therefore, in terms of the actions -- or failed actions -- of the physicians. In keeping with the predominant technical bias of biomedicine, "dying" becomes a cultural metaphor which symbolizes treatment failure. The focus of medical work is on "doing," and when everything fails, the patient is "dying":

There is this tendency to just sort of press ahead, ignoring the obvious. And then finally reaching a point where you've done everything you can do, and the patient is not turning the proverbial corner, and you're sort of sitting there, going, "What can I do next?" And if there's nothing obvious to do next, then sort of throw up your hands and say, "Well, I guess they're dying."

The words of these residents illustrate how the context of dying becomes defined in terms of the physicians' own treatment agenda. As long as they interpret the situation in terms of there being "more to do" medically for the patient, even if it is considered a "long shot," considerations that death will be the probable final outcome are not significant in their approach to the patient. The patient is not even defined as dying until the clinicians determine there are no further interventions they can make that will improve the patient's condition. Therefore, acknowledgement of a patient's dying status may not be made until death is imminent or, in some cases, has already occurred. As Cassell has observed with cardiac patients, whose care can involve many therapeutic procedures, "the patient with the numerous myocardial infarcts around whom the many physicians and their machines are crowded *is not dying until he is dead!*" (italics in text) (1972: 532).

Implications for the Care of Dying Patients

Integrated as we are within the Western biomedical model, the implications of how and when patients are defined as dying by medical practitioners may seem self-evident. At first glance it would seem exemplary, rather than problematic, that physicians should try to do everything possible for their patients in order to forestall death. Indeed, there are many circumstances in which the application of intensive biomedical interventions is entirely appropriate. If, however, patients who are very ill and likely to die are not recognized as dying until death is imminent, they will not be treated as dying persons. The point at which actors involved in the care of seriously ill patients make the determination that patients are dying influences how these patients are handled along their trajectories to death. Expectations surrounding patients' deaths are a key determinant both in how medical practitioners approach those patients as well as how the patients approach their own impending deaths (Glaser and Strauss 1968).

When patients continue to be defined as if they are seriously and acutely ill, rather than dying, a therapeutic tenacity often drives physicians to continue with "aggressive" interventions beyond the point of therapeutic benefit. While effective and imperative in many situations where patients suffer from acute illnesses, biomedical interventions and therapies may be neither appropriate nor necessary for patients who face death in the near future. If the predominant focus of care remains one of medical intervention, terminally ill patients may be subjected to costly, painful procedures from which they do not benefit and indeed, may suffer needlessly. Since these interventions frequently have the effect of prolonging the lives of those who would otherwise die rapidly, their use raises serious ethical questions about the prolongation of the difficult and often very painful process of dying (Fox 1974).

In addition, the propensity to label patients as dying only when they have failed to respond to attempted therapies means that the shift from a medical-technological orientation stressing intervention to a non-aggressive "palliative" orientation which stresses pain and symptom control and emotional support may not be made until late in the patient's trajectory to death or, in some cases, not at all. Current medical rhetoric supports the proposition that different levels of care are appropriate at different stages in the disease process. There is general agreement among medical practitioners about the need to shift to palliative care when confronted with a dying patient (Bayer et al. 1983; Wanzer et al. 1984). However, if patients are not defined as dying until the last moments of their disease trajectory, they may not receive the type of care that is deemed appropriate for dying patients.

In certain situations, the ambiguity surrounding the recognition of the dying status may be compounded by conflicting interpretations among doctors and nurses of the "reality" of the status. As we have seen in the cases discussed above, different readings of a patient's actual or potential interactive ability, for example, led to different assessments of the patient's prognosis. Disagreements are generated about the appropriate course of action to take if some participants feel the patient is clearly in the process of dying, while others believe that more interventions can still be attempted in an effort to save the person. These differing definitions of the situation influence how practitioners approach their tasks concerning the patient. While housestaff continue to pursue activities in accordance with their medical treatment agenda, for example, nurses may be eager to begin the activities associated with aiding patients until death. If conflicting perceptions of the dying trajectory among hospital personnel become too pronounced, they can create disturbances in existing work relationships.

Finally, it is important to note that within this biomedical perspective which defines certain terminally ill patients as dying only where there are no more medical interventions to make, it is frequently the case that patients are no longer perceived as a *medical* problem after they have been labelled as dying. Although housestaff in American teaching hospitals have overall responsibility for the medical care of patients, once patients are seen to be on an irreversible course toward imminent death and diagnostic or therapeutic procedures are no longer necessary, residents may withdraw from active engagement in patients' treatment. At this point other personnel, usually nurses, take over the palliative care necessary to support patients until death. Thus, once defined as dying, patients are thrust to the periphery of medical work. "Dying" becomes an exit label symbolizing patients' passages beyond the physicians' professional domain.

In summary, this chapter has illustrated how physicians do not rely only on biomedical knowledge of pathophysiology or statistical data about disease outcomes in their assessment of a seriously ill patient's status, treatment, and future. They also evaluate and interpret information regarding the patient's capacity for interaction and available therapeutic possibilities. Through these interpretative activities, a clinical reality of a patient as having "a chance" is constructed. Once a patient is identified as having a chance of life, physicians focus their energies on seeking medical interventions which will forestall death and, in the process, delay the definition of the patient as dying.

Although the approach of physicians to the dying has been criticized in recent years (cf. Aries 1974; Illich 1976; Kübler-Ross 1969), we have attempted to demonstrate in this chapter that their behavior is not the result of pernicious intent. Rather, it is rooted in the biomedical world view which guides their construction of patients' clinical realities and their actions towards these patients. Biomedicine now provides the cultural framework for death in contemporary American society: "Today we cannot think about death except in a language informed by medicine" (Arney and Bergen 1984:37). While it was once the dying person who acknowledged his or her own impending death, in many situations today, as this discussion has illustrated, it is the physician who determines an individual's admission to the category of dying and controls the timing of death. Whereas at one time physicians could do little more than offer comfort and moral support to seriously ill patients while attending their transition to death, they now can intervene, often dramatically, in this transition: not only can they define the circumstances of dying, but death itself can be challenged, fought, and sometimes overcome.

When death does occur, it increasingly has come to be viewed as a technical matter, a failure of technology to rescue the body from a threat to its functioning (Cassell 1975). This conception of death is epitomized in the attending physician's comment on Tommy Sheldon's death: "I think he had the right medicine; he just didn't have the right disease . . . he was just a little bit beyond our technology."

ACKNOWLEDGMENTS

We would like to thank Margaret Clark, Albert Jonsen, Anselm Strauss, Judith Barker, Deborah Gordon, and Patricia Bell for comments on earlier drafts of this chapter.

NOTES

1. The research on which this study is based was conducted with support from the Maureen Church Coburn Charitable Trust and the James Picker Foundation. The analysis of the data was supported in part with funds from the Academic Senate and from the School of Medicine, University of California, San Francisco. This support is gratefully acknowledged. In addition, the first author received partial support for data analysis from the National Institute on Aging Training Grant to the Medical Anthropology Program, University of California, San Francisco (#5 T32 AG00045-07). The first author collected and analyzed the data on which this chapter is based and wrote this manuscript. The second author also collected data for this study and directed the larger research project of which this study is one part.

2. This research was based on 100 interviews with housestaff, medical students, and attending physicians, as well as eighteen months of observation of housestaff "teams" as they carried out their day-to-day medical work on hospital wards. This included direct observation of their care of 45 dying patients.

3. It is significant to note that because this research was conducted exclusively with medical residents working in a university teaching hospital, the cultural patterns described here may vary somewhat with different medical settings and physician groups. Surgeons, for example, may exhibit other ways of handling the care of the dying. Likewise, more experienced clinicians in non-teaching community hospitals may approach dying patients somewhat differently from the physicians described here. Nevertheless, we believe that the arguments advanced in this paper have general significance for the care of the dying by physicians in the United States. The majority of American physicians spend at least part of their training in medical settings similar to the one studied here and it is in these settings that the dominant values, attitudes, and premises of biomedicine are transmitted to the young physicians. These socialization experiences help to shape the expectations and practices associated with "good medical care" that physicians take with them when they leave their residency.

4. All names are pseudonyms.

REFERENCES

Aries, Philippe
    1974 Western Attitudes Toward Death: From the Middle Ages to the

Present. Baltimore: The Johns Hopkins Press.

1981   The Hour of Our Death. New York: Knopf.

Arney, William Ray and Bernard J. Bergen
1984   Medicine and the Management of Living: Taming the Last Great Beast. Chicago: University of Chicago Press.

Atkinson, Paul
1984   Training for Certainty. Social Science and Medicine 19:949-956.

Bayer, Ronald et al.
1983   The Care of the Terminally Ill: Morality and Economics. New England Journal of Medicine 309:1490-1494.

Cassell, Eric
1972   Being and Becoming Dead. Social Research 39(3):528-542.

1975   Dying in a Technological Society. In P. Steinfels and R. Veatch (eds.), Death Inside Out. Pp. 43-38. New York: Harper and Row.

Comaroff, Jean
1984   Medicine, Time and the Perception of Death. Listening: Journal of Religion and Culture 19:155-171.

Comaroff, Jean and Peter Maguire
1981   Ambiguity and the Search for Meaning: Childhood Leukemia in the Modern Clinical Context. Social Science and Medicine 15B:115-123.

Crane, Diana
1975   The Sanctity of Social Life: Physicians' Treatment of Critically Ill Patients. New York: Russell Sage.

Fox, Renée
1974   Ethical and Existential Developments in Contemporaneous American Medicine: Their Implications for Culture and Society. The Milbank Memorial Fund Quarterly/Health and Society 52:445-483.

Fox, Renée and David Willis
1983   Personhood, Medicine and American Society. The Milbank Memorial Fund Quarterly/Health and Society 61:127-147.

Glaser, Barney and Anselm Strauss
1965a   Awareness of Dying. Chicago: Aldine.

1965b   Temporal Aspects of Dying as a Non-Scheduled Status Passage. American Journal of Sociology 71:48-59.

1968   Time for Dying. Chicago: Aldine.

Good, Byron and Mary-Jo Delvecchio Good
1981   The Meaning of Symptoms: A Cultural Hermeneutic Model for Clinical Practice. In L. Eisenberg and A. Kleinman (eds.), The Relevance of Social Science for Medicine. Pp. 165-196. Dordrecht, Holland: D. Reidel.

Hingson, Ralph, Norman Scotch, James Sorenson and Judith Swazey
1981   The Dying Patient. In R. Hingson et al. (eds.), In Sickness and in Health: Social Dimensions of Medical Care. Pp. 181-208. St. Louis: C.V. Mosby.

Illich, Ivan
1976   Limits to Medicine. London: Marion Boyars.

Kleiman, Dena
1985   Probing Life: The New Dilemma. New York Times. January 18, p. 10.

Kleinman, Arthur
    1980   Patients and Healers in the Context of Culture. Berkeley: University of
            California Press.
Kübler-Ross, Elisabeth
    1969   On Death and Dying. New York: MacMillan.
    1975   Death: The Final Stage of Growth. Englewood Cliffs, NJ: Prentice
            Hall.
Parsons, Talcott
    1964   Social Structure and Personality. New York: The Free Press.
Parsons, Talcott, Renée Fox and Victor Lidz
    1972   The "Gift of Life" and its Reciprocation. Social Research 39:367-415.
President's Commission for the Study of Ethical Problems in Medicine and
Biomedical and Behavioral Research
    1983   Deciding to Forego Life-Sustaining Treatment: A Report on the Ethi-
            cal, Medical and Legal Issues in Treatment Decisions. Washington
            D.C.: U.S. Government Printing Office.
Schutz, Alfred
    1967   The Phenomenology of the Social World. Evanston, Illinois: Northwest-
            ern University Press.
Sudnow, David
    1967   Passing On: The Social Organization of Dying. Englewood Cliffs, NJ:
            Prentice Hall.
Wanzer, Sydney et al.
    1984   The Physician's Responsibility Toward Hopelessly Ill Patients. New
            England Journal of Medicine 310:955-959.

PART VI

BIOMEDICAL KNOWLEDGE AND PRACTICE ACROSS
CULTURES

MARGARET LOCK

A NATION AT RISK: INTERPRETATIONS OF SCHOOL REFUSAL IN JAPAN

Opposition

In my youth
I was opposed to school.
And now, again,
I'm opposed to work.

Above all it is health
And righteousness that I hate the most.
There's nothing so cruel to man
As health and honesty.

Of course I'm opposed to "the Japanese
spirit"
And duty and human feeling make me
vomit.
I'm against any government anywhere
And show my bum to authors' and artists'
circles.

When I'm asked for what I was born,
Without scruple, I'll reply, "To oppose."
When I'm in the east
I want to go to the west.

I fasten my coat at the left, my shoes
right and left.
My *hakama* I wear back to front and I
ride a horse facing its buttocks.
What everyone else hates I like
And my greatest hate of all is people
feeling the same.

This I believe: to oppose
Is the only fine thing in life.
To oppose is to live.
To oppose is to get a grip on the very
self.

Mitsuharu Kaneko

M. Lock and D. R. Gordon (eds.), Biomedicine Examined, 377–414.
© *1988 by Kluwer Academic Publishers.*

INTRODUCTION

It is well known that Japan now has the second largest economy in the world, the greatest longevity for women, and the second longest for men. It also has a highly literate population, relatively little poverty, and a long established socialized medical system. All ingredients for a smoothly running, salubrious society one would assume, but a glance through a few recent newspaper headlines indicates otherwise: "More Girls, House-wives Becoming Drug Addicts"; "Schools Reverting to Corporal Punishment"; "More Middle-Aged Men Killing Selves"; "Stress: Serious Problem of Japanese Work Force"; "Number of Suicides By Youths Up 30%"; "School Phobia Caused by High-Pressure Education"; "Japanese Youth Unhappiest [among eleven industrialized countries polled], in Spite of Economic Growth." This last article concludes with the statement: "they are like a floating generation, without any sense of purpose. *And the real problem lies in the family*" (emphasis added, Asahi Evening News 1984).

When reading newspapers and magazines or watching television in Japan one frequently gains the impression that many Japanese believe their society to be in crisis and that one manifestation of this crisis is the plethora of syndromes and social disorders said to be rampant today. Any reasonably well-stocked bookstore will have a section devoted to health related matters, and conspicuous among the books are numerous volumes which contain the following topics: "school-refusal syndrome," "apartment neurosis," the "kitchen syndrome," "moving day depression," "salary-man depression," sleep, eating, and sexual disorders, and family violence (by which is meant children attacking their parents). Some commentators believe that the emotional price that the Japanese people are paying for modernization is perhaps too great; that what is described as the post-war embrace of Western values, and the more recent rapid move into the international market has left the country without any clear direction as to morals and purpose, and its people "floating" and alienated.

I have selected for analysis one of these "new" disorders which is creating particular concern and which is widely discussed in both professional and popular literature, the "school refusal syndrome." After presenting five case studies I will discuss current ideas in Japan about the "meaning" and importance of children. The problem of school refusal will then be situated in the larger current cultural debate in Japan known as *nihonjinron* (essays on being Japanese), and will be followed by a brief discussion of the school system, its perceived function and its on-going deficiencies. The final part of the paper will be devoted to a discussion of causal explanations given by involved Japanese medical professionals and educators in connection with school refusal. It will be shown how these explanations reflect current values in Japanese society at large.

Passive resistance is a characteristic form of communication in Japan, a quiet but persistent plea for some changes in ongoing social relations and expectations placed upon individuals. It is my contention that school refusal children embody a dual representation to those around them: the particular conjunction of variables in a child's biological makeup, life history, and family relationships are used to account for individual cases of school refusal and these factors can often be manipulated successfully in therapy to relieve debilitating distress. At the same time, school refusal children are living manifestations of a potent anxiety which is current in Japan about youth in general, the impact of modernization upon them, and the future of Japan as a nation.

The suffering of each child takes on an ideological dimension, a massive shadow self which everyone who refuses to go to school shares with all other children in the same situation. This phantom ideological child, a creation of political and professional rhetoric, has characteristics, a few of which do indeed correspond to features of some of the real children. The phantom, however, is much more predictable than a real child; a stereotype, prosaic and mundane, with little resemblance to a living child at all. The parents of the phantom child are equally predictable, also stereotypes; together this mythical family represents what is feared could happen to Japan's New Middle Class, itself a cultural construction, if the slide into the abyss of individualism and freedom from social constraints is not checked as soon as possible.

The mythological, symbolic child, creation of contemporary Japanese consciousness, is a very visible creature despite its withdrawn nature, and the actual experience of real children, and the behavior of family members, health professionals, school teachers and peers towards children who refuse to go to school, is marked by the rhetoric about the phantom. By examining the rhetoric in historical and cultural context it is possible to gain some insights as to why school refusal is classed as a medical problem -- a syndrome -- and why, although the absolute numbers appear to be relatively small, it is causing great concern at present in Japan. The relationship of the real, scared, sad and sometimes very sick child to its political shadow is not however simple, neither one of a neat dialectic nor of a blatant opposition.

DEFINING "SCHOOL REFUSAL SYNDROME" (TŌKŌKYOHI)

As in the West, the label of school refusal is clearly distinguished from those of truant or delinquent (Inomata 1983). Unlike a truant, the school refusal child apparently wants to go to school and frequently packs his school bag in the evening saying that he will go the next morning but when the time actually comes cannot make himself leave the house. Alternatively, the child will go half way to school and then come home, or return from school after just an hour or two. The age range in both the West and Japan is from primary school through to college level, but whereas it is generally agreed that the majority of cases in the West occur in young children (Coolidge et al. 1960),[1] the problem is said to occur five times more frequently with junior high school students than with primary school students (Bungei Shunjū 1978:341). In both Japan and the West, school refusal in young children is believed to be triggered primarily by fears about maternal separation and is generally regarded as a different and less severe phenomenon than that observed in older children (Hersov 1985). In this paper I will focus on school refusal at the middle (junior high) school level, in which children aged between twelve and fifteen are involved.

Many commentators in Japan claim that school refusal in older children is a post-war phenomenon and that it was either non-existent or a very minor problem before the war (Hirata 1980:35). The problem was first given official attention in Japan in the 1950's and the label "school refusal syndrome" first applied in the 1960's (twenty five years later than in the West, Broadwin 1932; Johnson et al. 1941). Statistics compiled by the Ministry of Education in which reasons for long-term (50 days or more per year) middle school absence between 1974 and 1982 are recorded indicate that school refusal is currently on the increase (see Table I). In 1974 there were 12,975 cases of long-term absence for reasons of sickness (55.1%), and in 1982 there were 12,943 cases, but these cases were only 33% of all long-term absences since the number of school refusal cases had risen dramatically from 7,310 in 1974 to 20,165 in 1982, representing a percentage change from 31.1% in 1974 to 52.7% in 1982. Absence due to sickness and economic reasons had remained at approximately the same level. Despite an apparent exponential increase in the number of middle school refusal cases the percentage of children is still small and comprises only 0.36% of all middle school students in 1982. The statistics are probably on the conservative side because many children drop in and out of school for shorter periods of time over a number of years, but since their absence falls outside the arbitrarily assigned formal definition of "long-term" absence they are not recorded in national surveys. It is estimated that in about 90% of the cases which are diagnosed as school refusal syndrome the problem first manifests itself as somatic complaints which are very frequently taken initially to a pediatrician (Bungei Shunjū 1978:341). Parents, child, and physician may all choose to treat the physical sickness as the primary problem for weeks

TABLE I

Percentages of "Long-Term" School Absence
in Middle School

Year	1974	1975	1976	1977	1978	1979	1980	1981	1982
Total number of middle school children	4,735,705	4,762,442	4,833,902	4,977,119	5,048,296	4,966,972	5,094,402	5,299,282	5,623,975
Sickness	12,957 (55.1)	12,731 (54.0)	13,137 (52.3)	13,134 (48.9)	11,997 (46.0)	11,960 (43.0)	12,150 (40.9)	12,327 (37.7)	12,943 (33.8)
Economic reasons	629 (2.7)	520 (2.2)	557 (2.2)	575 (2.1)	496 (1.9)	516 (1.9)	527 (1.7)	616 (1.9)	656 (1.7)
School refusal	7310 (31.1)	7704 (32.7)	8362 (33.3)	9808 (36.5)	10,429 (40.0)	12,002 (43.2)	13,536 (45.6)	15,912 (48.7)	20,165 (52.7)
Other	2597 (11.1)	2629 (11.1)	3060 (12.2)	3353 (12.5)	3153 (12.1)	3326 (12.0)	3440 (11.6)	3824 (11.7)	4481 (11.7)
% of children absent due to school refusal	0.15	0.16	0.17	0.20	0.21	0.24	0.27	0.30	0.36

Adapted from: Tōkōkyohi mondai o chūshin ni: chūgakko, kōtōgakkōron, Monbushō, p. 19, 1983. (Figures in brackets are percentages of the total number of absentees).

or even months and this too introduces a bias into statistical estimations. It has been pointed out that if one includes both short term cases and cases diagnosed as somatic complaints of unknown origin, then the total number of children involved would be much larger (Bungei Shunjū 1978). The problem of school refusal is usually assumed to be much more frequently associated with boys than with girls, and this has been confirmed by studies done in Shimane and Aichi prefectures (Wakabayashi 1983:819).

School refusal is a topic which has attracted a large amount of attention from the mass media over the past ten years in Japan. Numerous popular books and articles have been written on the subject (see, for example, Kanezawa and Maruki 1983; Iino 1980; Takahashi 1984; Takuma and Inamura 1980), several television programs have been devoted to it, specialists from various professions have given lectures and held public discussions, and the Ministry of Education has put out an official booklet on the subject (Monbushō 1983). School refusal is clearly a phenomenon which creates a great deal of anxiety at the national as well as the family level. It is difficult to judge just how frequent a problem middle school refusal is, partly because the statistics are unreliable (and tend to encourage an underestimation of the problem) and partly because the media attention directed towards the subject no doubt inflates its significance considerably (and probably leads to an overestimation of the problem). Individual Japanese admit that school refusal certainly did happen before the war, but the actual number of cases does appear to be on the increase at present, and clearly a large number of children are suffering because they cannot cope with their expected role in school. The complex reasons which contribute to the plight of individual children are presented below, but the greater symbolic significance which each school refusal child also embodies should not be underestimated.

FIVE EXAMPLES OF SCHOOL REFUSAL[2]

Case #1

Kenichi[3] is 15 years old and was diagnosed over a year and a half ago as a case of school refusal syndrome. Two years ago he started to complain of stomach aches quite regularly (the physical complaint most usually associated with school maladjustment problems in Japan (Ikemi and Ikemi 1982)). His mother took him to the family pediatrician who ordered a series of laboratory tests and prescribed medication. The results of the tests were negative and even though Kenichi's symptoms intensified his parents insisted that he go to school since they were very concerned about him dropping behind the other students. Kenichi started to stand in the corridor outside his classroom during some of his lessons and the teacher responded by ridiculing him in front of the class. Kenichi's father says that it was at this time that his son became violent at home; he hit his mother and one of his sisters on several occasions and threw furniture and dishes at the walls. Kenichi's father tried to talk to his son and asked him why he was feeling so upset, and the reply, repeated on many occasions, was *"shiranai"* ("I don't know"). The stomach aches continued, and his parents decided to allow Kenichi to stay home from school. He remained on his futon (sleeping mattress) for several days, and on the third evening he announced that he was better and packed his books ready for school the next day, but when morning came he refused to leave the house. His mother, afraid of violence again, allowed her son to stay home. This cycle was repeated many times and settled into a pattern which lasted for more than a month. After consultation with the school counsellor the parents were recommended to a nearby hospital. Kenichi was admitted as an in-patient where he remained for fourteen months, receiving medication and psychotherapy which focused initially on getting up early and eating meals at regular times. Gradually school studies were introduced, and then Kenichi started going to school while still an in-patient. After this routine was well established he was finally discharged and has been going to school since that time. His parents take part in a monthly support group run by the hospital. They say that they have learned to lower their expectations of their only son -- he can work in a *sushi* shop if he wants to, and he will probably finish his schooling after junior high school. His father says that he now realizes that he was more distant with his son than his daughters and that he showed him very little affection.

Case #2

Junko is 13 years old. From the age of 8 until she was nearly 12 she lived in England with her parents where her father was engaged in scientific research at a London hospital. Together with her younger brother, Junko went to a large public comprehensive school on the outskirts of London for the duration of her stay in England. Initially Junko had problems due to lack of facility with English and during the first few weeks of school she was teased by some of her classmates but, says her mother, the teacher soon put a stop to that, and made every effort to have Junko integrated properly into the class. The teacher was apparently completely successful and Junko's mother reports that despite the stressful beginning her daughter did very well at school in London and remembers it as a very happy time in her life.

Upon her return to Japan Junko started going to the local junior high school near her home but from the second day she complained of feeling unwell in the mornings. Her mother allowed her to stay home for a few days but then insisted that Junko go to school and accompanied her as far as the school gates for the next few days. Junko's mother noticed that in contrast to the situation in London where, when she returned from school each day Junko was very keen to tell her family about her day at school, in Tokyo she had nothing to volunteer, and when questioned answered in a flat and perfunctory fashion. Her mother talked to Junko's teacher by telephone and was told that her daughter was having difficulties because she was "different," that her Japanese was no longer "natural," and that she stood out from the other children in the class so that she was not liked by her classmates.

Junko's parents decided to insist that their daughter go to school despite the difficulties she was facing since they believed that it was essential that she be fully integrated back into Japanese society. Junko spent days and sometimes weeks at a
time away from school continually complaining of stomach pains and headaches.
Her mother took her to the pediatrician several times and was told each time that
there was nothing seriously wrong with her child although she was given medication
for her. She telephoned the teacher two or three more times and was led to
believe on each occasion that her daughter was still not behaving in an acceptable
fashion in school and therefore was liable to bullying by the other children. The
teacher suggested that Junko's mother take her to one of the special "treatment"
centers which have been set up in major metropolitan areas of Japan especially to
deal with "returnee students" (*kikoku shijo*) as they are collectively labelled. This
suggestion was not taken up, and the following academic year Junko was transferred by her parents to a private school in Tokyo where many returnee students
study together in a somewhat ghettoized situation. Gradually Junko settled into
her new school and once again started to take pleasure in her life. She later confessed to her mother that she could not bear the teasing that she had been subjected to at the other Tokyo school and added that her friends in England had
never treated her like the outsider she was made to feel when she was in Japan.
Although Junko's family described her as a school refusal child, her case probably
has not been recorded in the official statistical records in connection with this
problem, firstly because Junko has never received counselling, and secondly
because she was never absent from school for more than fifty consecutive days, the
arbitrary length of time designated as constituting an official case.

Case #3

Akira is 12 years old. He was diagnosed as having school refusal syndrome when
his mother went with him for consultation at a child guidance center at the recommendation of his social teachers. Akira has a scar on the inner side of his upper
left thigh which he sustained when he was 10, as the result of an accident while
playing with his friends. Akira has always been regarded as rather selfish (*waga-
mama*) and lazy (*ōchaku*) by his parents and "unconcerned with what other people
feel about him." Akira's eldest brother is praised by his parents in front of Akira
for his sensitive temperament (*shinkeishitsu*), because he is exacting (*kichōmen*),
and because he likes to do things "just right" (*chanto*).

Akira's father thinks that his son's problems started when he was badly bullied by
the other boys while in the fifth grade. His father says that Akira is still angry
about that experience and that he, the father, probably should have complained to
the teacher at the time, but that he thought a *boy* should be able to handle these
matters on his own. Akira's mother says that they should not have been so strict
with him and shown more warmth, but that his father was frightened that Akira's
bullying attitude towards his mother would turn into violence. His mother summed
up her view of the situation thus: "He was beaten up at school, his father was
extremely strict, and his own mother [referring to herself] was afraid of him."

Akira's teacher says that Akira used to say a lot of nasty things to the other children, and that he was made to take part in *hansei-kai* (self-examination meetings) in front of the rest of the class. Akira was repeatedly made to apologize for his behavior at these meetings, but the teacher says that this did not lead to improved behavior.

Akira's mother says that her son has complained every summer since his accident about his leg being painful and swollen. Akira's father and grandmother thought that it might be cancer and persuaded his mother to take him to the doctor, who found nothing wrong.

When Akira was 11 years old his homeroom teacher told his parents that he thought that their son was frustrated, that he did not get along with his classmates, and that he should go to a counsellor. An appointment was made, but Akira refused to go, saying that he was not ill. His mother persuaded him to take some patent medicine called *kiogan* which is normally given to babies to keep them calm and healthy. She also went to a shrine and took part in a ritual to eliminate bad fortune. Akira's mother believes that these measures were effective since her son became more "co-operative and obedient," but then the homeroom teacher reported that Akira was not dressed in an acceptable way; he had started to take a pair of *hakama* (a loose garment like a divided skirt which is worn to practice the martial art of *kendo* (stick fighting)) in his schoolbag on some mornings. He changed into *hakama* on the way to school and stayed in them all day despite the derision of his classmates.

At about this time Akira's mother noticed that he came home from school several times with scotch tape on the front of his underpants and sometimes tightly taped around his upper thigh near the scar from his accident. Akira says that the other children put it there.

Akira moved on to junior high school with many of his elementary schoolmates. His parents say that he never was a dedicated student and that he started to find the lessons extremely difficult at this point. It was again suggested that Akira should go to a counsellor with his parents; the teacher reported that he was being bullied and that he never ate his lunch.

After Akira's father saw his son "walking as though he had polio" on his way to school one day, he decided that Akira had a mental illness (*kokoro no yamai*) and agreed that he be taken by his mother to a counsellor.

Akira has never completely stopped going to school; he misses one day every week or two and when he goes he does not participate in group activities; he is embarrassed about changing for physical education and he will only go to school dressed in *hakama* (he also wrapped his lower torso up in bandages under the *hakama* for awhile and then covered himself with cushions (*zabuton*) on his lap for a few days until the teacher hid them and told Akira that he had burnt them).

Akira's father was gradually persuaded to come to family counselling sessions with his wife and son which sometimes also included the teacher. After several sessions Akira's mother burst into tears and confessed that her husband was an alcoholic and had on occasion beaten up both Akira and her, but not his eldest son. The ambivalent feelings of both mother and son towards the father are very apparent in the counselling sessions.

Akira had been going, on his father's insistence, to karate and *kendō* classes for many years. He has stopped these now and says he disliked them. He has also stopped going to *juku* (a supplementary afternoon school, see below) which he says he liked because he had friends, and which his mother says he disliked.

To an outside observer Akira appears to be a pleasant child, not withdrawn and not particularly shy, especially when compared to other Japanese adolescents. Counselling sessions are very goal-oriented and also very authoritarian by Western standards. A psychodynamic interpretation of his case has not been discussed with the parents to my knowledge, and the counselling focuses at present on getting Akira to participate in school activities and on getting him to co-operate in tidying his things up at home. The next goal is to have him give up the *hakama*. Akira's mother is also being counselled to allow her son to do a few things for himself. So far, no assistance has been offered to his father.

Case #4

Yasuyuki is 16 years old and was diagnosed as school refusal syndrome several months ago. He is an in-patient in a large teaching hospital where he receives medication and regular psychotherapy. He was admitted to hospital after his mother came, at the insistence of her husband, to ask for help. Yasuyuki had stopped going to school 2 years previously, and when his mother tried to persuade him to go he had retreated, carrying his *futon*, into the space between the ceiling and the roof of the family house. According to the psychiatrist in charge of the case, Yasuyuki's mother carried food to her son's hiding place everyday for 2 years and entreated him day after day to get up and go to school. Yasuyuki remained indifferent to these pleas, and his mother, full of guilt about what she perceived to be a reaction to her inept mothering, endlessly put off asking for help. After she finally went to the hospital, a psychiatrist came with her to the house, administered medication to Yasuyuki, and had him removed to hospital. (This is clearly a very extreme example; I have not met the family in question, but am reporting the account as it was given to me by the psychiatrist at present in charge of the case.)

Case #5

The death of a junior high school student in the winter of 1986 was reported at great length in the Japanese newspapers. Hirofumi Shikagawa, aged 13, hanged himself rather than facing the bullying of his classmates which he described in his suicide note as "hell on earth." Hirofumi named two of his classmates of many years' standing in his note and asked them to stop bullying other students. Hirofumi who was described in the newspapers as "quiet" had apparently been made into an "errand boy" by these two boys, both bigger than him in physique, and they had required him to fetch and carry all their belongings. He had also had his face written on with marking ink and been forced to make a fool of himself in public (Japan Times 1986). At the police inquiry subsequent to the suicide it was revealed that Hirofumi had been made to attend his own mock funeral set up by some of his classmates and condoned by the homeroom teacher (Mainichi Shinbun

1986).

During the final months of his life Hirofumi had often excused himself from class with complaints of stomach pains or headaches. In January, he attended school for only eleven days and often hid in the school washroom rather than go to class, until he finally committed suicide in early February.

In 1985 a total of nine children in Japan were recorded as having committed suicide because they were allegedly the victims of school bullies, and it is estimated that of all "school phobic" children, 40% stay at home in order to avoid bullying (Japan Times 1986).

The values and attitudes towards children and their education in contemporary Japan will be briefly considered next in order to better understand the basis on which ideas about causation and therapy for school refusal are constituted.

THE MAKING OF A NATION

The stated objectives of schooling in Japan are "to assist in the harmonious psychological and physical development of the child and to nurture the foundations for a healthy mind and body" (Early Childhood Education 1979:105). These foundations are, of course, laid down during early socialization in the home and are largely the responsibility of the child's mother. A newborn child is regarded as pure, and is sometimes described as pure white – *masshiro* there are no associations made between an infant and either sin or pollution. The connection between purity and the newborn has been interpreted by some to mean that in Japan one starts with a *tabula rasa* for the purposes of socialization (Hendry 1986:17), but this runs contrary to the very old concept of *umaretsuki* (what is attached at birth). The terms *seishitsu* and *kishitsu* (which are very similar to each other), are used to indicate a disposition or temperament with which one is endowed at birth, including certain emotional tendencies and aptitudes for specific tasks and skills. Another term, *taishitsu*, is used to express the idea of a physical endowment at birth. The *umaretsuki*, comprised of these latent physical, emotional, and motor qualities, are the basic building blocks with which a Japanese person is created; included are elements which potentially can put one at a distinct disadvantage in terms of a healthy life, either physical or social (Lock 1980).

I know of no Japanese who does not accept these traditional concepts but, naturally, ideas about genetics have been grafted onto them in recent times. So too have European derived post-Darwinian beliefs about

innate instincts and drives which have been adopted in many cases as
though they were scientifically established. It is very common, for example, for social scientists in Japan to accept as fact the idea that aggression is innate: "Youthful violence, now a serious social problem, is in no
way a sign of mental abnormality. It is merely a manifestation of the
aggression inherent in the structure of the human brain . . . However
adorable a baby may be, aggression, that is, the killer instinct, is innate"
(Suzuki 1983:21).

Concepts derived from genetics and innate drives, however, have by
no means replaced traditional ideas about *umaretsuki*. One is certainly
not therefore starting with a Lockean *tabula rasa*, but it is, nevertheless,
generally accepted that despite the limitations imposed by endowed characteristics, a young child is "good" material, and that the task ahead of
the parents (most especially the mother) and later the child's teachers, is
to work together with the child to bring out its full potential, largely
through the application of perseverance and endurance. When this is
accomplished successfully an individual grows up to learn and accept
their limitations, and to adjust to the constraints of their inherited
endowment and given social environment (Lebra 1976; Lock 1982).

It is commonly believed in Japan that although humans share with the
animal world the features of the lower brain, what is unique about them
is the potential for social and moral development. In *this* sense the child
is "white," in that it is a *tabula rasa* as far as morals are concerned; the
process of becoming a "good Japanese" is essentially a lifelong exercise
in learning how to behave appropriately both socially and morally. Moreover, since mind and body are not dichotomized, temperament and even
physical disposition must be allowed for and built upon in the creation
(*sōzō*) of the socially acceptable adult.

An ancient and much quoted adage in Japan states that the "soul of
the three year old must last for one hundred years," implying that early
socialization is of the greatest importance and something which cannot
easily be undone if handled incorrectly. Even today Japanese mothers
are very reluctant to hand the early training of their children over to any
outside caretaker; their own mother is usually the only acceptable substitute for themselves (Hendry 1986). During the first few years of life a
small Japanese child is taught *shitsuke*, a diffuse, but key concept which
incorporates the idea of "putting into the body of a child the patterns of
living, ways of conduct of daily life and a mastery of manners and correct
behavior" (Hara and Wagatsuma 1974:2). Other meanings covered by
this term include the idea of making the child "fit into a mould," and of
learning habits before one can reason why (Hendry 1986:13). The raising
of children is regarded as a vital investment in the future and as the only
effective means of ensuring continuity of the Japanese culture. "The

creation of people" is therefore "seen as [a] skill to be cultivated with a good deal of time and careful attention" (Hendry 1986:14).

As soon as the Japanese infant shows signs of talking, it is taught (as part of *shitsuke*) how to greet people properly and a little later it will be shown how to bow, eat food in the correct fashion, keep both its surroundings and itself as neat and tidy as possible, and how to practice elementary hygiene. The aim of much of this socialization is to encourage the child to do things for itself but equally important is the idea that being responsible for oneself is ultimately for the benefit of those around one and for society at large. The child is reminded in many different ways to think of others, to be kind and sympathetic, and to avoid causing trouble for other people. From a very early age small children are encouraged to stick at things, to persevere (*gaman*), to be patient, and to be tolerant of others in their group (Hendry 1986:83). This early "moulding" of the behavior of the child is designed to lead ultimately to the internalization of a basic moral style in which one is made highly sensitive to both the verbal and nonverbal cues of other people and in which individuals then bend (like a bamboo) in order to accommodate themselves to the needs of others, especially those designated as more mature and experienced.

Certain temperamental qualities are especially fostered in children in order to enhance moral development. The quality of *sunao* is particularly prized, another difficult concept to gloss, and covering a range of meanings from meek, submissive, passive, compliant, and honest to guileless. This term is often translated into English simply as obedient, but the positive, spontaneous behaviors associated with it are then lost. Two other qualities held in high regard are those of *yasashii* (gentle, kind) and *omoiyari* (sympathy, fellow-feeling). Women reported to Hendry that the temperamental styles of small children which are particularly difficult to deal with include being selfish, a cry-baby, impatient, obstinate, or of a strong-will (1986:90). Throughout early socialization a mother will consciously identify the unique characteristics of her child, and will work to encourage those perceived as good, and to modify those viewed as in conflict with a co-operative group life.

There is a slogan currently very visible on large bill-boards in some parts of Japan: *kodomo wa kuni no takara* (children are the nation's treasure). This sentiment, emblazoned at great expense, no doubt by nationalistic groups for the edification of the minds of the masses, provides a glimpse of the larger, political meaning of children of Japan. When a mother trains her child in connection with greeting other people, she is not merely training it how to use the Japanese language correctly, nor to be polite, but to be a "proper" Japanese. The future of the nation is invested in the fruits of her dedicated and diligent child-training and the bill boards indicate a concern that this function is no longer being properly carried out by today's parents, the first post-war generation to raise children.

DISSENT IN THE CIRCLE OF NURTURANCE

In Japan, the training of a child is generally accomplished by gentle persuasion. Children are praised for their success much more than they are chastised for their shortcomings. Physical separation and ostracism rather than direct punishment or displays of anger are used in order to gain compliance in a willful child. The early environment is, in general, extremely nurturant, and one in which actual physical contact is frequent (Caudill and Weinstein 1974). It has been pointed out many times that the goal of socialization is not one of striving for independence, but rather of encouraging a state in which dependence on primary groups is accepted as natural and inevitable (DeVos 1973:47). The term *amae* is used to indicate the feelings of dependence and a desire to be passively loved that all normal infants experience in the primary relationship with their mother. Takeo Doi, a well-known Japanese psychiatrist, has pointed out how dependency and the "desire to presume upon another's love" are not merely confined to the experience of an infant and are particularly encouraged and fostered in Japan throughout the life cycle (Doi 1973). The relationship between a mother and her child, especially her son, is a particularly close one in which it is difficult for a clear differentiation of self to be established, and where the child learns to manipulate the behavior of others through the demonstration of *amae*. DeVos has suggested, on the basis of thematic apperception test results, that as the result of the intense socialization process that the Japanese child undergoes, considerable guilt is experienced when one does not live up to adult expectations (1973:148).

In a culture such as Japan where harmony in primary groups is revered, it is particularly difficult for people to express open dissent to the wishes of those around them, most especially their mother. Any kind of behavior which is thought of as different or unusual is likely to produce a response of parental and peer group ostracism (Kiefer 1980:436). Even the wearing of clothes or the use of school bags which are not like those of everyone else can instigate jeering and taunting. There are several characteristic styles of behavior for the indirect expression of inter-personal conflict, ranging from "negative communication" (remaining silent when a response could be expected) to the use of self-directed aggression (Lebra 1984). The most dramatic of these, of course, is suicide.

Given an environment of sensitivity to non-verbal communication, and the preference for indirect forms of expressing dissent, a child learns that one way to control the behavior of others is through its own suffering. In extreme cases, such as that of Hirofumi cited above, suicide becomes the last resort, but for many Japanese, both adults and children, somatization (the expression of distress as physical symptomatology) is a very common means of experiencing and expressing dissent (Lock 1987; Reynolds 1976). As a form of protest, somatization may be unconscious, or partially or fully conscious, and represents a very common (perhaps universal) and potent style of self-aggression, made use of especially by those who are lacking in power (Comaroff 1985; Kleinman and Kleinman 1985; Nichter 1982). Somatization and the presentation of non-specific complaints have been noted as very frequent reasons for visits to the doctor's office in Japan (Lock 1980; Reynolds 1976; Vogel 1978) and it is not surprising to find that 90% of school refusal children first complain of physical symptoms. Their physical discomfort represents a plea to those around them, especially their mother or their school friends, or both, for a change in relationships or behavior. If this message is not noticed, then a more dramatic form of protest can ensue, sometimes violence, or more often, the passive withdrawal associated with school refusal. Unfortunately, the combination of the competitive environment of the classroom and the determination of many mothers that their children will have academic success at all costs means that the quiet message of the frightened child, although noticed, often goes unheeded. This situation is particularly acute because of the intense way in which the Japanese mother is involved in the eduction of her child. As part of this relationship, from kindergarten until the first years of middle school, the majority of Japanese mothers take it upon themselves to help their children intensively with their homework. This means that a child who refuses to go to school represents a failure on the part of his mother, not only himself. His decision not to persevere is perceived as a harsh indirect criticism of his family and especially his mother and as such cannot be treated as a personal crisis.

THE PATHOLOGICAL FAMILY

There has been recent discussion in Japan on the question of whether a new national holiday should be created known as "family day." Those supporting the idea state that despite a "healthy" economy, the "spiritual health" of the country is poor (Mochida 1980). Almost all commentators agree that recent changes in demography, urbanization, and the transformation of the majority of extended families into nuclear households has brought about some fundamental changes in family relationships. For some analysts one of the biggest concerns is that the nuclear family is no longer equivalent to the spiritual corporate body that the extended family used to be. The traditional Japanese family, apart from its social and economic functions, also served as the center of religious activities since the ancestors were enshrined in the home, and family members regularly communicated with them. The traditional family was, therefore, the source of moral and spiritual education (Smith 1974:151) and parents made use of the presence of the ancestors to legitimatize their authority beyond mere parental rights into a sacred mission.

Some families, notably in rural areas, continue to function this way, but the majority have become "modernized." Compared with the traditional family, the nuclear household is often regarded as a potentially fragile and even "pathological" system in which sexual equality, liberalization of parent/child relationships, and egalitarian inheritance laws have disrupted traditional forms of control, leading to the deviance that is said to be rampant in modern Japan (Eto 1979; Mochida 1980). There are many dissenters to this view (see Yuzawa 1980 for example), but there is, even among progressive segments of the Japanese elite, a grave concern over the actual functioning of the nuclear family in Japan today (Higuchi 1980; Imazu et al. 1979). It has been widely reported over the past twenty years both inside and outside of Japan that 90% of the Japanese are middle class. The concept of the "New Middle Class" composed of nuclear families has been described by Kelly (1986) as "folk sociology." It is an idealization, an abstraction which bears only a superficial resemblance to reality, but one around which much current rhetoric is generated about the mass educated, consumer society of modern Japan. The fragility of the New Middle Class, is also part of the rhetoric.

In most modern Japanese families, roles are still rather strictly segregated by gender. The image of a father is of someone who is devoted to his work group, who spends up to 15 hours a day, 6 days a week commuting, working and socializing on work related matters, and who only takes a small portion of his allotted vacation. It is widely believed that to be less committed will provoke slow promotion and peer group ostracism (Atsumi 1979). It is also believed that if one refuses to be re-located at the company's suggestion, this too will affect promotion and pay increases. A 1983 survey of more than 1100 salaried workers revealed that 50% of them moved without their families, usually for periods of several years, and sometimes up to 6 or 7 years. It is very rare for a company to pay for the expenses of moving an entire family, especially since it is generally agreed that it is inappropriate for children to be made to change schools[4] (Japan Times 1984). This phenomenon of families living with an absent father known as *tanshin funin*, is regarded by many as an inevitable part of company life and something to be adjusted to;

however, it is claimed by school counsellors, physicians, and government officials that school refusal syndrome and parent abuse are particularly prevalent in families where the "father's shadow is thin" and where he is frequently absent (Hirata 1980; Monbushō 1983).

Of course, not all Japanese fathers are company employees and many live what appears to be a "normal" life. Moreover, statistics on the actual relationship of single parent and absent father families and the incidence of school refusal are not available -- there may not be particularly close association between these variables, (the fathers of Kenichi, Junko and Akira all live at home) but the "Absent Father," who, paradoxically is often behaving like a model company employee, represents a symbolic threat to family harmony and continuity. His enforced absence, which has been described as the "feminization" of the family (Mainichi Shinbun 1985), becomes part of the pathology lurking beneath the surface of the paradigmatic modern nuclear family.

The role of professional housewife (sengyō shufu) is taken very seriously in Japan: not only is child raising important, but the health, physical and spiritual, of the entire family rests upon her shoulders. However, women raised in post-war Japan are often accused of being selfish, too individualistic and spoilt; they no longer know how to raise children and over-indulge them (Kyūtoku 1979), and they are said to be more concerned with their own interests than with caring for those of their husbands (Katsura 1983). Some authors are explicit that the desire of women to work and to fulfill their own aspirations is "wrecking" the children of Japan (Eto 1979), but therapists and counsellors often point out that they rarely see cases of school refusal in families where the mother is employed (Lock 1986:108) and none of the mothers of the children whose cases are described above are working. As in the case of the modern father, there is paradox and ambiguity associated with the roles and expected behavior of the wife/mother in Japan today (see Lock 1988 for a full development of this theme). Even though the relationship of the mythological Selfish Mother to the actual incidence of behavioral problems and pathology in children is not at all well established, she stands symbolically for a wide-spread fear that the nation is not guarding its "treasure" well -- that children are not being raised properly, and that the future of Japan as a distinct and carefully preserved cultural entity is at stake.

The discussions and literature produced on these symbolic parent figures represent part of the much larger ongoing cultural debate (Parkin 1978), known in Japan as nihonjinron, essays on being Japanese. Central to nihonjinron is a concept of racial (genetic) homogeneity which is thought to lead "naturally" to language and cultural unity. A distinction between insiders and outsiders is sharply delineated, and ideas about the separateness and uniqueness of the Japanese are taken for granted in this discourse (Kawai 1976; Miller 1982). This type of rhetoric is not unique to the Japanese, of course, nor is it new to Japan (Kawamura 1980), but it forms the basis for what Hobsbaum calls the "invention of tradition" (Ranger and Hobsbaum 1983) and is recast with different emphases at different historical moments. With accelerating impetus, the latest variation in this genre of writings has taken up the question of loss of traditional values such as group interdependence, tenacity, and endurance, and their replacement by values associated with modernization and especially Westernization, most notably that of individualism. Almost 700 monographs in a 32 year period have been officially identified as part of this current genre, and if journal articles were added to this number the total would be many thousands (Befu 1983).

The nuclear family and the New Middle Class are often selected out for special critical attention in this type of rhetoric, and are usually held responsible for the "neuroses" which are said to be rampant in modern Japanese society. The Selfish Mother and the Absent Father raise children who are weak in both body and spirit and who become ready victims to the diseases and evils of modern society. The school system is generally acknowledged as having problems, but many commentators go on to ask why so many children apparently enjoy school or at least pass through it in a satisfactory fashion and only a relative few exhibit behavioral problems (Kumagai 1981; Tamai 1983). In order to account for this apparent anomaly, the fragile nuclear family is usually invoked as causal. Before examining explanations given by clinical practitioners for school refusal in more detail, a brief discussion of the school system is in order.

THE MAKING OF A MERITOCRACY

Japanese education is theoretically designed to be egalitarian. The majority of children go to public schools where everyone in the same age-grade proceeds at exactly the same pace; there is no streaming or acceleration of certain students, and textbooks, course materials and examinations are standardized by the Ministry of Education. Despite the superficial uniformity of the system everybody is well aware that some educational paths are more likely to lead to the top than are others. Certain prestigious universities for example have kindergarten, primary, middle, and high schools attached to them and the successful three year old who passes the required examinations can, with reasonable application, remain in the safe stream to the very top. Graduation from a good university is essential in order to procure a rewarding career, but university places are limited. In 1980, for example, 58% of high school seniors decided to apply for advanced education, but it was estimated that only 1 in 4 succeeded at the first attempt in finding a place in college or university (Rohlen 1983). At each stage, from kindergarten through to university, one must pass demanding examinations which require an enormous amount of rote memory work in order to proceed to the next level. The examinations at the end of middle school are especially important because compulsory education ends at the point, and competition to enter a good high school is especially fierce.

Pressure is compounded by the existence of *juku*, cram schools, where as many as 50% of middle class school children go several days a week after regular school (Rohlen 1980:212). *Juku* are private institutions where pupils aspiring for a good university place are usually taught extra math, Japanese, and English. Parents who are willing to settle for less for their children enroll their offspring in a *juku* where there is less emphasis on academic prowess. Children frequently have to commute a considerable distance to *juku* and by the time they have finally arrived home at night, completed two sets of homework, and often piano practice as well, it is usually very late. They are frequently up at five thirty or six o'clock each morning in order to get to school early for a sports practice. Children go to school for 5 1/2 days a week and the academic year is one of 240 days.

Despite its commitment to equality, the school system is intensely competitive and in recent years appears to have become even more so. The Ministry of Education has recently officially complained about the "schoolization" of kindergarten in which children are made to sit for long stretches of time at desks and to start learning the 3 R's in earnest (Mainichi Shinbun 1984). On the other hand, the infamous "examination hell" remains unmodified by the Ministry. Paradoxically, at the same time, the National Council of Educational Reform and some school teachers and parents are calling for the introduction of more individuality into the school system. Nevertheless, the use of school uniforms, standardized hair cuts, lack of heating in schools to "toughen" children up, punishment by one's group if an individual is late, rude, untidy and so on, and other requirements such as no unseemly behavior outside of school including the buying and eating of food, still go almost totally unquestioned.

School in Japan is not merely a place for learning academic skills; it remains as it always was, an institution tuned to continue the moral training begun during early socialization in the home. Correct behavior is taught formally in a subject called moral education from primary school onwards (Cummings 1980). Qualities of diligence, endurance, an ability to decide to do the hard thing, wholehearted dedication at all times, and cooperation, are the specific objectives that the teachers in the moral education classes should try to achieve with their students (White 1987:45), qualities which a diligent and dedicated mother has in her turn tried to foster. Children also learn in these classes to put group needs before their own, and this training is reinforced, for example, as they clean their school together, and work regularly in all their classes in small groups on projects for which they are graded as a group. These groups also act as disciplinary agents and the teacher may delegate considerable responsibility and authority to them. The preferred mode of curbing the behavior of a non-conforming peer is through ostracism.

In kindergarten and primary schools children tend to be compliant with the training to which they are subject, but in middle and high schools in recent years serious discipline problems have been documented with great frequency, including cases of delinquency, inter-personal violence, bullying, and vandalism. The National Council on Educational Reform describes this as "a state of desolation in education" and calls for an enrichment of moral education both in the home and the schools to counter the trend (1986:16).

The current school-related problem which is creating the most media attention is labelled as "returnee children" and this issue reveals, as does no other, what the Japanese perceive education to be all about. These students are children like Junko who have received a portion of their education outside of Japan while their fathers are posted overseas on business. Returnee children, it is said, cause trouble in the classroom because they are different, ask too many questions, show too little discipline, and do not work well cooperatively. Many teachers are reluctant to have them in their class (although these children are often from well-educated and influential families), and state in effect that the child is too much trouble because he or she must be taught first of all how to be a Japanese, and, at the same time, how not to be foreign (Goodman 1986). The mandate for a school teacher is clearly one of training "proper" Japanese, someone who not only has their head filled with "facts," but who has been moulded into the correct *persona*.

Returnee children and children labelled as having school refusal syndrome are thought to have several qualities in common: they are unpopular and not part of any group; they may try to compensate for this by being pushy, after which they often withdraw and become introverted; they are frequently the victims of bullying. To an outsider what is rather horrifying are accounts where the teacher has apparently condoned the bullying. Obviously this would not be allowed to happen in most Japanese classes, but a few teachers apparently feel little sympathy for the "outsider." Presumably these teachers believe that the introverted child is not merely expressing a personal quality, but is in effect posing a threat to the entire group; that ostracism and even violence is an acceptable way to deal with them. In an individually oriented society, refusal to go to school causes concern largely about the future academic success and hence employment opportunities of the child. In a group oriented society such as Japan there is a similar concern, but it is magnified enormously by a potent fear on the part of parents and teachers that they have in some way failed to train the child adequately in some very basic social qualities. Although it is popular in Japan today to talk about encouraging more individualism (Moeran 1984), this quality is so closely associated with selfishness and even anarchy that if it manifests itself at the expense of group interests it then becomes a source of great concern. Children who do not fit into groups are *de facto* exhibiting excessive individualism. The school refusal child, therefore, becomes a living symbol of malaise in the system at large and is not merely a sad kid. He is living proof that the nuclear family is fragile, and that the entire country is potentially in jeopardy because of its willful abandonment of traditional values.

CAUSES OF SCHOOL REFUSAL SYNDROME: PROFESSIONAL EXPLANATIONS

It is agreed by medical professionals that school refusal syndrome can subsume a number of psychiatric diagnostic categories. In a few extreme cases (such as that of Yasuyuki) a diagnosis of psychosis will be given, others will be labelled secondarily as having depression, and a large number are considered to be neurotic or border-line neurotics (Akira has been given this last label). Well over half the children do not, however, receive a psychiatric diagnosis of any kind and are labelled as "pure" (*junsui*) school refusal cases. The majority of all children refusing to go to school will be taken initially by their mother to see a primary care physician since the child usually complains about various somatic aches and pains. The primary care physician is unlikely to ask about the child's behavior -- he usually has a very full waiting room, and the mother may choose not to volunteer this information even though she may well believe that the physical complaints and the school refusal are in some way linked. At this stage medication is usually prescribed. If the child continues to miss school, is causing trouble or is very isolated in the classroom, then the teacher will call in a school counsellor if one is available, who may well suggest that the parents solicit psychiatric help. This process can take weeks, months, or even years and, since many schools have no counsellors, little pressure is put on the parents to seek profes- sional help. Media exposure and popular medical books on the topic have helped to break down some of the guilt and humiliation which many parents feel, and advice is more readily sought out than was formerly the case, although one still hears of extremely sick children who have been hidden in their homes by their parents, sometimes for years on end (as was apparently the case with Yasuyuki).

Virtually all interested medical professionals and educators in Japan believe in a multi-causal explanation for school refusal, but those who are actually involved with therapy, almost without exception, opt for one or more specific variables as key factors for manipulation in the therapeutic process. Since there is general agreement that school refusal is a post-war phenomenon, the majority of professionals attribute the problem in part to changing environments, physical and social, since the war. Modernization and urbanization are the two biggest changes which are most often cited in a negative fashion.

It is said that due to urbanization children no longer have contact with nature, and especially the passing seasons. They are therefore not able to associate their changing moods with the changing seasons as was formally the case, and so experience a sense of alienation (Hirata 1983). It is also said that children no longer participate in traditional Japanese regional festivals; growing up in an urban environment severs their sense of belonging to a particular region, and also diminishes their ability to play. School refusal children, it is claimed, are bad at playing (Hirata 1980:36).

It is also pointed out that modernization has meant that the daily lives of most people has become one of convenience. It is claimed that mothers prepare pre-packaged food for their children, and that in addition children get very little exercise because of the demands of school work, television, and isolation and lack of opportunity to play in groups, enforced by an urban environment. These changes have led to a weakening of the physical bodies of children, poor brain functioning, and often to overly sensitive children who are at risk for school refusal (Masaki 1982; Hirata 1983:37).

Everyone agrees that the school environment is not ideal for all children and that those who "stick out" in some way are likely to feel unhappy. There is a proverb which is made much use of to illustrate this point: *Deru kui wa utareru* (a nail which protrudes must be hammered down). Some professionals claim that children with low intelligence are likely to have the most trouble in an equalitarian system and that they make up the bulk of school refusers (Lock 1986:104), others state that it is the bright, bored children who drop out (Bungei Shungū 1978). Still others focus on the fact that some children are especially nervous about tests (Hirata 1984:40).

Members of the nuclear family, notably those who have their homes in urban high rise apartments, are also thought to contribute to school refusal. Mothers living in nuclear families are said to be lonely and lacking in confidence as to whether they can raise a child by themselves.

They enter into a symbiotic relationship with their offspring and become the source of its neuroses (Hirata 1983:42; Monbushō 1983; Suzuki 1983; Takahashi 1984). Mothers are accused by several writers of having lost their "natural" childrearing instincts (Hirata 1983; Kyūtoku 1979). Absent fathers are similarly labelled as a contributing factor and are accused of not projecting a "father-image" (Monbushō 1983). The personality "types" of both mothers and fathers who are likely to produce school refusal children are frequently "analysed" and reported in the literature (see Table II). Mothers are described as anxious, worried, turned inwards, unsociable, infantile, immature, unmotherly, overly methodical, unexpressive, dependent and so on. Fathers are characterized as lacking in confidence, poor at making decisions, turned inwards, unsociable, not fatherly, verbose, dependent and so on.[5] Adults with personalities such as these are not good role models for their children; like

TABLE II
Personality Types of Parents

Investigators	Fathers	Mothers
Ro	Lacking in self confidence, Avoidance of contact with other people.	Tendency to instability, excess of internal tension.
Manita	The psychological representative of the mother.	Faint hearted, lacking in self-confidence, a worrier.
Murata	Faint hearted, poor at making decisions, emotionally immature.	Unexpressive, talkative, socially and psychologically immature.
Satō	Taciturn, turned inwards, antisocial.	Infantile, nervous, turned inwards.
Yamabashi	Antisocial, turned inward lacking in a masculine image, lacking in a positive attitude.	Juvenile, unsocialized, dependent, distressed about relationships with other people, reserved, exacting.
Yamamoto	Too egotistical, verbose, warped personality.	Lacking in a motherly image, image, lacking in motherly warmth.
Tamai	Dependent, faint hearted, not good socially.	Dependent, withdrawn, unsociable.

Translated from Shūsaku Satō, "School Refusal Children" Kokudosha, 1968, reproduced in T. Taketoshi and H. Inamura (eds.), *Tōkōkyohi Dōshitara Tachinao-eruka*. Tokyo: Yuhikakusensho, 1980.

their children, they are warped products of modernization who because of their personality types have not been able to adjust to post-industrial life as well as the rest of their generation.

The personality of the child, the result of poor socialization and training, is also at fault. School refusal children are antisocial, withdrawn, egotistical, immature, of nervous temperament, overly sensitive, timid, stubborn, repressed and poor problem solvers (see Table III). Their *umaretsuki* (inherited qualities) have not been adequately nurtured and schooled by their parents or teachers, and their self-centered approach to life, symbolic of a general blight in modern Japan, is apparently not unduly surprising to many commentators. Since school refusal children are basically regarded as morally deficient it is not surprising that the more chronic and "difficult" cases are referred to the modern gate-keepers of good behavior -- it is the psychiatrist who has the final word about who is to be labelled as suffering from this shameful syndrome.

TABLE III

Special characteristics and personalities of juveniles

Investigators	Personality and Special Characteristics
Ro	Retiring, inclined to be egotistical, inclined to be excessively tense.
Yamamoto	Rambunctious in the house but quiet outside (*uchi benkei*), retiring, disruptive temperament, takes refuge easily, antisocial, sensitive, timid, obstinate.
Wakabayashi	Antisocial, turned inwards, egotistical, uncooperative, poor emotional development.
Manita	Poor problem solver, feels pressure because of wish to run away from things, strong craving to be perfect.
Tamai	Faint hearted, withdrawn, rambunctious in the house but quiet outside (*uchi benkei*).
Murata	Basis of the problem is emotional immaturity, tendency to be a weakling and of nervous temperament, dependent and hence very tense around other people.
Satō	Tendency to have a nervous temperament, socially immature, turned inwards, egotistical, sensitive to the point of irritability.

Translated from Shūsaku Satō, "School Refusal Children" Kokudosha, 1968, reproduced in T. Taketoshi and H. Inamura (eds.), *Tōkōkyohi Dōshitara Tachinaoeruka*. Tokyo: Yuhikakusensho, 1980.

HAMMERING DOWN THE PROTRUDING NAILS – THERAPY FOR SCHOOL REFUSAL

Since there is a shared belief in multi-causality one might expect to find a multi-faceted approach to the management of the problem of school refusal. At first glance this appears to be the case, since treatment ranges from dietary changes through to electro-shock therapy.[6] However, a closer look reveals that virtually all of the treatment modalities focus on the "identified patient" – the child, and to a lesser extent, on his mother. Very occasionally fathers will be involved. Although the larger issues of modernization and urbanization associated with physical pollution and social anomie, the latent competition in the school system and the current rash of bullying in many schools, the oppressive work load of many Japanese men and their apparent enforced absence from their families, the supposed isolation and boredom of many Japanese women and sex discrimination in the work force, are all recognized and discussed in abstract terms, these factors do not enter directly into the therapeutic encounter.

It is not surprising, of course, that counsellors, clinical psychologists, and psychiatrists focus on bringing about change in individual behavior – this is their mandate, what they have been trained to do. It is neither reasonable nor appropriate to expect clinical therapists to sit and discuss at length with their clients the dilemmas of society at large, nor to commiserate with them in general terms about the pressures of the school system. It *is* possible, however, to incorporate a flexible stance in therapy in which one does not start from the assumption that the patient and his family are morally inadequate, misfits in a society where there is a rather narrow definition of what is correct and incorrect, and must therefore be socialized into taking on a more appropriate *persona*. My impression from observing therapeutic encounters, and from reading reports on clinical cases and popular books on the subject is that, although kindness and sympathy are very often extended to the child, a moralistic stance is usually central and therapists take upon themselves the task of pointing out and correcting poor behavior in the child and sometimes in his mother. The ultimate objective of the exercise is to help parents and child adjust to what is taken as inevitable and virtually unquestionable in the organization of social relationships and working environments in Japanese society.

The psychodynamic language of Western psychiatry is absent in these encounters, one does not hear a discussion in terms of feelings of loss, anger, fear, or sadness. Family relationships are explored largely in terms of concrete behavior and roles, and a moralistic tone often pervades the atmosphere. For example the psychiatrist says to Akira:

"You used to say that you were afraid of your father. How about now?"
"Sometimes I'm afraid."
"That's because you don't do what he says" (*"iu koto o kikanaikara"*).

This last statement is reiterated by both psychiatrist and mother almost in unison.

The "unconscious" does not have a role to play at all in these encounters and behavior modification, whether it be in terms of diet, habits, or in social encounters with parents and peers is by far the most popular approach taken to therapy.[7] Some children are hospitalized, the majority remain at home; some receive anti-depressants or other medications, most do not; virtually every child (apart from those few who are completely withdrawn or psychotic) will be trained into a routine where sleep and eating habits, tidiness and order in daily life, are the immediate goals to be achieved. The creation of this routine will be established using the kind of language and value orientation that a "good" mother should use in socialization. The qualities of *sunao* (compliance), *yasashii* (gentleness) and *omoiyari* (thoughtfulness for others) will be encouraged in association with the formation of routines.

Murase has described the concept of *sunao* as central in many types of Japanese therapy. He points out that to say that someone is not *sunao* is to pass a decidedly negative judgement about them, and lists several psychological states which are associated with a lack of this quality. Included are situations where some forms of resistance, conflict, and suppression are found both within oneself and in one's relation to the environment; secondly where there is in a person's mind or behavior, a *muri* or quirk which prompts one to behave unnaturally in certain situations; thirdly, a somewhat tense or anxious state of mind; fourthly, a state where one's view of the world is rather onesided and predisposed towards one-way communication; and lastly a state where one's perspective is very limited and narrow (1982:321).

Murase goes on to point out that there are thought to be two different levels of *sunao*. The prototype of *sunao* is found in the infant before the phase of ego development (1982:324). The natural trusting attitude of a baby towards its mother is associated with this pristine *sunao* and so too is a "naive, submissive and nondefensive state of mind" (1982:324). In certain kinds of traditional Japanese therapeutic systems, notably *naikan* therapy, a regression to this early state can sometimes be achieved and used to advantage in the healing process. Regression to infantile *sunao* is thought to aid in re-establishing a sense of trust and profound security in human relationships which is considered to be the basis for a mature adult life as a Japanese.

At the secondary level of socialized *sunao*, individual identity is clearly established but with an acute sensitivity to the way in which one's own desires and behavior affect other people. The ego tends to be viewed rather negatively, and is equated with selfishness. There is a striving to "repress oneself in order to *realize* oneself" (Murase 1982:325), implying an attainment of a state of "greater ego" (*taiga*), where others come before self, and harmony in human relationships is the ideal. One interpretation of the behavior of a school refusal child is that he is jeopardizing group harmony with his own desires and neurotic anxieties, and that once he learns to think of others, then he will start to function better as a human being: "The key point is that the relationship *per se* is far more important than the individuals themselves" (Murase 1982:325).

Once regular routines and some signs of moral development are established, then the child will be slowly eased back into the school system and set once more onto the path of meritocracy. The parents may well have been persuaded to lower their academic expectations for their child, and in some instances the child will be placed in a special class or school, in which case everyone accepts in effect the fact that he/she is not a "normal" Japanese, and is therefore likely to have chronic problems. This appears to be especially true for returnee children and children who are very slow academically.

Many therapists also work to modify the behavior of the mother, most often by giving her instructions as to how to allow her child more independence and autonomy while at the same time providing structure and order to his life. Alternatively, "selfish" mothers will be encouraged to provide better care for their children in the form of more home cooking or family activities and "immature" mothers will be encouraged not to treat their children "like pets" (Lock 1986:104). Insight (in the Freudian sense of that word) is not the objective, but a gradual modification of behavioral habits and styles. Because of inborn characteristics, neither parents nor children can be totally re-created, and therapy is always constrained by the limits which biology imposes on human beings as cultural products.

Meanwhile the debates about school reform, equal employment opportunities, trade deficits, the New Middle Class, and Japan as a "consumer culture" out of which its traditions are draining at a rapid rate, provide the backdrop against which the little dramas of each school refusal child is acted out. In the minds of therapist and parents this backdrop is not unimportant, but it is irrelevant to the more pressing issue of making an adequate citizen and human being out of the child who will not go to school. The child may feel sad, unloved, guilty, angry, or frightened and may also be ill, and is very frequently the victim of bullies. The complex reasons, historical and contemporary, biological, individual, and social, for the patient's feelings are acknowledged, but remain

unexplored in more than a very superficial way, and are relegated to second place while the hard "work" of creating a moral person takes prece-
dence.

Children such as Kenichi, Junko, Akira, Yasuyuki and Hirofumi are
making a plea for some attention to be given to the psychic pain which
they can no longer bear on their own. Those who end up in therapy usually receive solicitous care, and sometimes effective solutions are found
for their problems for at least the immediate future. The treatment of
these children in therapy is culturally constructed and furnishes an example of how, for a therapeutic system to be successful, it must reflect and
draw upon values taken from the society at large (Lebra 1981; Reynolds
1976). At the same time the behavior of all of the children represents for
many Japanese a frightening omen for the future, created through an
extrapolation into a symbolic world in which associations of school
refusal with selfishness, disorder and the decline of the nation are dominant. While some therapists contribute whole heartedly to the fueling of
the symbolic specter (notably by writing books on the subject), many do
not, and some even try to counter it.

Although statistics are not available, it seems highly unlikely that in
real life all school refusal children come from the New Middle Class and
have Absent Fathers, Selfish Mothers, and are themselves antisocial,
stubborn and so on. Moreover, their numbers are indeed rather small,
although the media gives the impression that the problem is epidemic and
a "Phobia Hotline" has been set up in Tokyo and is apparently kept very
busy. The imaginary school refusal child which represents a general
political and cultural anxiety is indeed troublesome. Its showiness,
coupled with its personality traits and behavioral style, probably inhibits
the development of a more flexible approach to its living counterpart.
The media version of school refusal syndrome is exactly the kind of
metaphorical creation which Susan Sontag (1978) has exposed in connection with illness, since it infuses public thinking with a dangerous rhetoric; dangerous because it helps to deflect attention away from larger
social issues of which school refusal is one small manifestation. The
necessity of continued economic growth and consumerism go virtually
unquestioned for example. The impact of rapid social change on real
individuals who have grown up in post-War Japanese society is another
topic which needs to be examined as honestly and openly as possible and
not merely through the use of survey data. The possible loss of the
essence of Japanese culture is hotly debated, but surely it is important to

ask if it is necessary to cultivate "the spirit of Japan" in the children of today who are also told that their country must promote internationalism. The fostering of concepts of insiders and outsiders, and of unique racial qualities, and the moral training of children to incorporate these national-istic ideas appears singularly paradoxical in a shrinking world. Values which seem more appropriate are surely tolerance and sensitivity to oth-ers -- qualities which were also fostered in traditional Japan and continue to be valued today, and which should perhaps be encouraged in the com-mentators who have helped to fuel the phantom of the school refusal child.

SOCIO-SOMATICS: CONFUCIAN IDEOLOGY AND PATIENT CARE

Modern Japan has been characterized as a society which "couples the organizational discipline of Confucian traditions with the unfettered energy of private enterprise" (Rohlen 1980:242). This dominant ideology of "Confucian capitalism" is reproduced in ideas and behavior in connec-tion with health and illness, and influences both health professionals and the general public. Although there has been exposure to and continuous contacts with European medicine from the 16th century onwards, and despite the fact that for over 100 years Western-style medicine (originally in its German form) has been the official medical system in Japan, there remain some profound differences with both European and North Ameri-can medical thinking. The origin of many of these differences lies in the respective attitudes which are held in connection with fundamental con-cepts about the inter-relationship of nature, the human body, and society and culture. Shinto, the indigenous folk religion of Japan, and the philo-sophical tradition of Buddhism, together with the ethical system of Con-fucian, have all contributed to modern Japanese ideas in connection with the relationship of the individual person to society and to the natural world. This pluralistic system of thought shapes the way in which mod-ern scientific and medical thinking has developed in Japan.

The ethical system of Confucianism is, of course, not acknowledged today as an ideology upon which modern Japanese society is structured. There are, however, thinly disguised signs of nostalgia for it in many aspects of social life in Japan, and its legacy is drawn upon regularly to uphold or restore order in the family (Vogel 1968), in business (Rohlen 1974) and in the schools (Lanham 1979:1-18).

Neither Buddhist philosophy nor Confucian ethics ever subscribed to the equivalent of a Cartesian split between psyche[8] (Simon 1978) and soma, and the issue of whether social events and emotional states have an influence on somatic function and vice versa has not been problematic

until very recent times. Even in modern Japan, biological reductionism
is subject to regular criticism (Watanabe 1974:279-282) and in medicine,
although basic research usually takes a form familiar to Western scientists, in clinical practice an extreme biomedical orientation is very rare.
But the long shadow of Confucianism is cast over this "holistic" orientation. The history of Confucianism is in a broad sense essentially a study
in the suppression (but not the elimination) of individual needs and aspirations for the sake of group harmony, cohesion, and advancement;
political order and the hierarchical structuring of society could only be
questioned at great risk by those designated as subordinates.

I have tried to show above how this type of ideology is passed on and
recreated through early socialization. Ideas derived from traditional medical theory and practice (which have evolved under the influence of Confucianism) also reflect a similar style of thinking. Medical classics state
that it is the duty of an individual to lead a balanced and ordered life
which in turn contributes to a healthy body and, more importantly, to a
harmonious social order (Lock 1980). Throughout much of Japanese history the state was "not subject to human control; humankind was subject
to the control of the state" (Smith 1983). Chang, in discussing Neo-Con-
fucianism in general, says that at the heart of Confucian ideology "was a
belief in the substantial unity of the central values and norms of the existing social order with the structure of the cosmic order. This belief could
generate tensions with the existing sociopolitical condition, but its more
important effect was to freeze the normative order of state and society
and render it absolute" (Chang 1980:267).

The chief function of a medical practitioner in this kind of political
climate becomes one of restoring patients to a condition in which they
can once again contribute energetically to society. Medical practitioners
in historical Japan, therefore, readily acknowledged the interaction of
environmental and social variables with the physical state of individuals,
but usually regarded the social order as given, and focused on furnishing
somatic treatments which, it was believed, would restore their patients to
good social standing (Lock 1980).

In contemporary Japan, although modern medical theory is based
upon the principles of science, the Confucian legacy continues to intrude
directly and indirectly. Individuals or families are usually held responsible for problems which clearly have a very broad range of etiological
components, such as sickness of the elderly (Lock 1984:65-73), behavioral and psychiatric problems in children (Kyūtoku 1983), depression in
housewives (Katsura 1983), problems at menopause (Lock 1987a) and so
on.

It has been stated that biomedicine, which claims that it is free of values and ideology, since it is purportedly based upon scientific principles,

"depoliticizes" the medical encounter (Habermas 1971). By manipulating biological processes, and viewing these processes as independent from environmental, social, and psychological variables, disease has come to be regraded as an amoral and asocial event (Comaroff 1982:49-68). I believe that in Japan this assumption is not generally accepted since an epistemology that "decontextualizes" the body has never been widely accepted.

The physical body can be regarded relatively mechanistically in Japan, but personal responsibility should be taken for preventive medicine which includes the maintenance of physical, emotional, and moral order. Today, psychotherapy, both indigenous and Western-derived, is used (often together with pharmacotherapy) for problems such as school refusal which are thought of as "behavioral," but traditional principles remain evident even in modern psychotherapy, and restoration of moral order is primary. This is accomplished primarily through the use of behavior modification techniques in which an appeal is made neither to the rational mind nor to the unconscious of the patient, but at the level of "socio-somatics." The undifferentiated mind/body (*shintai*) is the focus for manipulation, in which the patient (and often his mother) are taught to embody new disciplines which will enable them to participate more fully as upright and morally mature members of their society. As members of Japanese society, physicians are predisposed to approach therapy in this way.

The role of the therapist as mediator between the real child and his or her ideological usurper, the phantom child, awaits further analysis. My belief is that medicalization of this problem does not necessarily serve as a weapon for the preservation of the *status quo*. The very exposure of it through medicalization and the mass media, replete with inaccuracies and ambiguity, can potentially lead to an awakening of political consciousness and hopefully to an uncoupling of the phantom child of ideology and the pernicious shadow which it casts, from the suffering of real children.

ACKNOWLEDGEMENTS

The research reported in this chapter was made possible by a Social Sciences and Humanities Research Council of Canada Grant, #410-83-0175R-1.

NOTES

1. Leventhal and Sills reported in their 1964 article that the greatest frequency

"seems to be in the 10 to 12-year-old group." Most other researchers emphasize entry into primary school as a particularly difficult time.

2. Cases #1, #3 and #4 have previously been described in a paper published in
    1986 in Social Science and Medicine 23:99-112. The data for this paper,
    which I regard as a preliminary survey, were obtained in the spring of 1984
    from interviews with five psychiatrists, a group of psychiatric residents, two
    social workers, two school teachers, and one school administrator. Five
    books and numerous papers written on this subject by health professionals
    and educators, and a booklet put out by *Monbushō* (Ministry of Education)
    also furnished useful information. I attended four sessions where counselling
    of children and/or parents took place, and was permitted to tape-record two
    of these sessions. I also interviewed two sets of parents and two children
    without the counsellor present.

3. All the names used in this article are fictions except where reports are taken
    directly from newspaper articles.

4. Compulsory education finishes at the end of middle school in Japan and high
    schools are not obliged to accept new students. There have been cases
    reported of high school children not finding a place in school after they were
    moved to a new district.

5. The personalities of parents are also regarded as contributory factors to
    school phobia in the West (see Atkinson et al. 1985 and Hersov 1985, for
    example). Some of the descriptions of these parents, notably those diag-
    nosed as having "characterological" disorders sound quite similar to the
    descriptions of Japanese parents. There are no marked differences between
    Japan and the West in terms of causal explanations given for school refusal,
    but the interpretation and meaning attributed to the explanations are cultur-
    ally constructed and therefore have different implications. School phobia in
    America, for example, does not generate the same national concern as it
    does in Japan and is not a subject which is given much priority.

6. A female member of the Japanese Diet is currently investigating the use of
    electric shock treatment for patients with "school phobia." An official of the
    Ministry of Health and Welfare responded to enquiries on the matter by say-
    ing that electric shock therapy is a supplementary procedure used on children
    with endogenous depression and other severe medical problems. He added
    that "it shouldn't be practiced as a punishment" (Japan Times Weekly 1985).

7. Behavior modification is reported as the most often used therapeutic tech-
    nique in the West (Atkinson et al 1985). A psychoanalytic approach is gener-
    ally regarded as too time consuming and less successful then behavior modifi-
    cation.

8. In traditional East Asian medicine the concept of *ki* (mind or spirit) is often
    glossed as psyche. It is important to note that *ki* is quite close to the Hom-
    eric concept of *psyche*, which is never portrayed as a thinking, feeling or
    reflecting entity, but that it does not at all resemble the Platonic idea where

psyche is capable of functioning as one's ethical and cognitive core.

REFERENCES

Asahi Evening News
  1984 Japanese Youth Unhappiest, in Spite of Economic Growth: Poll. February 22nd.
Atkinson, L., B. Quarrington and J.J. Cyr
  1985 School Refusal: The Heterogeneity of a Concept. American Journal of Orthopsychiatry 55:83-101.
Atsumi Reiko
  1979 Obligatory Personal Relationships of Japanese White-Collar Company Employees. Human Organization 38:63-70.
Befu Harumi
  1983 Internationalization of Japan and *Nihon Bunkaron*. *In* H. Mannari and H. Befu (eds.), The Challenge of Japan's Internationalization: Organization and Culture. Pp. 232-266. Tokyo: Kwansei Gakuin University.
Broadwin, Isra
  1932 A Contribution to the Study of Truancy. American Journal of Orthopsychiatry 20:599-607.
Bungei Shunjū
  1978 *Tōkōyohiji te Nandaro* (What are School Refusal Children?) *Zadankai* (Round table discussion) 56:340-347.
Caudill, William and Helen Weinstein
  1974 Maternal Care and Infant Behavior in Japan and America. *In* T.S. Lebra and W.P. Lebra (eds.), Japanese Culture and Behavior: Selected Readings. Pp. 225-276. Honolulu: University Press of Hawaii.
Chang, Hao
  1980 Neo-Confucian and Moral Thought and its Modern Legacy. Journal of Asian Studies 39:267.
Comaroff, Jean
  1982 Medicine: Symbol and Ideology. *In* P. Wright and A. Treacher (eds.), The Problem of Medical Knowledge. Pp. 49-68. Edinburgh: Edinburgh University Press.
  1985 Body of Power, Spirit of Resistance: The Culture and History of a South African People. Chicago: University of Chicago Press.
Coolidge, J.C., M.L. Willer, E. Tessman and S. Waldfogel
  1960 School Phobia in Adolescence, a Manifestation of Severe Character Disturbances. American Journal of Orthopsychiatry 30:599-607.
Cummings, William K.
  1980 Education and Equality of Japan. Princeton: Princeton University Press.

Doi, Takeo
  1973   The Anatomy of Dependence. Tokyo: Kodansha International Ltd.
DeVos, George
  1973   Socialization for Achievement. Berkeley and Los Angeles: University of California Press.
Early Childhood Education of Japan
  1979   Early Childhood Education and Care in Japan. Tokyo: Child Honsha.
Eto, Jun
  1979   The Breakdown of Motherhood is Wrecking our Children. Japan Echo 6:102-109.
Goodman, Roger
  1986   Japan's Returnee School Children: A Threat to the System or Threatened by the System? Proceedings of the British Association of Japanese Studies, II (in press). Sheffield: Sheffield University.
Hara, Hiroko and Wagatsuma Hiroshi
  1974   Shitsuke (The Rules of Child-rearing). Tokyo: Kobundo.
Habermas, Jürgen
  1971   Towards a Rational Society. London: Heinemann.
Hendry, Joy
  1986   Becoming Japanese: The World of the Pre-School Child. Manchester: Manchester University Press.
Hersov, Lionel
  1985   School Refusal. In M. Rutter and L. Hersov (eds.), Child and Adolescent Psychiatry: Modern Approaches. (Second Edition) Pp. 382-399. Oxford: Blackwell Scientific Publications.
Higuchi, Keiko
  1980   Changing Family Relationships. Japan Echo 7:86-93.
Hirata, Keiko
  1980   Tōkōyohi no Genin (The Causes of School Refusal Syndrome). In T. Takuma and H. Hamura (eds.), Tōkōkyohi: Dōshitara Tachinaoreruka (School Refusal: How Shall We Get Over It?). Pp. 35-46. Tokyo: Yuhikakusensho.
Iino, Setsuo
  1980   Tōkōkyohi no kokufukuhō (How to Conquer School Refusal). Tokyo: Bunrishoin.
Ikemi, Y. and A. Ikemi
  1982   Some Psychosomatic Disorders in Japan in a Cultural Perspective. Psychotherapeutics and Psychosomatics 38:231-238.
Imazu, Kōjino, Hamaguchi Esyun and Sakuta Keiichi
  1979   Shakai Kankyō no Henyō to Kodomo no Hattatsu to Kyōiku (Strategic Points in the Social Environment and the Development of Children) In Kodomo no Hattatsu to Kyōiku I (Child Development and Education, Vol. 1). Pp. 42-94. Tokyo: Iwanami Shoten.

Inomata, Taketsugi
  1983 *Tōkōkyohi to Hikō* (School Refusal and Delinquency) *Rinshō Seishin Igaku* (Clinical Psychiatry) Special Issue 12:857-871.
Japan Times
  1984 Employment, Seniority Systems Cause Mental Depression. February 16th.
  1986 Principal Finds Lesson in Boy's Suicide. February 4th.
Japan Times Weekly
  1985 Truants Sent to Mental Hospitals: Diet Woman. December 14th.
Johnson, Adelaide M., Eugene I. Falstein, S.A. Szurek and Margaret Svendsen
  1941 School Phobia. American Journal of Orthopsychiatry II:701-711.
Kanezawa, Kaichi and Masaomi Maruki
  1983 *Tōkōyohi: gakko kirai no shinsō shinri.* (School Refusal: An Enquiry into a Dislike of School). Tokyo: Rodokuhosha.
Katsura, Taisaku
  1983 *Daidokoro Shōkōgun* (The Kitchen Syndrome). Tokyo: Sanmaku Shuppan.
Kawai, Hayao
  1976 *Bosei Shakai Nihon no Byōri* (The Pathology of Japan as a Maternal Society) Tokyo: Chūō Kōronsha.
Kawamura, Nozomu
  1980 The Historical Background of Arguments Emphasizing the Uniqueness of Japanese Society. Social Analysis 5/6 (December): 44-62.
Kelly, William
  1986 Rationalization and Nostalgia: Cultural Dynamics of New Middle Class Japan. American Ethnologist 13:603-618.
Kiefer, Christie
  1980 Loneliness and Japanese Social Structure. *In* J. Hartog, J.R. Audy and Y.A. Cohen (eds.), The Anatomy of Loneliness. Pp. 425-450. New York: International Universities Press.
Kleinman, Arthur and Joan Kleinman
  1985 Somatization: Interconnections Among Culture, Depressive Experience and the Meanings of Pain. *In* A. Kleinman and B.J. Good, (eds.), Culture and Depression. Pp. 429-490. Berkeley: University of California Press.
Kumagai, Fumie
  1981 Violence: A Peculiar Parent-Child Relationship in the Japanese Family Today. Journal of Comparative Family Studies 12:337-350.
Kyūtoku, Shigemori
  1979 *Bogenbyō* (Mother-Induced Illness). Tokyo: Sanmaku Shuppan.
Lanham, Betty
  1979 Ethics and Moral Precepts Taught in Schools of Japan and the United States. Ethos 7:1-18.

Lebra, Takie
  1976   Japanese Patterns of Behavior. Honolulu: University of Hawaii Press.
  1981   Self-Reconstruction in Faith Healing: A Hyperbolic Version of Japanese Morality and Womanhood. *In* G. White and A. Marsella (eds.), Cultural Conceptions of Mental Health and Therapy. Pp. 269-283. Dordrecht, Holland: D. Reidel Publishing Co.
Leventhal, Theodore and Malcolm Sills
  1964   Self Image in School Phobia. American Journal of Orthopsychiatry 34:685-695.
Lock, Margaret
  1980   East Asian Medicine in Urban Japan: Varieties of Medical Experience. Berkeley: University of California Press.
  1982   Traditional and Popular Attitudes Towards Mental Health and Illness in Japan. *In* A. Marsella and G. White (eds.), Cultural Conceptions of Mental Health and Therapy. Pp. 215-233. Dordrecht, Holland: D. Reidel Publishing Co.
  1984   East Asian Medicine and Health Care for the Japanese Elderly. Pacific Affairs 57:65-73.
  1986   Plea for Acceptance: School Refusal Syndrome in Japan. Social Science & Medicine 23:99-112.
  1987   Protests of a Good Wife and Wise Mother: The Medicalization of Distress in Japan. *In* E. Norbeck and M. Lock (eds.), Health and Medical Care in Japan: Cultural and Social Dimensions. Pp. 130-157. Honolulu: University of Hawaii Press.
  1988   The Selfish Housewife and Menopausal Syndrome in Japan. *In* Women in International Development Publication Series. Michigan State University.
Mainichi Shimbun
  1984   Kindergarten: How Many Kanji? March 13th.
  1985   Dad Playing Weak Role in Bringing Up Offspring. March 14th.
  1986   Teacher Knew About "Funeral" For Subsequent Suicide Victim. February 8th.
Masaki, Takeo
  1979   *Kodomo no Kokoro to karada* (The Mind and Body of Children). *Jurisuto*. Autumn, 75-80.
Miller, Roy
  1982   Japan's Modern Myth: Language and Beyond. New York: Weatherhill.
Mochida, Takeshi
  1980   Focus on the Family. Editorial Comment. Japan Echo 3:75-76.
Moeran, Brian
  1984   Individual, Group and *Seishin*: Japan's Internal Cultural Debate. Man (n.s.) 19:252-66.

Monbushō
  1983   *Tōkōkyohi mondai o chūshin ni*: *chūgakko, kōtōgakko ron* (A Discussion of Junior and Senior High School: Focus on School Refusal). Tokyo.
Mouer, Ross E. and Yoshio Sugimoto
  1983   Internationalization as an Ideology in Japanese Society. *In* H. Mannari and H. Befu (eds.), The Challenge of Japan's Internationalization: Organization and Culture. Pp. 267-297. Tokyo: Kwansei Gakuin University.
Murase, Takao
  1982   *Sunao*: A Central Value in Japanese Psychotherapy. *In* A.J. Marsella and G.M. White (eds.), Cultural Conceptions of Mental Health and Therapy. Pp. 317-329. Dordrecht, Holland: D. Reidel Publishing Co.
National Council on Educational Reform Government of Japan
  1986   Summary of Second Report on Educational Reform (provisional). April 23rd.
Nichter, Mark
  1982   Idioms of Distress: Alternatives in the Expression of Psychosocial Distress. Culture, Medicine and Psychiatry 5:379-408.
Parkin, David
  1978   The Cultural Definition of Political Response. London: Academic Press.
Ranger, Terence and Eric Hobsbaum
  1984   The Invention of Tradition. Cambridge: Cambridge University Press.
Reynolds, David K.
  1976   Morita Therapy. Berkeley: University of California Press.
Rohlen, Thomas P.
  1974   For Harmony and Strength: Japanese White-Collar Organization in Anthropological Perspective. Berkeley: University of California Press.
  1980   The *Juku* Phenomenon: An Exploratory Essay. Journal of Japanese Studies 6:242.
  1983   Japan's High Schools. Berkeley: University of California Press.
Simon, Bennett
  1978   Mind and Madness in Ancient Greece: The Classical Roots of Modern Psychiatry. Ithaca: Cornell University Press.
Smith, Robert J.
  1974   Ancestor Worship in Contemporary Japan. Stanford: Stanford University Press.
  1983   Japanese Society: Tradition, Self and the Social Order. Cambridge: Cambridge University.
Sontag, Susan
  1978   Illness as Metaphor. New York: Farrar, Straus and Giroux.
Suzuki, Shigenobu
  1983   What's Wrong with the Education System. Japan Echo 10:17-23.

Takahashi, Y.
   1984   *Tōkōkyohi no Ruutsu* (The Roots of School Refusal). Tokyo: Yuhika-
          kushinsho.
Takuma Taketoshi and Hiroshi Inamura
   1980   *Tōkōkyohi Dōshitara Tachinaoreruka* (School Refusal: How Can We
          Overcome It?) Tokyo: Yuhikakusensho.
Tamai, Nobuaki
   1983   *Tōkōkyohi no Gainen* (Conceptions about School Refusal) *Rinshō
          Seishin Igaku* (Clinical Psychiatry) Special Issue 12:809-813.
Vogel, Ezra
   1968   Japan's New Middle Class. Berkeley: University of California Press.
Vogel, Suzanne
   1978   Professional Housewife: The Career of Urban Middle Class Japanese
          Women. The Japan Interpreter 12(1-2):16-43.
Wakabayashi, Shinichirō
   1983   *Tōkōkyohi no Genkyō to Haikei* (The Background and Present State of
          School Refusal). *Rinshō Seishin Igaku* (Clinical Psychiatry) Special
          Issue 12:815-823.
Watanabe, Masao
   1974   The Conception of Nature in Japanese Culture. Science 183:279-282.
White, Merry
   1987   The Japanese Educational Challenge. New York: The Free Press.
Yuzawa, Yasuhiko
   1980   Analysing Trends in Family Pathology. Japan Echo 7:77-85.

SETHA M. LOW

MEDICAL PRACTICE IN RESPONSE TO A FOLK ILLNESS: THE DIAGNOSIS AND TREATMENT OF *NERVIOS* IN COSTA RICA

INTRODUCTION

This paper examines the dialogue which occurs between physicians and patients in response to *nervios* (nerves), the presenting complaint in 30% of general medicine consultations and 50% of psychiatric consultations observed in urban outpatient clinics of San José, Costa Rica (Low 1981). *Nervios* is both a common, culturally acceptable distress symptom, that is, a sign of psychosocial discomfort, and a syndrome, that is a local medical concept which identifies and gives meaning to a group of anxiety-related symptoms. It is a particularly interesting example of a folk illness that has no clear biomedical diagnosis or description, yet is normally presented to a physician in the clinic setting by a variety of patients searching for a cure or symptom alleviation. Physicians whom I observed respond to these requests positively in so far as they rarely question the use of the term *nervios* and usually take part in a discussion of the medical, social and family concerns of the patient. The physician often prescribes a pill, a tranquilizer or anti-depressant as part of the treatment. This "treatment" of talking and medication reflects cultural agreement on the part of both physicians and patients as to their expectations of the consultation and the cultural meaning of *nervios*.

Nervios provides a socially sanctioned vehicle for the expression of distress, and in some cases mental illness, that in other social situations would be stigmatized. The concept of *nervios* helps order a diverse series of complaints into a culturally meaningful unit that physicians recognize and treat and that people around the patient accept as a regular, although not normal, reaction to the stress of family disruptions and other disturbances of everyday life. Barlett and Low (1980) argue that *nervios* reinforces cultural conformity by facilitating the expression of cultural deviance through sickness and by socially sanctioning or controlling such deviance through medical treatment.

The treatment of *nervios* within the medical consultation presents a special opportunity to examine the creation of a medical discourse when the language and actions of physicians and patients focus on the cultural category, *nervios*. Biomedical theory and medical education lead to biological reductionism in both knowledge and practice (Wright and Treacher 1982; Turner 1984), yet in the Costa Rican example, physicians respond

415

M. Lock and D. R. Gordon (eds.), Biomedicine Examined, 415–438.
© *1988 by Kluwer Academic Publishers.*

to a diffuse "psychosomatic" symptom/syndrome which ignores mind-body dualism and biomedical diagnostic categories.

This paper specifically focuses on the medical consultations in which the physician is presented with a culturally acceptable symptom/syndrome that has specific cultural meanings for both patient and physician, but which does not fit neatly into the usual diagnostic categories of biomedicine. The observation of a medical consultation that affirms the meaning of *nervios* through the integration of sociocultural and biomedical knowledge illustrates the inherently social basis of the production of medical knowledge. In Costa Rica, physician/patient interaction during the medical consultation for the treatment of *nervios* mediates popular and biomedical clinical models in three different domains: (1) etiology, in which physicians tend to adopt popular explanations for the illness when speaking to the patient; (2) diagnosis, in which physicians use popular models when speaking to the patient and biomedical models when speaking to other physicians, the anthropologist or when recording the diagnosis in the patient's medical chart; (3) treatment, in which physicians again use both the biomedical and the popular models, this time the former for prescribing medication to treat physical symptoms, and the latter, for treating the illness as an idiom of distress which the physician validates in a culturally meaningful way. These models are produced by what I am calling "interactive knowledge sets"[1] that are integrated in practice through the behavior and language of the physician/patient interaction. Knowledge sets include: (1) the physician's learned biomedical model of psychosomatic illness; (2) the physician's understanding of the relationship between social stress and the expression of illness that has developed through his clinical practice and experience; and (3) the physician's cultural appreciation of the particular meaning of *nervios* as a signal of social and personal distress.

In the following sections this argument is developed through a selected review of the theoretical background for this analysis, a summary of the expression and meaning of *nervios* in Costa Rica, a presentation of selected cases from the 122 observed *nervios* medical consultations, and a discussion of the findings in terms of medical knowledge and practice with reference to Mishler's (1985) concept of medical control.

THEORETICAL APPROACHES TO MEDICAL KNOWLEDGE, PRACTICE AND TREATMENT

There are many approaches to understanding dialogue between a physician and a patient (Fisher and Todd 1983). Some of these concentrate on "talk," that is, the word-by-word analysis of the medical consultation

(Mishler 1985); some on the effects of the physician's clinical reasoning and practice, and the patient's cognitive model on language use (Gaines 1982, 1987; Kornfield 1986); and some on how social structural arrangements are replicated in medical discourse (Turner 1984; Wright and Treacher 1982). All of these approaches, however, acknowledge the influence of context on medical discourse (Fisher and Todd 1983). Medical anthropological studies of medical language and practice, for example, have often focused on the use of language, the relation of language to medical knowledge, and the sociopolitical and cultural contexts of language use and knowledge production (Comaroff 1983, 1985; Good and Good 1980; Lock 1982).

Language mediates the production of knowledge; it is through the assigning of special labels and meanings that new knowledge is created (Foucault 1970). Young (1978) objects to the common assumption of a separation of "scientific" and "cultural" knowledge. He points out that science too is ordered by language and thus scientific knowledge should be thought of as a cultural construction. Medical knowledge has claimed legitimacy through its "scientific" status and its supposed objectivity, effectiveness, autonomy, and the naturalness of its object of study (Wright and Treacher 1982). Yet medical knowledge also depends on language for the creation of its object of analysis: "Thus it may often happen that medical categories are merely transposed social categories rooted in the daily social experience and ideological practice of the period" (Wright and Treacher 1982:12); medical knowledge is necessarily social. The study of the production of medical knowledge therefore should focus on the meanings that are attached to medical categories and should concentrate on observing practice and interpretive work in medicine (Atkinson, this volume; Gaines 1979; Bourdieu 1977; Good et al. 1982; Lock 1982).

In an ideal system, knowledge is produced by a feedback loop of events that links existing knowledge to observations and practice, practice to outcomes, and then returns to knowledge (Young 1982:100). Knowledge is usually produced through either the reaffirmation of existing knowledge, or by an observed or practical challenge to existing knowledge, in which case new knowledge is produced. Other influences, such as the conditions of the material environment and normative beliefs, effect practice and the expression of knowledge (Young 1978). Western medical knowledge production, however, differs from other scientific knowledge production systems by (1) client-dominated and patron-dominated practices in which the practitioner's action may be based on career or personal advancement; (2) the difficulty of monitoring the outcome of medical practice in order to correct or change that practice; (3) the reliance on experience and "authorities" for knowledge without a system for evaluating them; and (4) situations where the necessary equipment or experience are not available so that practitioners are limited in their practice (Young 1978:108-109). Medical knowledge, while it is "not restricted to what is represented in textbooks and journal articles but is a product of a dialectic between knowledge and practice/experience" (Young 1978:107), is also constrained by other dimensions of the social, political and economic realities (Low 1985).

Medical knowledge production in developing countries with pluralistic medical traditions such as Costa Rica (Hill 1985; Low 1985) and in industrial countries with historically significant alternative health care systems (Maretzki and Seidler 1985; Lock 1980) is complicated by the presence of multiple medical care systems and practice settings. The categories that Kleinman (1980) articulated of popular, folk, and professional sectors of health care and the concept of clinical "explanatory models" are attempts to analyze knowledge and practice relationships within therapeutic encounters. These well-used concepts provide an heuristic model to explain how knowledge is organized for clinical use and have been applied to communication problems and patient compliance studies in clinical settings.

Medical anthropologists, concerned with issues of medical practice and explanations of that practice, have explored the social and cultural dimensions of the therapeutic role and their impact on the medical encounter (Gaines 1979; Gaines and Hahn 1982; Hahn and Gaines 1985), different interpretations of the patient by members of a medical team and the resulting treatment (Good, Herrera, Good and Cooper 1982), and clinical knowledge as an interpretive and meaning-oriented process (Gaines 1982; Good and Good 1980). One of the clearest statements of this reformulation of a model of practice is presented by Lock (1982) in which she argues that physicians' folk models are used in clinical practice for dealing with the gap that always exists between biomedical knowledge and practice. Physicians' own life experiences, social and political relationships, and personal ideology play an important part in the determination of practice and treatment modes. Helman (1985) suggests that physicians often have a poor knowledge of the patients' clinical models, which further limits communication.

In the search for an understanding of the relationship between medical knowledge and practice, medical anthropologists have reinterpreted medical discourse (Turner 1984; Armstrong 1983; Lock 1982; Wright and Treacher 1982) and critiqued the separation of "medical" and "cultural" realities (Comaroff 1983). The distinction between "illness" as the social expression, and "disease" as the biological expression of sickness has been reassessed such that "sickness is defined as the process whereby both disease and illness are made social" (Lock 1987:131; Frankenberg

1980; Young 1982). It is frequently assumed that disease is a "natural" object rather than a social construction based on social circumstances and cultural norms (Wright and Treacher 1982). Disease, however, is not a fact, but is a relationship and "a system of signs which can be read and translated in a variety of ways" (Turner 1984:208). The body, then, is a cultural object, a representation of the social world, and disease is a form of "language" that must be interpreted. When physicians give meaning to the array of symptoms and regulate the body independently of the patient's discourse (Mishler 1985), then the therapeutic approach is perceived as inappropriate and the treatment invasive (Quill 1985). The biomedical treatment of *nervios* in Costa Rica (Low 1981, 1982, 1985), psychiatric diagnosis of *embrujado* among Mexican-Americans in the United States (Koss and Canive n.d.) and Ayurvedic therapy in India (Nichter 1981), for example, represent alternative approaches in which the physician negotiates the meaning of illness as a somatic as well as a social event and, during treatment, makes some attempt to re-establish the patient's sense of control and self-worth.

Medical anthropological research which takes this "meaning-centered approach" (Good and Good 1982) examines both the relationship of the sociocultural beliefs about illness causation and symptomatology (Good and Good 1980), and the discursive nature of the medical consultation and doctor-patient interaction. These studies demonstrate that medical knowledge and practice is always culture-dependent. The process of labelling an illness enhances treatment by providing "an interpretive context within which allusive and partially hidden 'texts' unfold" (Rhodes 1984:49) and can facilitate communication and rapport between the physician and patient (Obeyesekere 1977; Leslie 1978, 1980).

The meaning-centered approach, when applied to cultural "idioms of distress" (Parsons 1984) such as *nervios*, uncovers sources of social and cultural conflict that are symbolically encoded in the body disturbance. In the Costa Rican case the "unpacking of performed meaning" (Geertz 1983:29) in the medical consultation elicits stories of family disruption, social isolation and interpersonal abuse which become part of the medical discourse. The physician's treatment strategy -- to provide medication based on a biomedically determined diagnosis and to offer advice, confirmation of feelings, reassurance, and behavioral directives based on a popular or personal appraisal of the patient's condition -- suggests that in practice Costa Rican physicians utilize both a patient-centered and a reductionistic type of approach. A culturally acceptable idiom of distress such as *nervios* is likely to invoke a broad base of interactive knowledge sets because of its social and cultural meaning, diffuse symptomatology, and psychosomatic character. Although physicians' ideas about etiology, diagnosis, and treatment of *nervios* may derive from different knowledge sets than those of the patient, and be expressed in both popular and biomedical clinical models, in practice these two different types of information are integrated and unified in the language and action of the consultation. The physician's behavior is a response to a cultural and professional imperative to control the symptom/syndrome and return the patient to a healthy state.

NERVIOS IN COSTA RICA

Research data for this Costa Rican urban health study were collected in outpatient clinics of four hospitals. These clinics serve a majority of urban patients living in San José and the surrounding suburbs and towns, but also are visited by rural residents who prefer the urban medical facilities. The data on *nervios* are derived from a variety of sources, including 457 cases of observed doctor-patient interaction in medical consultations, 117 interviews with patients before and after their medical consultations, 12 team-conducted family interviews in the patients' homes, and twenty months of participant observation in three neighborhoods of San José. The observed consultations varied in length from a few to twenty minutes, the interviews, both before and after consultations, took approximately fifteen minutes to complete, and the home interviews ranged from two hours to two days depending on the number of the family members and the complexity of the health information. The clinic sample included 305 patients from general medicine clinics and 152 from psychiatric and psychosomatic clinics. Seventy percent of the patients in general medicine and 63 percent of those in the psychiatric clinics were female, with a total patient mean age of 33.5 years.

Of the 457 patients in the sample, 122 complained of *nervios* it was the most commonly presented complaint in the psychiatric clinics (50 percent of all patients observed) and the second most common complaint in general medicine (15 percent of all patients observed). *Nervios* was presented to the physicians both as a symptom, a single complaint or statement of discomfort, and as a syndrome, a culturally labelled grouping of related symptoms. Both *nervios* as symptom and as syndrome were culturally shaped conceptualizations of the patients' experience, expressed in terms of multiple causes and associated symptoms of varying duration and severity.

Patients experienced an interesting pattern of other symptoms associated with their *nervios*: headache, insomnia, lack of appetite, depression, fears, anger or bad character, trembling, disorientation, fatigue, itching, altered perceptions, profuse sweating, lifelessness, vomiting and hot sensations (in that order of frequency). The common attribute given to these symptoms is that the patient feels "out of control," or separated from body and self. Patients report that the sensations are not part of their normal behavior, but are experienced as undesirable body responses over which they have no control. The body is seemingly objectified by the patient since he/she views the self as feeling and acting inappropriately.

The description of the patient as "out of control" is reiterated in data collected from interviews carried out before and after medical consultations:

A 24 year old man from the rural highlands complained of a noise in his head, *nervios*, constant dizziness, sweats, fear, pressures, and neck pain. "I lost control," the patient told the interviewer, "and blacked out twice."

A 32 year old woman from a southern valley of Costa Rica, with nine children complains of headache, dizziness, crying and temporary blindness. A week before she had an attack of *nervios* during which she was unconscious for two hours. She thought it might be due to family worries.

A single man from a working-class neighborhood employed in a laboratory department of a large hospital complains of *nervios*, anxiety, desperation, and being disoriented; he is not sure of his acts or words. He says that he has developed an impersonal attitude, doesn't care about other people, and doesn't feel himself. He reports that his problems come from his family who do not appreciate him and stay apart from him. His work is very lonely, and he lives alone.

Nervios patients differ from other people in the general medicine sample in that they experience more family disruption. They are more often single, separated, or widowed, and more frequently mention abuse or abandonment by their spouses, or death of a relative in connection with their symptoms. *Nervios* appears as a symptom in association with family interactional discord, disruption of family structure, and past family disturbance. As further evidence of the etiological link between family disruption and *nervios*, it was found that the clinic patients who present *nervios* differ from those who do not in that they attribute the cause of their symptoms more often to family, social reproductive or sex-related factors.

Physicians also attribute *nervios* to family factors, but suggest that economic problems, spouse inattention and abandonment, lack of friendships outside the family, overdependency, boredom and sexual problems are other causes. Physicians responded positively to patients who presented *nervios* and often asked them about their problems and concerns. The use of *nervios* elicited attention to social and emotional aspects of the patient's illness and enhanced physician-patient interaction through communication about social and familial concerns.

The usual pattern in seeking therapy for the treatment of *nervios* typically begins at home with an herbal remedy, such as dandelion tea, known to the sufferer or suggested by a family member. If the home remedy is not effective, then the sufferer might ask the local pharmacist or herbal practitioner for a non-prescription medication which would alleviate the discomfort. The local pharmacist may suggest some kind of tranquilizer or may direct the sufferer to the medical clinic. The next step in health-seeking behavior is to see a physician, either a general practitioner, or in some cases, a psychiatrist at the local Social Security clinic. Finally, if the medical treatment is not effective, the sufferer may continue to search for a physician in a specialty clinic -- such as psychiatry, neurology, or psychosomatic medicine -- within the Social Security system or may pay to see a private physician. (When the health care data were initially collected, patients could use both the Social Security and Public Health hospitals and clinics; however, currently all clinics are directed by Social Security and only the very poorest patients are not covered by this national health care, pension and death benefit plan. For a more extended discussion see Low 1985:44-48.)

Nervios can be interpreted as a communication about oneself and one's relation to the social system expressed through a disturbance or "discontinuity" of the body perception. The importance of the symptom/syndrome is that it reflects both the relations of an individual in the social system and represents their participation in Costa Rican culture; it allows for help-seeking behavior of individuals or families attempting to re-establish a "tranquil" (*tranquilo*) or "healthy" (*sano*) sociocultural state. But *nervios* appears to have a more specific cultural meaning as well. Somatic symptomatology and sensations of disturbed body boundaries refer symbolically to broken family ties and relations which are the basis of Costa Rican identity. In both the urban clinic sample and the rural communities (Barlett and Low 1980), family change, formation, and breakdown are correlated with the incidence of *nervios*. In the rural setting *nervios* occurs in large families with many problems and economic burdens; in the urban setting, the presented problems focus on family disruption, poor relationships, death and abandonment.

The presentation of *nervios* absolves the individual of responsibility and provides a culturally acceptable idiom of distress. The symptom/syndrome elicits a positive response from family members and significant others usually in the form of increased attention, help with daily tasks, and expressions of concern and support which may work to resolve the underlying conflict. The medical practitioner also reaffirms the social and cultural message of the symptom/syndrome by engaging the patient in further discussion of the illness or by giving the patient the requested and often validating medication. Culturally, *nervios* indicates an inability to fulfill the cultural ideal of maintaining a proper attitude toward life and of behaving in culturally acceptable patterns. Treatment by family, friends and physicians works to reintegrate cultural expectations with social performance. If successful the treatment of *nervios* returns the patient to the cultural ideal: *vivir tanquilo* (to live tranquilly), to live in harmony with family, community, and friends.

NERVIOS FROM THE PHYSICIANS' PERSPECTIVE

The following data on physician-patient interaction is based on 122 cases of *nervios* observed in seven outpatient clinics in the four major hospitals of San José: Hospital San Juan de Dios, Hospital Mexico, Hospital Capui and Hospital Calderon Guardia; 46 *nervios* patients were observed in general medicine clinics (Calderon Guardia and San Juan de Dios) and 76 were observed in psychiatric (Calderon Guardia and Chapui) or psychosomatic (Mexico) clinics. The interaction between physicians and patients with *nervios* was recorded in the context of the physicians' ongoing consultations. The clinics were sampled by the day of the week, hour of the day, and by the physicians who were working on various schedules. The 122 observed cases include observations of eleven out of a total sample of seventeen physicians observed during the entire health study. The eleven physicians were all white males between the ages of 27 and 55, middle to upper-middle class family men who said that they enjoyed their work but often complained about the poor pay and the disorganization of the medical care system (see Low 1985 for details of the backgrounds and conflicts of physicians in the Social Security and Public Health medical institutions).

From the 122 observed cases of *nervios* a typical physician-patient interaction pattern emerges when *nervios* is presented. The patient often begins by saying, "Doctor, I feel bad all over; I have nerves" (*tengo nervios*), and then goes on to describe the other accompanying symptoms and their characteristic configuration. The physician typically responds in one of four ways:

(1) He engages the patient in a discussion of *nervios*, seeking to understand its cause. For example:

An older man enters with a frown on his face. He has terrible pain and has felt this way since he left the hospital. He says that he has *nervios*. The doctor asks him, "How are your *nervios*?" The patient replies that he cannot work because of the pain, especially when he is around strangers. He lives with uncertainty. The doctor moves closer to the man and asks him to continue. The man continues

saying that he feels criticized and never feels complimented. The doctor asks him
which is more important -- the pain or the problem. The patient says the he wants
to change the pain.

(2) The physician responds by reaffirming that the patient has *nervios*,
but emphasizes other aspects of the symptom report:

A young girl, quite heavy, enters smiling. She says that she has *nervios* and difficulty breathing. The doctor asks about her *nervios* and how she is doing. She replies that she is much better, but still has the pain in her head. "And the pain in
your interesting parts?" asks the doctor, "is it gone?" "Yes," giggles the patient, "it
is gone." The doctor goes on: "Don't eat so much!" "But I don't eat!" the patient
responded emphatically. "Then eat less, eat nothing," continued the doctor, and
"behave yourself." (The reference to "behave yourself" has a sexual meaning in
Spanish which the doctor explained to me). The patient looked at him out of the
corner of her eye, smiled, and left the office.

(3) The physician does not verbally reaffirm the diagnosis *nervios* but
gives the patient a tranquilizer and comments to me (the observer) that
the patient is "just neurotic" or "psychosomatic" or "having problems":

A young man with eyes wide open and rather strange behavior enters the office.
He has had headaches for two months and does not want to work. He is dizzy and
suffers from *nervios*. He has been in the hospital for parasites recently and feels
that his mind is weak. (This refers to a folk concept in that the mind can become
weak without adequate nourishment or from too many parasites which take your
strength. This may be a quasi-medical translation of doctor explanations of para-
site-related illness.) The doctor asks about what he has eaten -- meat, eggs, or
milk. The patient replies that he has eaten all of these. The doctor orders a laboratory exam and sends the patient out with a prescription for a tranquilizer. He
then turns to me and says that the man was a "typical neurotic."

(4) The physician does not reaffirm the patient's diagnosis of *nervios* and
seeks instead an "organic" cause for the symptoms:

A 33 year old woman enters and says that she is bad. "Good," replies the doctor,
"that is when we want to see you so we can help." The patient says that she is concerned about her mother who is in a coma and that she must care for both households. The patient says that her only problem is this pain in her breast and she
understands that it is *nervios*. She also has a problem with sleeping. "My role,"
says the doctor, "is to investigate the organic possibilities and there are two: a her-
nia or a tumor of the esophagus." The doctor listens to her heart and orders laboratory exams. The patient continues telling the doctor that she is from a small
town, and that her husband cannot eat or drink. The doctor tells her to sleep with
two pillows and orders her to diet. He tells her to come back when she can get
another appointment.

From the patients' perspective the first three responses are experienced as positive and reinforcing since their symptoms are validated either verbally or through the receipt of medication. The fourth type of response is unusual because the physician appears to ignore the emotional and social factors in the illness. The majority of physician responses indicate that they share a similar paradigm as that of their patients; the clinical encounter was socially meaningful in that physicians behaved within the expected pattern of social interaction and because the patients appeared satisfied with the consultation and treatment.

A contextual analysis of the recorded interactions, however, produced a more complex description of the physician-patient interaction. The following three cases illustrate the range of physician-patient interaction patterns in the 122 patient sample and were selected as representative of the observed *nervios* consultations.

Case 1: Psychiatric clinic

A twenty-four year old woman enters the office. She is married with two children and lives in a small town on the Pacific coast. Her problem, she says, is with her husband and other women. She has come because of her *nervios* which she has had for a year. She does not eat, trembles, has heart palpitations, and *descomposiciones* (disorders, usually means a fainting spell or falling down). She talks and is angry all day long. The doctor asks her if she feels like crying? She replies, "Yes," and continues to say that she cannot sleep well and wakes at whatever hour of the night with a headache. The doctor asks if it is worse when she has her menstrual period, and she replies that it is a lot worse, that even the back of her neck (*la nuca*) feels strange. The doctor asks if she had a problem a year ago that might be the cause of her *nervios*. She replies that she had four years of colitis and has suffered because of her daughter. She asks the doctor to give her something for her appetite. The doctor agreed and gave her a B12 vitamin shot (quite popular with Costa Rican patients) and prescribed a tranquillizer (Valium) and an antidepressant (Tofranil). Recorded diagnosis: depressive neurosis.

Case 2: General medicine clinic

A middle-aged woman enters the office and the doctor turns to me and says that she has the same problem (referring to a previous patient with *nervios*). He smiles and begins by asking the patient: "How are your *nervios*?" She replies that she feels very badly and cannot do anything because of her dizziness. The doctor continues, "and how is your chest?" The woman replies that it is fine, but not really good. The doctor asks if she cooks with a woodstove. The woman replies that she does and that it has become very difficult for her to walk, as it hurts so much. She feels *susto* (fright), bad all over, and hot at times: "My *nervios* are very bad." The doctor turns to me and says, "We do a kind of hypnosis and tell the patient that this is the very best medicine for this and that she will do well with this medicine." He then turns to the patient and says that he would prescribe her a pill for her *nervios* which would cure her (prescribes Valium).

Case 3: General medicine clinic

As small, well-dressed woman from the countryside enters the office. She is 38 and works as a domestic. She apologizes to the doctor and to me for subjecting us to her story and all her subsequent remarks are prefaced by this statement. She lives with a black man who humiliates her. She earns money for her children and this man treats her badly – the way that he treats her gives her *nervios*. The doctor responds by saying, "And why not, with such problems?" She has been sterilized and feels that she is not a woman. The father of her children wants to marry her but she cannot do anything with the man she lives with who hits her. She has head-aches, cannot drink coffee and cannot eat. When her boyfriend comes in from drinking it is worse. The doctor sends her out with some comforting words and a prescription for an anti-depressant (prescribes Tofranil). Recorded diagnosis: hysteric with depression.

Based on an analysis of the total 122 cases, physicians accept *nervios* as a valid symptom and respond positively to the patient's request for attention in the presence of the patient. On the other hand physicians record a case of *nervios* as a medical diagnosis of "anxiety," "anxious depression" or "conversion reaction"; they treat *nervios* with a conversation about its cause; and they validate and treat the distress experienced by the patient with a prescription of Valium (tranquilizer) or Tofranil (antidepressant). Costa Rican patients, in fact, expect some kind of explanation of their illness and medication not only for *nervios* but in all medical encounters; they requested medication in 29.3% of all the observed doctor-patient interactions and requested an explanation in 14.6% (Low 1985:320). Data on medication prescribed in the Costa Rican Social Security health care system confirm that tranquilizers and psychotropic drugs are ranked among the most frequently prescribed medications in medical clinics (Vargas et al. 1978). These cases illustrate how Costa Rican medical practice incorporates both a biomedical and popular model in the diagnosis and treatment of *nervios*, while the discussion of etiology follows the popular cultural explanations of the illness.

Another way to look at the 122 *nervios* cases is to tabulate physicians' diagnoses (Table I) and treatments (Table II). Of the 100 *nervios* patients who were diagnosed, over a third were diagnosed as "psychoneu-rotic" and half were diagnosed as either "psychoneurotic" or "psychoso-matic." Twenty-two patients were diagnosed as having *nervios* and only 6 were diagnosed as having "psychosocial" problems (see Table I). Of the 95 patients treated, most patients received two treatments – "talking" and some form of medication (tranquilizer, antidepressant, antipsychotic or other "medication," see Table II). Cases 1 and 3 presented above illustrate the simultaneous use of a popular and a biomedical diagnosis and treatment – the prescription of tranquilizers and antidepressants (biomedical) – and a popular or cultural treatment of "talking." The frequent use

TABLE I

Physician diagnoses of observed *nervios* cases[a]

PHYSICIANS' DIAGNOSES[b]	SUBTOTAL	TOTAL
Psychoneurotic		35
Depression	12	
Neurosis	11	
Anxiety	7	
Hysteria	3	
Conversion reaction	2	
Psychosomatic		15
Psychosomatic illness	6	
Asthma	3	
Diabetes	2	
Anorexia	1	
Allergy	1	
Ulcer	1	
Hypochondriasis	1	
Organic/Developmental		8
Brain damage	5	
Mental retardation	2	
Tumor/hernia	1	
Psychosocial		7
Family problems	5	
Alcohol	2	
Psychotic		6
Schizophrenic	5	
Psychosis	1	
Women's Problems		6
Menopause	3	
Infection	2	
Birth control pills	1	
Nutrition		3
Poor nutrition	1	
Goiter	1	
High blood pressure	1	
Nervios (confirmed)	22	22
Total number of physician diagnoses	102	
Total number of patients diagnosed	100	

a. This analysis is based on 122 observed *nervios* cases reported in Low (1981) and collected as part of a health care study completed 1972-1974, 1976, and 1979 in San José, Costa Rica. The 122 cases reflect all presented cases of *nervios* observed during the medical consultation in seven urban clinics.

b. The designation "physician diagnosis" indicated that the physician *stated* a diagnosis either to me or to the patient during the interaction. This total includes the few cases in which the physician gave me the patient's chart after the consultation. In other words, these designation are all the possible diagnoses I was able to determine. Some patients were given more than one diagnosis so that the total number of diagnoses reflects all separate statements and the total number of patients diagnosed reflects the number of interactions in which a diagnosis was made.

TABLE II
Physician treatments of *nervios* cases[a]

PHYSICIAN TREATMENTS[b]	TOTAL
Talking	65
Tranquilizer (Valium, Librium)	32
Medication (drug name not specified)	29
Antidepressant (Tofranil)	17
Vitamins, tonics and diet	13
Antipsychotics (Stelazine)	9
Appointment, referral	6
Psychotherapy	3
Stop medication	3
Cream, soap (for skin irritation)	2
Change mode of contraception	1
Laboratory examination	1
Recommend a change of job	1
Shock therapy	1
Total number of treatments	182
Total number of patients treated	95

a. This analysis is based on 122 observed *nervios* cases reported in Low (1981) and collected as part of a health care study completed 1972-1974, 1976, and 1979 in San José, Costa Rica. The 122 cases reflect all presented *nervios* cases.

b. The designation "physician treatments" indicates the kind of treatment ordered in the observed medical consultation. Some physicians indicated only that they gave "medication" while others specified the type of drug or drug name.

of medication is summed up by the attitudes of two physicians, one who explains that his favorite drug is Tofranil and that he has had good results with this drug; and the other who said that "taking a tranquilizer is like taking an aspirin -- if you have a headache, you can wait until it goes away or you can take an aspirin. It is the same thing with stress -- you can take pills quite safely to be comfortable."[2]

The range of diagnoses and treatments by the physicians includes a biomedical emphasis on symptom removal with "active" pharmacological agents, but also includes popular modes of diagnosis and treatment, which emphasize understanding, explanation, and talking. The professional "action system" of medical practice relies on a limited number of learned diagnostic categories, organized and maintained through the social organization and structure of the profession -- medical schools, medical texts, licensing exams -- and the sociocultural relations of practice (Low 1982). The significance of the *nervios* example is that metaphorical and local cultural meanings of *nervios* are used by the physician with the patient during the consultation in conjunction with the medication, a product of the biomedical culture.

Examples of the physicians' explanations of *nervios* discussed during a representative sample of 16 medical consultations illustrate the diversity of social, psychological, and cultural explanations for the etiology, diagnosis and treatment of *nervios*. This diversity reflects the interactive knowledge sets that form the basis of the physicians' available knowledge for clinical practice. Through eliciting symptoms and life histories and discussing them with the patient and with the observing anthropologist, the physician arrives at a clinical explanation that draws upon both his own knowledge of possible explanations for *nervios* and knowledge of *nervios* derived from the medical consultation. Following are the 16 examples:

(1) The physician comments to me before the patient enters that it would be better if the patient did not have more children. He says that her children are very sad and that she is giving them her *nervios*. The physician proceeds to ask her more questions about where she works and how much money she makes. The patient seems confused by this questioning.

(2) The physician tells the patient to control herself, that what she has (*nervios*) is not so strange. She also has something in her lungs. The physician discusses the patient, saying to me that he has an existential conflict between the necessity of treating her potential tuberculosis and increasing her fear of the hospital.

(3) The patient's husband says that she should not drive because of her *nervios*, but the physician points out that very often to give a person with *nervios* a responsibility and the satisfaction of having control over something acts as a good therapy. The physician asks me if I agree.

(4) The doctor tells me that this is a psychiatric case, not *nervios*, and does not belong in the psychosomatic clinic. The patient was admitted for an overdose of sleeping pills, and was placed in the psychiatric hospital. He does not want to see a psychiatrist. The doctor suggests that the man does not want to see himself as crazy (*loco*) and that is why he is using the label of *nervios*.

(5) The patient replies to a question that she does not have any sugar; the physician answers that sugar is bad for her *nervios*. The patient responds, "Oh, then I will not have any sugar." After the patient is gone, the physician explains that what is really wrong with the patient is that she has a son that left the house and came back with a mistress who tries to run the house. This son is an alcoholic and it is not clear if the patient did not like the mistress or if the mistress treated the family in a certain way. The physician said that he would have asked more about the mistress but that he just did not want to; he did not have time to get the whole story.

(6) The woman was born out of wedlock, the physician tells me after she has left. She was adopted by another family and lived with them, but she does not see then now. At a very young age she went to live with a man who was ten years older than her and had a child with him. The man left her. She then started living with another man and had another child, and he left too. She is a cleaning lady and works very hard to support her son. These difficult conditions have caused her *nervios*.

(7) The physician uses a metaphor of a sponge that cannot take any more water to explain the patient's *nervios*.

(8) The physician tries to convince the patient not to take pills for his *nervios*. He is fat and is taking pills and vitamins. His feeling of fatigue is psychic and vitamins have nothing to do with it. The weakness is the state of his mind, not the fault of the *nervios*. Nonetheless, the physician orders the pills that the patient requested. The patient asks about his lowered sex drive and the physician says that this is also part of the influence of the pills.

(9) The physician tells me that the patient could not "absorb" his problems when he was young and that his current problem of *nervios* is based on his earlier unresolved conflicts.

(10) The physician turns to me and says that very often many months after an accident the patient will experience many pains which do not have any organic root and are simply a nervous reaction. Sometimes this reaction is the basis of *nervios*.

(11) The physician explains that the patient had an accident which left him very upset and that is what is causing his *nervios*. The shock of the accident triggered the patient's *nervios*.

(12) The physician tells the patient this his hallucinations are part of his *nervios*.

(13) The physician tells the patient that her *nervios* is due to a problem with her thyroid. The physician then says to me that this was a person from an upper-class family who married a poor man and her family abandoned her.

(14) The physician says to me that the husband drinks a lot and only works occasionally. "Family problems make it worse for your *nervios*," he explains to the patient.

(15) The physician gives the patient an appointment and suggests to me that she

has influenza and that *nervios* is just a complication.

(16) The physician suggests that the patient has troubles with her husband who goes around with other women. Her husband does not pay attention to her as she would like; the more he rejects her the worse she gets.

As these examples indicate, physicians recognize *nervios* as a "real" illness and explain the symptom/syndrome through popular Costa Rican explanatory models of motivation and appropriate behavior. They use the same reasons as patients when they explore the "real" explanations of why *nervios* occurs. The attributed causes range from social explanations of family conflict, alcoholism, and abandonment; to psychological explanations of a difficult childhood; and include cultural and economic explanations of an impoverished lifestyle, concepts of poor nourishment, and fright or shock (*susto*). The diversity of explanations reflects the different backgrounds and gender of the physicians as well as whether the physician was speaking to the anthropologist, to other physicians or to the patient.

Verbal explanations are part of the physician-patient or physician-observer interaction and are influenced by the emotional and social tone of the consultation, and the rapport, or lack of it, within the medical context. Physicians' explanations, like their clinical models, are responsive to the actual social and medical situation, and as such, include local cultural categories as well as more universal biomedical categories of attributed etiology, diagnosis and treatment. These models are expressions of interactive knowledge sets which include medical knowledge obtained from both medical education and from experience, and cultural knowledge derived from the local cultural context, social institutions, and interpersonal setting. Interactive knowledge sets, such as the physicians' understanding of the relationship of psychological, social, and cultural bases of stress-related illness, and their appreciation of the cultural meaning of *nervios* as an idiom of distress, provide the basis for a popular, or meaning-centered model of clinical practice. Both models and the knowledge sets are integrated in the language and practice of the doctor-patient interaction.

DISCUSSION: *NERVIOS* AS SOCIAL AND MEDICAL CONTROL

In Mishler's perceptive study of the discourse of medicine he begins by stating that,

the illness "discovered" in a medical consultation is "constructed," not found. A diagnosis is a way of interpreting and organizing observations ... Since the discovered illness is, in this sense, partly a function of the talk between a patient and a physician, the study of this talk is central to our understanding of both illness and clinical care (1985:11).

In order to analyze this talk, Mishler introduces the concept of "voices" to specify the relationship between the actual language and pattern of speech and the underlying "frameworks of meaning" (1985:14). Two voices are distinguished, the "voice of medicine" and the "voice of the lifeworld," "representing, respectively, the technical-scientific assumptions of medicine" (1985:14) and assumptions of everyday life. His analysis of physician-patient interaction explores the differences of these voices and their conflict within the medical consultation (Mishler 1985). Through his examination of transcripts of medical interviews, he identifies the effect of physician interruptions as dominating the talk and controlling the domain of meaning. The struggle between the "voice of medicine" and the "voice of the lifeworld" represents the control of the interview and the control of the domain of meaning within which the illness will be understood and treated (Mishler 1985:120-121).

The data on physician-patient interaction in Costa Rica are consistent with Mishler's analysis in as much as the illness is partly constructed by talk, the consultation is made up of the "voices" of medicine and of the lifeworld, and the talk of the physicians in some sense controls the definition of the illness and treatment. However, the data from the *nervios* consultations suggest that the use of the "voices" and the control of meaning is more complex than the analogy of a conflict or struggle between the physician and the patient. In the Costa Rican data the physicians use both "voices" in that they talk with the patient in terms of the lifeworld meaning of *nervios* as well as in the "voice of medicine." Further, both the "voice of medicine" and the "voice of the lifeworld" are composed of interactive knowledge sets which include social, cultural, technical-scientific, and everyday assumptions. The talk about *nervios* observed between the physician and the patient includes popular conceptions of etiology, biomedical and popular or stress-related diagnoses, and biomedical and meaning-centered treatment. The imposition of power on the part of the physician is not necessarily in the control of the talk through interruptions or questions, but is present in the affirmation, redefinition and/or mode of treatment of *nervios*.

The idea of control is central in *nervios* as a cultural concept. Sufferers experience *nervios* as feeling "out of control" and disoriented. Treatment, from the patients perspective, whether suggested by family, or friends, or a medical practitioner, includes developing self-control and/or control over bodily functions and sensations. When the sufferer goes to the clinic the expectation is that the illness will be "controlled" in the sense that: (1) it will be explained and/or a cause will be identified; (2) the physician will tell the sufferer what to do; and (3) the physician will prescribe medication which will alleviate the symptoms or cure the illness. The control offered by the physician includes talking about *nervios*, either reaffirming the patient's diagnosis or redefining the patient's perception of the illness and attribution of cause; behavioral control in the form of medical directives to eat less or more, go back to work, stop work, or relax; and pharmacological control in the form of psychotropic drugs.

Nervios at the level of social and cultural meaning is also an idiom of distress which indicates family disruption and breakdown. It signals to family and significant others that there is some role conflict, mental disturbance, or family problem that needs to be resolved, and it can be assumed that many of these problems are solved in the context of the home and the community. However, in a medicalized health culture such as Costa Rica, many family problems are somatized, labeled as *nervios*, and presented in the readily available urban medical clinics. The physician's control through talk, behavioral directive or medication, then, becomes a substitute for family or social control. This process is complicated in that many of the *nervios* cases are biomedically diagnosable as mental illnesses that may have been ineffectively treated in the community. In cases such as these, physicians' affirmation of a diagnosis of *nervios* becomes a socially acceptable way to label mental disturbance. The diagnosis is a form of control but it enables the patient to continue living in the community as a "normal" rather than "crazy" citizen.

In this sense then, the "medicalization" of *nervios* may be beneficial for patients whose social, psychological and medical problems have not been adequately dealt with by family and friends. Further, the physician becomes an important resource for *nervios* sufferers who do not have a community or family to help them with the resolution of these problems. This positive therapeutic outcome, however, is also due to the cultural knowledge about the meaning and causes of *nervios* shared by both the physician and the patient. In a typical *nervios* medical consultation the "voice of medicine" does not dominate the interaction, but is integrated through dialogue with the "voice of the life-world" shared by both the physician and the patient. The physician, in treating and "controlling" *nervios*, participates in the construction of a medical metaphor that reinforces culturally appropriate behavior and lifestyle through the medium of the medical consultation. Medical practice thus transforms a symptom/syndrome into a metaphor for cultural conflict and control that draws its meaning from both the social interaction of physician and patient and from the cultural context of shared ideals.

ACKNOWLEDGEMENTS

This research was funded over a period of years by a National Institute of Mental Health Combination Research Fellowship, travel grant from the Center for Latin American Studies at the University of California, Berkeley, and a faculty research award from the University of Pennsylvania, Philadelphia. The author would like to thank the editors and reviewers of this volume for their helpful comments, Lucile Newman for her encouragement in the study of *nervios*, and Erve Chambers for his critical reading of this manuscript.

NOTES

1. Erve Chambers of the University of Maryland suggested this term during a discussion of knowledge production as a cognitive process.
2. Treatment of nerves by "nerve pills" has been noted by Davis (1983) for women in Newfoundland, and by Arny (1955), Ludwig and Forrester (1981) and Van Schaik (1984) for patients in Eastern Kentucky. Davis comments that "nerve pills" may include both hormones and tranquilizers because of the association of nerves with menopause in her study area. Davis' designation of "nerve pills" is based on informants' reports and not on the observations of physicians' treatments. However, these other studies suggest that there may be some cross-cultural similarities in the biomedical model for the treatment of nerves.

REFERENCES

Agar, Michael
    1986   Review of M. Stubbs, Discourse Analysis, and G. Brown and G. Yule, Discourse Analysis. International Journal of Pragmatics 5:710-716.
Armstrong, David
    1983   Political Anatomy of the Body. Cambridge: Cambridge University Press.
Arny, M.
    1955   My Nerves are Busted. Mountain Life and Work 31:24-29.
Barlett, Peggy and Setha Low
    1980   *Nervios* in Rural Costa Rica. Medical Anthropology 4:523-564.
Bourdieu, Pierre
    1977   Outline of a Theory of Practice. Cambridge: Cambridge University Press.

Casas, A. and H. Vargas
    1980 The Health System of Costa Rica: Towards a National Health Service.
          Journal of Public Health 1:258-279.
Comaroff, Jean
    1978 Medicine and Culture: Some Anthropological Perspectives. Social Sci-
          ence and Medicine 12:247-254.
    1983 The Defectiveness of Symbols or the Symbols of Defectiveness? On
          the Cultural Analysis of Medical Systems. Culture, Medicine and Psy-
          chiatry 7:3-20.
    1985 Body of Power, Spirit of Resistance: The Culture and History of a
          South African People. Chicago: University of Chicago Press.
Conrad, Peter
    1985 The Meaning of Medications: Another look at Compliance. Social Sci-
          ence and Medicine 20:29-37.
Davis, D.L.
    1983 Blood and Nerves: An Ethnographic Focus on Menopause. St. John's,
          Newfoundland: Institute of Social and Economic Research, Memorial
          University of Newfoundland.
Fisher, Susan and A.D. Todd
    1983 Introduction: Communication and Social Context — Toward Broader
          Definitions. In S. Fisher and A.D. Todd (eds.), The Social Organiza-
          tion of Doctor-Patient Communication. Pp. 3-17. Washington D.C.:
          Center for Applied Linguistics.
Foucault, Michel
    1970 The Order of Things: An Archaeology of the Human Sciences. New
          York: Random House.
Frankenberg, Ronald
    1980 Medical Anthropology and Development: A Theoretical Perspective.
          Social Science and Medicine 14B:197-207.
Freidson, Eliot
    1970 Profession of Medicine: A Study of the Sociology of Applied Knowl-
          edge. New York: Dodd, Mead and Company.
Gaines, Atwood D.
    1979 Definitions and Diagnoses: Cultural Implications of Psychiatric Help-
          seeking and Psychiatrists Definitions of the Situation in Psychiatric
          Emergencies. Culture, Medicine and Psychiatry 3:381-418.
    1982 Knowledge and Practice: Anthropological Ideas and Psychiatric Prac-
          tice. In N.J. Chrisman and T.W. Maretzki (eds.), Clinically Applied
          Anthropology. Pp. 243-273. Dordrecht, Holland: D. Reidel Publishing
          Co.
    1987 Culture and Medical Knowledge in France and America. In A. Gaines
          and A. Young (eds.), The Social Origins of Biomedical Knowledge.
          Special Issue of Medical Anthropological Quarterly. (forthcoming).

Gaines, Atwood D. and Robert A. Hahn (eds.)
   1982   Physicians of Western Medicine: Five Cultural Studies. Culture, Medi-
          cine and Psychiatry 6:215-322.
Geertz, Clifford
   1983   Local Knowledge. New York: Basic Books.
Good, Byron J. and Mary-Jo DelVecchio Good
   1980   The Meaning of Symptoms: A Cultural Hermeneutic Model of Clinical
          Practice. In L. Eisenberg and A. Kleinman (eds.), The Relevance of
          Social Science for Medicine. Pp. 165-196. Dordrecht, Holland: D.
          Reidel Publishing Co.
   1982   Toward a Meaning-Centered Analysis of Popular Illness Categories. In
          A.J. Marsella, G. White (eds.), Cultural Conceptions of Mental Health
          and Therapy. Boston: D. Reidel Publishing Co.
Good, Byron J., H. Herrera, Mary-Jo DelVecchio Good and J. Cooper
   1982   Reflexivity, Countertransference and Clinical Ethnography: A Case
          from a Psychiatric Cultural Consultation Clinic. Culture, Medicine and
          Psychiatry 6:281-303.
Hahn, Robert A. and Atwood D. Gaines (eds.)
   1985   Physicians of Western Medicine. Dordrecht, Holland: D. Reidel Pub-
          lishing Co.
Helman, Cecil
   1985   Communication in Primary Care: The Role of Patient and Practitioner
          Explanatory Models. Social Science and Medicine 20:923-931.
Hill, Carole
   1985   Local Health Knowledge and Universal Primary Health Care: A Behav-
          ioral Case from Costa Rica. Medical Anthropology 9:11-24.
Kornfield, Ruth
   1986   Dr., Teacher, or Comforter?: Medical Consultation in a Zairian Pedia-
          trics Clinic. Culture, Medicine and Psychiatry 10:367-387.
Koss, Joan D. and J. Canive
   n.d.   Embrujado: The Clinical Impact of a Cultural Diagnosis. Manuscript.
          Department of Psychiatry, University of New Mexico.
Leslie, Charles (ed.)
   1978   Special Issue on the Theoretical Foundations for the Comparative Study
          of Medical Systems. Social Science and Medicine 12B(2).
   1980   Special Issue on Medical Pluralism. Social Science and Medicine
          14B(4).
Lock, Margaret
   1980   East Asian Medicine in Urban Japan: Varieties of Medical Experience.
          Berkeley: University of California Press.
   1982   Models and Practice in Medicine: Menopause as Syndrome or Life
          Transition? Culture, Medicine and Psychiatry 6:261-280.
   1987   Protests of a Good Wife and Wise Mother: The Medicalization of Dis-

tress in Modern Japan. *In* E. Norbeck and M. Lock (eds.), Health, Illness, and Medical Care in Japan: Continuity and Change. Pp. 130-155. Honolulu: University of Hawaii Press.

Low, Setha M.
  1981   The Meaning of *Nervios*: A Sociocultural Analysis of Symptom Presentation in San José, Costa Rica. Culture, Medicine, and Psychiatry 3:25-47.
  1982a Dr. Moreno Canas: A Symbolic Bridge to the Demedicalization of Healing. Social Science and Medicine 16:527-531.
  1982b The Effect of Medical Institutions on Doctor-Patient Interaction. Milbank Memorial Fund Quarterly/Health and Society 60:17-50.
  1985   Culture, Politics and Medicine in Costa Rica. South Salem, New York: Redgrave.

Ludwig, A.M. and R.L. Forrester
  1981   The Condition of "Nerves." The Journal of the Kentucky Medical Association 79 (6):333-336.

Maretzki, Thomas W. and E. Seidler
  1985   Biomedicine and Naturopathic Health in West Germany. Culture, Medicine and Psychiatry 9:383-421.

Mesa-Lago, Carmels
  1985   Health Care in Costa Rica: Boom and Crisis. Social Science and Medicine 21:13-21.

Mishler, Elliot C.
  1985   The Discourse of Medicine: Dialectics of Medical Interviews. Norwood, NJ: Ablex Publishing Corporation.

Nichter, Mark
  1981   Negotiation of the Illness Experience: Ayurvedic Therapy and the Psychosocial Dimension of Illness. Culture, Medicine and Psychiatry 5:5-24.

Obeyesekere, Gananath
  1977   The Theory and Practice of Psychological Medicine in Ayurvedic Tradition. Culture, Medicine and Psychiatry 1:155-181.

Parsons, Clare
  1984   Idioms of Distress: Kinship and Sickness Among the People of the Kingdom of Tonga. Culture, Medicine and Psychiatry 8:71-93.

Quill, T.E.
  1985   Somatization Disorders: One of Medicine's Blind Spots. Journal of the American Medical Association 254:3075-3079.

Rhodes, Lorna AmaraSingham
  1984   "This Will Clear Your Mind": The Use of Metaphors for Medication in Psychiatric Settings. Culture, Medicine and Psychiatry 8:49-70.

Turner, Bryan S.
  1984   The Body and Society. Explorations in Social Theory. Oxford: Basil

Blackwell.

Van Schaik, Eileen
  1984   "My Nerves Bother Me, But That's Not All." Paper presented at the
          Department of Psychiatry Colloquia, University of Kentucky, Spring.

Vargas M., H., R. Gutierrez S., O.R. Fallas C., Y J. Gainza E.M.S.
  1978   *Estudio Sobre el Consumo de Medicmentos Por Parte de la Poblacio'n
          Asegurada. Revista Centoamericana de Ciencias de la Salud* 4:121-154.

Woodward, M.R.
  1985   Healing and Morality: A Javanese Example. Social Science and Medi-
          cine 21:1007-1021.

Wright, Peter and Andrew Treacher (eds.)
  1982   The Problem of Medical Knowledge. Edinburgh: Edinburgh University
          Press.

Young, Allan
  1978   Mode of Production of Medical Knowledge. Medical Anthropology
          2:97-122.
  1982   The Anthropologies of Illness and Sickness. Annual Review of Anthro-
          pology 11:257-285.
  1985   Scientists' Beliefs and Scientists' Facts in the Discourse on Post-trau-
          matic Stress Disorder. Paper presented at the Annual meetings of the
          American Anthropological Association. Washington D.C., December 6.

PART VII

CONSTRUCTING THE "ORDINARY" OUT OF THE "EXTRAORDINARY"

KATHRYN M. TAYLOR

PHYSICIANS AND THE DISCLOSURE OF UNDESIRABLE INFORMATION

INTRODUCTION

Studies of physicians who care for patients with a potentially fatal illness often include a discussion of "disclosure" (Quint 1965; Fitts et al. 1953; Krant 1976). This term refers to the event of telling patients their diagnosis and the probable outcome of their disease (Weisman 1967). It is a complex, difficult and frequently misunderstood element of any doctor-patient interaction (Christman 1965). This is particularly true in the treatment of cancer (Crayton 1969).

In the past, many physicians chose to withhold the diagnosis of cancer, even when it was clear that the patient was going to die. For example, in a study by Glaser and Strauss, cancer physicians agreed that it was in their patients' best interest not to be told of their impending death (Glaser and Strauss 1968). In a similar study, ninety-five percent of Oken's physician respondents reported that they routinely withheld the diagnosis from their cancer patients (Oken 1961). Scheff pointed out that "when there are conditions that both encourage and discourage the physicians to tell patients, the combined effect makes it easier not to act" (Scheff 1977:433). Freidson argued that since the physicians' authority is based on the esoteric nature of medical knowledge, physicians are inclined to restrict the flow of information rather than expose themselves to evaluation and criticism (Freidson 1970:127).

A dramatic change in disclosure policy occured in the late 1960s and early 1970s. These changes are attributed to the demands of patients who had become better informed. This is due in part to: increased media coverage (Abrams 1966), new laws requiring that patients understand the nature of their disease before consenting to experimental therapy (Barber 1980), and a shift toward holistic medical care, based on the ideal of open communication between doctor and patient (American Cancer Society 1982).

In more recent studies, many physicians claim to disclose all relevant information to all patients (Bird 1973; Braver 1960). The literature contains numerous articles describing physicians who advocate the new policy of total disclosure. For example, Novack reissued Oken's questionnaire and revealed that 90 percent of his respondents said they disclosed all information -- a complete reversal of Oken's earlier findings (Novack

M. Lock and D. R. Gordon (eds.), Biomedicine Examined, 441–463.

et al. 1979). Indeed, it has become arduous for physicians NOT to tell patients. Identifiable disease-specific clinics (Germain 1979), detailed informed consent forms required before experimental medical treatment (Elkeles 1982), and continual media exposure of the inadequacy of current therapy (Dunn 1978) make it difficult to withhold the exact nature of the disease.

It is interesting to note, however, that studies still cite patient dissatisfaction with the amount of information they are given. Patients describe physicians as continuing to withhold and manipulate information that patients want (Cassileth et al. 1980). In particular, McIntosh clearly describes this incongruence (1977). It appears that while physicians believe they are informing patients, their patients continue to regard the flow of information as inadequate.

To address the issue of what physicians actually tell patients and how they arrive at that decision, this study focuses on a single episode in breast cancer treatment -- the event of telling the patient for the first time that she has cancer. This study is a participant-observation and interview account of the episode, framed from the physicians' perspective. It describes how physicians organize the event and what strategies they develop for routinizing a task they define as difficult and unpleasant. There are few ethnographic accounts of this sensitive and complex dilemma from the doctor's viewpoint. Difficulty in accessing the event, physicians' reluctance to acknowledge the issue and the tendency of social science research to focus on the patient's perspective contribute to the paucity of data.

THE METHODS

The Setting

I was an observer of the 17 surgeons affiliated with the Breast Cancer Clinic at Garfield General Hospital[1] (Canada) from 1978 to 1981. The Clinic provides initial screening, detection and diagnosis for approximately 2000 women each year. Of these, over 200 would be found to have breast cancer and would receive treatment. Half of the patients were referred by their family physician, while the rest were self-referred. The women ranged in age from 20 to 85 years, and the average was between 50 and 65 years old. They were white, lower and middle class women with an average of 15.5 years of schooling and a stated average income of $21,000. The 17 surgeons affiliated with the Clinic were all male, and ranged in age from a junior associate (29 years old), to the hospital's senior surgeon (67 years old). The median was 47 years. Four surgeons had their offices within the Clinic. They were salaried, full-time staff, hired by the General Hospital. The rest were part-time associates, invited to take part in the Breast Cancer Clinic. They saw patients in the Clinic, as well as in their private offices, outside the Hospital. All seventeen doctors shared a common Clinic patient pool.

The 17 Clinic surgeons followed a predictable routine when examining women who came to see them, complaining of a lump in their breast. First, they took a detailed family history. They then palpated the mass, and often ordered a mammogram. If the physicians were suspicious that the lump might be malignant, they recommended an immediate biopsy. Small amounts of tissue removed from the lump were carefully examined under a microscope. If the biopsy was negative, the women were asked to come back to the Clinic for a check-up every six to twelve months.

For those women whose biopsy showed evidence of cancer cells, surgeons planned further treatment. Before proceeding, the surgeons had to explain to the patient the results of the biopsy, and impress on her that immediate surgery was imperative. The elapsed time between a patient's first visit to the Clinic and surgery was often as little as 48 to 72 hours.

Collecting the Data

The focus of this article is on the 118 "events" I observed during which the Clinic surgeons disclosed the results of the biopsy to women who had breast cancer. The term "event" refers to the first time the physician tells a patient she has cancer. This disclosure "event" may be followed by a communication process over time wherein additional details and explanations continue as part of the treatment process. However, this study is restricted to the first, and one-time occurence. The 118 events took an average of 47 minutes each within a range of 10 to 91 minutes. I attended an average of 7 events per Clinic surgeon.

I remained in the room while surgeons talked to their patients. I sat facing the doctor, so that the patient had her back to me. In this way I was able to observe the physicians' facial expressions and gestures, and could clearly hear the conversation. I was introduced to each patient, and all women were asked if they preferred that I leave the room -- none did. I made brief notes during the event and dictated the details immediately after. The notes were later transcribed verbatim, and coded and analyzed in the manner of Glaser and Strauss (1968).

Three characteristics of the events I observed differed from other reports of physician disclosures (Kutner 1958). First, the popular belief that a diagnosis of breast cancer is a declaration of an incurable, fatal illness defines disclosure of malignancy in the breast as the delivery of "catastrophic news". Unlike surgeons in other clinical settings who occasionally convey news of a terminal illness, Clinic surgeons routinely gave patients bad news. Second, the Clinic surgeons were aware that the patients' prognosis was not entirely linked to the surgeon's technical skill. That is, a surgeon may effectively remove the entire cancerous lump with cosmetically-pleasing results, yet the patient could die within a few months. Although patients were often unaware of the relative ineffectiveness of breast surgery, Clinic surgeons knew that 65 per cent of their patients would die within 5 years, regardless of how well they did their job. Third, there was no single treatment of choice. Clinic surgeons varied greatly on what they believed was the best operation to perform. For example, four argued that extensive amputation resulted in improved survival time, three believed that it was best not to remove the lymph nodes, and five felt that removal of the lump alone was sufficient. No single operation was accepted as best by all Clinic surgeons.

Clinic surgeons were obliged to tell women, most of whom they had never seen before, that they had breast cancer, that they must have immediate surgery with possible removal of their breast, that there was uncertainty as to which operation was best and uncertainty regarding the effectiveness of any surgery, and that no one could guarantee a permanent cure. Here, I describe how the physicians staged the event.

The Clinic surgeons were divided into two groups: the experimenters and the therapists (Strauss et al. 1964). The surgeons I labelled "experimenters" assumed that therapeutic innovation is a continuum. They believed it should move from experiments on animals in the laboratory, to clinical trials with terminally ill patients beyond the help of conventional therapy, to the use of experimental treatments on less critically ill patients. Clinic surgeons who adopted this philosophy defined their core task as generating scientific data by placing their eligible patients into experimental programs. Those surgeons I call "therapists" adopted a more traditional role, emphasizing the routine application of standard therapy. They saw their primary responsibility as recommending a particular treatment for each patient. The therapist philosophy is based on the individuality of each case; the experimenter philosophy on its generalizable features.

Of the 17 Clinic surgeons, fourteen (85%) fit the category of therapist. Their first allegiance was to their patients, and clinical expertise governed most of their decisions. Among this group, clinical acumen was see as a charismatic possession, with an almost mystical quality. In contrast, the three experimenters (15%) emphasized the rational and technical dimensions of their practice. Although fewer in number, they were a powerful force at the Clinic. They published extensively, held a high professional profile both within and outside the Hospital, and were responsible for large research budgets. This gave them a significant power base which allowed them to determine many of the Clinic's policy decisions.

Both groups of surgeons regarded telling patients they had breast cancer as a thankless, potentially exhausting, unpleasant, but central part of their job. Studies have shown that physicians confronted with what they believe to be stressful, ambiguous, or distasteful situations try to routinize the task (Siegler 1975). This allows them to regain some sense of control, reducing the impact of the event on subsequent behaviour. Each Clinic surgeon developed a predictable, rigid approach to routinizing the disclosure process. It varied according to whether the surgeon was an experimenter or therapist.

In this report I describe how the experimenters and therapists staged the event. I suggest that there are three phases of each episode, and delineate four strategies the doctors used to routinize this process.

STAGING THE EVENT

When a woman discovers a lump in her breast, her thoughts -- and those of her physician -- immediately turn to cancer. Although 80% of all breast masses are benign (Canellos et al. 1982), for those whose tumors proved to be malignant, Clinic surgeons were obliged to confirm the women's worst fears. The surgeons regarded the disclosure of the diagnosis as a routine part of their job, but they also considered it one of the worst.

I really hate this part of the work. I feel terrible . . . always . . . it never seems to get easier. Knots in my stomach a mile wide in that walk from the lab to the patient . . . I know it's not my fault . . . I didn't give the patient the disease . . . but I always somehow feel that it is . . . and besides . . . I never really know what to say. They sure never had a course on this part of medicine when I went to school, but let me tell you, I could sure use it now . . . and I usually don't ever get to know these patients who think their lives are in my hands. How am I supposed to deal with it? I hate doing it . . . but I can't avoid it!

Several times each week the physicians disclosed the information to patients either in the Clinic examination cubicle or in their offices. They often reserved two or three adjacent appointment slots for each event. The physician faced his patient across a desk and began to address the unspoken questions. Patients whose tumors proved to be benign – about 80 percent – reacted to the news with tremendous sighs of relief, often echoed by the physician. For the other 20 percent -- those with a diagnosis of cancer -- the physician anticipated the inevitable -- shock, tears, disbelief, and questions of "what am I going to do now?"

I know when I first tell my patients, they stop listening after I use the word "cancer". I know I will have to repeat what I say over and over again, no matter how
informed they think they are, or how much they say they figured it was cancer anyways. I tell you it's always a shock for them. This really is the rotten part of the
job. You try to tell them as little as possible in the first talk, because just hearing
the word "cancer", it's really hard for them. It really screws up your patient flow
for the rest of the afternoon, you know . . . Well, maybe that's not completely
true . . . I'm usually so screwed up, all hell breaks loose the rest of the afternoon
anyways . . .

Once the physician had told the patient that the diagnosis was cancer,
and reassured her that no mistakes had been made by the laboratory, he
attempted to shift the conversation to choice of treatment.

I feel much better when all the crying and that sort of stuff is over. It happens all
the time . . . there's not much you can do, just have to wait until they calm down
and pull themselves together. You'd be surprised how long that can take, sometimes. Since they can't do anything about having cancer, but CAN do something
about removing it, I try to start talking as soon as I can about the surgery. You
have to work fast, or they would all take forever to decide . . . and with my ladies,
we just don't have forever . . .

THE PHASES

Disclosure has been described as "a short, blunt announcement made by
the surgeon after the operation. He walked into the patient's room, and
made his speech, and turned and walked out, closing the door behind
him" (Glaser 1966:83). This did not happen at the Clinic. Biopsies were
done before, and not after, any operation. Unlike the traditional, "short,
blunt announcement", each of the 118 events had three predictable, discrete and mutually exclusive components. The three phases of disclosure
(Davis 1960:41) were designed to facilitate the surgeons' task of initiating
a status change in women they did not know, from "healthy female" to
"breast cancer patient with a life-threatening illness requiring immediate,
but not always effective, surgical intervention".

The first phase was the PREAMBLE, the conversation immediately preceding the delivery of the diagnosis. Although physicians agreed that
patients remembered little of this initial conversation, the physicians
themselves put great emphasis on it. Anticipating the patients' overwhelming sense of hopelessness which inevitably accompanied a cancer
diagnosis, the physician used this opportunity to set the stage. With each
new patient, the surgeon began to introduce such terms as "serious

illness" and "difficult decision" and spoke in grave tones of his role as a specialist treating patients with a life-threatening disease.

The preamble was often abruptly terminated when the patient asked, "Doctor, what did the biopsy show?" Most patients assumed, "the news wasn't good, otherwise he would have said so right away," but awaited confirmation of their fears. The CONFRONTATION was the phase during which the surgeon disclosed the diagnosis and potential prognosis. Since most patients had a painless breast mass and were asymptomatic, the news came as a great shock. Many had hoped that the lack of disabling sensations were an indication that they were among the 80 percent whose tumors would prove to be benign growths.

The confrontation was quickly followed by physicians' efforts to soften the blow. The DIFFUSION phase was used by the surgeons to reduce the impact of their words. Discussion shifted away from the diagnosis to what type of medical interventions might be effective in the particular case.

THE STRATEGIES

During each of the three phases of disclosure, each Clinic physician adopted a favorite technique and used it routinely. The first technique, "communication", is described by Davis as, "when the physician can, in accordance with the state of medical knowledge and his own skill, make a reasonable, definite prognosis of the condition, and communicate it to the patient in terms sufficiently comprehensible to him" (Davis 1960:41). Many of the Clinic surgeons were certain that this was their approach to disclosing information, but in fact, few did. Only 12 of the 118 disclosures (10%) could be classified as "physician communication".

The second, "admission of uncertainty", was used when "no prognosis is clinically justified, and is by its nature a difficult and unstable mode of communication" (ibid). There were many cases in which the patients' prognosis was less than certain. In addition, there was often a lack of consensus regarding the treatment of choice. For example, the entry of eligible Clinic patients into experimental programs was based on officially acknowledged uncertainty about which treatment is best (Fox 1980). However, few Clinic physicians chose to share this uncertainty with their patients. In only 18 of the 118 events (15%) did the surgeon clearly explain that there was uncertainty regarding the best way to treat the disease, and uncertainty regarding the outcome of any medical intervention.

The third approach, "dissimulation", the rendering of a diagnosis and prognosis which the physician knows to be unsubstantiated clinically, was closely correlated to the lack of "admission of uncertainty". If Clinic

surgeons were reluctant to share the extent of uncertainty with their
patients, they were obliged to pronounce a prognosis which they realized
could not be medically substantiated. This strategy was used in 35 of the
118 events (30%).

The final technique, "evasion" or the "failure to communicate a clinically substantiated prognosis", was used by surgeons who preferred not to
respond directly to patient questions. For those surgeons whose patients
asked direct questions to which the appropriate technical response might
reveal a low chance of long-term survival, repressing information was a
favored policy. Responding to specific questions with general statistics,
not easily applicable to the individual case, was not unusual. In 53 of the
118 events, Clinic surgeons used "evasion" as a method of coping with
direct patient questions (45%).

EXPERIMENTERS, THERAPISTS AND DISCLOSURE

Each Clinic surgeon adopted one of the four techniques of disclosing
information to patients and was consistent in utilizing the policy he
adopted. The most significant predictor of which approach an individual
clinician selected was his "disease philosophy". This term refers to an
integrated set of attitudes and beliefs which define physician behaviour
(Strauss et al. 1964).

The three experimenters were very articulate with patients. They used
medical jargon and statistics to explain the prognosis. They relied on
published studies as proof of their recommendations, and justified their
openness in terms of patient's rights. They believed that full disclosure
was an efficient means of obviating the need to weave intricate fabrica-
tions. They used the disclosure process to initiate discussions of the ran-
domly-assigned treatments they hoped would resolve the current controversy regarding the best treatment for breast cancer.

The fourteen therapists, on the other hand, often used complex
euphemisms to explain the diagnosis. They offered less scientific information, and referred to their clinical experience rather than published
data. They justified this evasive policy claiming that "patients didn't
really want to know". They preferred to preserve the traditional physician role, shielding patients from bad news and assuming full responsibility for all decisions.

The Experimenters' Disclosure

The following is typical of the experimenters' disclosure policy.

Experimenter :   Enormous strides have been made in cancer research. We are making terrific headway. You know, if you had had this diagnosis a few years ago, there would not have been a whole lot that could have been done. We now have techniques for conquering all sorts of problems we could not have touched, even ten years ago . . .

Patient :   Did you get my report back?

Experimenter :   Yes, as a matter of fact, I have the [pathology] report here somewhere on my desk. Let me see [reaching over to pick up a slip of paper from a pile]. Ah yes, it says, "infiltrating and intraductal lobular carcinoma, well encapsulated in . . ."

Patient :   You mean -- it is -- I've got -- cancer?

Experimenter :   Yes.

Patient :   Are you sure?

Experimenter :   Yes, I went down and checked in the lab and . . .

Patient :   Am I going to die?

Experimenter :   We have a lot of tools to fight with. Let's talk about the options. You see there is no real "best therapy" as I am sure you are aware. As a matter of fact in the latest issue of the Journal [New England Journal of Medicine], there was an interesting article this week . . .

Patient :   You mean I am going to die? [Patient cries and shakes]

Experimenter :   Now listen, I was saying that there is a lot we can do. We have to make some serious decisions. Pull yourself together . . . please.

Patient :   You mean I am going to die . . . my God . . . my . . . what was that you said about surgery? I know I don't have to lose my breast . . . right? . . . right? . . . Oh my God.

Experimenter :   We really don't know which surgery is best. We do not have any real answers. We are collecting data to help us with these questions. Let me tell you about this clinical trial . . .

Patient :   Doctor, I am asking YOU what you think is best for me. For God's sake you are a doctor . . . I don't want my breast taken off . . . but then . . . I want to live . . . I've got three kids you know . . .

Experimenter :   You know many women have breast disease. Why I was reading somewhere that the new figures show one out of every 11 women in North America will have some trouble during her lifetime. A lot of them will do well . . . and of course, some won't. But as you know, unfortunately, medicine has not always been considered an exact science . . . All those women who develop breast cancer now have a unique opportunity to help us get scientific answers to very old questions . . . To help their daughters

somewhere down the road. YOU can really help us get more
accurate information about the disease, and help us test the
ammunition we are now using. We can't be sure of anything until
we have hard proof, and you can be very important in helping us
do that. Entering a clinical trial is one positive step patients can
take in a very unfortunate situation.

The experimenter used the preamble to set the stage. He shifted the
conversation to his point of view -- improved potential survival for breast
cancer patients due to advances made by current experimental procedures. He then confronted the patient's question, "is this cancer?"
directly, using the terms, "breast cancer" and "survival" rather than
euphemisms. Finally, he attempted to diffuse the impact by suggesting
that the patient shift from the undesirable status of cancer patient to one
he considered more desirable: "research subject capable of making a significant contribution to medical science".

Experimenters were also predictable in the way they arranged the discussion. For example, if patients were sent to the physician's Clinic
office, rather than the busy examination rooms, the physical distance
between doctor and patient was prearranged. These sparsely furnished
offices were devoid of personal ornamentation and littered with slides,
charts and colored liquid in test tubes. Most of the textbooks and manu-
scripts had complex titles, such as, "Protocol III-AVB for Adenocarci-
noma With or Without Skeletal Metasteses using CMFVP". The experimenters sat behind their large desks during the disclosure, often shuffling
sheets of paper, looking for exact figures to support their statements.
They used medical jargon and technical terms to underline the seriousness of the conversation. For example, a surgeon told his patient:

Your lesion is in the inner upper quadrant, so it may have already infiltrated the
internal mammary chain of the lymph node system, making reliance on the results
of an axillary dissection only tentative. The protocol for your stage of disease suggests radiating the internal mammary chain, prophylactically, of course . . .

For the Clinic experimenters, disclosure usually lasted between 30 and
45 minutes. It ended with the physician's recommendation that his
patient consent to having her therapy randomly assigned as part of a clinical trial.

Let's get this show on the road. The sooner we get your therapy chosen, organized
and started, the better it will be for all concerned.

The experimenters were explicit, even when the prognosis was not
certain. In addition, they often slipped into technical jargon, although

they initially tried to deliver their speech in terms they thought patients could understand.

Yes, the test did show cancer. It is a malignant tumor. We call it an "early stage one," since it seems to be small, well-circumscribed, and I can't feel any lymph nodes under you arm. But, I can't be totally sure. After the operation I can be more specific, after I've had a look in there. It is in the inner upper quadrant, easily accessible, and once removed, you should be able to return to a completely normal life. Of course, you'll have to come back for routine checks, but new evidence shows that women in your menopausal state, with these characteristics, seem to do really well. Of course, there are no guarantees, no one knows for sure how you'll do . . . but from the new studies, it seems to be safe to say that you might do very well.

When the prognosis was less optimistic, the experimenters used uncertainty to offer help.

Well, the news is certainly not as good as it could be . . . but then, we really don't have all the information yet. It seems there are subsets of patients that we previously lumped together, who seem to be doing just fine. Why, just the other day I was reading that a subgroup of patients with advanced disease, but no real evidence of active involvement, were doing better than anyone expected. In your case, well, we just can't be sure.

However, when confronted directly, experimenters responded with accurate and detailed information that was often painful and unwelcome.

Yes, you are right, the survival rate for premenopausal women is lower than for postmenopausal. There seems to be some feeling that it has to do with the expanded prolactin secretions in younger women . . .

The experimenters also volunteered information that their patients had not requested. They agreed that few patients know what questions were appropriate, and assumed the responsibility of responding to implicit queries.

You can be sure I'm going to tell them everything. Most times I have to help them along, because they don't even know what questions to ask. Any guy who tells you he waits till his patients ask, before giving them the word, is crazy, because women don't know what the hell to ask. How can they? I tell them all, and then if they come back months later and say, "you didn't tell me about this", I can, with a clear conscience, show them I did.

The experimenters included many references to the previously published results of experimental studies when explaining the basis for their recommendations.

Our latest information shows that doing any tests before surgery is really not necessary. Might as well wait a couple of months. The group tested before surgery showed no more detectable tumors than those tested two months after surgery, so I say wait . . .

Finally, experimenters emphasized their expertise in interpreting the complex, confusing and often contradictory findings of the latest research.

In this month's Journal [New England Journal of Medicine], they were describing the new data on cases that resemble your problem very closely. Not too much to say, because the treatments are new, but it seems that Hopkins, Kettering and Anderson [large American cancer centres] are all saying the same thing . . .

The experimenters offered a variety of justifications for their disclosure policy. Some argued that legal requirements made total disclosure inevitable. One surgeon explained:

The law says that participation in any kind of clinical research must be preceded by pages of informed and documented consent from all patients. It is technically impossible to offer patients a chance to try good treatment options without their knowing what the hell they are being treated for. Besides, those committees sure check to make sure the damned sheet is signed!

Others said they feared possible malpractice suits if a patient subsequently discovered that she had cancer and had not been informed of all possible treatment options. The senior surgeon gave an example:

Mrs. H. is sure making me nervous. She had her operation in Switzerland a couple of years ago. They never told her. They just said she had some sort of cyst. Anyways, in she comes to see me with it spread all over. She's a goner, but boy is she out to get those guys, before she goes. She wants me to sign a statement that if she had known, she would have gotten better, or at least different care. Who the hell can tell if it would have made any difference anyway? She's out for blood. I really feel for the doctor in Switzerland, though . . . I think to myself . . . "there but for the grace of God" . . . if you know what I mean . . .

The experimenters also hoped that full disclosure would keep patients within the traditional medical system. They defined what was "scientifically valid experimentation", in contrast with "those crazy clinics set up to kill patients, and make a bundle doing it." The experimenters did not attack such complementary therapies such as diet, special exercise, or focused psychotherapy. They simply viewed them as useless, but not harmful, as long as the patient continued the treatments recommended by the Clinic. However, the experimenters regarded therapies designed to replace surgery -- such as unknown drug mixtures administered in foreign countries – as particularly dangerous. They were concerned about the tendancy to replace, rather than supplement, the surgeons' advice. The experimenters hoped that an open discussion of cancer, and its fatal nature, might prove to be an effective counter-strategy against what they called the "growth of medical quackery."

Offering the patient legitimate, academically sound, and scientifically valid research opportunities should be preferred to the false hopes and promises offered by medical quacks who prey on ignorance.

In addition, experimenters argued that offering a patient an opportunity to join an experiment whose outcome might increase medical knowledge was the best way to alleviate the patients' inevitable mental anguish. They often told patients:

You know how lucky you are to be here in a large hospital with the best of everything; the latest and the greatest, I always say . . . and to have the good fortune to be able to participate in the most recent types of therapy . . .

Finally, the experimenters said that the organization of the Breast Cancer Clinic made it impossible to conceal information from their patients.

It's impossible to not tell patients about their disease. In part, it's because the media tells them what to look for, and what to do, if it's cancer. They know before we do half the time. Who is fooling whom, but it is just because we are suspicious? Come on, before a woman will consent to that type of surgery, you can be damn sure she knows what the score is . . .

The experimenters viewed disclosure as an integral part of the experimental nature of their work. This policy had some interesting advantages. First, physicians who give their patients the worst possible prognosis cannot fail (Siegler 1975). If the patient dies, the physician is considered a talented prognosticator; if the patient lives, the doctor can take credit for saving a doomed life. Second, full disclosure freed the experimenters from an intricate web of lies.

I refuse to talk to any patient unless I can be completely honest. I tell them the way it is. After the initial shock, they always get over it. Look, can you imagine how complicated my life would be trying to remember what lies I said to whom and when? What would they think of me? No, I have to have their total trust. I promise each patient total honesty at the first visit, and I ALWAYS keep my promise . . .

Third, full disclosure was necessary to obtain the patient's signature on the detailed informed consent form required before randomly assigning the patient in a clinical trial. (The standard surgical consent sheet did not include details describing the exact nature of the disease.) Thus, a policy of full disclosure for all patients relieved the experimenter of having to acquire informed consent from a patient who did not know she had cancer. Finally, experimenters claimed that disclosure reduced their sense of personal responsibility when treatments failed.

> I really do prefer this way of doing things. I hope and pray that things will go well for each patient, but if they don't, I feel easier about it since the patient was told all the options, and hopefully made an informed choice. That really makes me feel off the hook when things start falling apart, as they always do . . . sooner or later!

However, there were risks involved in the experimenter's policy of full disclosure. There was the possibility that his patients would seek the advice of another, more traditional surgeon, and the possibility that patients and colleagues would view the experimenter as cold and unfeeling.

> Yes, I'll admit, I'm always a little concerned that if I appear to be really ambivalent while I explain the need for a clinical trial, or when they ask me to tell them something "for sure" and I say "I really don't know," that the woman will go down the street to a guy who doesn't have any more answers that I do . . . but tells his patients HE knows what's best for her particular case . . . who wouldn't worry . . . and another thing . . . I know that I'm called a real son-of-bitch behind my back, but, I guess the bearers of bad tidings for years . . . I guess that's for life . . .

The three experimenters used both "communication" and "admission of uncertainty" as preferred strategies for disclosing information to breast cancer patients. The fourteen therapists took a very different approach.

The Therapists' Disclosure

The difference in the therapist approach is evident in the following episodes.

Episode 1

Therapist :	You know, my dear, I can see you are upset, but, whether I like it or not, I have lots of women like you come into this office over the last 30 years . . . Why you're certainly not alone . . .
Patient :	Doctor, tell me . . . will you have to . . . operate?
Therapist:	Well, now, as you have probably guessed, we saw something in your breast that really shouldn't have been there, now

Patient :

Therapist :

should it? Well, I didn't like the looks of it under the microscope, and I must say . . . it could have looked better.

You mean . . . I guess I really have to have an operation, oh God no . . .

Yes, just to clean up the mess you know . . . make sure that the nasty lump doesn't give us any more problems. By the way, just today, I had the sweetest old lady in my office. Why she sat there, just where you are . . . anyhow . . . I had operated on her . . . well it must be 25 to 30 years ago, and she was in for her annual check-up . . . well, I said to her, "how have you managed all these years to look so terrific and to feel so chipper?" "Well," she said ". . . my mother always told me . . . don't get mixed up in things you know nothing about. So I didn't . . . just left everything up to you, Doctor." She has done very well. That's what I want you to do . . . just relax and leave everything to me . . . okay?

Episode 2

Therapist :

Patient :

Therapist :

Patient :

There was something we didn't like about your [pathology] report . . .

Was it cancer . . . no . . . don't tell me it was cancer . . . my mother died of it when I was 16 [Patient shakes and cries]. I knew it . . . I knew it was as soon as I felt the lump. . . Everyone said don't worry, but I knew . . . I know I am going to die . . . just like she did . . . it was awful . . . just awful.

Well, it's sort of a tumor, but mind you . . . not like she had . . . we can call it infiltrating ductal carcinoma, just your everyday garden variety type. To die as quickly as she did, your mother probably had some really rapidly invasive cells . . . Now, if you listen to me and follow my instructions . . .

Oh, thank God . . . I thought I had cancer . . . thank you, doctor . . . I'm so grateful . . . I'm sorry I got so upset . . . but, you know . . . I was scared that I had . . . you know . . . cancer.

Therapists used the preamble to focus on the importance of the physicians' personal clinical experience. In the confrontation phase, they conveyed the potential seriousness of the disease while maintaining the illusion of possible cure. Finally, in the diffusion phase, they used euphemisms and veiled messages to soften the blow, and believed that patient trust in the physician was an important part of the treatment process.

Therapists sometimes asked patients to come to their private office, outside the hospital, for the disclosure. A traditional doctor's office, it included personal mementos on the shelves behind the desk, pictures of family on the walls, background music, and a waiting room of patients, many of whom did not have cancer. The therapist often sat on the same side of the desk as his patient, facing her directly. This close proximity allowed him to offer a Kleenex or a quick pat on the arm as he spoke. He often used his patients' first names, preferred euphemisms to hard facts, and divulged only that information which patients might interpret as "good news".

Your lump is in a good spot. Perhaps a little x-ray therapy treatment to tidy up after your operation would be helpful . . . Just to be sure . . . now, it never hurts to be extra careful, that's what I always say . . . Don't you agree?

The discussion could last from one hour to one hour and a half. The therapist usually reserved his last appointments for disclosure, anticipating the extra time he needed. At the end of the conversation, the physician often suggested another meeting the following day, to finalize any decisions.

Well, now, why don't you go home and think it over, if you feel you really can't make up your mind right away. I'm sure you'll feel better tomorrow and decide to leave the worrying to me.

Therapists often gave a prognosis that they knew was not necessarily substantiated by medical fact. In the following case, the surgeon feared that this was a precancerous lesion, but chose to withhold this information from his patient.

Don't you worry your pretty little head about anything. Sure, it has to be watched, and I want you to come in every three months so I can have a good look at you . . . but the mammogram was negative, so let's just keep an eye on this little fellow, shall we? But, no, at the moment, I see absolutely no cause for concern.

Therapists routinely evaded direct questions, describing the disease in hazy terms. This tactic was used to impress a patient with the seriousness of the disease, without alarming her by using the word, "cancer".

Physician :   You have something in your breast we don't like.
Patient :   Is it serious . . . like is it . . . you know . . .?
Physician :   Well now, it doesn't look too bad . . . but we better take care of it before it starts to give us trouble, eh?

Another therapist told his patient:

You do have cancer in that breast. With surgery I'll try to get most of it out. It's spread around a little bit. Some radiotherapy in the area after the operation should clean up any cells I miss. Then, we'll just have to wait and see. You can never really tell in cases like this. Why, sometimes the body just seems to readjust after the bulk of the tumor is removed. We'll just have to wait and see . . . at the moment it's anyone's guess . . .

Even when patients asked direct questions, therapists continued to be evasive and optimistic.

Come now, where did you hear such nonsense? You know, you are not a statistic. That's for others. Let's talk about you and your problems, and leave the numbers to the statisticians.

Therapists said they preferred not to offer any information, unless the patient specifically asked.

The patient lets you know very clearly how much they can handle. Knowing that timing, getting a feel for it, well, that comes with experience, I guess. For God's sake, you sure as hell don't hit them over the head with it, or ram unwanted information down their throats. You just watch and listen, they'll tell you. They're the ones who really orchestrate the agenda . . .

When patients requested substantiation of a therapist's recommendation, he usually emphasized his personal clinical experience.

You know that any lump in the breast, well, it always can be a serious matter, of course, but you know, those early lesions, the ones that come out nice and clean, well, if women would just listen to their doctors and come in as soon as they feel anything . . . get their problem seen to by a specialist . . . well, just look at the hundreds of women I've cared for in the last 30 years . . . If they would just behave . . . things CAN work fine . . . just fine. My dear, in the last many years, I have seen lots of cases, just like yours. I call them regular, garden-variety type. I'm sure I can't even begin to count the number of patients I've taken care of . . . just like you.

Therapists believed that patients needed the reassurance that their physician had extensive clinical experience with similar cases.

Now, look, as I see it, you have two choices. You can wear sack cloth and heap ashes on your head for the rest of your life and make yourself and those around you nuts . . . Or, you can say to yourself, "well, yeah, I had an abcess, it's gone . . . I'm fine" and forget it. Just relax . . . I've seen patients a lot worse off than you, and I've managed to take really good care of them. So relax . . .

The therapists often justified their policy of not telling patients everything that was known by saying that it is the way things have always been done, or that patients DIDN'T really want to know even when they asked, or that it was the physicians' professional responsibility to shield women from the overwhelming feelings of depression and grief they assumed would accompany full disclosure.

What do you think I'm going to say to her . . . "Look lady, you're going to be dead in a few months. Your kids are going to be orphans. Forget it, I'm not going to operate on it at all. The stuff has spread and it's no use?" At least with my approach, the poor women will have a few months of peace. I know I'll probably have trouble convincing her to have the "cyst" radiated. She doesn't really have to know . . . That way she'll have some peace, even if it's only a short time . . . God knows she deserves it . . . It's the least I can do . . . It's my job, even if I don't always take the easiest way out.

All therapists agreed that telling a patient her diagnosis was a difficult task, but they accepted it as an integral and important part of their duty as physicians. Some felt it had been easier when . . .

. . . the patient did not have to be told the entire truth. We used to get that nice sense of confidence and trust from the patient who believed in the physician's sincere interest in her case. I'm not sure that it wasn't a large part of many of our earlier successes. Like this old doctor I used to visit at home once a day to change his dressing on his colostomy [surgery for bowel cancer]. He used to ask me, every day, "So, I guess my colitis [a benign intestinal disease] is causing all this trouble, eh?" "Yeah, I guess so," I used to reply. We went on like this till the day he died. He needed me to tell him that everything was going to be okay. He was a doctor. He knew he had cancer, but even HE wanted some comfort. What good would it have done to tell the old guy all the details of his case? He would just have been more miserable. It was kind of like a sacred convenant with patients in those days. I know you can't do that anymore, but I'm not totally convinced that this new way is right. But, what can I do? I have to go along with everybody at the Clinic.

Another therapist observed:

I never really know what to say, and then, what the hell can you say? You see this couch in my office? It's there for the patients, when they faint or get really weepy. I try to tell them as little as possible, and THEY DON'T ASK. I am telling you, patients do not want to know, even when they insist they do. I remember this lady I had as a patient years ago . . . I never told her . . . Oh, she'd sort of ask, "well, doctor, have my cysts . . . got worse?" You know . . . But, I'd always say, "Now you know you've got bilateral fibroplasia — lumpy breasts, as it is better known — we've talked about this." I will never forget, on her death bed seven years later, she looked up at me and said, "Thank you for not telling me," and then she died. The intern was in the room and he fainted. I'll bet he learned a valuable lesson that he will never forget. Anyway, I never forgot her. Even when she said she wanted to know, you see, she didn't. I often think of her when I decide not to tell the young ones. At least not in so many words.

Therapists disclosed the diagnosis and prognosis reluctantly, sometimes concerned that their tactics may not always be in the patient's best interest, and often aware of the difficulties in withholding information.

It's getting harder and harder to have to tell patients these days . . . They all know so much . . . or at least they think they do . . . When I try to tell them something they all say, "Yes, we understand," and fly back into my office the next day with some mixed-up tale of woe some neighbour saw on TV and they think I'm lying, or at least not telling them everything. I always try to tell just enough so they realize this is no joke, but not enough to scare them out of their wits. I'm telling you the consumer movement . . . well, they are making for very terrified patients . . . for no reason . . . and making the job harder and harder . . . you can be sure.

Another therapist declared:

There's no arguing with reality. Practical obstacles to withholding the unwelcome news from cancer patients are often insurmountable. What I object to are the philosophic considerations put forth in defense of completely frank conversations. Most of their arguments are based on the mistaken notion that "an aura of sanctity surrounds the truth." In actual fact, it makes a magnificent cop-out for the oncologist who doesn't want to help his patient bear the burden of the disease.

The fourteen Clinic therapists used "evasion" and "dissimulation" to inform their patients. This was in sharp contrast to the Clinic experimenters.

CONCLUSIONS

In contrast to many sociological studies of telling bad news, the focus of this report has been on the person who gives rather than receives the news. Three phases -- preamble, confrontation and diffusion -- of telling bad news, four strategies used by physicians -- evasion, dissimulation, admitting uncertainty and truth-telling -- and two physician policies -- experimenter and therapist -- are described. This report suggests that the event of telling bad news is perceived as stressful by physicians. In response, they find complex yet predictable ways to routinize this task.

Since improvement in any didactic relationship ideally includes the perspective of both parties, one way in which this information may be useful is for those seeking to improve doctor-patient communication. For example, patient education, informed consent regulations, and public lectures have taken on particular significance in North American health care delivery in the 1980s. Increasing emphasis is being placed on the importance of patients' rights and patient involvement in medical decision-making. At the same time, this report suggests that physicians may

be unconvinced that it is always in the patient's best interest to be given
full details of their case. If this is true, and physicians remain the gate-
keepers of information, the efforts to more fully inform patients may be
less than fruitful.

Difficulties in encouraging physicians to acknowledge and address this
issue is a serious concern. At the Clinic, the topic of what and how physicians inform their patients was never formally discussed, had not been
included in medical training, and was not considered a topic warranting
serious research. Clinic physicians adopted a policy of disclosing information based on a personal comfort level, a selective review of their past
experiences and anecdotal tales. Despite continuing personal discomfort,
the Clinic physicians preferred to develop techniques and policies to
reduce their tension rather than confront the issue directly. The result of
their actions makes disclosure a very sensitive topic. If routinizing the
process reduces the tension, suggestions that they review and evaluate
their actions may be difficult. This report suggests that physicians are
unconvinced of the value of fully informing patients in all cases. Unless
physicians believe otherwise, it is doubtful that any significant change in
the quality of communication between doctor and patient will occur.

If open communication between physician and patient is the goal,
then physicians' concerns cannot be ignored. It may be comforting and
familiar for some sociologists to explain the withholding or information
from patients solely in terms of physician power and see it as a result of
professional non-accountability. Surely this plays a part. However, this
study suggests that this reason may be insufficient to explain why physicians adopt, maintain and transfer a disclosure policy.

All Clinic physicians found effective methods to routinize what they
defined as a difficult task -- telling their patients that they had breast cancer. The experimenters were exact and gave detailed information,
whether patients wished it or not. In contrast, the majority of physicians, the therapists, defined their role as selecting and interpreting information for patients.

In addition to theoretical implications, these findings also raises
empirical questions. First, it appears that all Clinic physicians, whether
therapist or experimenter, found ways of bypassing the individuality of
each case. That is, by routinizing the task of disclosing information, they
were able to reduce their need to make an individual decision based on
the characteristics of the case. Physicians were then able to overlook
individual patient's variables, enabling them to operationalize a rigid, predetermined, yet efficient disclosure policy. This behaviour has important
implications for any analysis of patient care. If physicians do not make a
unique choice for disclosing information to each patient, the impact on
the ensuing doctor-patient relationship must be closely examined.

Second, the way in which physicians choose a policy and the social systems that encourage and support each philosophy require careful review. Medical education, physician role-modeling and the reward system in teaching hospitals each provide valuable clues. In addition it would be interesting to speculate on the full implications of encouraging physicians not to adopt a defined disclosure policy.

Finally, the physicians' policies should be examined, evaluated and assessed. Surgeons at the Clinic often contended that one or another of the approaches was most effective. However, they based their arguments on anecdotal data and personal belief. These disclosure policies could be examined in a more rigorous way, with special attention to assessing the impact on both physicians and patients.

This report suggests that there may be a need to redefine the term "disclosure" in ways that more clearly reflect the complex demands of current medical practice. Although this study uses breast cancer as a model, physicians in virtually all settings are obliged to disclose bad news. The generalizability of these findings to other clinical settings is an important question for future research.

ACKNOWLEDGEMENTS

The research reported in this chapter was supported by Health and Welfare (Canada) Fellowship (6605-1681-47). This article is reprinted with the permission of the journal *Sociology of Health and Illness*.

NOTES

1.   This is a pseudonym.

REFERENCES

Abrams, R.D.
    1966   The Patient with Cancer -- His Changing Pattern of Communication. New England Journal of Medicine 274:317-322.
American Cancer Society
    1982   American Cancer Society Working Conference -- The Psychological, Social and Behavioural Medicine Aspects of Cancer: Research and Professional Education Needs and Directions for the 1980's. Cancer 50:1919-1970.
Barber, B.N.
    1980   Informed Consent in Medical Therapy and Research. State University of New Jersey: Rutgers University Press.

Bird, B.
    1973 Talking with Patients. 2nd Edition. Philadelphia: J.B. Lippincott.
Braver, P.
    1960 Should the Patient be Told the Truth? Nursing Outlook 8:672-676.
Canellos, G.P., S. Hellman and V. Veronesi
    1982 The Management of Early Breast Cancer. New England Journal of
          Medicine 306:1430-1432.
Cassileth, B.R. et al.
    1980 Information and Participation Preferences among Cancer Patients.
          Annals of Internal Medicine 92:832-836.
Christman, L.P.
    1965 Patient-Physician Communication. Journal of the American Medical
          Association 194:151-156.
Crayton, J.
    1969 Talking to Persons Who Have Cancer. American Journal of Nursing 69.
Davis, F.
    1960 Uncertainty in Medical Diagnosis: Clinical and Functional. American
          Journal of Sociology 66:41-47.
Dunn, W.J.
    1978 What to Tell the Public. In Proceedings of the Fourth National Cancer
          Communications Conference held in Chicago, June, 1977. National
          Cancer Institute/American Cancer Society, Chicago, Illinois: HEW
          Publication No. 78-1463.
Elkeles, A.
    1982 Informed Consent in Clinical trials. Lancet 1:1189.
Fitts, W.J. and I.S. Raudin
    1953 What Philadelphia Physicians Tell Patients with Cancer. Journal of the
          American Medical Association 153:901-904.
Fox, R.C.
    1980 The Evolution of Medical Uncertainty. Milbank Memorial Fund Quar-
          terly Vol. 58:1-49.
Freidson, E.
    1970 The Profession of Medicine. New York: Harper and Row.
Germain, C.P.
    1979 The Cancer Unit: An Ethnography. Wakefield, Mass.: Nursing
          Resources.
Glaser, B.
    1966 Disclosure of Terminal Illness. Journal of Health and Human Behav-
          iour 83-91 (Summer).
Glaser, B.G. and A.L. Strauss
    1968a The Discovery of Grounded Theory: Strategies for Qualitative
          Research. London: Weidenfeld and Nicolson.
    1968b Time for Dying. The University of California: Aldine Publishing Com-

pany.

Krant, M.J.
   1976   Problems of the Physician in Presenting the Patient with the Diagnosis. *In* J.W. Cullen, et al., (eds.), Cancer: The Behavioural Dimensions. Pp. 269-274. New York: Raven Press.

Kutner, B.
   1958   Delay in the Diagnosis and Treatment of Cancer: A Critical Analysis of the Literature. Journal of Chronic Diseases 7:95-120.

McIntosh, J.
   1977   Communication and Awareness in a Cancer Ward. London: Croom Helm.

Novack, D.J., R. Plumer, P.L. Smith, H. Ochitill, G.R., Morrow, J.M. Bennett
   1979   Changes in Physicians' Attitudes Toward Telling the Cancer Patient. Journal of the American Medical Association 241:897-900.

Oken, D.
   1961   What to Tell Cancer Patients -- A Study of Medical Attitudes. Journal of the American Medical Association 175:1120-1128.

Quint, J.C.
   1965   Institutionalized Practices of Information Control. Psychiatry 28:119-132.

Scheff, T.
   1977   The Distancing of Emotion in Ritual. Current Anthropology 18:433-505.

Siegler, M.
   1975   Pascal's Wager and the Hanging of Crepe. New England Journal of Medicine 293:853-857.

Simonaites, J.E.
   1972   Recent Decisions on Informed Consent. Journal of the American Medical Associations 221 (4):441-442 (July).

Strauss, A., et al.
   1964   Psychiatric Ideologies and Institutions. New York: The Free Press of Glencoe.

Taylor, K.M.
   1985   Doctor's Dilemma: Supporting Clinical Trials. Cancer Treatment Reports October: 1282-89.

Weisman, A.
   1967   The Patient with a Fatal Illness: To Tell or Not to Tell. Journal of the American Medical Association 201:152-154.

BARBARA A. KOENIG

THE TECHNOLOGICAL IMPERATIVE IN MEDICAL PRACTICE: THE SOCIAL CREATION OF A "ROUTINE" TREATMENT

INTRODUCTION

While leaving the operating room after assisting with an early implantation of an artificial heart into a human being, Dr. Robert Jarvik, one of the inventors of the new device, commented to the press that the surgery had gone so smoothly it seemed "routine." The *New York Times* reported on February 18, 1985:

> Though it was only the second time the Humana team performed an artificial heart implant, there was a sense of the routine. "Boy this is a dull operation," one of the nurses who had participated in Mr. Schroeder's operation said, according to Dr. Jarvik. "That was great," Dr. Jarvik said, "because nothing exciting is going on, there didn't seem to be any danger, any great risk here."

On reflection these remarks seem truly extraordinary. To describe the physical removal of a man's ailing heart and its replacement with a mechanical substitute as "routine," indeed, "just a day's work," expresses something of the power of medical technology over the modern imagination. Images of dramatic technological progress dominate our understanding of modern medicine (Reiser 1978). As a culture we are fascinated with the details of medicine's most recent miraculous advance. The limits of technology seem boundless. Although increasingly aware that progress sometimes occurs at significant cost, both social and economic, we await eagerly news of the latest test-tube baby or liver transplant.

What accounts for the rapid development and quick adoption of new equipment in western medical practice? "The landscape of modern health care is filled with machines" (Reiser and Anbar 1984:3). Clearly, some new medical procedures (vaccines, for example) offer such significant improvements over existing practices that there is little question about their use. More often, however, the true nature of new procedures and equipment is extremely difficult to evaluate. In an attempt to explain the seeming primacy of technology in modern medicine many health economists and policy analysts have postulated the existence of a "technologic imperative" in medical practice; these commentators believe that the mere existence of a dramatic new medical device provides a mandate for its continued use. Victor Fuchs, who first coined the phrase

465

M. Lock and D. R. Gordon (eds.), Biomedicine Examined, 465–496.
© *1988 by Kluwer Academic Publishers.*

"technologic imperative," attempted to account for this preference for the latest machine by discussing the power of "tradition" among physicians; he speculated that physicians had been "imprinted" during training to provide the best possible medical care, generally meaning the newest and most technological care (1968; 1974). The idea of a technological imperative is both powerful and captivating.

In this paper I will examine the social processes which contribute to the operation of a technological imperative in medical practice. I will be asking how people construct understandings of the complex technologies which enter their lives. My focus will be on the meaning of medical technology, specifically on the changed meaning which new machines develop as they are used. In order to explain the significance of this changed meaning I must define one additional idea, that of the "experiment-therapy continuum" in medicine. This concept, developed by Fox and Swazey (1978), speaks to an inherent tension in the use of new medical technologies. What is their meaning? Are they experiments or are they treatments? New technologies must traverse this continuum, changing from a status of pure experiment to the standard of care. I believe that this transformation of meaning is an inherently social process which sustains the technological imperative. Using a case study of one new technology, therapeutic plasma exchange, I will describe the social "routinization" of a technical medical procedure. Evidence of routinization is found in the varying social relationships found in "experimental" versus "standard" uses of this technology and in the creation and maintenance of treatment rituals. The new routines are sustained by physicians' overriding interest in the research applications of the technology.

This analysis focuses on the social "hows" of the technological imperative, leaving aside for the moment a complete analysis of cultural "why" questions. Many commentators have discussed the fascination with technology in western, particularly American culture. I take for granted this basic cultural infatuation with technology as well as the engineering "know-how" required to produce new medical miracles. My subject is the social processes which support the technological imperative.

In the case of medicine this underlying cultural preoccupation with mastery of the body via ever more sophisticated equipment is supported by important political and economic considerations. Both the system of financing and delivering health care in American society have favored the hospital as the setting for providing services and the use of high-technology devices. For example, American physicians are reimbursed for their services in a fashion which is dramatically skewed toward the use of machines and procedures as opposed to more cognitive services, such as interviewing patients or providing counseling about health behavior (Schroeder and Showstack 1979). Although of great relevance in furthering our understanding of the technological imperative, these issues will not be discussed in great detail here.[1]

The practical ramifications of the technological imperative are serious. New treatments commonly diffuse into widespread clinical practice before evidence is available about their actual usefulness. "Marginally useful, expensive technologies are developed, while unmet needs abound" (Banta 1983:1365). Examples can be cited in the areas of both surgical procedures and therapeutic equipment. Two particularly notorious cases are the eventually discredited but for a time widespread use of a "gastric freezing" machine during the fifties to treat stomach ulcers (Fineberg 1979) and the quick acceptance and extensive use of lobotomy for the treatment of mental illness in the thirties, forties, and fifties (Valenstein 1986). Hindsight makes it clear that these procedures were not only ineffective, but in many cases were actually harmful. Remarkably, the physician who developed the lobotomy procedure received the Nobel Prize for medicine in 1949, indicating the transient nature of medical orthodoxies.

Examples are not limited to the infamous or to the distant, non-scientific past. Many more recent innovations have been adopted and diffused widely prior to conclusive evidence of effectiveness. The efficacy of coronary care unit technology in improving the outcome of acute myocardial infarction has been questioned in numerous studies (reviewed in Waitzkin 1979). Likewise, electronic fetal monitoring, although now believed to be useful in the management of high-risk pregnancy, was widely adopted in the situation of normal deliveries despite evidence that it failed to improve the survival of low-risk infants (Banta and Thacker 1979). This list could be extended but the examples cited are sufficient to clarify the nature of the problem: once a new technology is developed, the forces favoring its adoption and continued use as a standard therapy are formidable.

SOCIAL STUDIES OF MEDICAL TECHNOLOGY

Traditional social science approaches to technology range from an extreme form of "technological determinism," in which technology is viewed as determinant of culture, to a view that the influence of technology can only be understood in terms of the meanings which people give to it. The reality, as is usual in the case of extreme formulations, is much more complex. Clearly, the use of technology cannot be independent of its social context. Especially in the case of medical technology, with its potential for evoking strong feelings carrying potent symbolic references to the body, life and death, the relationship between the machine as object and its user is multifaceted.

Despite the great potential which social studies of medical technology hold, Aiken and Freeman (1984) and Banta (1983) have documented the lack of interest in the science and technology of medicine among social scientists concerned with health. With the exception of the work of Fox and Swazey (1978) and Fox (1959), the field is barren. Earlier social science studies dealt with more narrowly defined questions, such as the study of medical innovation and diffusion (Greer 1977; Rogers 1981). These works were concerned with pragmatic topics such as identifying "barriers" to the diffusion of new techniques. Most importantly, early studies accepted the inherent value of new medical techniques without question (ibid.). In the past few years more critical studies have begun to appear (Plough 1986; Guillemin and Holmstrom 1986).

The relative lack of interest among social scientists is compensated for by a large literature in economics, health policy, and in the emerging field of health care technology assessment (McKinlay 1981, 1982; National Research Council 1979; Institute of Medicine 1985; McNeil and Cravalho 1982). This keen interest in technology use in medicine is accounted for primarily by concerns about the rising costs of medical care, although other issues, such as clinical safety, are also important considerations.

Anthropologists have long understood the absolute importance of social and cultural knowledge in comprehending the medical systems of their traditional subjects, tribal or peasant societies in the third world. Ethnomedicine has been a common focus of research. What has not been undertaken is a study of biomedicine itself. We have not taken "biomedicine as an ethnomedicine" (Gaines and Hahn 1982). The reasons for this are complex. Our own society and its sophisticated technologies, part of a powerful scientific paradigm, have been "off limits" to the kind of investigation routinely performed on the medical systems of other societies. Lock has stated the issues clearly:

Because contemporary medicine in industrialized societies is based upon a scientific foundation, certain assumptions have been made about this type of medical knowledge which have led to its exemption from social analyses. In the first place, it has been assumed that the biomedical model is *the* representation of reality, clearly not complete, but nevertheless slowly but surely moving towards a final explanation of the causes, diagnosis, and treatment of diseases (1984:121).

I begin with the assumption that western medical knowledge is as open to social and cultural interpretation as any other medical system (Comaroff 1982). As social studies of modern biomedical science have demonstrated, even research into the realm of "pure science" on topics such as the biochemical structure of hormones or genetic disease can be viewed as at least partially social constructions and not simply as biological fact uninformed by cultural considerations (Latour and Woolgar 1979; Yoxen 1982).

THE ROUTINIZATION OF TECHNOLOGY USE

In medicine a new therapy begins as an unusual, perhaps dramatic, but certainly out of the ordinary event. This is true whether the therapy is a new medication, a complicated surgical procedure, or an impressive machine-based technology. Although administering a new tablet is quite different than orchestrating a new cardiac surgery technique, they share the basic characteristic of novelty. The use of a new therapy upsets the everyday routines of clinical practice. As with the artificial heart story mentioned above, they are front page news.

Over time, however, this upset must resolve. The news of subsequent heart implants is buried in later newspaper pages and gradually recedes from public attention and notice, attracting no attention. Medical procedures which are not abandoned in the earliest stages of use eventually proceed to a state of "ordinariness." Their use is no longer perceived by hospital staff and patient as unique and new. They become the standard of care.

The core of my argument is that the meaning of a new biomedical technology changes as participants become habituated to its use. Through a social process which I will call simply "routinization," the meaning a technology holds for its participants becomes redefined. Routinization, of course, lies "at the very basis of social life itself" (Goody 1977:28). Social life would be impossible without the familiar patterns of the routine. ". . . Habitualization makes it unnecessary for each situation to be defined anew, step by step" (Berger and Luckman 1966:53-54). Since illness is by its very nature chaotic and unpredictable the fact that social settings in medicine tend to be highly routinized is not surprising.

I believe that it is through a process of social routinization that this changed meaning takes place. An initial "experimental" designation gives way to a new interpretation, that of "standard therapy." The meaning of the technology for the participants, either experiment or standard, is derived from the social setting itself. An experimental situation is drastically different from the everyday hospital routine.

As Barley (this volume) shows, the use of sophisticated new imaging devices in medicine demonstrates the importance of the social context in comprehending the effects of technology. Although the CT scanner clearly causes change in the social organization of the radiology departments in which they are employed, the scanner's impact is felt only through the constructed understanding of the equipment which is developed by the users as they struggle with a novel technique. The meaning of a new technology is not automatic, but evolves gradually. In the case of therapeutic plasma exchange I will argue that one key facet of this meaning – the placing of the machine along an axis of experimental versus standard therapy – occurs as participants in the use of the technology struggle with its application and gradually tame the machine through a process of routinizing its use in everyday practice. Originally, the CT scanner's "adoption by a local hospital [entailed] a sudden rent in the tissue of day to day experience" (Barley, ibid.). Before too long, however, the machine users had developed strategies to keep up at least "the appearance of normal operations" (ibid.).

A CASE STUDY OF THERAPEUTIC PLASMA EXCHANGE

An examination of the initial development of one recent medical technology, therapeutic plasma exchange (TPE), will illustrate the social forces supporting a technological imperative in medicine.[2] The case study is drawn from a two year investigation of TPE using ethnographic research methods. I conducted extensive observational fieldwork in eight cities in the U.S. and U.K. over the two year period, working in TPE treatment units, research facilities, and in related settings such as medical equipment manufacturing companies and scientific research meetings. I focused attention on hospitals and investigators who were known to be innovators in the TPE field. During the research I observed numerous TPE treatments as they were carried out, conducted a detailed review of the medical literature concerned with the development of the procedure, and formally interviewed many of the key participants in TPE research and treatment. Those interviewed included primarily research physicians, nurses, technicians, patients, and medical equipment manufacturers.[3] In the interviews participants were asked to describe their personal involvement in the development of TPE, their reactions to the machine, and views of the procedure and its effectiveness.

Although when first observed TPE appears intimidating and complicated, in reality the basic principles are quite simple. A large centrifuge machine (called a cell separator) is used to remove one part of the patient's blood, the liquid component, or plasma. This is replaced with new plasma, processed from donors. At the same time the patient's own red and white blood cells are returned. The machine itself appears daunting because of the large array of tubes, bottles, blood pumps and flow regulating devices necessary to accomplish the task of plasma exchange. TPE is used primarily for a class of diseases called the autoimmune disorders in which the body produces harmful substances, generally either immune complexes or antibodies to its own tissues. Disorders which have been treated with TPE include myasthenia gravis, systemic lupus erythematosus, rapidly progressive glomerulonephritis, multiple sclerosis, rheumatoid arthritis, plus many others, some obscure and some quite common. The frequency of treatment varies in different diseases with patients receiving TPE daily for some forms of acute renal disease or monthly for neurological conditions. Therapy is generally not as intensive or long term as renal dialysis. The antibodies (or other harmful substances) collect in the plasma and are removed by the machine along with the old plasma. The new, "clean" plasma is free of disease causing proteins.

The innovation of TPE provides an instructive case through which to examine the social and cultural elements sustaining a technological imperative in medicine. The procedure, used for serious and debilitating disorders for which few treatments are available, is costly and dramatic. Yet at the same time its actual contribution to the treatment regimes of various autoimmune diseases is very difficult to evaluate. The high cost comes from the expense of plasma substitutes, the cost of the machine and disposables, and the labor intensive nature of the therapy. Estimates of the cost *per procedure* range from $400 to $1500 (ECRI 1985) with patients often requiring many treatments over a prolonged time period. Although some of the disorders for which TPE is used are rare, others, such as rheumatoid arthritis, are quite common and thus the potential impact on the health care system as a whole is significant. The dramatic nature of TPE is difficult to describe in words; one must imagine a patient connected to a large and complicated looking machine by means of which the patient's blood is removed by circulating it through many feet of tubes. A number of medical personnel are in constant attendance, checking the machine, the patient, and adding bottles or bags of new plasma. One physician informant described TPE as "the ultimate Walter Mitty-like experience. A patient's lying there in bed and attendants move levers back and forth . . . the machine goes toponka, toponka, toponka . . . It's all a sort of medical fantasy."

In spite of its dramatic appearance the efficacy of TPE and its ultimate place in medical therapy remain unknown (AMA 1985; Shumak and Rock 1984). Early in the development of TPE its use increased very rapidly.[4] Although it is impossible to determine the exact number of TPE procedures performed in the early stages, it is possible to get an indirect estimate of its use by examining the number of articles dealing with TPE published in the world medical literature. As can be seen, the number of articles about TPE increased geometrically during the study period (see Figure 1).

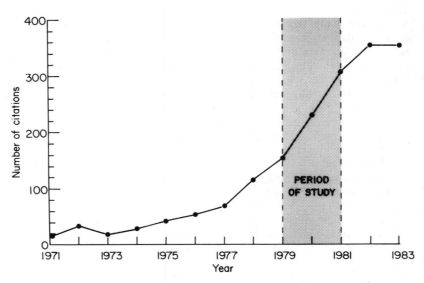

Fig. 1. Case reports and articles discussing TPE in the medical literature[5]

It is important to emphasize that this increase in use occured *prior* to the availability of "scientific" information about TPE's effectiveness for the many diseases treated. The scientific evaluation of efficacy is enormously difficult even under the best of circumstances. In the case of TPE this difficulty was confounded by the fact that many of the diseases treated, such as multiple sclerosis with its characteristic natural remissions and exacerbations, make evaluation notoriously difficult. In addition, many of the diseases treated were believed theoretically to be of autoimmune origins, but the exact noxious agent in the plasma was unknown. Because of the lack of clearly defined medical indications for the procedure it provides an ideal setting in which to study the social and cultural forces at work in the technological imperative.

Contrasting Views: "Experimental" versus "Routine" Treatment Settings

During fieldwork I was able to observe a "natural experiment," the introduction and early clinical trials of a new type of TPE equipment. Although it accomplished the same task as the earlier model, this equipment was based on a different principle: filtration of the blood rather than centrifugation. As I observed this equipment in use it immediately became apparent that something out of the ordinary was happening. An examination of the first use of this new procedure provides an illustration of the experimental treatment setting.

First of all, the "newness" of the machine was celebrated by a constant string of visitors; each person who entered the treatment room focused immediate visual attention on the machine itself:

They always asked technical questions and were given an elaborate explanation. The visitors seemed intrigued with the technology. One physician came in, asked a question, and received a detailed description of the size of the pores in the filter membrane (fieldnotes).

This picture of a circle of white-coated figures gathered around the machine, attracted as to a fire on a cold night, was repeated over and over in the course of my observations. Second, many of these visitors were physicians. In my experiences with the use of TPE in everyday treatment settings, physicians were rarely present as the therapy was being administered. By contrast, with the new machine it was physicians who were most intimately involved with both setting up and running the novel device.

The physician primarily involved, Dr. Parker (all proper names used here are pseudonyms), was constantly in attendance. The new machine was to be used daily for three days, including the weekend, when no nursing assistance was available. A patient with an unusual and damaging blood protein was the subject. The tone of the procedure was one of moment by moment uncertainty and hesitation, beginning with setting up. Dr. Parker had to negotiate with a nurse from the kidney dialysis department (experienced in the extracorporeal manipulation of blood) in order to get help in setting up the system. "A number of trips had to be made back and forth to the dialysis unit, which was as a distance away . . . [and] upstairs to the lab for filters and notebooks. They kept needing additional little bits of equipment that they had forgotten or didn't know they would need" (fieldnotes). After a struggle that lasted for a number of hours (and was repeated to some extent each day the machine was used) the treatment got under way.

This general level of confusion continued throughout the treatment. Every decision point was problematic and required additional work and negotiation. Dr. Parker was on hands and knees with the nursing staff (or when they were not present, with the anthropologist) piecing together bits of equipment and drawing up medication in syringes. Also present throughout was a representative of the manufacturer of the new machine. The representative removed his jacket and gave "hands-on" assistance with the equipment. He helped collect samples of the patient's blood every few minutes in order to gather data on the machine's efficiency in removing the suspect protein. He loaned the medical team a special infusion pump to make the procedure easier and made numerous telephone calls to gather needed information on technical questions, such as the proper amount of anti-coagulant medication needed. Dr. Parker attempted to reduce the tension level in the unit by joking, "This could be done by one person *if* they knew what they were doing."

Eventually the team managed to complete the series of treatments in spite of major complications during one procedure when the patient's blood clotted in the filter. In summarizing the three days of treatment I noted:

Dr. Parker kept repeating throughout that he didn't know what he was doing. The treatment room was in a constant state of chaos, with staff members laughing, IV poles falling to the ground and physicians running in and out of the room. How the patient managed to remain calm is a mystery. When the procedures were finished the room looked like a deserted battle field. There were empty boxes everywhere, the bed was covered with blood and there was spilled plasma all over the floor (fieldnotes).

The social scene was the complete opposite of the calm and orderliness characteristic of a "routine" TPE procedure. You had only to walk into the room to discern that something out of the ordinary was going on.

The contrasting images of a "routinized" treatment setting are strikingly different. The most powerful visual image of routinization is a picture of the plasma exchange nurse sitting calmly at the machine controls which have only half of her attention, the other half devoted to the morning newspaper. All the needed supplies are available; the equipment is set up and ready to go. The nurse has spent over an hour carefully preparing the machine and other equipment before the patient arrives. This orderly atmosphere was summed up by a patient who said, "It's almost an assembly line." Another veteran patient who had been receiving treatment for a number of years noted the great differences between the TPE unit "in the old days" and currently. Whereas previously she had felt like, "an oddity" while receiving treatment, she now described her TPE therapy as "almost like a tablet." By defining her TPE therapy as feeling as routine as swallowing an aspirin for a headache, this patient graphically illustrates the changed tone present on the unit.

Although every unit has busy and chaotic days when crisis follows upon crisis, especially when acutely ill patients are treated, the overall tone of the TPE unit that is well established is one of calm and orderliness. The "assembly line" feeling is pervasive. Patients come and go, supplies are ordered, routine blood tests are carried out, all managed efficiently by TPE nurses. Nurses dominate the setting. Their interaction with the equipment itself is matter-of-fact, embedded in the everyday hospital routines rather than set apart as in the scene described above when

Dr. Parker was in charge. The following observations are exemplary: "Edith read the *Guardian* and did paperwork during the exchange . . . The nurse heated her lunch in the microwave and ate while seated at the machine controls" (fieldnotes).

But this seemingly calm exterior can be deceptive. In order to maintain an appearance of the routine in the face of the realities of clinical practice in medicine, that is the constant problems of applying a technology to particular human beings, a major activity of the nurse on a TPE unit was that of "managing trouble." It is important to note that the underlying chaos does not disappear completely as a new therapy gradually loses its experimental designation. Rather, by managing trouble the nurse shields the physician investigator from all but the most serious kinds of aggravation (or "aggro" as the English nurses called it). The nurse makes sure that the healing rites of TPE are always performed in the "correct" fashion.

Although an exact estimate of the frequency of untoward events is impossible, information from my own observational data plus records of TPE procedures kept by some treatment units indicate that they are an everyday occurence. Numerous fieldnote entries include an example of a nurse coping with a disaster or smoothing over a procedural difficulty. In one unit over a two week period I noted that one out of three treatments (approximately one per day) was complicated by some difficulty. Although problems with vascular access were the most common, other difficulties include failed transport arrangements for patients, machine failures, or shortage of supplies, including plasma replacements. Some problems are predictable in that they occur regularly (even if this is often not acknowledged by the staff). Other problems are more unique, such as an entire shipment of plasma freezing -- and thus becoming unusable -- during a January blizzard in the American midwest. The following example is illustrative, describing a procedure that was troublesome from the beginning when staff had difficulty getting access to the patient's veins.

From here on things got even worse. There was a serious problem with the machine and blood leaked all over. It spurted all over the nurses, the machine itself, and the floor; it was a terrible mess. A physician looked in the room incredulously, made a face, and left. There was blood everywhere . . . Eventually the nurses were able to get control of the situation and clamp off tubes, etc. They then dissolved in almost uncontrollable laughter (fieldnotes).

This scene was reenacted over and over: sometimes air would mysteriously enter the system, sometimes the patient's blood would clot in the centrifuge bowl, occasionally the blood flow from the patient would be so slow that the procedure would drag on for hours longer than expected. Nurses were constantly engaged with controlling the effects of untoward events.

SOCIAL RELATIONS AND THE DIVISION OF LABOR

On the surface the social *milieu* of the TPE treatment episodes I have described appears alike; confusion and disorganization seem the order of the day. Yet as one examines the social relationships among key actors, these similarities disappear. Significant differences exist in the actual division of labor among TPE participants as well as in the hierarchical ordering of relationships. The changing roles of physicians, nurses, patients, and equipment manufacturers are revealing. These differences are crucial in interpreting each actor's understanding of the meaning of the TPE therapy. Of particular importance is the meaning of the therapy for physicians because of their dominant position in making decisions about the use of TPE.

Perhaps the most important change which occurs during routinization is the change in who actually *performs* the TPE procedure. During the experimental, uncertain stage of development physicians are in constant attendance, struggling with the new procedure and working alongside the other staff. A pioneering TPE patient told me, "Dr. Daniels used to sit by the machine with a book trying to figure out what to do next." In fact, most of the physicians who were pioneers in developing TPE were expert technicians, often conducting all treatments unaided. Nursing staff familiar only with the routinized stage of TPE's development expressed amazement that physicians had once done all the work themselves. The nurses' incredulous reactions speak to the fact that in the routinized treatment setting actions and patterns appear inevitable.

By contrast, the later, non-experimental stages of TPE's development are dominated by nurses. Notable was the physician fleeing from the room as trouble began in the scene described above. One nurse claimed the physician she worked with had never been in the plasma exchange room. Others complained of the lack of interest of physicians in the routinized treatment setting, stating that they were no longer helpful or willing to explain things. Whereas when the machine is new all eyes focus upon it, in the later stages a request for help is met with the physician's uninterested reply, "It's not my patient." Thus, important signs of routinization are the changes in the actors' level of involvement and in who actually performs the procedure.

Another sign of increasing routinization can be seen in the type of relationship which exists between physicians and nurses. Early on, in the experimental, non-routinized phase of TPE's development, nurses and physicians act as partners in the research endeavor. Tasks are shared, not rigidly divided. A nurse might appear as the co-author of a scientific paper. As time passes, however, these egalitarian relationships are transformed into the more traditional nurse/doctor relationship, hierarchical in nature. In this phase the physician simply "orders" the procedure which is then carried out by the nursing staff. Trouble, which is shared mutually in the experimental phase, becomes in the later stage an annoyance to be handled by the nurses. It is almost as if the experimental tone of the early phase allows a violation of the traditional superordinate/subordinate relationship. With increasing routinization the barriers to egalitarian cooperation go back up. No longer research partners working within the same uncertain treatment setting, the actors are now clearly identifiable as nurses who skillfully carry out a procedure (and manage trouble) and physicians who simply issue orders. Strauss and his co-workers also found changes in work relationships among professionals in their study of the use of medical equipment. They report that, "When the technology is quite new, especially when a new unit is being set up that embodies novel technology, then there is more likelihood of a blurring in the division of labor" (1985:157).

Other key actors, primarily patients and equipment manufacturers, are also involved in the transformation of roles and responsibilities which occurs as a treatment becomes routinized. Patients play a role in creating the "standard therapy" meaning of TPE as it develops in individual treatment units.[6] In highly experimental settings patients are much more likely to be treated as partners in the research endeavor rather than as passive recipients of treatment. When the patients' TPE procedure tended toward the new or unusual, their treatment by hospital staff members was much more casual and informal, more collegial. The patients were known as individuals, often called by first names. Their social distance from staff members was negligible. Indeed, some clinicians recalled their early use of the procedure in terms of specific patients and their treatment. The patients who had been involved in important "firsts" were remembered in great detail. A new therapy was recalled as, "the night that Mary almost died." Likewise the patients are very much involved in and aware of the research being conducted by their doctors. The patient "subject" in the experimental scene described earlier asked many questions which were answered seriously by the physicians. Fox found a similar relationship between pioneering patients and physicians in the early years of experimentation with kidney transplantation and related therapies (1959). She described patients being treated almost as if they were professional equals, calling them "patient-colleagues" (Fox 1959:89).

As the patient's social relationships with staff alter there also occurs a change in the patient's involvement in the actual TPE procedure. As routinization progresses they develop standardized ways of participating in the treatment itself. As they became familiar with the procedure for doing TPE, patients (who were not dangerously ill) began to play a role in monitoring the TPE equipment. Their role in watching the equipment occured primarily at "dangerous" moments in the treatment process.

One of these moments occurs when it is necessary to add a new bottle of plasma replacement fluid as the previous bottle is emptied. If not timed properly air can enter the system, creating the potential for a life-threatening bolus of air to the patient or a major delay in the treatment. After "learning" the machine's functions some patients created a role for themselves by monitoring the level of fluid remaining in the bottle or bag. The patient would watch the bottle carefully and then signal the nurse or technician when it was time to hang a new bottle. This was often done while marking the progress of the treatment by talking about which "bottle" you were on. Hanging the last bottle meant the treatment was almost over.

Another dangerous moment occurs at the very end of treatment (with one variety of TPE machine). It is necessary to return the patient's own blood from the centrifuge bowl as well as the blood remaining in the elaborate series of connecting tubing. One method of doing this is to flush the blood back to the patient by allowing air to enter the tubing from the machine side to force the blood back to the patient. This is an inherently dangerous procedure because of the risk of inadvertently infusing a large air embolus. Because of the danger it is essential that the tubing be clamped before the air nears the patient's vein. The nurse or technician usually stands at the patient's side holding a large pair of clamps. In routinized treatment settings, I observed patients take part in this procedure by carefully scrutinizing the exact position of the air in the tubing and signalling to the nurse when to clamp the tubing off. This patient initiated routine is accompanied by a great deal of stylized joking behavior, reflecting the high level of tension present at this critical moment in the treatment. A staff member joked that if a patient "misbehaved" the nurse caring for the patient should "give him the air." The jokes disguise the tension felt by both patients and staff. They become part of the ritual of ending a treatment. The process of patients developing roles in the actual technical procedures of TPE is an important part of routinization.

Similarly, the role of the medical equipment manufacturer is very different as routinization progresses. As described above, in experimental settings it was not uncommon to see the company representative, with shirt sleeves rolled up, working alongside the professionals, solving problems, taking samples from the machine or assisting with calculations. The company representative present during the early stages of development is likely to function as an equal partner, actively working with the clinicians to solve an engineering difficulty. Role relationships are highly reciprocal at this point, with each side providing needed input into machine design and specification. Social relationships between physicians and representatives were quite egalitarian; first names were often used and, most significantly, representatives had easy access to the clinicians. Company representatives jokingly referred to this phase of development as the "cocktail napkin design" period, alluding to designs sketched out over drinks in a pub or restaurant near the hospital. All parties need each other; the exchange of ideas, data, equipment, and technical support sustain the relationship.

However, as with the easy relationship between nurses and physicians which is characteristic of the first stages of machine development, the nature of the relationship between clinicians and company representatives is also transformed over time. The reciprocity (based on mutual need) which typifies the experimental stage gives way to a more rigid and hierarchical set of social relationships. The social distance between physicians and manufacturers increases; company phone calls are not returned and representatives are shunted off to lower level hospital functionaries, such as purchasing agents or technical staff. As the machine is used more frequently, social barriers based on the established differences in the hierarchical position of representatives and clinical investigators return. At this time a likely pose for the company representative was patiently (at least in appearance) waiting outside the physician's office for a moment in which to extoll some new feature of the company's equipment. At this point the representative was likely to be mistaken for a "detail man" from a drug company, giving out free samples and engaging in classic business marketing techniques. By this stage of development the TPE machine is an accepted part of the hospital environment, described by one physician as "part of the furniture."

THE DEVELOPMENT OF WARD RITUALS

At the heart of the routinization process is the creation and maintenance of treatment rituals. Traditional anthropological explorations of ritual have focused on the religious/magical referrents of collective ceremonies (Moore and Myerhoff 1977). By definition, rituals are viewed as separate from their "technological consequences" (Helman 1984:123). Although discussing the place of ritual in the everyday world of medical work may seem incongruous, in reality the distance between ritual and technical task may not be great. In modern medical practice, the division between ritual and technical aspects of healing is not absolute; the two are often interwoven (ibid.:34). Turner reminds us that, "In tribal and archaic societies what people do in ritual is often described by terms which we might translate as 'work' . . . the ritual round in tribal societies is embedded in the total round of activities . . ." (1977:39).

The concept of "secular ritual" has been employed in the analysis of many non-religious settings (Moore and Myerhoff 1977; Gusfield and Michalowicz 1984). Some commentators have objected to this broadened scope of the concept of ritual, claiming that extending it to the description of everyday, secular activities makes it so all-inclusive as to become meaningless (Goody 1977). I use the concept here because it helps to point out the importance of purely social elements of behavior in a highly technical setting.

The minute by minute technical tasks involved in carrying out a TPE procedure are the substance from which the rituals of a "standard" medical procedure are created. An accepted function of ritual is "traditionalizing new material" (Moore and Myerhoff 1977). The development of ward rituals in TPE treatment units contributes to the changed meaning of the therapy; these rituals create order and a sense of certainty where none had existed before. I observed two types of rituals in operation: the actual *tasks* of carrying out the TPE procedure were turned into highly structured series of actions, repeated in exactly the same manner each time a treatment was carried out. Likewise, the *social interaction* surrounding the treatment developed ritualistic features; activities and verbal exchanges between patients and staff were highly patterned. The key elements of the treatment process which were subject to ritual formation were the tasks of setting up the equipment and getting the patient "on" the machine, actually carrying out the repetitive tasks associated with conducting the plasma exchange, and finally, removing the patient from the machine and restoring the patient to normalcy.

The existence of idiosyncratic treatment rituals became evident in the course of comparing TPE treatment across different settings. There were innumerable variations, both major and minor, in the way the procedure was carried out. What was constant was the rigid way in which the procedure was always done in the same manner in a particular unit. Some variation can be accounted for by the use of different models or types of equipment. However, even when the same machine was used in different units there was variation in the configuration of the I.V. tubing, the type of replacement fluids used, the way records were kept, and the way the patient was monitored. In one unit, for example, all the bottles of replacement fluid were carefully prepared (with medications added) and placed on a windowsill before the patient arrived for treatment. This procedure was invariably followed even though it occasionally meant the waste of hundreds of dollars worth of plasma if the treatment had to be cancelled. One might assume that these highly standardized routines were made necessary by technical requirement of the procedure itself. However, the significant variation among treatment settings and the rigorous adherence to routines suggest that something beyond the pragmatic is in operation.

A curious "starting up" ritual had evolved in one unit. In this highly routinized setting the physicians were only present at the very start of a procedure, when they were called to perform the venipuncture and connect the patient to the machine. After gaining access to the patient's vein the physician would wait until the machine had been "primed" with normal saline solution before doing the final connecting. A by-product of the priming procedure was the accumulation of excess saline in a waste bag. In this unit, after completing the task of hooking the patient up to the machine, the physician would invariably pick up the waste bag filled with saline and toss it across the room, attempting to make a successful "basket" into the sink. When questioned about this curious behavior – which they sometimes even half-jokingly referred to as a "ritual" – one physician laughed and stated that it was necessary to make a good throw in order to guarantee a successful, speedy procedure.[7] Other starting-up rituals evolved around the difficulties of connecting the patient to the machine, since, as discussed above, vascular access is often problematic. Asking the patient to soak his or her arm in a bucket of hot water immediately before treatment was a well established procedure. Although theoretically based on physiological principle, e.g. heat will cause veins to dilate, I did not see any observable benefit from this procedure other than allowing the participants to feel that they were doing everything possible to get through the difficult task of hooking up the patient to the machine. The complexities of "connecting work" are discussed in Strauss, et al. (1985).

Another unit had developed a particular "ending" ritual. As the last bottle of replacement fluid was administered, the nurse would send someone to the kitchen to make the patient a cup of tea. As the treatment was ending the patients would be served tea and biscuits as they recovered and rested quietly before going back to normal activities. The cup of tea marked an important transition: the TPE recipient moved from the status of patient, utterly dependent on the machine and staff to fulfill his every want, to an independent status, able to freely move around the unit, no longer under the direct control of the staff.

The middle period of the TPE treatment, which ranged in length from two to over four hours, was also subject to highly patterned behavior and the creation of small rituals. Many of the activities required of the staff in operating the equipment are highly repetitive, so some of the need for regularization is pragmatic. These types of activities include gathering and recording information about the patient's condition and the machine's functioning. However, other behavior was also subject to routinization. In some units, the verbal exchanges between patients and staff were highly patterned. For example, in one treatment setting there would invariably be an elaborate discussion, with comments from all present, about how many TPE treatments a patient had undergone. A patient would comment, "Today is my 105th plasma exchange." This statement would be seconded by a staff member who would exclaim that the patient had undergone an impressive number of procedures. Similarly, in another unit the patient and staff would always begin a treatment by making an informal wager about when the treatment would end, a subject of some importance because if a procedure was speedy it usually meant that it was painless and free of technical trouble, also that the patient and staff could leave the hospital.

Rituals in medicine serve many useful functions (Helman 1984:123-140; Bosk 1980). "Rituals may disguise realities, portray fictions, save face, and convince all parties that matters are in order and in their own control" (Myerhoff 1977:217). TPE treatment rituals, those actions which elaborate and embellish the technical task of plasma exchange, first normalize and then stabilize the meaning of TPE as an accepted, taken for granted therapy. Performing the TPE treatment in a particular fashion, in the same way over and over again, fulfills some vital ritual functions. Perhaps of most importance, the treatment rituals I have described function to reduce the omnipresent uncertainty of clinical encounters.

When a new therapy is used key questions remain unanswered: Is it safe? Will there be unexpected side effects? Will the machine perform as expected? And, of primary importance, will the treatment ameliorate the patient's disease process? Procedural routines and rituals reduce this uncertainity. They disguise the reality that the patient's condition might not be treatable with any known method, and that the therapy is only a chance which may or may not work. Likewise, the rituals of treatment allow the professionals to believe that they are in control of the patient's debilitating disease, that they have some power, can take some action against its ravaging effects. Patients' fears of the procedure are minimized if the treatment is performed in a competent and crisply efficient manner by professionals who carry out their tasks with authority and seeming security. The "starting up" rituals described above address this uncertainty directly. These actions are like exhortations to "protect" the participants -- patient and staff alike -- from the potential misfortunes of treatment. The "ending" ritual symbolizes the patient's safe transition away from the danger of the machine and back to ordinary life.

With the high degree of clinical uncertainty inherent in the work of medicine and nursing, the need to create a situation of "normal operations" as Barley (this volume) has described it, is acute. The disorder and lack of routines characteristic of the experimental setting cannot be tolerated for long. The social aberrations of the experimental setting, for example the temporary dissolution of normal hierarchical relationships, must be resolved. As routine procedures are developed and treatment rituals evolve, a new social cohesiveness forms among the staff. The inevitability of the new treatment, and the commitment of the staff to continue treating this patient, is celebrated by the ritualized discussion of the large numbers of procedures which the patient has undergone. In spite of the high level of excitement generated by the use of new therapies, over the long term it creates too much anxiety and disorder for patients and staff alike to be constantly faced with situations for which predetermined actions do not exist. Treatment rituals develop to fill this void and help create a climate of certainty.

As discussed above, the nurse's role as manager of clinical trouble is crucial to the process of routinization. Likewise, as ward rituals develop and solidify into everyday events, the nurse, fulfilling the role of "ritual specialist," is of paramount importance. It is the nursing staff who manage and guard these ward routines, making sure that they are carried out in the "correct" fashion and passing on the idiosyncratic ritual knowledge of the TPE unit to newly hired staff. After their central involvement during the experimental stage, physicians disappear into the background. As physician investigators search for the next new therapy, nurses run the unit, manage trouble, and carry out the therapeutic rituals of TPE.

PHYSICIANS AND TPE: RESEARCH OR THERAPY?

Nurses contribute to the changed meaning of TPE by "managing the trouble" of carrying out the procedure in clinical settings. Physicians, shielded from these more mundane tasks, are thus able to concentrate on other issues. The issue of primary importance to the physicians I interviewed was research. It became clear very early in my study of TPE that the meaning which different actors derive from their work with TPE varied greatly. Patients, nurses, and industry representatives viewed their work as primarily oriented toward the use of TPE as a *treatment* for individual patients. Physicians, on the other hand, were much more oriented toward the use of TPE as a method of advancing *basic research* in the biomedical sciences.[8] Although the meaning of the TPE experience for non-physicians is also of importance in understanding the technological imperative, it will not be discussed in detail here. The physicians' perspective is of most importance in explicating the shift in meaning of a new therapy from experimental to standard therapy. This importance derives from their position of economic primacy in determining patterns of resource utilization (Relman 1980) as well as from their cultural authority (Starr 1982). Physician emphasis on research adds a dynamic element to the routinization process. As each therapy moves toward acceptance, physicians change their focus to something new. The division of labor among health professionals allows physicians to devote themselves more exclusively to the investigative elements of their work with new technologies.

An indication of the importance of research in the meaning ascribed to TPE by physicians comes from examining the actual language clinical investigators used in discussing their work. One physician, involved in using TPE from a very early date in its development, said in a published review article: "Plasma exchange has recently been used in the *investigation* and treatment of immunologically mediated disease [emphasis added]" (Pinching 1978). Later in the same article the author described TPE as "a valuable clinical research and therapeutic tool" (ibid.), clearly revealing the dual meaning of TPE. The title of another article is even more explicit in defining TPE as a research tool. The physician called the paper, "Function of Circulating Antibody to Acetylcholine Receptor in Myasthenia Gravis: Investigation by Plasma Exchange."

Noteworthy in these illustrations is the clear indication that one "investigates" a disease using plasma exchange. TPE is not simply a therapy but a means to learn more about a disease process. In both instances quoted above the research application of TPE *precedes* the discussion of its therapeutic use. This ordering reflects the values of the physicians interviewed. When asked to discuss their involvement with TPE, physicians invariably began by talking about their research interests. Research is a key part of their thoughts about their work, even when patient treatment is an offshoot of the research activity.

Therapeutic efforts may be useful in elucidating the cause of a disease or in furthering scientific understanding of a perplexing complex of symptoms. When TPE was first used in connection with myasthenia gravis (a debilitating disease of extreme muscle weakness) the exact pathophysiology of the ailment remained unclear. Scientists had postulated the existence of an auto-antibody in the plasma which interfered with transmission of nerve impulses, hence causing muscle weakness. When the first few patients with myasthenia gravis actually became stronger after having plasma exchange, this fact in itself became an element in building the scientific case for documenting the etiology of the disease. More recently, the use of TPE in treating thrombotic thrombocytopenic purpura (TTP) provides another illustration of how therapy itself may help "solve" the mysteries of a disease. The disorder, a life-threatening defect of the blood coagulation system, is often successfully treated with TPE. And yet, as an editorial in the *New England Journal of Medicine* observes, "it is not yet clear why plasma transfusion and exchange, which have become the cornerstones of treatment, are beneficial" (Aster 1985:986). The editorialist speculates that the disease may turn out to have multiple causes. The implicit assumption is that therapeutic efforts may themselves help sort out the true nature of TTP. He states, "It is to be hoped that studies aimed at defining the mechanism by which plasma exerts its apparently life-saving effects in some patients will lead to a fuller or perhaps even complete understanding of this remarkable syndrome" (ibid.).

Most physicians are cognizant of this overlap between research and therapy, realizing that the two activities are constantly muddled together in the real world of everyday practice. It is the primacy of research goals which supports a technological imperative in medicine. A physician who supervised a large number of patients undergoing TPE described his major purpose as learning the *cause* of auto-antibody mediated disease. Although in reality TPE is used for both therapy and research, the division of labor which occurs as routinization progresses frees physicians from everyday involvement and allows them to focus primarily on research into the etiology of disease or on the development of another generation of medical equipment. The career structure of academic medicine, with its emphasis on new discovery, further supports this dynamic.

For the physicians, even actual clinical encounters with patients and troublesome machines sustain the meaning of TPE as research. While the patients and technicians remain in the chaos of the experimental treatment setting, the physician is able to escape to the world of "the lab" and "data." Even in the chaotic experimental treatment situation described above the end result for the physician, at work in the lab a few weeks later, is far removed from the scene described. When this physician was interviewed he enthusiastically displayed the data obtained from the treatment. The actual experience of treating the patient had been turned into a series of graphs and charts showing the machine's performance and the changes in levels of substances in the patient's plasma measured against clinical improvement. The entire event was neatly recorded and graphically represented, reduced to a few pieces of paper and in the process "transformed." The data, representing the fruits of research, remain long after the patient has gone home and the scene of the treatment has returned to normal. The data *become* the meaningful reality. For the physician, routinization occurs as clinical encounters are turned into research results. The imperative to make use of new machines derives, at least in part, from the equipment's data-generating capabilities.

A MORAL IMPERATIVE FOR TREATMENT

The technological imperative is sustained by inherently social forces which result in a new meaning for TPE: the meaning of standard therapy. New routines are created, rituals develop, and physicians continue their focus on research. The most interesting question is the relationship between the changed meaning of a new therapy and the nature of medical decision-making by individual physicians for patients who might benefit from a new machine. Or, how does the technological imperative become transformed into a moral imperative to provide a new therapy? The notion of an imperative implies constrained choice. I would argue that a moral imperative to provide treatment is experienced by physicians when they are faced with a decision about whether to prescribe a therapy which has begun to "feel" routine. The moral tone derives from the sense of social certainty experienced by health professionals. The standard of care becomes a moral, as well as a technical, obligation.

When a therapy clearly belongs in the experimental camp – and everything about the treatment situation reveals this to the staff – there is no obligation to provide it. Yet when a new procedure has crossed over the mysterious boundary into the territory of a standard therapy, it cannot be denied. As the use of TPE became ever more routinized, the clinicians I observed would break their own rules of scientific excellence, often providing treatment to patients not part of established controlled trials or in the face of contradictory evidence.

As with TPE, a new treatment may or may not be efficacious; it might be risky. The moral imperative for treatment overrides these concerns. It becomes unthinkable for the physicians not to perform the treatment. The social inevitability of therapy takes on a moral tone; the experience of a technological imperative becomes a moral imperative for action. Once a new therapy is available it becomes extremely difficult, if not impossible, to forego its use. Experimental procedures, those which upset the everyday hospital routine, are not morally indicated. Their use is not mandated. It remains optional, at least until a later stage of development when the social forces of the technological imperative have resulted in a changed meaning.

Hence, the creation of a moral imperative is a social process, the end result of the routinization and consequent acceptance of a new medical technology. The moral meaning of a technology, its perception as a standard therapy, is embedded in and expressed through changes in social organization. I do not mean to suggest that the ongoing scientific evaluation of new medical technologies is unimportant. Rather, I believe that scientific "facts" about efficacy are informed by a careful reading of the social *milieu* in which treatment takes place. The evaluation of a new technique as a standard therapy derives from a complicated "reading" of the social setting in which it is used and not a simple assessment of the results.

IMPLICATIONS: TECHNOLOGY IN HEALTH POLICY AND BIOETHICS

The key element in this analysis is the suggestion that even in a highly rational, scientific setting the meaning of actions and events evolves, at least in part, from the underlying social and cultural organization. The meaning of a new technology as standard therapy crystallizes, over time, from many sources. Scientific studies of effectiveness, assessments of economic costs and benefits, and political considerations all have their place. Although often ignored by policy analysts, an understanding of the social context of technological innovation can inform our knowledge of the meaning of new technologies. There are a number of significant practical ramifications of the analysis presented here. The power of the technological imperative in medical practice, particularly the strong tendency for routinization to speed the perceived meaning of a new therapy from experimental to standard therapy, has broad implications for the fields of health policy and bioethics. Two major implications of the technological imperative in medical practice will be discussed: (1) the effect of a technological imperative on policy debates about the appropriate use of costly health care technologies, and (2) the challenges of preserving individual patient autonomy in the face of a moral imperative to make use of new techniques.

First, the speed at which expensive, new therapies can gain the label of standard therapy is of major concern in the area of technology assessment. How best to guide and control the growth of medical technology is debated extensively in the health policy arena in both the U.S. and the U.K. (Banta et al. 1983; Council for Science and Society 1982; Jennett 1985; McKinlay 1982). Formal technology assessment programs are increasingly called for as a means of controlling the use of dangerous or marginally effective new technologies (Institute of Medicine 1985). One assumption behind the idea of conducting rigorous evaluations of new technologies early in their development is that new machines or techniques which are inefficacious or unsafe can be easily eliminated from the clinician's technical repertoire. Ironically, the very process generally considered necessary to evaluate new therapies in a rigorous and scientific way, that is, the use of randomized controlled clinical trials conducted in a number of medical centers, itself contributes to the social process of routinization. It should be fairly obvious that the act of setting up clinical trials can (and often does) begin the process of routinization. Once a new machine is in use, even if in a limited way, it is very difficult to change course and stop using the machine; its use becomes entrenched.

Plans to control the rapid proliferation of marginally effective, expensive technologies by resorting to "better science" -- usually meaning more elaborate clinical trials -- are seriously flawed because they ignore (or discount) the social forces which support the technological imperative. In some cases, for example in lobotomy, historical analysis reveals that the evidence of the new treatment's lack of effectiveness was available very early (Valenstein 1986). Unfortunately, this did not hinder the wide use of lobotomy in treating mental disorders. The procedure was not abandoned until it was supplanted by another new therapy -- chlorpromazine, the first psychoactive drug. Valenstein cautions that the social and political forces responsible for the wide acceptance of lobotomy are not behind us, but are "part of the bone and marrow of the practice of . . . medicine" (1986:291).

In the case of TPE there is also evidence of uncritical acceptance of early reports of treatment success. Searching for an explanation of this phenomenon, one recent analysis suggests that the problem of overly enthusiastic endorsements of new technologies could be solved if editors of scientific journals would be more circumspect about publishing conclusions that go beyond the scientific evidence presented (Shapiro and Shapiro 1987). Although laudable, calls for more rigorous science ignore the fact that many of the influences on the meaning of a new technology are outside the realm of scientific "facts." Also ignored is the fact that the ultimate decisions about appropriate use of medical technologies are social and political -- as with U.S. government funding of treatment for end-stage renal disease (Plough 1986) -- and not strictly medical.

Most of the recent political debate about technology use in health care has revolved around issues of economic cost. The adoption of a new technique as a standard therapy has a significant and direct impact on the overall cost of health services. Most American insurance companies and government programs provide reimbursement for health services based on the status of the service. A procedure which is considered to be the standard of care must generally be reimbursed while experimental procedures are specifically excluded. Hence the meaning of a new treatment, its perception as experimental or standard therapy, has serious economic consequences. Who determines this meaning? Often, a panel of experts is assembled. The expert physicians called upon to evaluate the status of a new technology, generally those using the new technique themselves, are in an inherent position of conflict of interest and highly likely to perceive the equipment as essential (McKinlay 1981). Using a new therapy contributes to its acceptance as a necessary part of routine treatment. An understanding of these physicians' involvement in the social process of routinization might lead to improved decision-making about reimbursement.

A final policy concern is the need to choose between competing demands for limited resources. There is general agreement that the dramatic advances in sophisticated diagnostic and therapeutic technology have added to the overall cost of medical care. More difficult is the question of whether these costs are justified by an accompanying improvement in health status. Although a controversial issue, it is not always clear that these additional expenses are justified. McKeown, among others, argues persuasively that the common perception of modern clinical medicine as the cause of improvements in health status is erroneous (1979). There is undoubtedly a point of diminishing returns on investment from expenditures on expensive equipment that benefits only a small proportion of the population. This is especially true if resources are shifted from public health measures in order to finance technological advances. Should governments sponsor artificial heart implants or research into the etiology of coronary artery disease?

This issue is of relevance in both the U.S. and the U.K. Decisions about the appropriate balance of high technology versus other services must be made regardless of the means of financing health care. It may be politically less feasible to say "no" to specific medical advances in the U.S. than in the U.K. (Miller and Miller 1986). Nonetheless, I would argue that a technological imperative operates in both systems. The social processes of routinization and physician emphasis on research occur on both sides of the Atlantic. The effects are, of course, modified by the health system as a whole, with primary care services receiving greater priority within the National Health Service (ibid.). Political and economic conditions may limit the pace – but not the overall trends of the technological imperative.

The second major implication of the routinization of technology use is in the realm of personal autonomy in health care decision-making. The technological imperative experienced by health care providers, alive with moral meaning, has the potential to wrest control of decisions about the use of technology from patients themselves. The patient's voice may not be heard, as Plough (1981) demonstrates for end-stage renal disease. The field of bioethics has emerged over the past fifteen years in tandem with the belief that respect for patients' wishes, even when difficult to realize, is of paramount importance (President's Commission 1982). Many of the clinical problems confronted in the late sixties and early seventies – problems which contributed to the growth of bioethics as a discipline – centered around when and under what circumstances an individual patient could make his own, independent decisions about medical care. The growth of a legal doctrine supporting the right of patients to provide "informed consent" for treatment has paralleled developments in bioethics in the U.S.

This large body of legal and ethical thinking has failed to grapple with the fact that, by and large, the clinical reality within which patients make their decisions about care is defined by providers. On a pragmatic level, patients are privy only to information provided by health care professionals. They have very few independent sources of information or means of interpreting the complex events involved in their care. But of much more significance then merely withholding or failing to disclose relevant "facts" about a case is the provider's ability to *create* those facts. As Arney and Bergen (1984) have noted, the meaning of health and illness, life and death, have come increasingly under the sway of a "medicalized" definition. In the case of decisions about technology, the meaning of what constitutes "standard therapy" versus an "experimental" treatment is created by the physicians themselves. It is a highly subjective assessment, not based exclusively on scientific facts gained by evaluation of the safety and efficacy of new techniques. As I have suggested, the standard therapy meaning comes, at least partially, from a "reading" of the social *milieu* of treatment settings, from the experience of altered social relations and treatment rituals.

The implication is that patients make decisions about whether to receive medical therapies based on an assessment of the technology's state of development that may be, if not seriously flawed, at least biased by the operation of a technological imperative. The forces of social routinization which push new therapies forward along the experiment/standard therapy continuum appear inexorable. Supported by the activist orientation of western physicians (Freidson 1970), the centrality of research goals, and reimbursement policies biased toward the use of equipment-embodied technologies, the decision to use a new technology seems inescapable. In the process, true patient autonomy is usurped. It may well be true, as Jonsen has suggested, that there is "an inverse relation between scientific, technologic medicine and freedom of therapeutic choice" (1975:126).

I began this paper with a sense of astonishment that a medical "miracle" like the implantation of an artificial heart could so quickly be perceived as a routine, taken for granted therapy. The exploration of therapeutic plasma exchange has revealed some of the social complexities which must be unravelled to understand medicine's technological imperative. To conclude, I return to the apparent contradiction of analyzing scientific medicine with the same tools one would employ in describing a healing ritual in a social setting very different from a modern hospital. Since high technology medicine seems to be a direct embodiment of scientific knowledge, we wish to believe that the application of these new machines to patients is objectively determined, comprehensible to all. This case study of TPE reveals that even in the seemingly rational world of medical science one cannot ignore the social realm -- encompassing the highly subjective experience of participants in medical innovation. A full understanding of the relentless advance of medical technology requires knowledge of the social world in which medical machinery is developed and used. As routinization occurs and a new meaning for a medical technique solidifies, policy options narrow. Without a broad understanding of the social and cultural roots of the technological imperative we will be unable to make fully informed decisions about the appropriate use of technology or comprehend the constricted choices for patients that this imperative implies. Our understanding of the technological imperative in medical practice is enhanced by the recognition that the meaning of a new therapy is, in large part, a social construction.

ACKNOWLEDGEMENTS

The research on which this paper was based was supported by a National Research Service Award from the Division of Nursing, U.S. Public Health Service, the Wenner-Gren Foundation for Anthropological Research, and the Graduate Division, University of California, San Francisco. I wish to thank Margaret Clark, George Foster, Sharon Kaufman, Philip Lee, Jessica Muller, and the editors of this volume for comments and suggestions on earlier drafts of this manuscript.

NOTES

1. An inclusive survey of the many factors supporting the development of a technological imperative for the use of TPE is long and complex, beyond the scope of this paper. A complete account requires that the analysis move beyond the bounds of the TPE treatment unit to include issues such as the role of the for-profit medical equipment industry, physician reimbursement, media accounts of TPE, the effects of government regulations and payment mechanisms, and patient demand for treatment (see Koenig 1987).

2. There are alternative names for this technology, including plasmapheresis, total plasma exchange, and therapeutic apheresis. They all refer to the same basic procedure -- removing plasma and returning a replacement fluid -- although technical variations exist in how the treatment is carried out.

3. A more complete account of the study methods can be found in Koenig (1987).

4. Since the completion of data collection for this project some of the initial enthusiasm for TPE has waned (ECRI 1985; Hamblin 1984; Shapiro and Shapiro 1987). This discussion is based on an analysis of the social forces in operation during the early period of rapid expansion of the technology. An understanding of the later phases would require further study.

5. Compiled using information obtained from a bibliographic citation search completed using the National Library of Medicine's "MEDLARS II" computerized data base.

6. Another way in which patients may influence the routinization of a new therapy is by exercising "demand" for new therapies, thus contributing to the technological imperative. In the U.S., many patients with chronic illnesses actively sought out TPE in response to desperate situations where few treatment alternatives were available. A ready supply of patients willing to try new therapies is itself an interesting cultural phenomenon which supports the use of new treatments. The situation in the U.K. was quite different because self-referral to medical specialists is much more difficult than in the U.S. Most patients I interviewed in the National Health Service had never heard of TPE until they themselves received it after referral to a hospital or

specialist.

7.. Fox describes similar, "magical" routines to guarantee the success of uncertain treatments (1959:111).

8. In this discussion of the meaning of the TPE physician's work, I am purposefully excluding the strong motivation to treat patients which is also a potent force in the research physician's role. As Fox and Swazey make clear in *The Courage to Fail* (1978), clinical investigators experience constant conflict because their role encompasses the duties of both physician and researcher. The strong motivation to treat patients is also an important force in the technological imperative (see Warner 1975). However, for the purpose of analysis, it is necessary to make a heuristic separation of the two physician roles. The emphasis here is on the clinical investigator as *researcher*.

˙REFERENCES

Aiken, Linda and Howard Freeman
  1984   Medical Sociology and Science and Technology in Medicine. *In* Paul T. Durbin (ed.), A Guide to the Culture of Science, Technology, and Medicine, (2nd edition). Pp. 527-582. New York: The Free Press.
American Medical Association (AMA) Council on Scientific Affairs
  1985   Current Status of Therapeutic Plasmapheresis and Related Techniques. Journal of the American Medical Association 253:819-825.
Arney, William R. and Bernard J. Bergen
  1984   Medicine and the Management of Living: Taming the Last Great Beast. Chicago: University of Chicago Press.
Aster, Richard H.
  1985   Plasma Therapy for Thrombotic Thrombocytopenic Purpura: Sometimes it Works, but Why? New England Journal of Medicine 321:985-987.
Banta, H. David and Stephen B. Thacker
  1979   Policies Toward Medical Technology: The Case of Electronic Fetal Monitoring. American Journal of Public Health 69:931-935.
Banta, H. David
  1983   Social Science Research on Medical Technology: Utility and Limitations. Social Science and Medicine 17:1363-1369.
Banta, H. David, Anne K. Burns and Clyde J. Behney
  1983   Policy Implications of the Diffusion and Control of Medical Technology. Annals of the American Academy of Political and Social Science 468:165-181.
Berger, Peter L. and Thomas Luckman
  1966   The Social Construction of Reality: A Treatise in the Sociology of Knowledge. Garden City, New York: Doubleday.
Bosk, Charles L.
  1980   Occupational Rituals in Patient Management. New England Journal of

Medicine 303:71-76.

Comaroff, Jean
    1982   Medicine: Symbol and Ideology. *In* Peter Wright and Andrew Treacher
           (eds.), The Problem of Medical Knowledge: Examining the Social Con-
           struction of Medicine. Pp. 49-68. Edinburgh: Edinburgh University
           Press.

Council for Science and Society
    1982   Expensive Medical Techniques: Report of a Working Party. London:
           Calvert's Press.

ECRI Technology Assessment
    1985   Therapeutic Apheresis: Treatment in Search of a Disease. Journal of
           Health Care Technology 1:279-298.

Fineberg, Harvey V.
    1979   Gastric Freezing -- A Study of Diffusion of a Medical Innovation. *In*
           Medical Technology and the Health Care System. Pp. 173-200. Wash-
           ington, D.C.: National Academy of Sciences.

Fox, Renée C.
    1959   Experiment Perilous: Physicians and Patients Facing the Unknown.
           Philadelphia: University of Pennsylvania Press.

Fox, Renée C. and Judith P. Swazey
    1978   The Courage to Fail: A Social View of Organ Transplants and Dialysis.
           2nd edition. Chicago: University of Chicago Press.

Freidson, Eliot
    1970   Professional Dominance: The Social Structure of Medical Care. New
           York: Atherton Press.

Fuchs, Victor R.
    1968   The Growing Demand for Medical Care. New England Journal of Med-
           icine 279:190-195.
    1974   Who Shall Live? Health, Economics and Social Choice. New York:
           Basic Books.

Gaines, Atwood D. and Robert A. Hahn
    1982   Physicians of Western Medicine: Introduction. Culture, Medicine and
           Psychiatry 6:215-218.

Goody, Jack
    1977   Against Ritual. *In* Sally F. Moore and Barbara G. Myerhoff (eds.), Sec-
           ular Ritual. Pp. 25-35. Amsterdam: Van Gorcum.

Greer, Ann L.
    1977   Advances in the Study of Diffusion of Innovation in Health Care Organ-
           izations. Milbank Memorial Fund Quarterly 55:505-532.

Guillemin, Jeanne Harley and Lynda Lytle Holmstrom
    1986   Mixed Blessings: Intensive Care for Newborns. New York: Oxford Uni-
           versity Press.

Gusfield, Joseph R. and Jerzy Michalowicz
    1984   Secular Symbolism: Studies of Ritual, Ceremony and the Symbolic

Order in Modern Life. Annual Review of Sociology 10:417-435.
Hamblin, Terry
    1984   Where Now for Therapeutic Apheresis? British Medical Journal 289
           (6448):779-780.
Helman, Cecil
    1984   Culture, Health and Illness. Bristol, England: John Wright and Sons.
Institute of Medicine, Committee for Evaluating Medical Technologies in Clinical
Use
    1985   Assessing Medical Technologies. Washington, D.C.: National Academy
           Press.
Jennett, Bryan
    1985   High Technology Medicine: Benefits and Burdens. London: Nuffield
           Provincial Hospitals Trust.
Jonsen, Albert R.
    1975   Scientific Medicine and Therapeutic Choice. New England Journal of
           Medicine 292:126-127.
Koenig, Barbara A.
    1987   The Technological Imperative in Medical Practice: An Ethnographic
           Study of Therapeutic Plasma Exchange. Unpublished Ph.D. Disserta-
           tion, University of California, San Francisco.
Latour, Bruno and Steve Woolgar
    1979   Laboratory Life: The Social Construction of Scientific Facts. Beverley
           Hills, CA.: Sage.
Lock, Margaret M.
    1984   Licorice in Leviathan: The Medicalization of Care for the Japanese Eld-
           erly. Culture, Medicine and Psychiatry 8:121-139.
McKeown, Thomas
    1979   The Role of Medicine: Dream, Mirage or Nemesis? 2nd edition.
           Oxford: Blackwell.
McKinlay, John B.
    1981   From "Promising Report" to "Standard Procedure:" Seven Stages in the
           Career of a Medical Innovation. Milbank Memorial Fund Quarterly
           59:374-411.
    1982   Technology and the Future of Health Care. Cambridge, Mass.: The
           MIT Press.
McNeil, Barbara J. and Ernest G. Cravalho
    1982   Critical Issues in Medical Technology. Boston: Auburn House.
Miller, Frances H. and Graham A.H. Miller
    1986   Why Saying No to Patients in the United States is so Hard: Cost Con-
           tainment, Justice and Provider Autonomy. New England Journal of
           Medicine 314:1380-1386.
Moore, Sally F. and Barbara G. Myerhoff
    1977   Secular Ritual. Amsterdam: Van Gorcum.

Myerhoff, Barbara G.
    1977   We Don't Wrap Herring in the Printed Page: Fusion, Fictions and Con-
           tinuity in Secular Ritual. *In* Sally F. Moore and Barbara G. Myerhoff
           (eds.), Secular Ritual. Pp. 199-226. Amsterdam: Van Gorcum.
National Research Council
    1979   Medical Technology and the Health Care System. Washington: D.C.:
           National Academy of Sciences.
Pinching, Anthony J.
    1978   Plasma Exchange. British Journal of Hospital Medicine 20:552-559.
Plough, Alonzo L.
    1981   Medical Technology and the Crisis of Experience: The Costs of Clinical
           Legitimation. Social Science and Medicine 15F:89-101.
    1986   Borrowed Time: Artificial Organs and the Politics of Extending Lives.
           Philadelphia: Temple University Press.
President's Commission for the Study of Ethical Problems in Medicine and
Biomedical and Behavioral Research
    1982   Making Health Care Decisions: The Ethical and Legal Implications of
           Informed Consent in the Patient-Practitioner Relationship. Washington,
           D.C.: U.S. Government Printing Office.
Reiser, Stanley J.
    1978   Medicine and the Reign of Technology. New York: Cambridge Univer-
           sity Press.
Reiser, Stanley J. and Michael Anbar
    1984   The Machine at the Bedside: Strategies for Using Technology in Patient
           Care. Cambridge, England: Cambridge University Press.
Relman, Arnold S.
    1980   The Allocation of Medical Resources by Physicians. Journal of Medical
           Education 55:99-104.
Rogers, Everett M.
    1981   Diffusion of Innovations: An Overview. *In* Edward B. Roberts, et al.,
           (eds.), Biomedical Innovation. Pp. 75-97. Cambridge, Mass.: The MIT
           Press.
Schroeder, Steven and Jonathon Showstack
    1979   The Dynamics of Medical Technology Use: Analysis and Policy
           Options. *In* Stuart Altman and R. Blendon (eds.), Medical Technology:
           The Culprit Behind Health Care Costs. Hyattsville, MD: National Cen-
           ter for Health Services and Bureau of Health Planning.
Shapiro, Martin F. and Stanley Shapiro
    1987   Apheresis in Renal Transplantation and Chronic Inflammatory Demyeli-
           nating Polyneuropathy: Case Studies in the Evolution of Enthusiasm for
           a Technology. International Journal of Technology Assessment in
           Health Care 3:148-171.
Shumak, Kenneth H. and Gail A. Rock
    1984   Therapeutic Plasma Exchange. New England Journal of Medicine

310:762-771.

Starr, Paul
1982 The Social Transformation of American Medicine. New York: Basic Books.

Strauss, Anselm, Shizuko Fagerhaugh, Barbara Suczek and Carolyn Wiener
1985 Social Organization of Medical Work. Chicago: University of Chicago Press.

Turner, Victor
1977 Variations on a Theme of Liminality. *In* Sally F. Moore and Barbara G. Myerhoff (eds.), Secular Ritual. Pp. 36-52. Amsterdam: Van Gorcum.

Valenstein, Elliot S.
1986 Great and Desperate Cures: The Rise and Decline of Psychosurgery and Other Radical Treatments for Mental Illness. New York: Basic Books.

Waitzkin, Howard
1979 A Marxian Interpretation of the Growth and Development of Coronary Care Technology. American Journal of Public Health 69:1260-1268.

Warner, Kenneth E.
1975 A "Desperation-Reaction" Model of Medical Diffusion. Health Services Research 10:369-383.

Yoxen, E.J.
1982 Constructing Genetic Diseases. *In* Peter Wright and Andrew Treacher (eds.), The Problem of Medical Knowledge: Examining the Social Construction of Medicine. Pp. 144-161. Edinburgh: Edinburgh University Press.

STEPHEN R. BARLEY

THE SOCIAL CONSTRUCTION OF A MACHINE: RITUAL, SUPERSTITION, MAGICAL THINKING AND OTHER PRAGMATIC RESPONSES TO RUNNING A CT SCANNER[1]

INTRODUCTION

That technology can disrupt the tissue of experience and overturn taken for granted assumptions is a fundamental premise in all science fiction involving time travel. Thrust suddenly into a future whose principles are unknown, time travelers are forced to unravel a new world's logic so as to act knowingly amidst strange socio-technical surroundings. The drama of the protagonists' sense making is heightened when they are unable to consult with knowledgeable informants, and yet must pass as competent insiders to avoid perils of detection. Such plots are fraught with dangers narrowly averted as heroes and heroines piece together the logic of the future by trial and error and serendipitous discovery. Whether time travelers ever fully comprehend the future's technical logic is irrelevant; they need only learn to act as if they might.

Our notion of how people should adapt to actual cases of technical change contrasts sharply with the equivocality we tolerate, even relish, among travelers in time. While we allow time travelers the luxury of interpreting bewildering technical orders in light of their own cultural and personal experience, we treat contemporary technical knowing as a matter of fact. Because technologies embody scientific knowledge, we believe they should be objectively comprehensible and that their mastery should involve no more than learning a discrete body of information that may be quite complex but nevertheless definite. Moreover, because we view technical knowledge as potentially universal, we expect technologies to be understood equivalently by all who use them skillfully. Hence, aside from ritualistic nods in the direction of adequate training, social scientists interested in the problems of technical change rarely ponder how people make sense of the technologies they use or what their sense making may imply for patterns of social organization.

However, our common sense notions about technical knowing depend on conditions that may rarely hold. We assume that so long as technical knowledge is available it can, in fact, be known. But since knowing is a social as well as an epistemological matter, what is epistemologically possible may be sociologically unlikely. For example, even if newly acquired technical information is meticulously archived in journals and other

497

M. Lock and D. R. Gordon (eds.), Biomedicine Examined, 497–539.
© 1988 by Kluwer Academic Publishers.

media, it may be naive to assume that knowledge has been distributed, much less consumed or understood. For example, transmission will be necessarily limited to those who have access to appropriate channels or who can consult others who do. Moreover, people generally have little incentive to acquire technical information unless it has immediate relevance for their daily life. Even in a narrow professional specialty, technical knowledge may remain unknown to most practitioners until they develop some need to know. Local knowledge is therefore always incomplete and unevenly distributed, even among people who are positioned to know.

Furthermore, we assume that technical information will flow freely once the need to know exists. But the flow of information from a technology's purveyors to its users may be constricted at numerous points. Those who need information may not know where to search or may be unable to search thoroughly for want of time or money (March and Simon 1958; March and Olsen 1976). Moreover, available information may be incomplete, and critical details may not be recorded. Such omissions are more likely the more complex the technology (Perrow 1984; Hirschhorn 1984). Even if all necessary information is transmitted accurately, communication may still be unsuccessful if knowledge is coded in a language others do not share. Finally, brokers in charge of transmitting information may have incentives to restrict the flow of information to enhance their own indispensability and power (Pettigrew 1973; Pfeffer 1981). To the degree that such phenomena operate, technical information becomes sociologically rather than epistemologically uncertain.

It would appear then, more often than we care to admit, that users of new technologies may confront pragmatic and epistemological difficulties akin to the troubles time travelers face. The analogy may be especially apt if change involves a fundamental transformation in a setting's technical infrastructure. On such occasions, knowledge gained in the past may transfer poorly to the future. We currently know next to nothing about how people with incomplete technical knowledge construct understandings of complex technologies so as to proceed with their work. Yet, the stakes of knowing under such conditions can be quite high, particularly when novices are expected to use a technology skillfully, in a short period of time, without serious mishap. That people do, in fact, construct such theories and that their construal has important implications can be demonstrated by examining recent events in radiology.

Central to the so-called technical transformation of medicine has been the rapid infusion of sophisticated computational technologies into the very infrastructure of diagnosis and treatment. Computation lies at the core of a wide variety of innovations from digital thermometers to expert programs that offer differential diagnoses. But perhaps in no area of medicine has computation triggered a more thorough metamorphosis than in radiology. Until the late 1960s the work of a radiology department consisted almost entirely of radiography and fluoroscopy. To be sure, both technologies had been substantially modified since the late 1890s, when x-rays were first used for medical purposes. However, by and large, radiological innovations during the first half of the Twentieth Century consisted of incremental improvements to existing machines and techniques (Dewing 1962). Consequently, radiology's knowledge expanded gradually, with the result that most radiologists and technologists found it relatively easy to remain up to date (Barley 1986a).

However, in the late 1960s the pace of technical change suddenly accelerated as medical researchers and engineers began to link computers to an array of old and new data sources to create new imaging techniques. By the mid 1980s, the list of "new modalities" has grown to impressive length: ultrasound, computed tomography (CT), digital subtraction angiography, nuclear magnetic resonance (NMR) and positron emission tomography (PET) being the most commonly known. Nor is the computerization of radiology likely to slow in the near future. NMR and PET have just begun to diffuse beyond the elite medical centers. Manufacturers are beginning to equip standard x-ray tables with minicomputers, and numerous observers predict that by the turn of the century digital radiographs will replace analogical films. Other visionaries argue that during the next decade large hospitals will begin to use integrated data bases to store, retrieve, and analyze all medical images which, in turn, will be accessible throughout the hospital on a network of graphics terminals.

Unlike most technical developments in medicine, radiology's transformation has not gone unnoticed. Medical journals have burgeoned with studies attesting to each modality's diagnostic utility. The popular press has also spread its own version of the good news by repeatedly portraying each modality as science fiction turned fact. But because the new technologies have carried hefty price tags, not all of radiology's publicity has been positive. For a time during the 1970s the CT scanner became the quintessential symbol of the rising cost of health care and, as a consequence, radiology departments were among the first to feel the brunt of the Federal government's plan to regulate the diffusion of medical technology (O.T.A. 1978). In fact, one might argue that radiology's transformation crystallized public awareness of the uncertain trade-off between economic efficiency and medical efficacy (Banta and McNeil 1978).

But as with most medical technologies, debate and research on the new imaging modalities rarely extends beyond medical and economic issues. While there are numerous studies of the diagnostic utility and economic consequences of the new technologies (see Weiner 1979), there has been little speculation (Stocking and Morrison 1978), and even less research, on what new modalities portend for radiologists and radiological technologists. However, if studies of technical change in industry are any guide, there is good reason to believe that the new modalities should dramatically alter the roles, the relations, and even the organizational structure of radiology departments (Barley 1984). Each new modality departs radically from traditional radiological practice and represents a disjuncture in radiology's cumulative knowledge. Thus, a modality's diffusion may appear orderly from the perspective of radiology as a whole (Baker 1977), while its adoption by a local hospital may entail a sudden rent in the tissue of day-to-day experience.

New imaging modalities are not simply more complex than their precursors, they operate by computation, a technical paradigm that differs substantially from the mechanical world of radiography and fluoroscopy. Moreover, each modality brings with it a system of signs, a language of images whose syntax and morphology are well structured, but whose semantics are obscure, especially for those without extensive training in the technology's use. Because the technology and its images diverge so radically from the past, large portions of the average radiologist's or technologist's practical experience may fail to generalize to the new situation. When radiology departments contemplate adopting a new modality, they therefore almost always confront the fact that most members understand neither the technology nor its images (Barley 1986b).

The problems that face such a department are acutely pragmatic. Since new modalities entail sizable sunk costs, radiologists and technologists are generally expected to utilize newly acquired technologies skillfully in a short period of time. Mistakes under such conditions can be particularly costly. At a minimum, downtime signifies substantial lost revenue and technical errors may lead to diagnostic errors which, in turn, threaten more serious repercussions including endangering patients and the possibility of malpractice suits. Radiologists and technologists, therefore, face the necessity of developing strategies of practical action in short order under conditions of uncertainty, while at all times maintaining at least the appearance of knowledge.

The remainder of this essay explores aspects of this practical problem by focusing on the strategies two groups of radiological technologists devised to assist them in operating the whole-body CT scanners their hospitals had recently acquired. At issue are the tactics and accounts that technologists constructed to make sense of a technology whose actual functioning they dimly understood. After outlining the nature of the study and the data on which the analysis is based, the exposition turns to the technology, the technologists' work, and the problems the former created for the latter. The stage is then set for discussing the strategies technologists employed to handle the technical problems they faced. The exposition comes to rest on several surprising attributes that character- ized the technologists' theories of practical action and concludes by draw- ing implications for the management of radical technical change in medi- cal settings.

THE SITES AND THE DATA

Urban and Suburban were two of four community hospitals in the state of Massachusetts whose radiology departments acquired body scanners in 1982. Each had a capacity of approximately 350 beds. Both radiology departments employed six radiologists and roughly fifty other individuals; both performed a similar range of radiological procedures; and both pur- chased identical scanners, Technicare 2060s. Although Urban had oper- ated a first generation EMI head scanner since 1977, the body scanner was Suburban's first experience with CT.

Since the radiologists at both hospitals were unfamiliar with body scanners and the images they produced, in the summer before the scan- ners went on line each hired as an internal expert a young radiologist who had recently completed a fellowship in CT. However, the departments pursued different strategies for assembling the technologists who would staff their scanners. Urban built its staff internally by drawing on tech- nologists who operated its head scanner, by rotating the department's specials technologists between CT and special procedures, and by pro- moting a technologist who had worked in the x-ray department. Subur- ban, on the other hand, hired two experienced technologists who had worked previously in one of the area's first body scanning installations and then rounded out its staff by transferring three techs from its x-ray department.

In June 1982, after spending three months learning CT technology at a well known medical center, I began a year long stint as a participant observer in both radiology departments as part of a larger project on how new modalities have altered the social organization of radiological work (Barley 1984). Although the research covered all aspects of radiological practice, after the scanner went online in late September observation centered heavily on the two CT operations. As throughout the study, I gathered data by attending individual examinations in their entirety. Dur- ing the course of each exam I recorded the occurrence and timing of events chronologically in small spiral notebooks to compile behavioral records for every procedure observed. Conversations between patients, radiologists, and technologists were either taped or written in a shorthand devised to quickly record setting -- specific turns of phrase. Observation was conducted in each department on alternate days to ensure an even flow of data from the two sites. Over the nine months I spent in the two CT departments, I observed approximately a hundred and fifty scans.

Early in the study, the technologists' sometimes frantic attempts to solve technical problems drew my attention. From that point forward, when I witnessed the beginning of a technical snag I noted in order, and in as much detail as possible, the actions technologists took to correct the malfunction. The duration of these technical problems varied tremendously. Some lasted less than a minute, others took several hours to correct, and a few dampened routine operations for days at a time. But regardless of their duration, all problems occurred unpredictably and were experienced as such by technologists and radiologists alike. Over the course of the research I observed CT techs confront technical difficulties on 65 separate occasions.[2]

Since coping with computer problems entails interaction between humans and a machine, it was crucial to capture the machine's response as well as the technologists' acts. Otherwise, it would be impossible to determine how the technologists shaped the machine's behavior and how the machine influenced the technologists. Documenting the machine's response usually involved noting the messages that appeared on the scanner's video monitor. In addition, I solicited participants' interpretations of the machine's response when such accounts were not spontaneously offered. From these sources of data I was able to construct problem solving histories that recount, step-by-step, the actions technologists took to diagnose and solve technical problems.

Several examples of such problem histories appear as Appendices. Each history depicts a problem as two parallel but entwined streams. One stream documents the technologists' actions, while the other displays the signals and responses emitted by the computer. The chronological flow of events over the course of a problem history can be traced by reading from left to right down the page. Problem histories comprised the core data for the following analysis. From time to time I will draw on the histories contained in the Appendix as well as my fieldnotes to illustrate how technologists approached the dual task of solving and accounting for problems they poorly understood while maintaining control of their work and the appearance of normal operations.

TRAINING FOR THE ROUTINE

CT scanners are technical hybrids that pivot on a mini-computer's capacity to convert data generated by x-ray equipment into cross-sectional images of internal anatomy. In programmed increments, a mechanical table moves a patient's body through the scanner's gantry, a device that resembles a large rectangular doughnut. Between movements, an x-ray tube mounted inside the gantry rotates around the patient's body firing tight beams of electrons at predetermined intervals. The electrons pass through the patient's body, strike detectors, and generate electrical impulses that flow into the scanner's computer. The computer enters the data into a series of simultaneous equations to produce triangulated estimates of the density of the tissues through which the beams passed. Ranges of density values are then correlated to shades of grey on a video monitor to produce detailed cross-sectional images of the body.

Except for when patients are placed on and taken off the table, the whole procedure is controlled from a computer console located in a room adjacent to the gantry. The consoles of the Technicare 2060's used at Urban and Suburban contained a keyboard, two cathode ray tubes (CRT's), and a bank of touch sensitive buttons. The first CRT functioned as a terminal for communicating with the scanner's computer; the second served as a monitor for viewing images. The touch sensitive buttons controlled the x-ray tube's parameters and the movement of the gantry's table. By typing commands at the keyboard and by selecting appropriate buttons at appropriate times, technologists guided the scanner through each study.

Routine scanning procedures were nearly identical at both hospitals. At the start of a typical scan, one technologist sat at the console while another positioned the patient in the gantry.[3] The technologist at the console entered the patient's demographics in response to prompts provided by the computer and then programmed the computer to process the data using parameters that were also made available on the CRT. The choice of parameters was set by protocol. After entering the patient, the technologist "set a technique" by pushing a series of buttons to determine the number of scans to be taken, the time of each exposure, the milliamperes and kilovolts at which the x-ray tube would operate, the scan's diameter, and the increments at which the table would advance. Although "techniques" were also dictated by protocol, in practice, technologists deviated from standard when they thought it necessary to adapt to a patient's unique characteristics.[4] With the computer programmed to collect and process scans, the techs activated the "pause" and "start scan" buttons sequentially so that the scan would immediately commence when the "pause" button was released.

Over the course of a scan, technologists might increase the number of slices, adjust the table's increments, or change the x-ray tube's settings. Most of the time, however, the scanner was allowed to proceed on its own accord until the required images were obtained. On completing the programmed   number   of   scans,   the   computer   automatically "reconstructed" the raw data into images. After reconstruction the technologist commanded the computer to load the images into temporary memory and display them one-by-one, on the video monitor. As each successive scan appeared, the technologist photographed the image using a camera designed to transfer images from monitor to films. After finishing the photography, the computer was told to store, or "archive," the data on magnetic tape and the scan was deemed complete.

Except for the four technologists who had previously operated Urban's head scanner and the two experienced techs hired by Suburban, none of the fifteen CT techs had more familiarity with computers than could be gained by playing video games or by banking at an automatic teller. As one might expect, the technologists with prior CT experience were able to transfer broad principles, especially those pertaining to the overall conduct of a scan, from their earlier jobs to the new setting. But since none had ever operated a Technicare 2060, their advantage was not as clear cut as it might, at first, seem. The Technicare scanner was far more complex and interactive than the EMI head scanner, and even Suburban's techs, who had operated a third generation Sieman scanner, had to learn new procedures and a different computer language to operate the Technicare machine. Moreover, there was evidence that previous knowledge occasionally inhibited learning. For example, at Suburban the technologist who had most difficulty learning the Technicare's "deltaview" software was an individual with extensive experience doing functionally similar "topograms" on a Sieman scanner.[5]

Since Technicare realized that most radiologists and technologists assigned to newly purchased scanners would be unfamiliar with CT, the company provided three resources for training new users. First, a series of manuals accompanied the equipment. Since the scanners arrived a month before they were brought on-line, personnel were encouraged to use the lag to acquaint themselves with the manuals' contents. In practice, however, the manuals were of limited assistance, especially in the early days of the operation. Since the company provided a single set of manuals, when one person studied others were excluded. Consequently, none of the technologists, and only one of the radiologists at each site, actively studied the manuals prior to the operation's inception.

However, had the manuals received wider attention it is still by no means clear that their reading would have enchanced understanding. Filled with computer jargon, the manuals were designed as reference rather than teaching texts. One could not read the manuals from front to back and obtain a progressively refined knowledge of the computer's language or the scanner's operation. Thus, even when technologists were finally able to peruse the manuals, they had difficulty determining what they should read. For example, one morning after the scanner began to operate I came into Suburban's control room and found two technologists examining a manual. They complained that the book said nothing about how to do a scan. A glance vindicated the technologists' assessment. The documentation referred to the computer's operating system and consisted of a technical description of storage parameters cast in terms of bits, bytes, and hexadecimals.

In addition to the manuals, Technicare also provided an "applications technologist" whose job was to train new users during the first week of an installation's operation.[6] Training at both sites consisted of a combination of lectures and hands-on experience at the scanner's console. The training program aimed to acquaint new users with the computer's most important routines and centered on imparting information about when to push what button and what the scanner's subsequent response should be. After the second day of training, both departments began to scan patients with the applications technologist acting as coach. During this period discussions of technical problems occurred in an *ad hoc* fashion: when a glitch occurred in the course of a scan the applications tech helped diagnose the problem and provided rules of thumb to forestall its reoccurrence. But the applications technologist offered little systematic information on how the computer operated, and explained commands only in terms of the scanner's surface response.

Though indisputably useful, training was further circumscribed by contextual constraints. First, the training period lasted less than a week. Second, training was done *en masse* so that each technologist had but brief opportunity for individualized instruction. Finally, the radiologists at both sites lobbied to begin scanning soon after the applications technologist arrived and usually attended sessions when actual scans were conducted. Since the radiologists sat at the console, their presence further diluted the actual amount of time that technologists could spend at the keyboard. Moreover, since their interests differed from the technologists', they tended to steer conversation to issues of medical diagnosis rather than to the technology itself.

Finally, Technicare assigned its customers an "engineer." Engineers were usually graduates of two year technical schools who were further trained by the company in the particulars of its machines. Each hospital contracted with Technicare for an engineer's services at the rate of $30,000 per year. In return, the engineer was to maintain the equipment and act as a troubleshooter when technical problems arose. Since the engineers were nominally available at all times by pager, technologists and radiologists at both sites viewed them as their main source of support and information in times of technical difficulty.

The radiologists expected that by interacting with the engineer, technologists would slowly learn the computer's constraints and techniques for solving its routine problems. While some information was, in fact, acquired in bits and pieces through such interaction, several factors militated against any systematic transfer of knowledge. The engineers hesitated to teach technologists about the computer's subroutines for diagnosing and correcting technical problems, even when problems recurred frequently. For example, during the third month of operation the shutter on Urban's x-ray tube repeatedly failed to close at the end of a scan. Only after directing the technologists through diagnostic software numerous times by phone did the engineer explain how to make decisions without his direction. Moreover, the information was not passed along until the radiologists demanded that the engineer fully inform the techs about the subroutine's use.

The engineers' reluctance to impart technical information partially reflected their belief in the adage: "a little knowledge is a dangerous thing." The engineers feared that because the technologists did not have an adequate technical background, they would fail to fully understand the ramifications of their actions and, thereby, unwittingly amplify problems by attempting to correct them. But, their reluctance also involved a desire to protect their own role. The engineers admitted both rationales. For example, when asked why he did not share earlier his knowledge of the subroutine that controlled the scanner's shutter, the engineer explained that if he were to educate technologists about the existence of diagnostic programs, they might "screw up the machine even worse." Later in the conversation the engineer added, "Besides it's my job to fix the scanner, not theirs." When engineers did provide explanations, they usually spoke in a technical jargon whose terms they rarely explained. Consequently, technologists comprehended only portions of what the engineers said, a fact about which the techs repeatedly complained. Since the technologist had little knowledge of the system's hardware and software, the engineers' explanations offered technologists labels for problems, but provided little information for solving them. The technologists complained that the engineers purposely kept their knowledge secret, and the radiologists argued that the engineers did not fully appreciate the intensity of the need to keep the scanner on-line. Nevertheless, most willingly acquiesced to the engineers' expertise. Except in the heat of a problem, few technologists sought detailed information from the engineers. Most felt they did not have the time to learn what the engineer knew. Consequently, when engineers repaired the scanner or conducted preventive maintenance, technologists rarely observed their work even though they often had no other duties to perform. The technologists thereby forfeited numerous opportunities to observe how the engineers diagnosed and corrected technical malfunctions.

Despite the fact that the technologists knew little about how the scanner actually worked, by mastering its surface responses they rapidly developed considerable finesse at producing high-quality scans. In fact, once they learned the basics of guiding the scanner through a procedure, the technologists usually took the scanner for granted and rarely mentioned the technology unless something out of the ordinary occurred. However, since the technologists came to computers as novices and since the manuals, the training, and their relations with the engineer imparted little knowledge of technical principles, their acumen was circumscribed by the boundaries of routine operations and their evolving experience. The techs' normally complacent and confident demeanor regarding the technology was therefore precariously balanced and liable to rupture at a moment's notice.

If, in the flow of events, an anomalous message appeared on the monitor or the scanner failed to respond as predicted, the well-mannered interaction between technologist and scanner degenerated as tension pervaded the otherwise calm control room. The intensity of the tension generated by the onset of a technical problem was proportional to the problem's duration and was reflected in the technologists' behavior. When technical problems occurred, joking and social conversation ceased as the technologists focused on the scanner's erratic behavior. The pace of activity in the control room quickened, and profanities became more common. But, the level of tension occasioned by a technical problem was perhaps most strikingly gauged by the technologists' behavior toward patients.

Under normal conditions technologists were exceptionally attentive and civil to patients, regardless of how troublesome they actually perceived the patient's behavior to be. Technologists regularly explained all facets of the scanning procedure, routinely checked on patients who seemed restless, and promptly responded to patients' complaints about being cold or uncomfortable.[7] The technologists' behavior towards patients during the heat of a technical problem, however, contrasted sharply with their conduct under normal conditions. The more persistent the problem and the more deeply absorbed the technologists became in attempting to correct it, the less they remained aware of the patient's situation. While the technologists scurried about the control room attempting to correct a malfunction, the patient lay unattended, surrounded by a gantry that had fallen conspicuously silent. When patients failed to complain about the delay, the technologists frequently forgot them. To be sure, all patients were eventually informed of the events that had transpired, but only when a technologist suddenly realized that the patient had not been kept abreast of the situation. Such realizations often occurred as much as twenty to thirty minutes after a problem initially began. Patients who complained about perceived delays were told immediately of the technical difficulty in a calculatedly civil manner. Continued complaints, however, brought curt replies. For instance, on one occasion a technologist told a patient who repeatedly expressed her dissatisfaction by whining to "stop acting like a baby." I never observed this technologist or any other condescend to a patient except under the duress of a persistent malfunction.

Inability to act, despite knowing what needed to be done, was another indicator of the tension provoked by prolonged technical problems. Paradigmatic were those instances when technologists knew that solutions resided in the computer manual, but were unable to use the manual effectively. Consider the following excerpt from my fieldnotes during the fifth month of operation at Urban. At the time the events transpired, the technologist had spent half an hour working through an unrelated problem:

> At this point I noticed that the name on the scan was not the patient's name. Sally had typed in the name and number of another patient who had missed his appointment and was scheduled for a completely different exam. I told Sally about my discovery . . . She immediately looked at the schedule and realized the mistake. She picked up the manual and tried to figure out how to change labels. After several seconds, however, she put the book down without discovering the correct command and began to try the ALTER command, which did not work. I looked up the format for ANNOTATE and read her the document as she typed the commands. Within seconds all the labels were changed. "When you're in a bind," Sally said, "you feel like you don't have the luxury to go hunting through the manual."

Inaction was partially predicated on the potentially realistic fear that taking action might amplify the problem. Ironically, however, technologists often eschewed the very act that later solved the problem. Such a case occurs in Problem History #1 (Appendix A, event 58) when the technologist tells the radiologist, "The system's down. It won't do anything. I'm afraid that if I reboot it, I'll screw up the mag tape. We might get another mount error." The technologist later makes a similar comment to the engineer (event 68), who tells her that rebooting the system will eliminate the malfunction.

The onset of technical problems threatened to push technologists to the edge of their understanding, where they confronted ill-defined conditions which they were ill-prepared to handle. Because technical problems were experienced as random aversive events whose causes were often unclear, they challenged the technologists' sense of control. The threat had several sources. Technical problems disrupted the flow of a scan and made the technologists' work more difficult. When persistent, problems also created backlogs, inconvenienced patients, and provoked unpleasant encounters with radiologists. The technologists considered such complications undesirable and sought to minimize their occurrence. Moreover, technical problems threatened to belittle the technologists' reputed grasp of the technology. Since knowledge of the scanner constituted the technologists' primary source of power in their day-to-day dealings with radiologists (Barley 1986a), the technologists were not eager to appear at the scanner's mercy. Finally, since the radiologists held the technologists (and the technologists considered themselves) responsible for keeping the the the scanner on-line, the techs felt they had little recourse but to attempt to meet technical problems head-on. It was during such encounters that the techs began to devise strategies and theories to help them combat the scanner's caprice.

STRATEGIES FOR COUNTERING MACHINE PROBLEMS

Normalizing the Situation

Ethnomethodologists have repeatedly shown that when events tear the tissue of common sense expectations, people often attempt to repair the rent and retain their sense of control by either denying that problems exist or by asserting that nothing drastic has in fact occurred (Garfinkel 1967; Emerson 1970a, 1970b). To "normalize the situation" is to bracket anomalous events as quirks in the flow of experience whose implications for practical action are nil. Because normalizing strategies conserve an interpretative system's viability, they protect actors from the vagaries of negotiating new understandings. Given that most people prefer assimilation over accommodation, it is unsurprising that the technologists also chose to handle most technical malfunctions by normalizing the situation. When inexplicables occurred during the course of a scan, technologists frequently attempted to bracket and nullify their importance by constructing accounts for why the problem need not be taken seriously, since its potential for marring the study was limited. Such an account occurs in Problem History #2 (Appendix B). When the radiologist notes an anomalous deltaview (event 2), the technologist explains (event 3) that the radiologist need not be concerned with the problem since she has already photographed the image.

Technologists most commonly sought to normalize technical malfunctions by eliminating traces of the problem's occurrence. If a scan's appearance and the scanning context could be returned to their expected states, then for all practical purposes one could argue that a malfunction had been of little importance. In the following scenario extracted from fieldnotes at Suburban, two technologists pursue a complicated and ingenious line of action to eliminate subtle signs left in the wake of a technical mystery:

Sam went to the scanner to comfort the patient who had been complaining that she was cold. While he was gone, the computer returned the following message: SCAN SELECT PROCESS CANCELLED BY OPERATOR. (Rita had been archiving images at the console in the radiologist's office. Her actions most likely aborted the deltaview subroutine Sam had been using. However, the techs did not appear to recognize the possibility that this had occurred.) When Sam came back and saw the message, he muttered, "I'm in trouble now!" Rita asked if he had put the scanner on manual, because sometimes this triggers the new patient routine which was now showing on the screen. Sam claimed that he did not touch the button.

To compensate for the problem, Sam again entered the patient into the computer. He then reset the techniques he had been using when the deltaview program was in control. Fortunately, he could complete the scan without having to change the gantry's angle, since only the last disc space was left to be scanned.[8] Since "new patient" reset the table's counter to zero, Sam had Rita move the patient into the gantry until the counter read 70; she then re-zeroed the counter and moved the patient out of the gantry until the counter read -70. The number on the next scan would now begin where the last scan in the sequence left off. Since deltaview moves the table in a direction opposite to the direction it normally moves, Rita had to go into the scanner room between scans and move the table manually so that the scans would occur at the right level and in the right order. With this elaborate tactic, the techs ensured that no one would be able to tell from reading the films that a problem had occurred.

Attempts to normalize the situation were fundamentally pragmatic. If the techs could eliminate a problem's ill-effects, then from the standpoint of producing a usable study, no problem existed. The strategy was eminently sensible given that technologists were primarily responsible for supplying radiologists with diagnostically adequate films. Since the radiologists evaluated the technologists on the quality of the images they produced and on their efficiency in handling the patient load, from the techs' perspective it was better to frame problems as glitches in the immediate study, rather than as indicators of generic mistakes or malfunctions that could be discovered, understood, and corrected once and for all. Although the strategy taught technologists little about a problem's cause, by it they salvaged scans, attained a sense of competence, and saved face in front of radiologists who judged their work.

However, normalizing the situation was effective only when malfunctions did not recur and when their effects did not ramify over time. Recurrent or persistent problems forced technologists to attempt to diagnose the malfunction. At this point, the technologists' circumscribed knowledge of the scanner began to handicap their problem solving. Unless the problem was one they had seen before and had successfully solved, the techs had little recourse but to resort to trial and error. Technical rituals were not only the key elements in the technologists' trial and error approach, they were one of the strategy's most significant products.

Ritual Solutions

When technologists faced problems that resisted normalization, their first line of action was to engage in behavior that had, on other occasions, successfully eliminated a technical malfunction. These ritual solutions represented the behavioral sediments of what technologists had learned during previous bouts with the scanner's computer and constituted an assortment of tactics for confronting the scanner's black box. When new tactics were discovered, they rapidly passed from one technologist to another to become part of the group's technical lore. Although ritual solutions did not always work, and although technologists did not always know when or why they were appropriate, the solutions succeeded frequently enough for the techs to deploy them as a first line of defense in their struggles with the machine.

Repeating the last command issued to see if the computer would accept its second entry was the most frequent ritual solution technologists at both sites employed when they encountered technical problems. In most cases, the second entry elicited the same error message. In fact, syntax errors were one of the few problems whose solution required reissuing a corrected command. It was because syntax errors were relatively common that the practice of reentering commands persisted. However, unless the syntax error involved misspelling, most technologists did not recognize the syntactical nature of the solution. Instead, when error messages occurred, technologists habitually reentered commands and justified their action by claiming that the computer "sometimes just doesn't take it."

Reentering a command was usually a harmless, if fruitless, act. Occasionally, however, the ritual complicated the problem. Such a case constitutes Problem History #3 (Appendix C) in which a technologist fails to notice a critical error message (event 1-2) and then interprets the computer's next response (events 3-4) as a case of where the computer "just didn't take it." The technologist did not discover the actual nature of the problem until the end of the scan (event 6). Although the tech eventually resolved the problematic condition (event 8-9), he never realized that reentering the information had complicated the situation.

A second ritual solution, common at both sites, consisted of "downloading" or "rebooting" the computer. The first was a command that broke the link between computer and console and cleared the console's microchips; the second was a sequence of actions that shut down the whole system and then brought it back on-line. Both actions required that the computer be "set up" before scanning could resume. At times, downloading or rebooting was the technologists' first response to a technical problem. At other times, technologists used the download or the reboot to culminate a futile series of problem solving attempts. Faced with a choice between the two tactics, downloads were always executed first, since they were considered the least drastic of the two alternatives.

Downloading and rebooting were like wiping the slate clean and starting over. Each cleared the computer's screen and its temporary memory. The procedures often seemed to eliminate the problem, but at the cost of retaining no data on the problem's cause since both erased the console's log of error messages. Moreover, when problems were recurrent these tactics could actually complicate matters. Downloading and rebooting broke persistent problems into discrete chunks. The machine's apparent calm after a download or reboot occasionally lulled technologists into falsely assuming that the problem had been solved. The resulting sense of security could, in turn, divert recognition that the problem remained. Consequently, by chunking the flow of action, downloads and reboots could foster the illusion that one faced a string of problems rather than different manifestations of the same malfunction.

Even when a ritual solution was immediately shown to be ineffective, technologists nevertheless repeated the solution several times before concluding it brought no progress. For instance, techs at both sites routinely reentered commands or downloaded the computer three or four times in succession before attempting another course of action. Even complex rituals taking several minutes to perform were reiterated before they were rejected. For example, the standard, and usually sufficient, response to a "mount error" was to ensure that the reels were mounted properly on the mag drive. However, mount errors could occur for reasons other than improperly mounted tapes. In Problem History #1 (events 7-16) two technologists respond to a mysterious mount error by persistently winding a magnetic tape back onto its original reel only to have the mag drive unravel their work once it is activated. The technologists initially repeat their actions three times before ceasing (a span of about ten minutes) and then return to the task some twenty minutes later (event 32) only to obtain the same results. Note that each time the technologists rewound the tape, they pushed the mag drive's buttons in a new sequence under the assumption that they had forgotten the correct sequence. Since the technologists had mounted and dismounted many tapes in the past, a simultaneous memory lapse on the part of both individuals was unlikely.

The incident demonstrates not only why such solutions were repeated, it also underscores their ritualistic flavor. Technologists repeated ritual solutions primarily to convince themselves that they had performed the act exactly. Certain solutions therefore assumed their ritualistic character not simply because they were at the top in the technologists' bag of tricks, but because the techs believed the solution's success was contingent on fastidious performance. Although precision was indeed important when working with the scanner's computer, in the technologists' evolving theory of practical action, a ritual's precision often overshadowed its function. Thus when ritual solutions failed, technologists almost always initially assumed that failure reflected sloppy execution rather than the solution's inadequacy. The fetish of precision not only contributed to the solution's ritualistic tinge, it slowed the technologists' problem solving.

Confirmatory Testing

When technologists were forced to battle technical problems by trial and error, they proceeded from action to action in what might at first appear to be random order. In fact, however, potential solutions were tried and tested according to a line of practical reasoning whose logic was systematically patterned. When problems arose and when ritual solutions were found wanting, technologists would propose a possible explanation, act as if the explanation was correct, and then allow the scanner's response to tell on the action's adequacy. If the action overcame the problem, the act was taken as appropriate and the explanation as valid. If the action failed, then a new explanation would be formulated and enacted. Problem solving sessions of any duration, therefore, consisted of a chain of cycles whose elementary links were composed of a postulate, an action based on the postulate, and acceptance or reformulation based on the scanner's response.

Problem History #1 illustrates such an extended chain of practical reasoning. At first (event 4) the technologists suppose the problem reflects a twisted tape. However, since the problem persists after the techs straighten the tape, they discard their first explanation in favor of a second: the tape should be rewound (event 7). When the second source of action also fails, a radiologist suggests a third explanation (event 21): two problems may have actually occurred, an open file failure and a problem with the tape drive. The radiologist therefore advises rebooting (event 28) and cleaning the drive's heads (event 34). Although the dual actions appear to work, one technologist proposes a fourth possibility in anticipation of the solution's failure: a key mysteriously turned in the wrong direction signals another plausible cause (event 44). After seemingly unrelated problems occur several minutes later, the technologists finally claim that the computer is "frozen." At this point, the technologists call the engineer and abdicate resentfully to his direction.

The distinction between confirmatory and disconfirmatory approaches to hypothesis testing is relevant for understanding the technologists' problem solving behaviour. Under a disconfirmatory approach, problem solvers formulate an hypothesis, propose implications that would force them to discard the hypothesis, and then seek disconfirming evidence. If no such evidence is found, the hypothesis is accepted. In a confirmatory strategy, problem solvers formulate an hypothesis and then seek data consistent with the hypothesis. If such data occur, the hypothesis stands until additional evidence forces reformulation. In the realm of technical troubleshooting, a disconfirmatory approach is typified by the use of an explicit diagnostic tree that progressively eliminates alternatives. The confirmatory method is nicely demonstrated by the technologists' approach.

Though generally more precise and more consistent with the ethos of technical rationality, disconfirmatory diagnosis requires problem solvers who possess extensively structured prior knowledge of the technology. It was precisely such knowledge that the techs lacked. In contrast, the *ad hoc* nature of the confirmatory approach suited the technologists' situation well. Not only did a confirmatory strategy require minimum prior knowledge, it enabled the technologists to act in an equivocal situation. By acting, the technologists could satisfy their own desire "to do something," to remain in control, while meeting the radiologist's expectation that the techs do their best to keep the scanner on-line.[9] Moreover, as a form of active exploration, the confirmatory approach enabled techs to accumulate experiential knowledge of the scanner's workings. However, the confirmatory approach had its unintended side effects.

When problems generate sets of symptoms that intersect but are not coextensive, and when a solution is effective for more than one but not all problems whose symptoms fall within the intersection, a confirmatory strategy may bolster misconceptions. Such conditions apparently gave rise to and reinforced the use of ritual solutions. When problem solvers focus on symptoms within the intersection and then choose a non-specific solution, resolution of the problem will reinforce the solution's use whenever the symptoms recur. However, since the symptoms map onto more problems than does the solution, the latter will work on some occasions and not others. Unless problem solvers subsequently discriminate between problems by their idiosyncratic signs, use of the global solution will persist even if it works only intermittently. As mentioned above, the ritual act of reentering commands was maintained as a global response to a variety of problems precisely because equivocal symptoms ensured the act was intermittently reinforced. Moreover, as will be argued below in the section on superstition, confirmatory strategies of problem solving also reinforced technically inaccurate conceptions of the machine itself.

LOCAL THEORIES OF THE TECHNOLOGY

Practical action by members of an occupation is a two-sided coin. In addition to developing strategies and tactics for acting on the world, practitioners must also devise interpretations to justify their behavior and account for the situations they face.[10] I have suggested that encounters with machine problems shaped the technologists' knowledge of the scanner. But in order for the techs to act, it was also initially necessary for them to frame the problem they faced. How technologists construed a particular malfunction influenced their subsequent line of action. Because the techs had but slight knowledge of how computers work, and because their training was oriented to the routine, the techs had necessarily to ground their evolving theories in prior experience. Since most of their technical training prior to CT had been with traditional x-ray equipment, and since the scanner's hardware was more accessible than its software, it is perhaps unsurprising that the technologists tended to cast technical difficulties in mechanical rather than computational terms.

Mechanical Metaphors

Technologists frequently attributed technical problems to burnt-out circuit boards, shorted wires, and faulty detectors. While such mechanical failures did in fact occur, their incidence was less frequent than their appearance in the techs' accounts of malfunctions. Even when technologists verbalized no explicit reason for a problem, their actions nevertheless attested to the centrality of mechanical metaphors in their theory of the machine. How the technologists' accounts and behaviors pivoted on mechanical notions is illustrated by the techs' approach to the technical difficulties encountered in Problem History #1.

The reader will recall that the technologists first attempted to resolve the mount error by untwisting a twisted tape, and that when this ploy failed, they tried to rectify the problem by remounting the reel itself. To act as if the problem were caused by a twisted tape or an improperly mounted reel clearly presupposed a mechanical, rather than a computational, failure. Even after a radiologist suggested that the difficulty may have partially resulted from the computer's software (an open file failure), the technologists continued to propose mechanical explanations: a wrongly turned key was indicted as a possible cause. Finally, when technical difficulties recurred, the computer was said to be "frozen," a term used in x-ray to describe moving parts jammed by pressure or fused by friction or other sources of heat.

Since the technologists lacked an understanding of how computation could fail, even when they knew indisputably that problems resided in the scanner's software, they continued to use mechanical metaphors to explain the problem. Open file failures were a case in point. The scanner produced an "open file failure" and its accompanying error message whenever the computer failed to close the file to which it had allocated the last image's data. Until Technicare updated the scanner's software, open file failures were common at both sites. After numerous encounters with open file failures, one of Suburban's techs hit upon the following explanation to account for the problem's source. The explanation pivots on the mechanical metaphor of a skipping phonograph:

Author : What do you do when you get an open file failure?

Tech : Nothing! There's something wrong with the way the file, the way it puts . . . puts it on the disk, the record. (The technologist now drew a circle on a piece of paper with a tiny circle at its center so that the drawing resembled a phonograph record.) I don't know what it looks like. A circle? Everytime it gets to this place (stabbing the center of the circle with her pencil), the problem happens. That's how I explain it anyway.

External Causality

The technologists' proclivity to seek sources of problems in the scanner's external environment was closely related to their preference for mechanical metaphors. For instance, the first time an artifact was seen at Suburban, technologists contended it was the image of an I.V. tube. The scanner personnel had been told that the computer would begin to malfunction if the temperature in the control room rose much above seventy-five degrees. Consequently, on several occasions when artifacts appeared, Urban's techs noted that similar manifestations had occurred at approximately the same time on the previous day and suggested that the problem was a function of how long the machine had been running. The machine was said to have raised the room's temperature. Similarly, chronic problems that occurred in the afternoon at Suburban were, for a period of time, blamed on the afternoon sun striking the plate glass windows of the CT department's waiting room. The sunlight was said to have raised the temperature to a point where the scanner began to "act up."

Technologists framed technical problems in mechanical terms and sought causes in the external environment because they were comfortable thinking in terms of a world whose principles were familiar. Had the techs been operating standard x-ray equipment, their framing would have served them well. But while scanner problems were occasionally mechanical, and at times even resulted from environmental conditions, as the engineers repeatedly verified, most malfunctions arose from the scanner's software rather than its hardware. Since technologists had no vision of how software might work, to offer an account they had little option but to fall back on models of machines and physical causes they understood. Mechanical metaphors and external causes provided a convenient trellis for structuring the technologists' evolving theory of how scanners worked.

Superstitions

While the techs drew broadly on their knowledge of mechanical technology to interpret the scanner's operation, other elements of their theories reflected previous interactions with the scanner itself. Many of the rules of thumb technologists inferred on such occasions constituted isolated, but nevertheless accurate, bits of technical knowledge: for instance, the heuristic belief that mount errors usually indicated an improperly mounted reel. However, other elements of local knowledge were technically inaccurate and resembled what one might call technological superstitions.

Most superstitions had relatively short life spans, but others endured over the course of the whole study. One particularly persistent superstition was Suburban's technologists' belief that even though archiving images slowed the computer's response, one could minimize CPU time by using the physician's console to issue the command to archive.[11] The second console clearly allowed technologists more flexibility in their work, since it enabled two techs to interact with the computer simultaneously. But the notion that archiving from the second console would minimize CPU time was unfounded. Since both consoles were peripherals attached to the same computer, and since the computer's central processing capacity was fixed, running the same two tasks simultaneously placed an identical burden on the computer regardless of the terminal from which the techs entered the commands.

Technological superstitions were the cognitive counterparts of ritual solutions. Like the latter, most superstitions arose from the technologists' interactions with the computer during bouts of confirmatory problem solving. The malfunctioning drive in Problem History #1 again provides a particularly good example of how confirmatory problem solving sired superstitions as situated by-products. Consider the conditions surrounding the act of cleaning the drive's heads. Before beginning to clean the heads, the techs rebooted the computer (event 33). However, they did not bring the computer on-line (event 39) until after they had finished the cleaning (event 36). When the scanner was finally brought up, it appeared to function correctly so that when problems commenced several minutes later, they were deemed unrelated to the drive. The techs consequently inferred that dirty heads must have caused the earlier problem.

In fact, however, the heads had nothing to do with the problem. Over the following weekend, identical difficulties plagued the drive, and despite the fact that the engineer tried repeatedly to correct the problem, the malfunction continued to occur five days later. Yet despite strong evidence that dirty heads made little difference, some of the scanner's personnel still contended the heads were at fault. For example, the following exchange took place a week after the events in Problem History #1.

This morning the problem with the drive recurred. The Chief Technologist asked Sally about the drives:

C. Tech :   Was the head dirty? When did you clean the heads?
Sally :   It was cleaned yesterday.
C. Tech :   Did you clean it this morning?
Sally :   No.
C. Tech :   The heads should be cleaned every day. I think it's the humidity. The tape has an emulsion on it and the humidity probably makes it gunky.

Belief in the importance of clean heads initially arose because the techs were unable to determine which of two actions, rebooting or cleaning, had eliminated signs of the original problem. However, since the problem ceased soon after the heads were cleaned, and since cleaning had been the tech's last act before bringing the computer on-line, from the tech's vantage point cleaning was most closely associated with the problem's cessation. Consequently, by the logic of temporal contiguity, dirty heads appeared to be the problem's most likely cause. The conclusion was bolstered by the fact that dirty heads fit the technologist's proclivity for mechanical metaphors better than did vague notions of bugs in the software.

A superstition's life span depended on how readily the technologists could acquire information that would invalidate their belief. Because problems with the drive continued despite the techs' actions, they abandoned belief in the efficacy of cleaning heads soon after the last exchange occurred. However, when feedback from the scanner was subtle or unavailable, superstitions could endure for weeks or months at a time.

This was precisely the case with the belief that archiving from the physician's console would speed the computer's operation. Since the techs had no knowledge of how to monitor CPU time and since the computer did not make measures of CPU time readily available, the technologists never confronted data that would challenge their belief.

Anthropomorphism

The mere ability to pose hypotheses and offer explanations, however inaccurate they might be, helps sustain faith in the comprehensibility of the world. Conversely, the repeated inability to formulate even inaccurate hypotheses can undermine confidence in one's capacity for knowing. In the long run it may therefore be more comforting, and even more practical, to be wrong about specifics than to be wrong about the assumptions that brace one's knowing. I submit that it was partially for this reason that technologists strove to construct plausible explanations for the scanner's operation even when they realized their explanations were likely to be wrong. The techs' mechanical metaphors and superstitions about the scanner's operation allowed them to presume a world whose causes and effects could be understood. Even though their theories might be invalid, the capacity to offer plausible accounts sustained the techs' belief that they could ultimately make sense of their technical milieu.

That the techs were aware their explanations could be faulty is demonstrated by the qualification that ends the previously cited analogy between open file failures and skipping phonographs. The technologist concluded her explanation with the caveat, "That's how I explain it anyway." Like most of us, the tech preferred a plausible, but possibly inaccurate, explanation to no explanation at all. However, because the scanner was essentially a computational machine, because the techs had little familiarity with computational notions, and because they could not ignore overpoweringly obvious disconfirming evidence, the techs frequently confronted events for which they could devise no suitable account. It was on these occasions, and only these occasions, that the techs engaged in forms of anthropomorphic talk that others have observed among novice computer users (see Turkle 1984).

It is important to remember that technologists were trained to treat machines objectively, to understand technologies as physical devices governed by physical laws. In fact, their certification as technologists required formal courses in the physics and mechanics of an x-ray machine's operation. That the CT techs generally adopted a "no nonsense" attitude toward the scanner was reflected by the fact that they typically referred to the scanner's computer by the technical phrase, "the system." But when, in the course of daily events, techs encountered malfunctions or anomalies for which they could ultimately devise no plausible explanation, they began to speak of "THE COMPUTER" rather than "the system." References to the system were made without emphasis. However, the phrase, "THE COMPUTER," was accentuated in the flow of talk. Whereas, "the system" was a neutral term of direct reference for a physical entity, COMPUTER" was not. As used by the techs, "THE COMPUTER" implied a mysterious force which, if not malevolent, was surely fickle.

When techs spoke of THE COMPUTER their conversations took on anthropomorphic qualities that contrasted sharply with their matter-of-fact approach to "the system" and other technologies. THE COMPUTER was said to be capricious: it had, in the techs' own words, "a mind of its own." THE COMPUTER was a sentient entity that "liked" or "did not like" commands, that acted "crazy," and that beeped when it wanted to say, "I'm hot." In the throes of a persistent problem, technologists beseeched THE COMPUTER to do as they desired, and the bold among them even insulted THE COMPUTER with word and gesture.

Most importantly, however, when events went irretrievably wrong, it was THE COMPUTER that was said to have caused the problem. Although THE COMPUTER always lurked in the background, the techs usually kept it at bay with their mechanical metaphors, their confirmatory strategy of problem solving, and the ritual solutions and superstitions that the confirmatory strategy engendered. It was only when these practical tools failed that techs resorted to anthropomorphic talk. To say THE COMPUTER was a cause was, in effect, to admit that one didn't know what was wrong. THE COMPUTER, then, was the techs' label for the heart of the system's darkness.

THE PRACTICALITY OF RITUAL, SUPERSTITION, AND MAGICAL THINKING

Given Western attitudes about the nature of technology and technical knowing, even the suggestion of anthropomorphism, superstition, and ritual in high technology settings seems incongruent, unacceptable, and somewhat embarrassing. Such behaviors and interpretations smack of magical thinking, a mode of thought we expect among the religious and among members of traditional societies, but one we believe to be singularly out of step with (and perhaps a threat to) the ethos of technical rationality (Malinowski 1948). We might therefore wish to dismiss Suburban's and Urban's technologists as foolish or ignorant. However, to brush aside the technologists' practical theories as simple-minded and wrong-headed would be to ignore the possibility that they served practical purposes, that magical thinking may be more common in technical settings that we normally admit, and that the technologists' way of knowing may tell us more about their situation than about the actors themselves.

With the scanners' arrival the techs confronted a technology whose fundamental principles differed from any they had previously encountered. Though they were trained to run the scanner under normal operating conditions, even seasoned technologists knew little about how computers work. So long as the scanner functioned smoothly its vagaries remained in the background and the techs proceeded with their work undaunted. However, because the scanners were complex systems liable to frequent and unpredictable malfunctions, the techs were unable to take the scanners' operation completely for granted. Not only were the techs responsible for keeping the scanners on-line, their local reputation and source of occupational power lay in their ability to control the technology. Therefore, when technical problems disrupted their work and threatened their control, technologists had strong incentives to confront a technology they poorly understood.

The techs' theories of the machine and their ploys for countering technical malfunctions can be understood only in light of the technology's complexity, the technologists' uncertainty, and the practical problems the technologists faced. As I have repeatedly emphasized, the techs were inclined to construe technologies as principled entities. Despite the fact that the scanner's underlying technical paradigm broke radically with the technologists' experience, they nevertheless endeavored to treat the scanner as a comprehensible object. To do otherwise would have been to admit little hope of mastery. Unfortunately, in their struggles with the machine, the techs' resources were limited to their past experience with mechanical devices, their prior interaction with the scanner itself, and the expertise of engineers whom they regarded ambivalently. Decidedly absent was even a rudimentary knowledge of computation, cybernetics, or the mechanics of microelectronics.

In their bouts with the scanner, technologists strove to remain on the solid ground of what they knew for as long as possible. Their preference for dealing with equivocality from the safety of the certain influenced both their strategies for countering malfunctions and the substance of the theories they built to make sense of the scanner's operation. The technologists' first response to a technical problem was to normalize the situation, to return the contours of the scanning context to their typical state as quickly and convincingly as possible. By normalizing the situation technologists could establish an even keel by drawing primarily on their firm knowledge of the scanner's routine. When normalization failed, the technologists turned next to ritual solutions. While a ritual solution's efficacy was more uncertain than methods of routine scanning, its spotted history of success offered techs more certainty than did full-fledged trial and error problem solving. Consequently, it was only when ritual solutions failed that technologists submitted to the open-endedness of confirmatory diagnosis. Finally, if confirmatory hypothesis testing brought no progress, the techs summoned the engineer.

Note that the techs' hierarchy of tactics for confronting technical problems escalated systematically from the most to the least certain, from routines over which techs had the most control to an admission of complete uncertainty in their abdication to the engineer. A similar progression from certainty to uncertainty characterized the accounts technologists constructed. When techs could not say precisely why events occurred, they resorted first to explanations that pivoted on mechanical metaphors or external causes. Such accounts allowed technologists to remain firmly grounded in a familiar technical paradigm. When mechanical metaphors and external causes proved insufficient, techs drew next on *ad hoc* rules of thumb which were accumulated piecemeal during previous bouts with the scanner. Some of these explanations were technically accurate, but of limited applicability. Others were completely inaccurate and resembled superstitious beliefs. Finally, when certain knowledge, mechanical metaphors, external causes, and even local superstitions could provide no plausible account then, and only then, did technologists slip into anthropomorphic talk of the computer as agent.

The parallel progression in deed and thought from the relatively certain to the seemingly ritualistic and magical can be seen as correlated with the amount of strain that escalating uncertainty placed on the techs' ability to control the technology. As the scanner's complexity became ever more obvious and recalcitrant, notions of causality based on non-computational paradigms became increasingly brittle. To continue to act as uncertainty escalated, the techs were forced to take stands on increasingly shaky ground. That the technologists would employ local superstitions and rituals before admitting ignorance and defeat attests to their need to keep the scanner on-line and to appear in control of the situation. Before rushing to chastise the technologists' "unscientific" behavior, one should hasten to remember that to act on superstition or to engage in ritual is, *a priori*, to have a better chance of success than if one were not to act at all.

Recent anthropological theory suggests that the Victorian legacy of casting magical thought as science's nemesis obscures important commonalties between the two modes of sense making. To be sure, scientific thought is fundamentally different from magical thinking and there can be no doubt that scientific reasoning has a far better track record at predictive control (Horton 1967b). Nevertheless, both systems of thought offer adherents similar and quite critical epistemological benefits. In particular, each imposes order on uncertainty by creating a causal context that goes beyond, and yet preserves, what normally passes in a society for solid common sense (Horton 1967a; Geertz 1983).

When common sense fails to illuminate, members of Western societies turn to scientists for higher order explanations. Under similar circumstances, members of traditional societies consult diviners. But just as the average member of an industrial society makes out quite nicely in day-to-day life without constant reference to scientific theory, so members of traditional societies need only rely intermittently on magical beliefs. Members of both live out most days in a world of practical activity whose principles they expect all competent members of society to know. It is only when events go awry, when predictions and explanations fail, that one needs either the uncompromising eye of science or the power of magic. Of course, magicians' divinations and the predictions of science are both open to disconfirmation. However, the critical difference between incompetent magic and faulty science is that consumers of the first remain trapped within the boundaries of their cosmology, while consumers of the latter have leeway to adopt new theories of events (Horton 1967b). Unlike magical thinking, science is predicated on rejecting paradigms that can be shown to no longer adequately predict and explain anomalies.

This essay began by drawing a distinction between what is known in a technical field and what is actually understood by its average practitioner. As ideal and disembodied members of an occupational community, practitioners of science and technology are granted the luxury, if not the obligation, of jumping a sinking paradigm in the wake of change. But practitioners with real identities are rarely so ideal as to be unconstrained by the particulars of their own organizational context or by the limits of their own (as opposed their field's) knowledge. To reiterate the argument made in the beginning, for numerous sociological reasons individuals may have only a dim grasp of what their field claims to know. It therefore seems important to ask what might actually happen when flesh and blood members of a technical specialty find themselves facing practical conditions for which their current paradigms are unsuited and for which they have no alternate paradigm as refuge.

Although adequate answers are likely to be multiple and require far more research in a variety of technical settings, the behavior of Urban's and Suburban's technologists suggests several intriguing possibilities. When pushed against the wall of understanding and yet forced to act knowingly, practitioners may seek to salvage and defend outmoded paradigms, even if they suspect the paradigms are inadequate. Such tenacity should come as no surprise given the history of resistance to new theories by members of formerly dominant scientific communities (Kuhn 1970). More critical and more unexpected is the possibility that practitioners may knowingly or unknowingly resort to superstition, ritual, and even faith before admitting confusion and lack of control. At least when technical complexity and extreme uncertainty combine with a pressing need to act, the rudiments of magical thinking may be far more pragmatic than we care to admit.

Technologists routinely enacted ritual solutions and proceeded on superstition not only because they had work to do and face to save, but also because they were victims of their technical environment. While the technologists' uncertainty surely reflected their scanty knowledge of the rules and principles by which the scanner functioned, the scanner itself enhanced the techs' uncertainty in ways that interacted with and amplified their incomplete knowledge. Like most complex technologies, the scanners were assemblies of tightly interrelated subcomponents that were themselves quite complex. Combinations of events in separate subcomponents could, and in fact did, lead to malfunctions that were unanticipated even by the scanner's designers. Because the scanner was a cybernetic device that monitored and responded to its own internal state, as well as the commands of it users, it frequently communicated in surprising ways at surprising times. Moreover, the scanner's software contained a large number of bugs which began to come to light only as the scanners were used. The scanner's complexity, its cybernetic design, and its faulty software combined to create a technical milieu which even the engineers, at times, found incomprehensible.

The technology's unpredictability united with the technologists' scanty knowledge to create situations in which the scanner's responses were but loosely coupled to the technologists' actions. In particular, problems would crop up unexpectedly, the technologists would attempt to solve them, and then during the course of action the problems would disappear for reasons that were independent of what the technologists had done. Such an environment created conditions ripe for what behaviorists termed "adventitious reinforcement": the learning of a response by the chance pairing of a behavior and a reward that immediately follows, but is otherwise unrelated to, the behavior (Skinner 1948, 1969; Morse and Skinner 1957). Ironically, Skinner called behaviors learned under such circumstances "superstitions."

Experiments have shown that adventitious reinforcement is most likely to occur when rewards involve the removal of conditions that actors experience as noxious (Mowrer 1939; Herrnstein 1969). As was suggested earlier, in perhaps more palatable terms, the sequence of events that lead Suburban's techs to their belief in the efficacy of clean heads followed precisely the paradigm of adventitious negative reinforcement.

Though behaviorist thought has fallen out of grace, primarily because its eschews a theory of mind, its tenets may in the long run prove useful for explaining how mindful people unwittingly develop rituals and superstitions in their attempts to cope with complex technologies whose principles they do not fully comprehend. Under conditions of extreme technical uncertainty and complexity, the sense making strategies of a novice computer user may be more at the mercy of the technology and the ordering of events than less circumstantial theories would predict.

CONCLUSIONS

If we assume that Urban's and Suburban's CT techs resemble most novice computer users, and if we also assume that an increasing number of health care workers will be asked to operate computational technologies whose principles they find foreign, then one might expect similar phenomena to surround the use of other complex, cybernetic technologies regardless of the context in which they are used. If these assumptions are correct and if the projection is plausible, then the foregoing discussion entails implications that are particularly critical for understanding the organizational implications of medicine's recent technical transformation.

Of primary importance is the degree to which medical practitioners may become increasingly dependent on technical experts outside the medical community. As was the case at Urban and Suburban, physicians and other health care workers may become quite adept at controlling a complex technology under normal circumstances. However, when technical problems occur, medical personnel may find themselves without sufficient knowledge to control the technology and, hence, dependent on outsiders to keep the technology in line. To the degree that such a division of knowledge becomes institutionalized, medical personnel may find themselves sacrificing no small degree of power and autonomy to outsiders who better understand the technology's mysteries.

Evidence of increased dependence on outsiders can be seen in Urban's and Suburban's reliance on the engineer. If the computer was sometimes cast as a mysterious force, the engineer was treated as its diviner. Technologists and radiologists relied on the engineers to control the scanner with their knowledge of diagnostic subroutines which were perceived by most scanner personnel as mysterious incantations. Moreover, since familiarity with the scanner's actual operation was limited to Technicare's personnel, the radiologists and technologists could purchase no alternative source of expertise. To summon the engineer was therefore not only to seek expertise, it was to abdicate technical control to the vendor. Consequently, the need for the engineer's expertise fostered an ambivalent combination of dependence and resentment.

On one hand, the engineer was viewed as a source of technical salvation. For example, during a particularly troublesome problem at Urban the engineer was seen putting on his coat to go to lunch. A radiologist, who was by comparison to his colleagues quite knowledgeable about the scanner's working, turned to the engineer and asked with obvious concern, "Are you leaving us?" When told that he would be returning, the radiologist exclaimed, "Good, I was afraid we'd have to face it on our own." Yet, the engineer was also suspect. Both technologists and radiologists believed engineers actively fostered their dependence, a state of affairs which members of neither group found acceptable and which engendered hostility. Consequently, when engineers did not respond immediately, scanner personnel became simultaneously anxious and indignant.

The reader may claim that the situation is no different than that faced by users of more conventional technologies, for example, those of us who must take our automobiles to mechanics for repair. While such a comparison is superficially apt, it overlooks several critical distinctions. Automobile owners enjoy numerous alternatives. Not only may they choose among several mechanics, but other means of transportation are generally available. For some diagnostic tasks there are simply no technical substitutes for a CT scanner, and at present, personnel who understand a scanner's operation well enough to effect repairs are scarce. Moreover, aside from the automobile owner, few people are inconvenienced by the failure of a car. But the failure of a CT scanner not only irritates technologists and radiologists, it can inconvenience a large number of patients and, in some cases, even make the difference between a patient's life and death.[12]

Reliance on a central technology and its small community of technical experts has several upshots. Because the designs of complex technologies are typically unique to their manufacturers, when hospitals purchase a complex system like a CT scanner they, in effect, agree to engage in a long term symbiotic relationship with a vendor capable of monopolizing technical expertise. The manufacturers' market strategies and the priorities of their technical representatives, the engineers, may conflict with the needs and goals of medical personnel. Such conflict may not always be resolved in favor of the latter. Moreover, when the two perspectives conflict, physicians may find themselves with few persuasive levers for bargaining. For example, physicians may even find market mechanisms a useless threat, since there are typically no alternate sources of expertise available and since few hospitals are able to absorb the cost of discarding a million dollar technology in order to do business with a more responsive vendor. Consequently, the purchase of complex, computational technologies may not only entail constraints on professional autonomy, it may introduce into medical settings non-medical sources of uncertainty by creating dependence on organizations outside the normative structure of the medical community and immune to the discipline of a competitive market.

Although it is unlikely that medical professionals will be able to completely rid themselves of such dependency so long as complex technologies remain unique to particular manufacturers, there are several strategies medical professionals might pursue to reduce their dependency. All pivot on recognizing and planning for the paradigm shock that computational systems may entail. First, new users must be given extensive training in the fundamental principles of computational technology prior to being forced to operate such a technology on short notice. The training must go beyond normal operating procedures to include strategies for reading technical literature, for thinking about computational systems, and for diagnosing system problems. In the case of CT techs, such training could be sponsored and required by the College of Radiology or the American Registry of Radiological Technologists.

Because hospitals are the major consumers of technologies like the CT scanner, because each sale represents a source of substantial revenue for a vendor, and because vendors are at least *initially* competitive, with an awareness of their market power and more effective collaboration hospitals could force manufacturers to assume the full burden of adequately educating new users. In short, guidelines for adequate resources and training could be devised by professional associations and the purchase of equipment could be made contingent on a vendor's ability to meet the guidelines. If it were to organize more effectively, the medical community is likely to have sufficient market power to force vendors to become more open about technical designs, to provide more useful documentation, and to offer long term training programs for their customers.

One might rightfully argue that in time, computational technologies will become so common that their underlying paradigm will be as widely understood as mechanical paradigms are today. While such a prediction is likely to be valid, in the meantime we should not ignore the possibility that during the transition individuals may be forced to assume a stance toward technical knowing similar to that which time travelers must take in order to pass muster in some future world. While medical personnel may be able to make do with equivocal technical knowledge, unless the social implications of "making do" are well understood, the resulting change in the distribution of knowledge may so radically alter the structure of medicine that any future return to present divisions of labor may be ruled unlikely. For once relations become institutionalized, they are notoriously difficult to change.

APPENDIX A: PROBLEM HISTORY #1

Hospital: Suburban
Date: 12/1/82

EVENT	TECHNOLOGIST/RADIOLOGIST ACTION	COMPUTER RESPONSE/MESSAGE

A SCAN IS IN PROGRESS. THE TECHNOLOGISTS ARE ARCHIVING THE LAST PATIENT'S SCANS.

1.	While Tech 1 is in the gantry room attending the patient, Tech 2 downloads the computer and proclaims, "I got an error message I don't understand."	The computer's screen goes blank. The patient's demographic data is erased. The table's counter is set to zero.
2.	Tech 2 enters the computer room and examines the magnetic tape on the drive unit.	
3.	Tech 2 fiddles with the tape.	
4.	Tech 2 discovers the tape is twisted and suggests that this may have caused the problem, but she is not sure since she thinks it is possible that she may have twisted the tape when adjusting it.	
5.	Tech 2 takes the reel off the drive and puts it back on.	
6.	Tech 1 brings the computer back up and reenters the patient's demographics.	
7.	Tech 2 decides the tape needs to be rewound and begins to push buttons on the mag drive.	No response from the drive.
8.	Tech 2 pushes the buttons in a different sequence.	The tape moves forward, rather than backwards, winding the tape completely off the original reel and back onto the take-up reel.
9.	Tech 2 manually winds the tape leader back on the original reel.	
10.	Tech 2 repeats step 8 with the same result and then repeats step 9.	

11.	Tech 1 says she'll hit the "dismount" button on the console and that Tech 2 should hit the rewind button on the mag drive.	
12.	Tech 1 hits the dismount button.	Error message.
13.	Tech 2 hits the rewind button.	The tape winds forward onto the take up reel.
14.	Tech 2 threads the tape through the heads onto the original reel.	
15.	Tech 2 uses a pencil eraser to hold the tape next to the reel while manually winding the leader onto it.	
16.	Tech 2 presses buttons on the mag drive in another sequence.	
17.	Tech 2 pushes the "load button."	The tape winds forward onto the take up reel.
18.	Radiologist 1 enters the control room and reads the error message repeating over and over on the terminal.	Computer displays error "message: OPEN FILE FAILURE . . . BAD TAPE FORMAT."
19.	Radiologist 1 asks, "What's going on?"	
20.	Tech 1 explains, "The tape got messed up and we couldn't get it to archive."	
21.	Radiologist 1 suggests that there may have been two separate problems; one 2 with rive and another with the file.	
22.	The techs argue that there is only one problem.	
23.	The radiologist points out that the error message is an open file failure; a software, not a hardware problem.	
24.	A secretary enters the control room and summons the radiologist to the telephone.	
25.	Tech 1 calls the engineer and leaves a message with the engineer's paging service.	
26.	Tech 1 hangs up phone and pushes the tape mount button on the console.	Error message: "MTO---SELECT ERROR."

27.   Radiologist 1 returns, looks at the
      screen and asks if the computer has
      been rebooted.

28.   Tech 1 says she tried "AT DN" (the
      command to download).

29.   Radiologist asks what "AT DN" means.

30.   Tech 1 replies, "A way to clear the
      machine."

31.   The radiologist suggests bringing the
      computer all the way down.

32.   Both techs go into the computer room
      and repeat steps 15-17 two more times.

33.   The technologists reboot the computer.   All scanner functions cease.

34.   The radiologist asks if the tape heads
      have been cleaned.

35.   Tech 1 says that they haven't, and the
      radiologist replies that they should be.

36.   The techs clean the heads.

37.   The techs put an empty tape on the
      mag drive.

38.   The radiologist apologizes to the
      patient for the delay.

39.   After cleaning the heads, Tech 1 brings   The computer comes on-line.
      the computer back on-line.

40.   Tech 1 pushes the mount button and   The computer accepts the
      initializes the tape.   command without error
                                                    messages.

41.   Tech 2 re-enters the patient's data.

42.   Tech 2 commands the computer to
      archive the last patient's scans.

43.   The techs discuss why the problem may
      have occurred.

44.   Tech 1 claims that Tech 4 dismounted
      the tape at the end of last night's work.
      A key on the side of the computer has

been turned. Tech 1 suggests that the
problem may have something to do
with Tech 4's action and the key.

*THE SCAN PROCEEDS. ARCHIVING IS UNDERWAY. ALL FUNCTIONS APPEAR
NORMAL. THE TUBE HEAT RISES TO ITS CUT OFF VALUE AND THE TUBE SHUTS
DOWN. THE TECHS WAIT FOR THE TUBE HEAT TO FALL SO THEY CAN RESUME
SCANNING.*

45.	(4 min. later) Tech 2 notes that the tube heat hasn't fallen.	Digital readout on tube temperature remains 75 degrees.
46.	Tech 2 pushes the "pause scan" button to resume scanning.	Light indicating that the pause is on does not go off.
47.	Tech 2 pushes the "start scan" button.	No response.
48.	Tech 2 claims the computer is "frozen."	
49.	Tech 2 notes she is puzzled by the fact that the timer is still running.	(The timer is controlled by a microchip in the console rather than by the computer).
50.	Tech 2 looks at the computer. Notes the mag drive has stopped moving.	
51.	Tech 2 proclaims, "Geez, this is too much."	
52.	Tech 2 hits the carriage return.	The system prompt that should appear, does not appear.
53.	Tech 2 calls Tech 1 to the control room.	
54.	Tech 2 explains events 45-52 to Tech 1.	
55.	Tech 1 walks to the computer room and says, "I wonder what those keys do."	
56.	Tech 2 apologizes to the patient for the delay.	
57.	Radiologist 1 happens by.	
58.	Tech 1 immediately says, "The *system's* down. It won't do anything. I'm afraid that if I reboot it, I'll screw up the mag tape . . . We might get another mount error."	

59. Tech 1 asks Tech 2 if the engineer isn't at another hospital. Tech 2 says he is.

60. Tech 2 calls the hospital, explains why she needs to talk to the engineer and asks if they will have him call immediately.

61. (5 minutes later) Engineer calls.

62. Tech 1 answers the phone, explains the problem, and relates the error messages that were received.

63. Tech 1 (to engineer), "I can't put in SHUT DOWN, so I don't know what that'll do."

64. Tech 1 downloads the computer and relates to the engineer the configuration of lights on the mag drive's panel.

Computer beeps, does not accept the download command. The cursor, but no input, can be seen on the terminal.

65. Tech 1 (to engineer), "It won't take it (e.g. download). I hit AT DN and nothing happened . . . I did it !!. . . I'll do it again."

66. Tech 1 types AT DN. Says to engineer, "Now, *the computer* does not respond. I've got two witnesses."

Computer beeps, does not accept the download command.

67. Tech 1 to engineer, "You listen!" Tech 1 holds the phone to the keyboard so engineer can hear the beep when she types AT DN.

68. Tech 1 walks to the computer and says to engineer: "It's not going to screw up the mag tape?"

69. Tech 1 reboots the system, returns to the console, and brings the computer back on-line.

70. Radiologist 1 returns to control room.

71. Repeating the engineer's words, "Take it off-line and hit rewind," Tech 1 walks into the computer room with the phone in hand and pushes the off-line and rewind buttons on the mag drive.

The tape begins to move.

APPENDIX B: PROBLEM HISTORY #2

Hospital: Suburban
Date: 10/20/82

EVENT	TECHNOLOGIST/RADIOLOGIST ACTION	COMPUTER RESPONSE/MESSAGE
1.	The tech loads the scans into the 512 imager (the images were taken with a 256 pixel matrix).	The deltaview (longitudinal image) appears on the monitor. The lines on the deltaview appear as dots rather than lines.
2.	Both the tech and the radiologist express concern that this does not appear normal.	
3.	The tech tells the radiologist not to worry: "That's OK, I've already matrixed" (i.e. filmed the image).	Scans begin to appear on the video monitor but they are out of sequence.
4.	The tech notes the sequencing problem and loads the scans into the imager a second time.	
5.	The tech begins to film the images as each appears on the monitor. The radiologist views the images while the tech films.	
6.		The second scan appears on the monitor after the tech has photographed the fourth.
7.	Both the tech and the radiologist note the sequencing problem. They become perplexed.	
8.	The tech loads the scans into the imager for the third time.	The first six scans appear in correct order.
9.		The sequencing problem again occurs.
10.	The tech states, "I don't know what just happened."	
11.	The radiologist replies, "You know what it may be doing? It may not have finished reconstructing, so it's taking 256's and plopping the reconstructions in when it's done."	

12.   The tech contests the hypothesis by
        noting that the computer has not
        given a message about a completed
        reconstruction for some time.

13.   At the radiologist's suggestion, the
        tech loads the images into the 256
        imager rather than the 512 imager.

14.   The application's tech enters the
        control room.

15.   The tech explains the problem to the AT.

16.   The AT explains that when the tube heat
        shut the computer down, reconstruction
        began. The reconstructed images were
        dumped into queue on top of the
        quicklooks that had already been taken.
        To eliminate the problem one must
        load only 512 pixel images or delete
        the quicklooks (256 pixel images)
        before loading.

APPENDIX C: PROBLEM HISTORY #3

Hospital: Suburban
Date: 11/12/82

EVENT	TECHNOLOGIST/RADIOLOGIST ACTION	COMPUTER RESPONSE/MESSAGE
1.	Tech 3 enters a new patient's data in the computer.	The name is longer than the alloted number of characters. The computer gives the message: ERROR ON INPUT CONVENTION.
2.		The computer automatically returns a line for the patient's name so that it can be re-entered in shorter form.
3.	Tech 3 assumes the patient's name was accepted and therefore enters the patient's ID number.	The computer accepts the input as the patient's name and prompts the operator for the patient's number.

4. Tech 3 notices the computer's prompt, claims the computer must not have accepted the number, and enters the number again.	The computer accepts the input. The patient's number now appears on the video monitor in the spaces where the patient's name as well as the patient's number, typically appear.

THE SCAN COMMENCES AND THE PROBLEM GOES UNNOTICED UNTIL THE TECH IS READY TO FILM THE IMAGES

5. Tech 3 prepares to film the first image.	The first image is displayed on the video monitor.
6. Tech 3 examines the image. Notes that the patient's number is where the patient's name should be.	
7. Tech 3 enters the ANNOTATE subroutine.	
8. Tech 3 enters the patient's complete name.	The computer responds with the error message: ERROR ON INPUT CONVENTION.
9. Tech 3 enters a shortened version of the name.	The computer accepts the input and the error is corrected.

NOTES

1. My thinking in this essay has profited by careful comments from several colleagues who served as early readers: Frank Dubinskas, Margaret Lock, John Van Maanen, and Edgar H. Schein. Since each duly pointed me toward several paths of redemption, it was clearly I who slipped. I therefore must bear sole responsibility for all sins of omission and commission either attributed or real. The research discussed here was funded, in part, by a Doctoral Dissertation Grant from the National Center for Health Services Research (HS-05004).

2. This number should be taken as no indication of the actual frequency of problems, since the count depends on my having been present at the time a problem occurred. I would, however, estimate that if a complete record were kept, one might discover that machine problems of one sort or another occur, on average, at least once a day during the first year of a scanner's operation.

3. In radiology departments the noun, scan, is used in two ways. A scan may refer either to an examination performed with a CT scanner or to a single image produced during such an examination. I will follow the practice in radiology departments and use the noun in both senses. Which meaning is intended should be clear from the context of use.

4. For example, if a patient is obese, the milliamperes per second, and/or the kilovolts at which the x-ray tube operates, are usually increased to ensure that sufficient radiation penetrates the patient's body to produce an adequate image. The practice reflects the fact that the number of electrons that penetrate an object is inversely proportional to the object's density.

5. Deltaviews and topograms are subroutines that produce longitudinal images of the body, rather than the cross-sectional images for which the scanner is famous. Once this "digital radiograph" is obtained, the technologist uses the image to program where each cross-sectional scan will occur. The programming is done by using a joystick and graphics software to draw lines through the longitudinal image at points where slices are desired. The operator determines not only the number of slices, but the plane at which the slice will occur. Setting slices at varying intervals and changing the plane of an image are crucial for effective spinal scans.

6. Application techs are usually former radiological technologists who once worked as CT techs in a hospital. Becoming an applications tech is one of the few clear career paths open to technologists after they become competent at running the technology and their job becomes boring.

7. As in all CT areas I have visited, there was a large leaded glass window in the wall between the gantry and the console room so that technologists could monitor the patient's behavior. In addition, the Technicare scanners were equipped with an intercom that allowed techs to monitor a patient's vocalizations.

8. During spinal scans it was necessary to reset the gantry angle for each disk space so that the image would show a plane parallel to the disk. Deltaview software was used to set these angles. Without the deltaview, angle setting would have been impossible. This particular problem permanently disrupted the deltaview software's control over the scan. If another angle had been needed, the techs would have had to start over again.

9. Traditionally, organizational theorists have described the logic of informed action as deliberately rational. Makers of good decisions are said successively to define the problem and their objectives, to generate alternate solutions, to weigh alternatives against objectives and then to choose the best course of action. Recently, however, a small number of organization theorists have argued that such rationality rarely characterizes organizational life, and that instead, people tend to act with little deliberation and then wait to see what the results of their action might be. The argument is that an orientation to action may be adaptive when people face equivocal situations under time pressure. The term for this evolving stance seems to be "action rationality" and is to be found in the work of such theorists as March and Olsen (1976), Brunsson (1982), Starbuck (1983), and Swieringa and Weick (1984).

10. Schön's recent (1983) research on professional knowing suggests that the behavioral and cognitive components of occupational action are reflexively

entwined. Practitioners' actions and their actions' consequences mold their evolving theories of work which, in turn, shape their action strategies in a cyclic process of knowing. Schön intriguingly labels this reflexive cycle of adjustment, a conversation with the medium. In their interactions with the scanner's computer, the CT techs engaged in an explicitly conversational relationship with the technology. The fact that such openly "conversational" relations objectify the latent semiotic relation between persons and all machines may be a distinguishing characteristic of computer use. The pragmatics of such interactions bear closer scrutiny than they have previously received. This paper merely touches on the potential yield of a full-fledged "conversational analysis" of computer use.

11. The Technicare scanners came with two consoles. One was located in the control room, the other in the radiologists' office. The notion was that a radiologist could review studies stored on disk without interrupting the technologists' work. While Urban's radiologists used the second console, Suburban's did not.

12. Based on the average number of patients scanned at Urban and Suburban each day, every hour of downtime necessitated the rescheduling of 1.5 patients. Since scanners were utilized fourteen hours a day at each site (8:00 am until 10:00 pm), a day of downtime meant a backlog of approximately 21 individuals. The scanner's making a difference between life and death is most clearly seen in the case of emergency scans conducted on people who have sustained serious head injuries or who suffer severe heart attacks. Scanners give conclusive evidence of such problems as massive subdural hematomas or large aortic aneurisims.

REFERENCES

Baker, S.R.
   1977   The Diffusion of High Technology Medical Innovations: The Computed Tomography Scanner Example. Social Science and Medicine 13:155-62.
Banta, David and Barbara J. McNeil
   1978   Evaluation of the CAT Scanner and Other Diagnostic Technologies. Health Care Management Review 3:7-19.
Barley, Stephen R.
   1984   The Professional, The Semi-Professional, and the Machine: The Social Ramifications of Computer Based Imaging in Radiology. Doctoral Dissertation. Sloan School of Management: Massachusetts Institute of Technology.
   1986a Technology as an Occasion for Structuring: Evidence from Observations of CT Scanners and the Social Order of Radiology Departments.

Administrative Science Quarterly 31:78-108.

1986b Role Changes in Radiology. Administrative Radiology 4:32-41.

Brunsson, Nils
    1982 The Irrationality of Action and Action Rationality: Decisions, Ideolo-
          gies, and Organizational Actions. Journal of Management Studies
          19:29-44.

Cicourel, Aaron V.
    1974 Cognitive Sociology. New York: Free Press.

Dewing, Stephen B.
    1962 Modern Radiology in Historical Perspective. Springfield, IL: Charles
          C. Thomas.

Emerson, Joan
    1970a Behavior in Private Places: Sustaining Definitions of Reality in Gyneco-
          logical Examinations. In H.P. Dreitzel (ed.), Recent Sociology 2. Pp.
          73-100. New York: Macmillian.
    1970b Nothing Unusual is Happening. In T. Shibutani (ed.), Human Nature
          and Collective Behavior. Pp. 208-223. New Brunswick, N.J.: Trans-
          action Books.

Garfinkel, Harold
    1967 Studies in Ethnomethodology. New York: Prentice Hall.

Geertz, Clifford
    1983 Common Sense as a Cultural System. In Local Knowledge: Further
          Essays in Interpretive Anthropology. Pp. 73-94. New York: Basic.

Herrnstein, R.J.
    1969 Superstition: A Corollary of the Principles of Operant Conditioning. In
          W.K. Honig (ed.), Operant Behavior: Areas of Research and Applica-
          tion. Pp. 33-52. New York: Prentice Hall.

Hirschhorn, Larry
    1984 Beyond Mechanization. Cambridge, MA: MIT Press.

Horton, Robin
    1967a African Traditional Thought and Western Science: Part I. From Tradi-
          tion of Science. Africa 37:50-71.
    1967b African Traditional Thought and Western Science: Part II. The
          "Closed" and "Open" Predicaments. Africa 37:155-87.

Kuhn, Thomas
    1970 The Structure of Scientific Revolutions. Second Edition. Chicago:
          University of Chicago Press.

Malinowski, Bronislaw
    1948 Magic, Science and Religion. New York: Free Press.

March, James G. and Johan P. Olsen
    1976 Ambiguity and Choice in Organizations. Bergen, Norway: Universitets-
          forlaget.

March, James G. and Herbert A. Simon
  1958 Organizations. New York: John Wiley and Sons.
Morse, W.H. and B.F. Skinner
  1957 A Second Type of "Superstition" in the Pigeon. American Journal of Psychology 70:308-11.
Mowrer, O.H.
  1939 A Stimuli-Response Analysis of Anxiety and its Role as a Reinforcing Agent. Psychological Review 46:553-65.
O.T.A.
  1978 Policy Implications of the Computed Tomography Scanner. Washington, D.C.: Office of Technology Assessment.
Perrow, Charles
  1984 Normal Accidents: Living with High Risk Technologies. New York: Basic Books.
Pfeffer, Jeffrey
  1981 Power in Organizations. Boston: Pitman.
Pettigrew, Andrew M.
  1973 The Politics of Organizational Decision Making. London: Tavistock.
Schön, Donald A.
  1983 The Reflective Practitioner: How Professionals Think in Action. New York: Basic.
Skinner, B.F.
  1948 "Superstition" in the Pigeon. Journal of Experimental Psychology 38:168-72.
  1969 Contingencies of Reinforcement. New York: Prentice Hall.
Starbuck, William H.
  1983 Organizations as Action Generators. American Sociological Review 48:91-102.
Stocking, Barbara and Stuart L. Morrison
  1978 The Image and the Reality. London: Oxford University Press.
Swieringa, Robert J. and Karl E. Weick
  1984 Action Rationality in Managerial Accounting. Paper delivered at the Symposium on the Roles of Accounting in Organizations and Society. University of Wisconsin. July, 1984.
Turkle, Sherry
  1984 The Second Self: Computers and the Human Spirit. New York: Simon and Schuster.
Wiener, H.
  1979 Findings in CT: A Clinical and Economics Analysis of Computed Tomography. New York: Pfizer.

LIST OF CONTRIBUTORS

David Armstrong, Unit of Sociology, Medical School, Guy's Hospital, London SE1 9RT, England.

Paul Atkinson, Department of Sociology, University College, Cardiff CF1 1XL, U.K.

Stephen R. Barley, 110 Treva Avenue, Ithaca, NY 14850, U.S.A.

Mary Boulton, Academic Department of Community Medicine, St. Mary's Hospital Medical School, Praed St., London W2 IPG, U.K.

Deborah R. Gordon, Via P. Toselli, 140, Florence 50144, Italy.

C. G. Helman, 38 Lynmouth Road, London N2 9LS, England .

Patricia Kaufert, Department of Social and Preventive Medicine, University of Manitoba, S110 – 750 Bannatyne Avenue, Winnipeg R3E 0W3, Manitoba, Canada.

Laurence J. Kirmayer, 4333 Côte Ste Catherine Rd, Montréal H3T 1E4, Québec, Canada.

Barbara A. Koenig, 1459 5th Avenue, San Francisco, CA 94122, U.S.A.

Joseph W. Lella, Humanities in Medicine, McGill University, 3655 Drummond Street, Montréal H3G 1Y6, Québec, Canada.

Margaret Lock, 3655 Drummond Street, 4th floor, Montréal H3G 1Y6, Québec, Canada.

Setha Low, 2940 Guilford Ave., Baltimore, MD 21218, U.S.A.

Jessica H. Muller, 57 Homestead Blvd., Mill Valley, CA 94941, U.S.A.

Dorothy Pawluch, Dept. of Sociology, Concordia University, 1455 DeMaisonneuve W., Montréal H3G 1M8, Québec, Canada.

Andrea Sankar, 1420 Jorn CT, Ann Arbor, Michigan 48104, U.S.A.

Kathryn Taylor, 75 Prince George Drive, Islington M9A 1Y5, Ontario, Canada.

Anthony Williams, Dept. of Clinical Epidemiology, London Hospital Medical College, Turner St., London E1, U.K.

P. W. G. Wright, 6 Royal Mens, Taswell Road, Southsea P05 2RQ, Hampshire, U.K.

AUTHOR INDEX

SUBJECT INDEX